Lecture Notes in Computer Science 4100

Commenced Publication in 1973
Founding and Former Series Editors:
Gerhard Goos, Juris Hartmanis, and Jan van Leeuwen

Editorial Board

David Hutchison
Lancaster University, UK

Takeo Kanade
Carnegie Mellon University, Pittsburgh, PA, USA

Josef Kittler
University of Surrey, Guildford, UK

Jon M. Kleinberg
Cornell University, Ithaca, NY, USA

Friedemann Mattern
ETH Zurich, Switzerland

John C. Mitchell
Stanford University, CA, USA

Moni Naor
Weizmann Institute of Science, Rehovot, Israel

Oscar Nierstrasz
University of Bern, Switzerland

C. Pandu Rangan
Indian Institute of Technology, Madras, India

Bernhard Steffen
University of Dortmund, Germany

Madhu Sudan
Massachusetts Institute of Technology, MA, USA

Demetri Terzopoulos
University of California, Los Angeles, CA, USA

Doug Tygar
University of California, Berkeley, CA, USA

Moshe Y. Vardi
Rice University, Houston, TX, USA

Gerhard Weikum
Max-Planck Institute of Computer Science, Saarbruecken, Germany

James F. Peters   Andrzej Skowron (Eds.)

Transactions on Rough Sets V

 Springer

Volume Editors

James F. Peters
University of Manitoba
Department of Electrical and Computer Engineering
Winnipeg, Manitoba R3T 5V6, Canada
E-mail: jfpeters@ee.umanitoba.ca

Andrzej Skowron
Warsaw University
Institute of Mathematics
Banacha 2, 02-097 Warsaw, Poland
E-mail: skowron@mimuw.edu.pl

Library of Congress Control Number: 2006933430

CR Subject Classification (1998): F.4.1, F.1, I.2, H.2.8, I.5.1, I.4

LNCS Sublibrary: SL 1 – Theoretical Computer Science and General Issues

ISSN   1861-2059
ISBN-10   3-540-39382-X Springer Berlin Heidelberg New York
ISBN-13   978-3-540-39382-5 Springer Berlin Heidelberg New York

Springer is a part of Springer Science+Business Media

springer.com

© Springer-Verlag Berlin Heidelberg 2006

Typesetting: Camera-ready by author, data conversion by Scientific Publishing Services, Chennai, India
Printed on acid-free paper   SPIN: 11847465   06/3142   5 4 3 2 1 0

Preface

Volume V of the Transactions on Rough Sets (TRS) is dedicated to the monumental life and work of Zdzisław Pawlak[1]. During the past 35 years, since the introduction of knowledge description systems in the 1970s, the theory and applications of rough sets have grown rapidly. This volume continues the tradition begun with earlier volumes of the TRS series and introduces a number of new advances in the foundations and application of rough sets. These advances have profound implications in a number of research areas such as adaptive learning, approximate reasoning and belief systems, approximation spaces, Boolean reasoning, classification methods, classifiers, concept analysis, data mining, decision logic, decision rule importance measures, digital image processing, recognition of emotionally-charged gestures in animations, flow graphs, Kansei engineering, movie sound track restoration, multicriteria decision analysis, relational information systems, rough-fuzzy sets, rough measures, signal processing, variable precision rough set model, and video retrieval. It can be observed from the papers included in this volume that research concerning the foundations and applications of rough sets remains an intensely active research area worldwide. A total of 37 researchers from 8 countries are represented in this volume, the countries being, Canada, India, P.R. China, Poland, Japan, Taiwan, UK and the USA.

A capsule view of the life and work of Zdzisław Pawlak is included in an article at the beginning of this volume. During his lifetime, the research interests of Pawlak were rich and varied. His research ranged from his pioneering work on knowledge description systems and rough sets during the 1970s and 1980s as well as his work on the design of computers, information retrieval, modeling conflict analysis and negotiation, genetic grammars and molecular computing. Added to that was Pawlak's lifelong interest in painting, photography and poetry. During his lifetime, Pawlak nurtured worldwide interest in approximation, approximate reasoning and rough set theory and its applications. Evidence of the influence of Pawlak's work can been seen in the growth in the rough-set literature that now includes over 4000 publications as well as the growth and maturity of the International Rough Set Society.

TRS V also includes 15 papers that explore the theory of rough sets as well as new applications of rough sets. In addition, this volume of the TRS includes a complete monograph on rough sets and approximate Boolean reasoning systems that includes both the foundations as well as the applications of data mining, by Hung Son Nguyen. New developments in the foundations of rough sets are represented by a number of papers in this volume, namely, Rough Truth, Consistency and Belief Change (Mohua Banerjee), Rough Set Approximations in Formal Concept Analysis (Yiyu Yao and Yaohua Chen), Rule Importance Measures (Jiye Li and Nick Cercone), Generalized Rough-Fuzzy Approximation Operators (Wei-Zhi Wu), Rough Set Flow Graphs (Cory Butz, W. Yan, and

---

[1] Prof. Pawlak passed away on 7 April 2006.

B. Yang), Vague Concept Approximation and Adaptive Learning (Jan Bazan, Andrzej Skowron, and Roman Świniarski), and Arrow Decision Logic (Tuan-Fang Fan, Duen-Ren Liu, and Gwo-Hshiung Tzeng). Applications of rough sets are also represented by the following papers in this volume: Matching 2D Image Segments with Genetic Algorithms and Approximations Spaces (Maciej Borkowski and James Peters), Rough Set-Based Application to Recognition of Emotionally-Charged Animated Characters Gestures (Bożena Kostek and Piotr Szczuko), Movie Sound Track Restoration (Andrzej Czyżewski, Marek Dziubinski, Lukasz Litwic, and Przemyslaw Maziewski), Multimodal Classification Case Studies (Andrzej Skowron, Hui Wang, Arkadiusz Wojna, and Jan Bazan), P300 Wave Detection Using Rough Sets (Sheela Ramanna and Reza Fazel Rezai), Motion-Information-Based Video Retrieval Using Rough Pre-classification (Zhe Yuan, Yu Wu, Guoyin Wang, and Jianbo Li), Variable Precision Baysian Rough Set Model and Its Application to Kansei Engineering (Tatsuo Nishino, Mitsuo Nagamachi, and Hideo Tanaka).

The Editors of this volume extend their hearty thanks to the reviewers of the papers that were submitted to this TRS volume: Mohua Banerjee, Jan Bazan, Teresa Beauboeuf, Maciej Borkowski, Gianpiero Cattaneo, Nick Cercone, Davide Cuicci, Andrzej Czyżewski, Jitender Deogun, Ivo Düntsch, Reza Fazel-Rezai, Anna Gomolińska, Jerzy Grzymała-Busse, Masahiro Iniguichi, Jouni Järvinen, Mieczysław Kłopotek, Beata Konikowska, Bożena Kostek, Marzena Kryszkiewicz, Rafał Latkowski, Churn-Jung Liau, Pawan Lingras, Jan Małuszyński, Benedetto Matarazzo, Michał Mikołajczyk, Mikhail Moshkov, Maria Nicoletti, Hoa Nguyen, Son Nguyen, Piero Pagliani, Sankar Pal, Witold Pedrycz, Lech Polkowski, Anna Radzikowska, Vijay Raghavan, Sheela Ramanna, Zbigniew Raś, Dominik Ślęzak, Jerzy Stefanowski, Jarosław Stepaniuk, Zbigniew Suraj, Roman Świniarski, Piotr Synak, Marcin Szczuka, Daniel Vanderpooten, Dimiter Vakarelov, Alicja Wieczorkowska, Arkadiuz Wojna, Marcin Wolski, Jakub Wróblewski, Dan Wu, Wei-Zhi Wu, Yiyu Yao, and Wojciech Ziarko.

This issue of the TRS was made possible thanks to the reviewers as well as to the laudable efforts of a great many generous persons and organizations. The editors and authors of this volume also extend an expression of gratitude to Alfred Hofmann, Ursula Barth, Christine Günther and the other LNCS staff at Springer for their support in making this volume of the TRS possible. In addition, the editors of this volume extend their thanks to Dominik Ślęzak for his help and suggestions concerning extensions of selected RSFDGrC 2005 papers included in this volume of the TRS. We anticipate that additional RSFDGrC 2005 papers now being reviewed will be included in future volumes of the TRS. We also extend our thanks to Marcin Szczuka for his consummate skill and care in the compilation of this volume. The Editors of this volume have been supported by the Ministry for Scientific Research and Information Technology of the Republic of Poland, research grant No. 3T11C00226, and the Natural Sciences and Engineering Research Council of Canada (NSERC) research grant 185986 respectively.

June 2006   James F. Peters
                                                     Andrzej Skowron

LNCS Transactions on Rough Sets

This journal subline has as its principal aim the fostering of professional exchanges between scientists and practitioners who are interested in the foundations and applications of rough sets. Topics include foundations and applications of rough sets as well as foundations and applications of hybrid methods combining rough sets with other approaches important for the development of intelligent systems.

The journal includes high-quality research articles accepted for publication on the basis of thorough peer reviews. Dissertations and monographs up to 250 pages that include new research results can also be considered as regular papers. Extended and revised versions of selected papers from conferences can also be included in regular or special issues of the journal.

Editors-in-Chief: James F. Peters, Andrzej Skowron

Editorial Board

M. Beynon	M. do C. Nicoletti
G. Cattaneo	H.S. Nguyen
M.K. Chakraborty	S.K. Pal
A. Czyżewski	L. Polkowski
J.S. Deogun	H. Prade
D. Dubois	S. Ramanna
I. Duentsch	R. Słowiński
S. Greco	J. Stefanowski
J.W. Grzymała-Busse	J. Stepaniuk
M. Inuiguchi	R. Świniarski
J. Järvinen	Z. Suraj
D. Kim	M. Szczuka
J. Komorowski	S. Tsumoto
C.J. Liau	G. Wang
T.Y. Lin	Y. Yao
E. Menasalvas	N. Zhong
M. Moshkov	W. Ziarko
T. Murai	

Table of Contents

Dissertations and Monographs

Zdzisław Pawlak: Life and Work
1926-2006

James F. Peters and Andrzej Skowron

In the history of mankind, Professor Zdzisław Pawlak, Member of the Polish Academy of Sciences, will be remembered as a great human being with exceptional humility, wit and kindness as well as an extraordinarily innovative researcher with exceptional stature. His legacy is rich and varied. Pawlak's research contributions have had far-reaching implications inasmuch as his works are fundamental in establishing new perspectives for scientific research in a wide spectrum of fields.

Preamble

Professor Pawlak's most widely recognized contribution is his incisive approach to classifying objects with their attributes (features) and his introduction of approximation spaces, which establish the foundations of granular computing and provide frameworks for perception and knowledge discovery in many areas. He was with us only for a short time and, yet, when we look back at his accomplishments, we realize how greatly he has influenced us with his generous spirit and creative work in many areas such as approximate reasoning, intelligent systems research, computing models, mathematics (especially, rough set theory), molecular computing, pattern recognition, philosophy, art, and poetry. This article attempts to give a vignette that highlights some of Pawlak's remarkable accomplishments. This vignette is limited to a brief coverage of Pawlak's work in rough set theory, molecular computing, philosophy, painting and poetry. Detailed coverage of these as well as other accomplishments by Pawlak is outside the scope of this commemorative article.

1 Introduction

This article commemorates the life, work and creative genius of Zdzisław Pawlak. He is well-known for his innovative work on the classification of objects by means

J.F. Peters and A. Skowron (Eds.): Transactions on Rough Sets V, LNCS 4100, pp. 1–24, 2006.

of attributes (features) [25] and his discovery of rough set theory during the early 1980s (see, e.g., [11, 22, 25, 27]). Since the introduction of rough set theory, there have been well over 4000 publications on this theory and its applications (see, e.g., [6, 35, 36, 37, 39, 71] and Section 12).

One can also observe a number of other facets of Pawlak's life and work that are less known, namely, his pioneering work on genetic grammars and molecular computing, his interest in philosophy, his lifelong devotion to painting landscapes and waterscapes depicting the places he visited, his interest and skill in photography, and his more recent interests in poetry and methods of solving mysteries by fictional characters such as Sherlock Holmes. During his life, Pawlak contributed to the foundations of granular computing, intelligent systems research, computing models, mathematics (especially, rough set theory), molecular computing, knowledge discovery as well as knowledge representation, and pattern recognition.

This article attempts to give a brief vignette that highlights some of Pawlak's remarkable accomplishments. This vignette is limited to a brief coverage of Pawlak's works in rough set theory, molecular computing, philosophy, painting and poetry. Detailed coverage of these as well as other accomplishments by Pawlak is outside the scope of this commemorative article.

The article is organized as follows. A brief biography of Zdzisław Pawlak is given in Sect. 2. Some of the very basic ideas of Pawlak's rough set theory are presented in Sect. 3. This is followed by a brief presentation of Pawlak's introduction of a genetic grammar and molecular computing in Sect. 8. Pawlak's more recent reflections concerning philosophy (especially, the philosophy of mathematics) are briefly covered in Sect. 9. Reflections on Pawlak's lifelong interest in painting and nature as well as a sample of paintings by Pawlak and a poem coauthored by Pawlak, are presented in Sect. 10.

2   Zdzisław Pawlak: A Brief Biography

Zdzisław Pawlak was born on 10 November 1926 in Łódź, 130 km south-west from Warsaw, Poland [41]. In 1947, Pawlak began studying in the Faculty of Electrical Engineering at Łódź University of Technology, and in 1949 continued his studies in the Telecommunication Faculty at Warsaw University of Technology. Starting in the early 1950s and continuing throughout his life, Pawlak painted the places he visited, especially landscapes and waterscapes reflecting his observations in Poland and other parts of the world. This can be seen as a continuation of the work of his father, who was fond of wood carving and who carved a wooden self-portrait that was kept in Pawlak's study. He also had extraordinary skill in mathematical modeling in the organization of systems (see, e.g., [20, 24, 28]) and in computer systems engineering (see, e.g., [16, 17, 18, 19, 21]). During his early years, he was a pioneer in the designing computing machines. In 1950, Pawlak presented the first project of a computer called GAM 1. He completed his M.Sc. in Telecommunication Engineering in 1951. Pawlak's publication in 1956 on a new method for random number generation was the first article in informatics

1.1: Interior of UMC1   1.2: UMC1 Prototype

Fig. 1. Snapshots of the UMC1 Computer System

published abroad by a researcher from Poland [13]. In 1958, Pawlak completed his doctoral degree from the Institute of Fundamental Technological Research at the Polish Academy of Science with a Thesis on *Applications of Graph Theory to Decoder Synthesis*. In 1961, Pawlak was also a member of a research team that constructed one of the first computers in Poland called UMC 1 (see Fig. 1).

The original arithmetic for the UMC1 computer system with base "-2" was due to Pawlak [14]. He received his habilitation from the Institute of Mathematics at the Polish Academy of Sciences in 1963. In his habilitation entitled *Organization of Address-Less Machines*, Pawlak proposed and investigated parenthesis-free languages, a generalization of polish notation introduced by Jan Łukasiewicz (see, e.g., [16,17]).

In succeeding years, Pawlak worked at the Institute of Mathematics of Warsaw University and, in 1965, introduced foundations for modeling DNA [15] in what has come to be known as molecular computing [3,15]. He also proposed a new formal model of a computing machine known as the Pawlak machine [21,23] that is different from the Turing machine and from the von Neumann machine. In 1973, he introduced knowledge representation systems [22] as part of his work on the mathematical foundations of information retrieval (see, e.g., [11,22]). In the early 1980s, he was part of a research group at the Institute of Computer Science of the Polish Academy of Sciences, where he discovered rough sets and the idea of classifying objects by means of their attributes [25], which was the basis for extensive research in rough set theory during the 1980s (see, e.g., [7,8,12,26,27, 29]). During the succeeding years, Pawlak refined and amplified the foundations of rough sets and their applications, and nurtured worldwide research in rough sets that has led to over 4000 publications (see, e.g., [39]). In addition, he did extensive work on the mathematical foundations of information systems during the early 1980s (see, e.g., [24,28]). He also invented a new approach to conflict analysis (see, e.g., [30,31,33,34]).

During his later years, Pawlak's interests were very diverse. He developed a keen interest in philosophy, especially in the works by Łukasiewicz (logic and probability), Leibniz (*identify of indiscernibles*), Frege (membership, sets), Russell (antinomies), and Leśniewski (*being a part*)). Pawlak was also interested in the works of detective fiction by Sir Arthur Conan Doyle (especially, Sherlock Holmes' fascination with data as a basis for solving mysteries) (see, e.g., [35]).

Finally, Zdzisław Pawlak gave generously of his time and energy to help others. His spirit and insights have influenced many researchers worldwide. During his life, he manifested an extraordinary talent for inspiring his students and colleagues as well as many others outside his immediate circle. For this reason, he was affectionately known to some of us as Papa Pawlak.

3   Rough Sets

> If we classify objects by means of attributes,
> exact classification is often impossible.
> – Zdzisław Pawlak, January 1981.

A brief presentation of the foundations of rough set theory is given in this section. Rough set theory has its roots in Zdzisław Pawlak's research on knowledge representation systems during the early 1970s [22]. Rather than attempt to classify objects *exactly* by means of attributes (features), Pawlak considered an approach to solving the object classification problem in a number of novel ways. First, in 1973, he formulated knowledge representation systems (see, e.g., [11, 22]). Then, in 1981, Pawlak introduced approximate descriptions of objects and considered knowledge representation systems in the context of upper and lower classification of objects relative to their attribute values [25, 26]. We start with a system $S = (X, A, V, \sigma)$, where X is a non-empty set of objects, A is a set of attributes, V is a union of sets V_a of values associated with each $a \in A$, and σ is called a knowledge function defined as the mapping $\sigma : X \times A \to V$, where $\sigma(x, a) \in V_a$ for every $x \in X$ and $a \in A$. The function σ is referred to as *knowledge function* about objects from X. The set X is partitioned into elementary sets that later were called blocks, where each elementary set contains those elements of X which have matching attribute values. In effect, a block (elementary set) represents a granule of knowledge (see Fig. 2.2). For example, for any $B \subseteq A$ the B-elementary set for an element $x \in X$ is denoted by $B(x)$, which is defined by

$$B(x) = \{y \in X \mid \forall a \in B \ \sigma(x, a) = \sigma(y, a)\} \tag{1}$$

Consider, for example, Fig. 2.1 which represents a system S containing a set X of colored circles and a feature set A that contains only one attribute, namely, *color*. Assume that each circle in X has only one color. Then the set X is partitioned into elementary sets or blocks, where each block contains circles with the same color. In effect, elements of a set $B(x) \subseteq X$ in a system S are classified as *indiscernible* if they are indistinguishable by means of their feature values for any $a \in B$. A set of *indiscernible* elements is called an *elementary set* [25]. Hence,

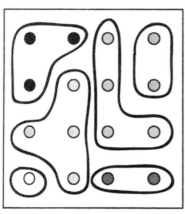

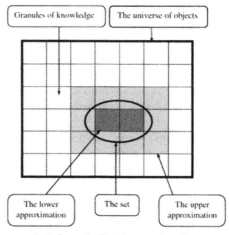

2.1: Blocks of Objects   2.2: Sample Set Approximation

Fig. 2. Rudiments of Rough Sets

any subset $B \subseteq A$ determines a partition $\{B(x) : x \in X\}$ of X. This partition defines an equivalence relation $Ind(B)$ on X called an *indiscernibility* relation such that $x Ind(B) y$ if and only if $y \in B(x)$ for every $x, y \in X$. Assume that $Y \subseteq X$ and $B \subseteq A$, and consider an approximation of the set Y by means of the attributes in B and B-indiscernible blocks in the partition of X. The union of all blocks that constitute a subset of Y is called the *lower approximation* of Y (usually denoted by B_*Y), representing certain knowledge about Y. The union of all blocks that have non-empty intersection with the set Y is called the *upper approximation* of Y (usually denoted by B^*Y), representing uncertain knowledge about Y. The set $BN_B(Y) = B^*Y - B_*Y$ is called the B-boundary of the set Y. In the case where $BN_B(Y)$ is non-empty, the set Y is a *rough (imprecise)* set. Otherwise, the set Y is a *crisp* set. This approach to classification of objects in a set is represented graphically in Fig. 2.2, where the region bounded by the ellipse represents a set Y, the darkened blocks inside Y represent B_*Y, the gray blocks represent the boundary region $BN_B(Y)$, and the gray and the darkened blocks taken together represent B^*Y.

Consequences of this approach to the classification of objects by means of their feature values have been remarkable and far-reaching. Detailed accounts of the current research in rough set theory and its applications are available, e.g., in [35, 36, 37] (see also Section 12).

4   Approximation

Some categories (subsets of objects) cannot be expressed exactly by employing available knowledge. Hence, we arrive at the idea of approximation of a set by other sets.

–Zdzisław Pawlak, 1991.

One of the most profound, very important notions underlying rough set theory is approximation. In general, an *approximation* is defined as the replacement of objects by others that resemble the original objects in certain respects [4]. For example, consider a universe U containing objects representing behaviors of agents. In that case, we can consider blocks of behaviors in the partition U/R, where the behaviors within a block resemble (are *indiscernible* from) each other by virtue of their feature values. Then any subset X of U can be approximated by blocks that are either proper subsets of X (lower approximation of the set X denoted $\underline{R}X$) or by blocks having one or more elements in common with X (upper approximation of the set X denoted $\overline{R}X$)[1]. In rough set theory, the focus is on approximating one set of objects by means of another set of objects based on the feature values of the objects [32]. The lower approximation operator $\underline{R}$ has properties that correspond closely to properties of what is known as the Π_0 topological *interior* operator [27,77]. Similarly, the upper approximation operator $\overline{R}$ has properties that correspond closely to properties of the Π_0 topological *closure* operator [27,77]. It was observed in [27] that the key to the rough set approach is provided by the exact mathematical formulation of the concept of approximative (rough) equality of sets in a given approximation space.

5   Approximation Spaces

> The key to the presented approach
> is provided by the exact mathematical formulation,
> of the concept of approximative (rough) equality
> of sets in a given approximation space.
>   –Zdzisław Pawlak, 1982.

In [32], an approximation space is represented by the pair (U, R), where U is a universe of objects, and $R \subseteq U \times U$ is an indiscernibility relation (denoted Ind as in Sect. 3) defined by an attribute set (i.e., $R = Ind(A)$ for some attribute set A). In this case, R is an equivalence relation. Let $[x]_R$ denote an equivalence class of an element $x \in U$ under the indiscernibility relation R, where $[x]_R = \{y \in U : xRy\}$.

In this context, R-approximations of any set $X \subseteq U$ are based on the exact (crisp) containment of sets. Then set approximations are defined as follows:

- $x \in U$ belongs with certainty to $X \subseteq U$ (i.e., x belongs to the R-lower approximation of X), if $[x]_R \subseteq X$.
- $x \in U$ possibly belongs $X \subseteq U$ (i.e., x belongs to the R-upper approximation of X), if $[x]_R \cap X \neq \varnothing$.

---

[1] In more recent years, the notation $R_* X$, $R^* X$ has been often used (see, e.g., Sect. 3) to denote lower and upper approximation, respectively, since this notation is more "typewriter" friendly.

- $x \in U$ belongs with certainty neither to the X nor to $U - X$ (i.e., x belongs to the R-boundary region of X), if $[x]_R \cap (U - X) \neq \oslash$ and $[x]_R \cap X \neq \oslash$.

Several generalizations of the above approach have been proposed in the literature (see, e.g., [35, 36, 37] and Section 12). In particular, in some of these approaches, set inclusion to a degree is used instead of the exact inclusion.

6   Generalizations of Approximation Spaces

Several generalizations of the classical rough set approach based on approximation spaces defined as pairs of the form (U, R), where R is the equivalence relation (called an indiscernibility relation) on the non-empty set U, have been reported in the literature. Let us mention two of them.

A generalized approximation space can be defined by a tuple $GAS = (U, N, \nu)$ where N is a *neighborhood function* defined on U with values in the powerset $\mathcal{P}(U)$ of U (i.e., $N(x)$ is the *neighborhood* of x) and ν is the *overlap function* defined on the Cartesian product $\mathcal{P}(U) \times \mathcal{P}(U)$ with values in the interval $[0, 1]$ measuring the degree of overlap of sets. The lower GAS_* and upper GAS^* approximation operations can be defined in a GAS by Eqs. 2 and 3.

$$GAS_*(X) = \{x \in U : \nu(N(x), X) = 1\}, \tag{2}$$
$$GAS^*(X) = \{x \in U : \nu(N(x), X) > 0\}. \tag{3}$$

In the standard case, $N(x)$ equals the equivalence class $B(x)$ or block of the indiscernibility relation $Ind(B)$ for a set of features B. In the case where R is a tolerance (similarity) relation[2], $\tau \subseteq U \times U$, we take $N(x) = \{y \in U : x\tau y\}$, i.e., $N(x)$ equals the tolerance class of τ defined by x. The standard inclusion relation ν_{SRI} is defined for $X, Y \subseteq U$ by Eq. 4.

$$\nu_{SRI}(X, Y) = \begin{cases} \frac{|X \cap Y|}{|Y|}, & \text{if } Y \neq \emptyset, \\ 1, & \text{otherwise.} \end{cases} \tag{4}$$

For applications, it is important to have some constructive definitions of N and ν.

One can consider another way to define $N(x)$. Usually together with a GAS, we consider some set F of formulas describing sets of objects in the universe U of the GAS defined by semantics $\| \cdot \|_{GAS}$, i.e., $\|\alpha\|_{GAS} \subseteq U$ for any $\alpha \in F$. Now, one can take the set the neighborhood function as shown in Eq. 5.

$$N_F(x) = \{\alpha \in F : x \in \|\alpha\|_{GAS}\}, \tag{5}$$

and $N(x) = \{\|\alpha\|_{GAS} : \alpha \in N_F(x)\}$. Hence, more general uncertainty functions having values in $\mathcal{P}(U)$ can be defined and as a consequence different definitions

---

[2] Recall that a *tolerance* is a binary relation $R \subseteq U \times U$ on a set U having the reflexivity and symmetry properties, i.e., xRx for all $x \in U$ and xRy implies yRx for all $x, y \in U$.

of approximations are considered. For example, one can consider the following definitions of approximation operations in GAS defined in Eqs. 6 and 7.

$$GAS_\circ(X) = \{x \in U : \nu(Y, X) = 1 \text{ for some } Y \in N(x)\}, \tag{6}$$

$$GAS^\circ(X) = \{x \in U : \nu(Y, X) > 0 \text{ for any } Y \in N(x)\}. \tag{7}$$

There are also different forms of rough inclusion functions. Let us consider two examples.

In the first example of a rough inclusion function, a threshold $t \in (0, 0.5)$ is used to relax the degree of inclusion of sets. The rough inclusion function ν_t is defined by Eq. 8.

$$\nu_t(X, Y) = \begin{cases} 1, & \text{if} & \nu_{SRI}(X, Y) \geq 1 - t, \\ \frac{\nu_{SRI}(X,Y)-t}{1-2t}, & \text{if} & t \leq \nu_{SRI}(X, Y) < 1 - t, \\ 0, & \text{if} & \nu_{SRI}(X, Y) \leq t. \end{cases} \tag{8}$$

One can obtain approximations considered in the variable precision rough set approach (VPRSM) by substituting in (2)-(3) the rough inclusion function ν_t defined by (8) instead of ν, assuming that Y is a decision class and $N(x) = B(x)$ for any object x, where B is a given set of attributes.

Another example of application of the standard inclusion was developed by using probabilistic decision functions.

The rough inclusion relation can be also used for function approximation and relation approximation. In the case of function approximation the inclusion function ν^* for subsets $X, Y \subseteq U \times U$, where $X, Y \subseteq \mathcal{R}$ and $\mathcal{R}$ is the set of reals, is defined by Eq. 9.

$$\nu^*(X, Y) = \begin{cases} \frac{card(\pi_1(X \cap Y))}{card(\pi_1(X))}, & \text{if } \pi_1(X) \neq \emptyset, \\ 1, & \text{if } \pi_1(X) = \emptyset, \end{cases} \tag{9}$$

where π_1 is the projection operation on the first coordinate. Assume now, that X is a cube and Y is the graph $G(f)$ of the function $f : \mathcal{R} \longrightarrow \mathcal{R}$. Then, e.g., X is in the lower approximation of f if the projection on the first coordinate of the intersection $X \cap G(f)$ is equal to the projection of X on the first coordinate. This means that the part of the graph $G(f)$ is "well" included in the box X, i.e., for all arguments that belong to the box projection on the first coordinate the value of f is included in the box X projection on the second coordinate.

The approach based on inclusion functions has been generalized to the *rough mereological approach*. The inclusion relation $x\mu_r y$ with the intended meaning x *is a part of y to a degree at least r* has been taken as the basic notion of the rough mereology being a generalization of the Leśniewski mereology [9,10]. Research on rough mereology has shown the importance of another notion, namely *closeness* of complex objects (e.g., concepts). This can be defined by $xcl_{r,r'}y$ if and only if $x\mu_r y$ and $y\mu_{r'}x$.

Rough mereology offers a methodology for synthesis and analysis of objects in a distributed environment of intelligent agents, in particular, for synthesis of objects satisfying a given specification to a satisfactory degree or for control in such a complex environment. Moreover, rough mereology has been recently used for developing the foundations of the *information granule calculi*, aiming at formalization of the Computing with Words paradigm, recently formulated by Lotfi Zadeh [42]. More complex information granules are defined recursively using already defined information granules and their measures of inclusion and closeness. Information granules can have complex structures like classifiers or approximation spaces. Computations on information granules are performed to discover relevant information granules, e.g., patterns or approximation spaces for complex concept approximations.

Usually families of approximation spaces labeled by some parameters are considered. By tuning such parameters according to chosen criteria (e.g., minimal description length), one can search for the optimal approximation space for a concept description.

7 Conflict Analysis and Negotiations

Conflict analysis and resolution play an important role in business, governmental, political and legal disputes, labor-management negotiations, military operations and others. To this end many mathematical formal models of conflict situations have been proposed and studied.

Various mathematical tools, e.g., game theory, graph theory, topology, differential equations and others, have been used for that purpose. In fact, as yet there is no "universal" theory of conflicts. Instead, mathematical models of conflict situations are strongly domain dependent.

Zdzisław Pawlak introduced still another approach to conflict analysis, based on some ideas of rough set theory [30, 31, 33, 34, 37]. Pawlak's model is simple enough for easy computer implementation and is adequate for many real-life applications.

The approach is based on the conflict relation in data. Formally, the conflict relation can be seen as a negation (not necessarily, classical) of the indiscernibility relation which was used by Pawlak as a basis of rough set theory. Thus, indiscernibility and conflict are closely related from a logical point of view. It turns out that the conflict relation can be used in conflict analysis studies.

8 Molecular Computing

> The understanding of protein structure and the processes of their syntheses is fundamental for the considerations of the life problem.
>
> – Zdzisław Pawlak, 1965.

Zdzisław Pawlak was one of the pioneers of a research area known as molecular computing (see, e.g., ch. 6 on Genetic Grammars published in 1965 [15]). He searched for grammars generating compound biological structures from simpler ones, e.g., proteins from amino acids. He proposed a generalization of the traditional grammars used in formal language theory. For example, he considered the construction of mosaics on a plane from some elementary mosaics by using some production rules for the composition. He also presented a language for linear representation of mosaic structures. By introducing such grammars one can better understand the structure of proteins and the processes that lead to their synthesis. Such grammars result in real-life languages that characterize the development of living organisms. During the 1970s, Pawlak was interested in developing a formal model of *deoxyribonucleic acid* (DNA), and he proposed a formal model for the genetic code discovered by Crick and Watson. Pawlak's model is regarded by many as the first complete model of DNA. This work on DNA by Pawlak has been cited by others (see, e.g., [3, 41]).

9   Philosophy

No doubt the most interesting proposal was given
by the Polish logician Stanisław Lesniewski,
who introduced the relation of "being a part"
instead of the membership relation between elements
and sets employed in classical set theory.
– Zdzisław Pawlak, 2006.

For many years, Zdzisław Pawlak had an intense interest in philosophy, especially regarding the connections between rough sets and other forms of sets. It was Pawlak's venerable habit to point to connections between his own work in rough sets and the works of others in philosophy and mathematics. This is especially true relative to two cardinal notions, namely, sets and vagueness. For the classical notion of a set, Pawlak called attention to works by Cantor, Frege and Bertrand Russell. Pawlak observed that the notion of a set is not only fundamental for the whole of mathematics but also for natural language, where it is commonplace to speak in terms of collections of such things as books, paintings, people, and their vague properties [35].

In his reflections on structured objects, Pawlak pointed to the work on mereology by Stanisław Leśniewski, where the relation *being a part* replaces the membership relation $\in$. Of course, in recent years, the study of Leśniewski's work has led to rough mereology and the relation *being a part to a degree* in 1996 (see, e.g., [38] cited by Pawlak in [35]).

For many years, Pawlak was also interested in vagueness and Gottlob Frege's notion of the boundary of a concept (see, e.g., [2,5]). For Frege, the definition of a concept must unambiguously determine whether or not an object falls under the concept. For a concept without a sharp boundary, one is faced with the problem of determining how close an object must be before it can be said to belong to a concept. Later, this problem of sharp boundaries shows up as a repeated motif in landscapes and waterscapes painted by Pawlak (see, e.g., Fig. 5.1 and Fig. 5.2).

Pawlak also observed out that mathematics must use crisp, not vague concepts. Hence, mathematics makes it possible to reason precisely about approximations of vague concepts. These approximations are temporal and subjective [35].

Professor Zdzisław Pawlak was very happy when he recognized that the rough set approach is consistent with a very old Chinese philosophy that is reflected in a recent poem from P.R. China (see Fig. 3).

The poem in Fig. 3 was written by Professor Xuyan Tu, the Honorary President of the Chinese Association for Artificial Intelligence, to celebrate the establishment of the Rough Set and Soft Computation Society at the Chinese Association for Artificial Intelligence, in Guangzhou, 21 December 2003. A number of English translations of this poem are possible. Consider, for example, the following two translations of the poem in Fig. 3, which capture the spirit of the poem and its allusion to the fact that rough sets hearken back to a philosophy rooted in ancient China.

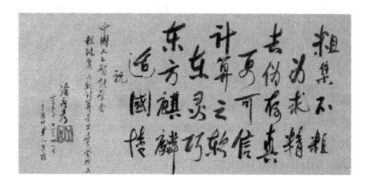

Fig. 3. Poem about Rough Sets in Chinese

Rough sets are not rough, and one moves towards precision. One removes the "unbelievable" so that what remains is more believable. The soft part of computing is nimble. Rough sets imply a philosophy rooted in China. Anonymous 8 January 2005

Rough sets are not "rough" for the purpose of searching for accuracy. It is a more reliable and believable theory that avoids falsity and keeps the truth. The essence of soft computing is its flexibility. [Rough Sets] reflect the oriental philosophy and fit the Chinese style of thinking. Xuyan Tu, Poet Yiyu Yao, Translator 21 December 2003

The 8 January 2005 anonymous translation is a conservative rendering of the Chinese characters in a concise way in English. The 21 December 2003 translation is more interpretative, and reflects the spirit of an event as seen by the translator in the context of the opening of the Institute of Artificial Intelligence in P.R. China.

Fig. 4. Zdzisaw Pawlak in Snow Country

10   Painting and Nature

Zdzisław Pawlak was an astute observer of nature and was very fond of spending time exploring and painting the woodlands, lakes and streams of Poland. A picture showing Pawlak during a walk in snow-covered woods is shown in Fig. 4. Starting in the early 1950s and continuing for most of his life, Pawlak captured what he observed by painting landscapes and waterscapes. Sample paintings by Pawlak are shown in Fig. 5 and Fig. 6.

A common motif in Pawlak's paintings is the somewhat indefinite separation between objects such as the outer edges of trees and sky (see, e.g., Fig. 5.3, Fig. 5.4 and 6.1). In Fig. 6.1, there is a blurring (uncertain boundary) between the tops of the trees and shrubs against the sky. Notice how the separation between the reeds in the foreground and the water on the far side of the reeds is rather indistinct in Fig. 6.1 (i.e., there is no sharp boundary between the reeds and water). This blurring the boundaries between tree shadows and water is also particularly pronounced in Fig. 6.1, Fig. 5.3 and Fig. 5.4. There is considerable charm in Fig. 5.4, where there is a colorful blending of the tree shadows, water and the surrounding land. The boundaries of objects evident in Pawlak's paintings are suggestive of the theoretical idea of the boundary between the lower and upper approximations of a set in rough set theory. There is also in Pawlak's

5.1: 1954 Landscape by Pawlak

5.2: 1999 Watercape by Pawlak

5.3: Treeline Painting by Pawlak

5.4: 1999 Tree Shadows by Pawlak

Fig. 5. Paintings by Zdzisław Pawlak

paintings an apparent fascination with containment of similar objects such as the roadway bordered by gorse in Fig. 6.3, line of haystacks in a field in Fig. 6.4, distant mountains framed by a border of evergreens and flora in the foreground in Fig. 5.3 as well as in Fig. 6.2 of the parts of a tree shadows shimmering in 'the water in Fig. 6.1 or the pixels clustered together to represent a distant building (see, e.g., Fig. 5.2). In some sense, the parts of a tree shadow or the parts of the roof of a distant building are indiscernible from each other.

The water shadows can be considered as approximations (substitutions) for the reflected objects in Fig. 5.3 and Fig. 5.4. To see this, try the following experiment. Notice that every pixel (picture element) with coordinates (x, y) has 4 neighbors at $(x+1, y)$, $(x-1, y)$, $(x, y+1)$, and $(x, y-1)$, which constitute what is known as a 4-neighborhood of an image. An image segment is a collection of 4-neighborhood connected pixels with the same color. Let U consist of the color segments in Fig. 5.4, and consider only the shape and color of the segments in U. The image segments making up the trees have "reflected" segments in tree shadows in Fig. 5.4. Mask or cover up the image segments contained in the trees

6.1: Reeds by Pawlak   6.2: Vista by Pawlak

6.3: Mountains by Pawlak   6.4: Hay Field by Pawlak

Fig. 6. Other Paintings by Zdzisław Pawlak

along the distant shoreline, then segments in the tree shadows can be used to approximate the corresponding segments of the trees shown in Fig. 5.4. To see this, go a step further, repaint the vacant space in the masked area of the painting with image segments from the tree shadows in Fig. 5.4. The new version of the painting will be approximately like the original painting. This approximation will vary depending on the time of day and the length of the tree shadows.

11   Poetry

In more recent years, Zdzisław Pawlak wrote poems, which are remarkably succinct and very close to the philosophy of rough sets as well as his interest in painting. In his poems, one may find quite often some reflections which most probably stimulated him in the discovery of the rough sets, where there is a focus on border regions found in scenes from nature. A sample poem coauthored by Pawlak is given next (each line of the English is followed by the corresponding Polish text).

Near To
Blisko

How near to the bark of a tree are the drifting snowflakes,
Jak blisko kory drzew płatki śniegu tworzą zaspy,
swirling gently round, down from winter skies?
Wirując delikatnie, gdy spadają z zimowego nieba?

How near to the ground are icicles,
Jak blisko ziemi są sople lodu,

slowing forming on window ledges?
Powoli formujące się na okiennych parapetach?

Sometimes snow-laden branches of some trees droop,
Czasami, gałęzie drzew zwieszają się pod ciężarem śniegu,

some near to the ground,
niektóre prawie do samej ziemi,

some from to-time-to-time swaying in the wind,
niektóre od czasu do czasu kołyszą się na wietrze,

some nearly touching each other as the snow falls,
niektóre niemal dotykają się wzajemnie, gdy śnieg pada,

some with shapes resembling the limbs of ballet dancers,
niektóre o kształtach przypominających kończyny baletnic,

some with rough edges shielded from snowfall and wind,
niektóre o nierównych rysach, osłonięte przed śniegiem i wiatrem,

and then,
i potem,

somehow,
w jakiś sposób,

spring up again in the morning sunshine.
Wyrastają na nowo w porannym słońcu.

How near to ...
Jak już blisko do ...

– Z. Pawlak and J.F. Peters,
Spring, 2002.

The poem entitled *Near To* took its inspiration from an early landscape painted by Pawlak in 1954, which is shown in Fig. 5.1.

12   Outgrowth of Research by Zdzisław Pawlak

This section briefly introduces the literature in that has been inspired by the Zdzisław Pawlak's research in rough set theory and applications.

12.1   Journals

Evidence in the growth in the research in the foundations of rough set theory and its many applications can be found in the *Transactions on Rough Sets* (*TRS*), which is published by Springer as a journal subline of the Lecture Notes in Computer Science [71]. The *TRS* has as its principal aim the fostering of professional exchanges between scientists and practitioners who are interested in the foundations and applications of rough sets. Topics include foundations and applications of rough sets as well as foundations and applications of hybrid methods combining rough sets with other approaches important for the development of intelligent systems. We are observing a growing research interest in the foundations of rough sets, including the various logical, mathematical and philosophical aspects of rough sets. Some relationships have already been established between rough sets and other approaches, and also with a wide range of hybrid systems. As a result, rough sets are linked with decision system modeling and analysis of complex systems, fuzzy sets, neural networks, evolutionary computing, data mining and knowledge discovery, pattern recognition, machine learning, and approximate reasoning. In particular, rough sets are used in probabilistic reasoning, granular computing (including information granule calculi based on rough mereology), intelligent control, intelligent agent modeling, identification of autonomous systems, and process specification. A wide range of applications of methods based on rough set theory alone or in combination with other approaches have been discovered in the following areas: acoustics, biology, business and finance, chemistry, computer engineering (e.g., data compression, digital image processing, digital signal processing, parallel and distributed computer systems, sensor fusion, fractal engineering), decision analysis and systems, economics, electrical engineering (e.g., control, signal analysis, power systems), environmental studies, digital image processing, informatics, medicine, molecular biology, musicology, neurology, robotics, social science, software engineering, spatial visualization, Web engineering, and Web mining. The journal includes high-quality research articles accepted for publication on the basis of thorough peer reviews. Dissertations and monographs up to 250 pages that include new research results can also be considered as regular papers. Extended and revised versions of selected papers from conferences can also be included in regular or special issues of the journal (see, e.g., [72, 73, 74, 75]). In addition, articles that have appeared in journals such as Communications of ACM [67], Computational Intelligence [95], Fundamenta Informaticae (see, e.g., [45], International Journal of Intelligent Systems [69, 70], Journal of the Intelligent Automation and Soft Computing [56], Neurocomputing [63], and Pattern Recognition Letters [83].

In the period 1997-2002, many articles on rough sets have been published in Bulletin of the International Rough Sets Society [90].

12.2   Conferences

The wide spectrum of research in rough sets and its applications can also be gauged by a number of international conferences. The premier conference of the International Rough Set Society (IRSS)[3] is the Internationl Conference on Rough Sets and Current Trends in Computing (RSCT) was held for first time in Warsaw, Poland in 1998. It was followed by successful RSCTC conferences in Banff, Canada (2000), in Malvern, U.S.A. (2002) and in Uppsala, Sweden (2004) [43,53,79,92,98]. RSCTC is an outgrowth of a series of annual International Workshops devoted to the subject of rough sets, started in Poznan, Poland in 1992, and then held alternatively in Canada, the USA, Japan and China (RSKD, RSSC, RSFDGrC, RSGrC series). The next RSCT will be held in Kobe, Japan in 2006. The aim of the RSCTC conference is to provide researchers and practitioners interested in new information technologies an opportunity to highlight innovative research directions, novel applications, and a growing number of relationships between rough sets and such areas as computational intelligence, knowledge discovery and data mining, intelligent information systems, web mining, synthesis and analysis of complex objects and non-conventional models of computation.

The IRSS also sponsors two other international conferences, namely, Rough Sets, Fuzzy Sets, Data Mining, and Granular Computing (RSFDGrC) (see, e.g., [51,82,86,87,91,93,95]) and Rough Sets and Knowledge Discovery (RSKD), held for the first time in 2006 in Chongqing, P.R. China [76]. RSFDGrC 2005 [86, 87] was a continuation of international conferences and workshops devoted to the subject of rough sets, held alternatively in Canada, P.R. China, Japan, Poland, Sweden, and the USA. RSFDGrC achieved the status of bi-annual international conference starting from the year of 2003 in Chongqing, P.R. China. This conference encompasses rough sets and fuzzy sets, granular computing, as well as knowledge discovery and data mining. RSKT 2006 [76] provides a forum for researchers in rough sets and knowledge technology. Rough set theory is closely related to knowledge technology in a variety of forms such as knowledge discovery, approximate reasoning, intelligent and multiagent systems design, knowledge intensive computations that signal the emergence of a knowledge technology age. The essence of growth in cutting-edge, state-of-the-art and promising knowledge technologies is closely related to learning, pattern recognition, machine intelligence and automation of acquisition, transformation, communication, exploration and exploitation of knowledge. A principal thrust of such technologies is the utilization of methodologies that facilitate knowledge processing. The focus of the RSKT conference is to present state-of-the-art scientific results, encourage academic and industrial interaction, and promote

---

[3] See http://www.roughsets.org/.

collaborative research and developmental activities, in rough sets and knowledge technology worldwide.

During the past 10 years, a number of other conferences and workshops that include rough sets as one of the principal topics, have also been taken place (see, e.g., [48, 58, 89, 85, 99, 100, 101, 102, 103]).

12.3 Books

During the past two decades, a significant number of books and edited volumes have either featured or included articles on rough set theory and applications (see, e.g., [44, 46, 47, 48, 49, 50, 52, 54, 55, 57, 59, 60, 77, 78, 80, 81, 84, 88, 94, 61, 62, 63, 64, 65, 66, 68]). For example, the papers on rough set theory and its applications in [80, 81] present a wide spectrum of topics. It is observed that rough set theory is on the crossroads of fuzzy sets, theory of evidence, neural networks, Petri nets and many other branches of AI, logic and mathematics. The rough set approach appears to be of fundamental importance to AI and cognitive sciences, especially in the areas of machine learning, knowledge acquisition, decision analysis, knowledge discovery from databases, expert systems, inductive reasoning and pattern recognition.

12.4 Tutorials

In 1991, Zdzisław Pawlak published a monograph that provides a comprehensive presentation of the fundamentals on rough sets [66]. The book by Pawlak on approximation of sets by other sets that is based on the study of a finite, non-empty sets of objects called *universe* where each universe is denoted by U, subsets of U called *concepts*, attributes (features) of objects, and an *indiscernibility* relation Ind that partitions U into a collection of disjoint equivalence classes (called *blocks* in Sect. 3). This seminal work by Pawlak was the forerunner of numerous advances in rough set theory and its applications. After 1991, a succession of tutorials have been published that capture the essentials of rough sets and exhibit the growth in research in the theory and applications (see, e.g., [6, 35, 36, 77]).

13 Conclusion

This paper attempts to give a brief overview of some of the contributions made by Zdzisław Pawlak to rough set theory, conflict analysis and negotiation, genetic grammars and molecular computing, philosophy, painting and poetry during his lifetime. Remarkably, one can find a common thread in his theoretical work on rough sets as well as in conflict analysis and negotiation, painting and poetry, namely, Pawlak's interest in the border regions of objects that are delineated by considering the attributes (features) of an object. The work on knowledge representation systems and the notion of elementary sets have profound implications when one considers the problem of approximate reasoning and concept approximation.

– James F. Peters and Andrzej Skowron

Acknowledgments

The authors wish to thank the following persons who, at various times in the past, have contributed information that has made it possible to write this article: Mohua Banerjee, Maciej Borkowski, Nick Cercone, Gianpiero Cattaneo, Andrzej Czyżewski, Anna Gomolińska, Jerzy Grzymała-Busse, Liting Han, Christopher Henry, Zdzisław Hippe, Bożena Kostek, Solomon Marcus, Victor Marek, Ryszard Michalski, Ewa Orłowska, Sankar Pal, Lech Polkowski, Sheela Ramanna, Grzegorz Rozenberg, Zbigniew Ras, Roman Słowiński, Roman Swiniarski, Marcin Szczuka, Zbigniew Suraj, Shusaku Tsumoto, Guoyin Wang, Lotfi Zadeh, Wojciech Ziarko.

The research of James F. Peters and Andrzej Skowron is supported by NSERC grant 185986 and grant 3 T11C 002 26 from Ministry of Scientific Research and Information Technology of the Republic of Poland, respectively.

References

I. Seminal Works

1. Cantor, G.: *Grundlagen einer allgemeinen Mannigfaltigkeitslehre.* B.G. Teubner, Leipzig, Germany (1883).
2. Frege, G.: *Grundgesetzen der Arithmetik, vol. II.* Verlag von Hermann Pohle, Jena, Germany (1903).
3. Gheorghe, M., Mitrana, V.: A formal language-based approach in biology. Comparative and Functional Genomics 5(1) (2004) 91-94.
4. Hazelwinkel, M. (Ed.): *Encyclopaedia of Mathematics.* Kluwer Academic Publishers, Dordrecht, 1995, 213.
5. Keefe, R.: *Theories of Vagueness.* Cambridge Studies in Philosophy, Cambridge, UK (2000).
6. Komorowski, J., Pawlak, Z., Polkowski, L., Skowron, A.: Rough sets: A tutorial. In: Pal, S., Skowron, A.: Rough Fuzzy Hybridization. A New Trend in Decision Making. Springer-Verlag, Singapore Pte. Ltd. (1999) 3-98.
7. Konrad, E., Orłowska, E., Pawlak, Z.: Knowledge representation systems. Definability of information, Research Report PAS 433, Institute of Computer Science, Polish Academy of Sciences, April (1981).
8. Konrad, E., Orłowska, E., Pawlak, Z.: On approximate concept learning. Report 81-07, Fachbereich Informatik, TU Berlin, Berlin 1981; short version in: Collected Talks, European Conference on Artificial Intelligence 11/5, Orsay/Paris (1982) 17-19.
9. Leśniewski, S.: Grungzüge eines neuen Systems der Grundlagen der Mathematik. Fundamenta Mathematicae 14 (1929) 1-81.
10. Leśniewski, S.: On the foundations of mathematics. Topoi 2 (1982) 7-52.
11. Marek, W., Pawlak, Z.: Information storage and retrieval systems: Mathematical foundations. Theoretical Computer Science 1 (1976) 331-354.
12. Orłowska, E., Pawlak, Z.: Expressive power of knowledge representation systems. Research Report PAS 432, Institute of Computer Science, Polish Academy of Sciences, April (1981).

13. Pawlak, Z.: Flip-flop as generator of random binary digits. Mathematical Tables and other Aids to Computation 10(53) (1956) 28-30.
14. Pawlak, Z.: Some remarks on "-2" computer. Bulletin of the Polish Academy of Sciences. Ser. Tech. 9(4) (1961) 22-28.
15. Pawlak, Z.: *Grammar and Mathematics*. (in Polish), PZWS, Warsaw (1965).
16. Pawlak, Z.: Organization of address-less computers working in parenthesis notation. Zeitschrift für Mathematische Logik und Grundlagen der Mathematik 3 (1965) 243-262.
17. Pawlak, Z.: *Organization of Address-Less Computers*. (in Polish), Polish Scientific Publishers, Warsaw (1965).
18. Pawlak, Z.: On the notion of a computer. Logic, Methodology and Philosophy of Science 12, North Holland, Amsterdam (1968) 225-242.
19. Pawlak, Z.: Theory of digital computers. Mathematical Machines 10 (1969) 4-31.
20. Pawlak, Z.: *Mathematical Aspects of Production Organization* (in Polish), Polish Economic Publishers, Warsaw (1969).
21. Pawlak, Z.: A mathematical model of digital computers. Automatentheorie und Formale Sprachen (1973) 16-22.
22. Pawlak, Z.: Mathematical foundations of information retrieval, Proceedings of Symposium of Mathematical Foundations of Computer Science, September 3-8, 1973, High Tartras, 135-136; see also: Mathematical Foundations of Information Retrieval, Computation Center, Polish Academy of Sciences, Research Report CC PAS Report 101 (1973).
23. Pawlak, Z., Rozenberg, G., Savitch, W. J.: Programs for instruction machines. Information and Control 41(1) (1979) 9-28.
24. Pawlak, Z.: Information systems–Theoretical foundations. Information Systems 6(3) (1981) 205-218.
25. Pawlak, Z.: Classification of objects by means of attributes, Research Report PAS 429, Institute of Computer Science, Polish Academy of Sciences, ISSN 138-0648, January (1981).
26. Pawlak, Z.: Rough Sets, Research Report PAS 431, Institute of Computer Science, Polish Academy of Sciences (1981).
27. Pawlak, Z.: Rough sets. International Journal of Computer and Information Sciences 11 (1982) 341-356.
28. Pawlak, Z.: *Information Systems: Theoretical Foundations*. (in Polish), WNT, Warsaw (1983).
29. Pawlak, Z.: Rough classification. International Journal of Man-Machine Studies 20(5) (1984) 469-483.
30. Pawlak, Z.: On conflicts. International Journal of Man-Machine Studies 21 (1984) 127-134.
31. Pawlak, Z.: *On Conflicts* (in Polish), Polish Scientific Publishers, Warsaw (1987).
32. Pawlak, Z.: *Rough Sets – Theoretical Aspects of Reasoning about Data*. Kluwer Academic Publishers (1991).
33. Pawlak, Z.: Anatomy of conflict. Bulletin of the European Association for Theoretical Computer Science 50 (1993) 234-247.
34. Pawlak, Z.: An inquiry into anatomy of conflicts. Journal of Information Sciences 109 (1998) 65-78.
35. Pawlak, Z., Skowron, A.: Rudiments of rough sets. Information Sciences. An International Journal. Elsevier (2006) [to appear].
36. Pawlak, Z., Skowron, A.: Rough sets: Some extensions. Information Sciences. An International Journal. Elsevier (2006) [to appear].

37. Pawlak, Z., Skowron, A.: Rough sets and Boolean reasoning. Information Sciences. An International Journal. Elsevier (2006) [to appear].
38. Polkowski, L., Skowron, A.: Rough mereology: A new paradigm for approximate reasoning. International Journal of Approximate Reasoning 15(4) (1996) 333-365.
39. Rough Set Database System, version 1.3:
    `http://rsds.wsiz.rzeszow.pl/pomoc9.html`
40. Russell, B.: *The Principles of Mathematics.* G. Allen & Unwin, Ltd, London (1903).
41. Wikipedia summary of the life and work of Z. Pawlak:
    `http://pl.wikipedia.org/wiki/Zdzislaw_Pawlak`
42. Zadeh, L.A.: A new direction in AI: Toward a computational theory of perceptions. AI Magazine 22(1) (2001) 73-84.

II. Rough Set Literature

43. Alpigini, J. J., Peters, J. F., Skowron, A., Zhong, N.(Eds.): Third International Conference on Rough Sets and Current Trends in Computing (RSCTC'2002), Malvern, PA, October 14-16, 2002, Lecture Notes in Artificial Intelligence, vol. 2475. Springer-Verlag, Heidelberg, 2002.
44. Cios, K., Pedrycz, W., Swiniarski, R.: *Data Mining Methods for Knowledge Discovery.* Kluwer, Norwell, MA, 1998.
45. Czaja, L., Burkhard, H.D., Lindeman, G., Suraj Z. (Eds.): Special Issue on Currency Specification and Programming. Fundamenta Informaticae 67(1-3), 2005.
46. Demri, S., Orłowska, E. (Eds.): *Incomplete Information: Structure, Inference, Complexity.* Monographs in Theoretical Computer Sience, Springer-Verlag, Heidelberg, 2002.
47. Doherty, P. Łukaszewicz, W., Skowron, A., Szałas, A.: *Knowledge Engineering: A Rough Set Approach.* Series in Fuzzines and Soft Computing 202, Springer, Heidelberg, 2006.
48. Dunin-Kęplicz, B., Jankowski, A., Skowron, A., Szczuka, M. (Eds.): Monitoring, Security, and Rescue Tasks in Multiagent Systems (MSRAS'2004). Advances in Soft Computing, Springer, Heidelberg, 2005.
49. Düntsch, I. , Gediga, G.: *Rough Set Data Analysis: A Road to Non-invasive Knowledge Discovery.* Methodos Publishers, Bangor, UK, 2000.
50. Grzymała-Busse, J. W.: *Managing Uncertainty in Expert Systems.* Kluwer Academic Publishers, Norwell, MA, 1990.
51. Hirano, S., Inuiguchi, M., Tsumoto, S. (Eds.): Proceedings of International Workshop on Rough Set Theory and Granular Computing (RSTGC'2001), Matsue, Shimane, Japan, May 20-22, 2001, Bulletin of the International Rough Set Society, vol. 5(1-2). International Rough Set Society, Matsue, Shimane, 2001.
52. Inuiguchi, M., Hirano, S., Tsumoto, S. (Eds.): *Rough Set Theory and Granular Computing.* Studies in Fuzziness and Soft Computing, vol. 125. Springer-Verlag, Heidelberg, 2003.
53. Inuiguchi, M., Greco, S., Miyamoto, S., Nguyen, H.S., Słowinski, R. (Eds.): Int. Conf. on Rough Sets and Current Trends in Computing. Lecture Notes in Artificial Intelligence. Kobe, Japan, 6-8 November, (2006) [to appear].
54. Kostek, B.: *Soft Computing in Acoustics, Applications of Neural Networks, Fuzzy Logic and Rough Sets to Physical Acoustics*, Studies in Fuzziness and Soft Computing, vol. 31. Physica-Verlag, Heidelberg, 1999.

55. Kostek, B.: *Perception-Based Data Processing in Acoustics. Applications to Music Information Retrieval and Psychophysiology of Hearing*, Studies in Computational Intelligence, vol. 3. Springer, Heidelberg, 2005.

56. Lin, T. Y.(Ed.): Special issue, Journal of the Intelligent Automation and Soft Computing, vol. 2(2). 1996.

57. Lin, T. Y., Cercone, N. (Eds.): *Rough Sets and Data Mining - Analysis of Imperfect Data*. Kluwer Academic Publishers, Boston, USA, 1997.

58. Lin, T. Y., Wildberger, A. M. (Eds.): Soft Computing: Rough Sets, Fuzzy Logic, Neural Networks, Uncertainty Management, Knowledge Discovery. Simulation Councils, Inc., San Diego, CA, USA, 1995.

59. Lin, T. Y., Yao, Y. Y., Zadeh, L. A.(Eds.): *Rough Sets, Granular Computing and Data Mining*. Studies in Fuzziness and Soft Computing, Physica-Verlag, Heidelberg, 2001.

60. Mitra, S., Acharya, T.: *Data Mining. Multimedia, Soft Computing, and Bioinformatics*. John Wiley & Sons, New York, NY, 2003.

61. Munakata, T.: *Fundamentals of the New Artificial Intelligence: Beyond Traditional Paradigms*. Graduate Texts in Computer Science, vol. 10. Springer-Verlag, New York, NY, 1998.

62. Orłowska, E. (Ed.): *Incomplete Information: Rough Set Analysis*. Studies in Fuzziness and Soft Computing, vol. 13. Springer-Verlag/Physica-Verlag, Heidelberg, 1997.

63. Pal, S. K., Pedrycz, W., Skowron, A., Swiniarski, R. (Eds.): Special volume: Rough-neuro computing, Neurocomputing, vol. 36. 2001.

64. Pal, S. K., Polkowski, L., Skowron, A. (Eds.): *Rough-Neural Computing: Techniques for Computing with Words*. Cognitive Technologies, Springer-Verlag, Heidelberg, 2004.

65. Pal, S. K., Skowron, A. (Eds.): *Rough Fuzzy Hybridization: A New Trend in Decision-Making*. Springer-Verlag, Singapore, 1999.

66. Pawlak, Z.: *Rough Sets: Theoretical Aspects of Reasoning about Data*. System Theory, Knowledge Engineering and Problem Solving, vol. 9. Kluwer Academic Publishers, Dordrecht, The Netherlands, 1991.

67. Pawlak, Z., Grzymała-Busse, J., Slowinski, R., Ziarko, W. (Eds.): Rough Sets. Special Issue of the Communications of the ACM 38(11), Nov. (1995).

68. Pawlak, Z., Polkowski, L., Skowron, A.: Rough sets and rough logic: A KDD perspective. In: Polkowski et al. [78], pp. 583–646.

69. Peters, J.F. (Ed.): Threads in Fuzzy Petri Net Research. Special Issue of the International Journal of Intelligent Systems 14(8), Aug. (1999).

70. Peters, J.F., Skowron, A. (Eds.): A Rough Set Approach to Reasoning About Data. Special Issue of the International Journal of Intelligent Systems 16(1), Jan. (2001).

71. Peters, J.F., Skowron, A.: Transactions on Rough Sets: Journal Subline, Lecture Notes in Computer Science, Springer
http://www.springer.com/east/home/computer/lncs?SGWID=5-164-6-73656-0.

72. Peters, J. F., Skowron, A. (Eds.): Transactions on Rough Sets I: Journal Subline, Lecture Notes in Computer Science, vol. 3100. Springer, Heidelberg, 2004.

73. Peters, J. F., Skowron, A. (Eds.): Transactions on Rough Sets III: Journal Subline, Lecture Notes in Computer Science, vol. 3400. Springer, Heidelberg, 2005.

74. Peters, J. F., Skowron, A. (Eds.): Transactions on Rough Sets IV: Journal Subline, Lecture Notes in Computer Science, vol. 3700. Springer, Heidelberg, 2005.

75. Peters, J. F., Skowron, A., Dubois, D.,Grzymała-Busse, J. W., Inuiguchi, M., Polkowski, L. (Eds.): Transactions on Rough Sets II. Rough sets and fuzzy sets: Journal Subline, Lecture Notes in Computer Science, vol. 3135. Springer, Heidelberg, 2004.

76. Peters, J. F., Skowron, A., Wang, G., Yao, Y. (Eds.): Rough Sets and Knowledge Technology (RSKT 2006), Chongqing, China, July 24-26, 2006. Lecture Notes in Artificial Intelligence vol. 4062. Springer-Verlag, Heidelberg, GermanyJuly (2006). [to appear] See http://cs.cqupt.edu.cn/crssc/rskt2006/.

77. Polkowski, L.: *Rough Sets: Mathematical Foundations.* Advances in Soft Computing, Physica-Verlag, Heidelberg, 2002.

78. Polkowski, L., Lin, T. Y., Tsumoto, S. (Eds.): *Rough Set Methods and Applications: New Developments in Knowledge Discovery in Information Systems.* Studies in Fuzziness and Soft Computing, vol. 56. Springer-Verlag/Physica-Verlag, Heidelberg, 2000.

79. Polkowski, L., Skowron, A. (Eds.): First International Conference on Rough Sets and Soft Computing RSCTC'1998, Lecture Notes in Artificial Intelligence, vol. 1424. Springer-Verlag, Warsaw, Poland, 1998.

80. Polkowski, L., Skowron, A. (Eds.): *Rough Sets in Knowledge Discovery 1: Methodology and Applications.* Studies in Fuzziness and Soft Computing, vol. 18. Physica-Verlag, Heidelberg, 1998.

81. Polkowski, L., Skowron, A. (Eds.): *Rough Sets in Knowledge Discovery 2: Applications, Case Studies and Software Systems.* Studies in Fuzziness and Soft Computing, vol. 19. Physica-Verlag, Heidelberg, 1998.

82. Skowron, A., Ohsuga, S., Zhong, N. (Eds.): Proceedings of the 7-th International Workshop on Rough Sets, Fuzzy Sets, Data Mining, and Granular-Soft Computing (RSFDGrC'99), Yamaguchi, November 9-11, 1999, Lecture Notes in Artificial Intelligence, vol. 1711. Springer-Verlag, Heidelberg, 1999.

83. Skowron, A., Pal, S. K. (Eds.): Special volume: Rough sets, pattern recognition and data mining, Pattern Recognition Letters, vol. 24(6). 2003.

84. Skowron, A., Pawlak, Z., Komorowski, J., Polkowski, L.: A rough set perspective on data and knowledge. In: W. Kloesgen, J. Żytkow (Eds.), Handbook of KDD, Oxford University Press, Oxford. 2002, pp. 134–149.

85. Skowron, A., Szczuka, M. (Eds.): Proceedings of the Workshop on Rough Sets in Knowledge Discovery and Soft Computing at ETAPS 2003, April 12-13, 2003, Electronic Notes in Computer Science, vol. 82(4). Elsevier, Amsterdam, Netherlands, 2003. http://www.elsevier.nl/locate/entcs/volume82.html

86. Ślęzak, D., Wang, G., Szczuka, M., Düntsch, I., Yao Y.Y. (Eds.): Proceedings of the 10th International Conference on Rough Sets, Fuzzy Sets, Data Mining, and Granular Computing (RSFDGrC'2005), Regina, Canada, August 31-September 3, 2005, Part I, Lecture Notes in Artificial Intelligence, vol. 3641. Springer-Verlag, Heidelberg, 2005.

87. Ślęzak, D., Yao, J. T.,Peters, J. F., Ziarko, W., Hu, X. (Eds.): Proceedings of the 10th International Conference on Rough Sets, Fuzzy Sets, Data Mining, and Granular Computing (RSFDGrC'2005), Regina, Canada, August 31-September 3, 2005, Part II, Lecture Notes in Artificial Intelligence, vol. 3642. Springer-Verlag, Heidelberg, 2005.

88. Słowiński, R. (Ed.): *Intelligent Decision Support - Handbook of Applications and Advances of the Rough Sets Theory.* System Theory, Knowledge Engineering and Problem Solving, vol. 11. Kluwer Academic Publishers, Dordrecht, The Netherlands, 1992.

89. Terano, T., Nishida, T., Namatame, A., Tsumoto, S., Ohsawa, Y., Washio, T. (Eds.): *New Frontiers in Artificial Intelligence, Joint JSAI'2001 Workshop Post-Proceedings.* Lecture Notes in Artificial Intelligence, vol. 2253. Springer-Verlag, Heidelberg, 2001.
90. Tsumoto, S. (Ed.): Bulletin of International Rough Set Society (IRSS) vol. 1-6, 1997-2002. http://www2.cs.uregina.ca/ yyao/irss/bulletin.html
91. Tsumoto, S., Kobayashi, S., Yokomori, T., Tanaka, H., Nakamura, A. (Eds.): Proceedings of the The Fourth Internal Workshop on Rough Sets, Fuzzy Sets and Machine Discovery, November 6-8, University of Tokyo , Japan. The University of Tokyo, Tokyo, 1996.
92. Tsumoto, S., Słowiński, R., Komorowski, J., Grzymała-Busse, J. (Eds.): Proceedings of the 4th International Conference on Rough Sets and Current Trends in Computing (RSCTC'2004), Uppsala, Sweden, June 1-5, 2004, Lecture Notes in Artificial Intelligence, vol. 3066. Springer-Verlag, Heidelberg, 2004.
93. Wang, G., Liu, Q., Yao, Y., Skowron, A. (Eds.): Proceedings of the 9-th International Conference on Rough Sets, Fuzzy Sets, Data Mining, and Granular Computing (RSFDGrC'2003), Chongqing, China, May 26-29, 2003, Lecture Notes in Artificial Intelligence, vol. 2639. Springer-Verlag, Heidelberg, 2003.
94. Zhong, N., Liu, J. (Eds.): *Intelligent Technologies for Information Analysis.* Springer, Heidelberg, 2004.
95. Ziarko, W. (Ed.): Rough Sets, Fuzzy Sets and Knowledge Discovery: Proceedings of the Second International Workshop on Rough Sets and Knowledge Discovery (RSKD'93), Banff, Alberta, Canada, October 12–15 (1993). Workshops in Computing, Springer–Verlag & British Computer Society, London, Berlin, 1994.
96. Ziarko, W. (Ed.): Special issue, Computational Intelligence: An International Journal, vol. 11(2). 1995.
97. Ziarko, W. (Ed.): Special issue, Fundamenta Informaticae, vol. 27(2-3). 1996.
98. Ziarko, W., Yao, Y.Y., (Eds.): Proceedings of the 2nd International Conference on Rough Sets and Current Trends in Computing (RSCTC'2000), Banff, Canada, October 16-19, 2000, Lecture Notes in Artificial Intelligence, vol. 2005. Springer-Verlag, Heidelberg, 2001.
99. IFSA/IEEE Fifth Int. Conf. on Hybrid Intelligent Systems, Rio de Janeiro, Brazil, 6-9 Nov. (2005). See http://www.ica.ele.puc-rio.br/his05/index.html
100. IEEE/WIC/ACM Int. Conf. on Intelligent Agent Technology, Complègne University of Technology, France, 19-22 Sept. (2005).
101. The 3rd Int. Conf. on Autonomous Robots and Agents, Palmerston North, New Zealand 12-14 Dec. (2006). See http://icara.massey.ac.nz/.
102. International Symposium on Methodologies for Intelligent Systems (ISMIS) http://www.informatik.uni-trier.de/~ley/db/conf/ismis/index.html
103. Information Processing and Management of Uncertainty (IPMU) http://www.informatik.uni-trier.de/~ley/db/conf/ipmu/index.html

Rough Belief Change*

Mohua Banerjee

Department of Mathematics and Statistics,
Indian Institute of Technology, Kanpur 208 016, India
mohua@iitk.ac.in

Abstract. The article aims at re-visiting the notion of *rough truth* proposed by Pawlak in 1987 [15] and investigating some of its 'logical' consequences. We focus on the formal deductive apparatus $\mathcal{L}_\mathcal{R}$, that is sound and complete with respect to a semantics based on rough truth. $\mathcal{L}_\mathcal{R}$ turns out to be equivalent to the paraconsistent logic J due to Jaśkowski. A significant feature of rough truth is that, a proposition and its negation may well be roughly true together. Thus, in [5], *rough consistency* was introduced. Completeness of $\mathcal{L}_\mathcal{R}$ is proved with the help of this notion of consistency. The properties of $\mathcal{L}_\mathcal{R}$ motivate us to use it for a proposal of *rough belief change*. During change, the operative constraints on a system of beliefs are that of *rough* consistency preservation and deductive closure with respect to $\mathcal{L}_\mathcal{R}$. Following the AGM [1] line, eight basic postulates for defining rough revision and contraction functions are presented. Interrelationships of these functions are also proved. The proposal is, therefore, an example of paraconsistent belief change.

1 Introduction

The notion of *rough truth* was introduced in [15] as a part of the first formal proposal on reasoning with rough sets. The work in [15], in fact, paved the way for much subsequent study on logics of rough sets, a good survey of which can be found in [11]. But rough truth seems to have escaped due attention, though it was developed to some extent in [5,3]. The present article investigates related issues and some further 'logical' implications of this notion.

Rough truth was proposed to reflect 'inductive' truth, i.e. truth relative to our present state of knowledge, and one that, with gain of knowledge, leads to total, 'deductive' truth. This sense of 'gradualness' finds an expression in, possibly, the only *qualitative* version of 'approximate' or 'soft' truth, as opposed to other quantitative definitions found in, e.g., probabilistic, multi-valued or fuzzy logics. Let us look at the definition formally.

It has generally been accepted that the propositional aspects of Pawlak's rough set theory are adequately expressed by the modal system $S5$. An $S5$ (Kripke) model (X, R, π) (cf. e.g. [9]) is essentially an *approximation space* [14]

---

* Part of work done while supported by Project No. BS/YSP/29/2477 of the Indian National Science Academy. Thanks are due to Pankaj Singh for discussions. I am grateful to the referees for their valuable comments.

(X, R), where $X \neq \emptyset$, with the function π interpreting every well-formed formula (wff) of $S5$ as a rough set in (X, R). If L, M denote the necessity and possibility connectives respectively, a *modal wff* $L\alpha$ ($M\alpha$), representing 'definitely' ('possibly') α, is interpreted by π as the lower (upper) approximation $\underline{\pi(\alpha)}$ $(\overline{\pi(\alpha)})$ of the set $\pi(\alpha)$.

Using this formalism, a wff α may be termed *roughly true* in (X, R, π), if $\overline{\pi(\alpha)} = X$. In [5,3], we extended rough truth to *rough validity*, and also introduced the notions of *rough consequence, rough (in)consistency*. These were further considered in the context of predicate logic in [4]. The rationale behind the introduction of the concepts was as follows.

Given the aforementioned syntax, one may wish to derive in it, roughly true propositions/beliefs from roughly true premises (in the same information system). In particular, one may look for interderivability of propositions that are both roughly true and logically equivalent in possibility. This led to the relation of rough consequence. It was also felt that the notion of (in)consistency needs to be relaxed. In the face of an incomplete description of a concept p, one may not always think that p and 'not' p represent conflicting situations. There could be two options to define consistency here – according as 'possibly' p is satisfied, and 'necessarily' p is satisfied. It is thus that we have the two notions of rough consistency and rough inconsistency.

In this paper, we focus on these features of rough reasoning again, and on the syntactic counterpart $\mathcal{L}_{\mathcal{R}}$ of a semantics based on rough truth. $\mathcal{L}_{\mathcal{R}}$ is built over $S5$, and is a modified version of the rough consequence logic of [5]. One observes that the logic is *paraconsistent*, i.e. if a set Γ of wffs in it contains two members, one of which is the negation of the other, then Γ does *not* yield all wffs as its consequence. In other words, it violates the principle of *ex contradictione sequitur quodlibet* (*ECQ*), viz. $\Gamma \cup \{\alpha, \neg\alpha\} \models \beta$, for all Γ, α, β. In fact, $\mathcal{L}_{\mathcal{R}}$ is seen to be equivalent to the paraconsistent logic J due to Jaśkowski (cf. e.g. [6], implicitly present in [10]). We present the system $\mathcal{L}_{\mathcal{R}}$ and its properties in Section 2. Proofs of the main results stated in the section are given in the Appendix.

Research on belief change has seen a lot of development through the years. A pioneering work, undoubtedly, has been by Alchourrón, Gärdenfors and Makinson in [1]. The formalisation propounded by the authors consists of three main kinds of belief change: *expansion, revision, and contraction*. In the first kind, a belief is inserted into a system S (say), irrespective of whether S becomes 'inconsistent' as a result. Revision and contraction represent, respectively, insertion and deletion of beliefs maintaining ('integrity' constraints of) consistency of the system, deductive closure and minimal information loss. The AGM 'rationality' postulates for defining revision and contraction functions were formulated in [1]. A to-and-fro passage between contraction and revision is established through the *Levi* and *Harper identities*.

The AGM postulates have since been questioned, modified, and alternatives like algorithms for computing the change functions have been proposed (cf. [7,19,18,20]). For instance, the AGM approach assumes representation of belief states by 'belief sets' – sets deductively closed relative to a (Tarskian) logic and so, usually, infinite. The agents are idealized, with unlimited memory and capacity of inference. Modification in defining belief change has been attempted (e.g.) by use of 'belief bases' (cf. [7,18]), or disbeliefs along with beliefs [8].

Another category of modification alters the base logic used to define belief change. Non-Tarskian logics have been considered, for the modelling of less idealized (more realistic) agents [22] or for the modelling of intuitive properties of epistemic states that the AGM approach fails to cover [12]. Investigations have been carried out, in particular, keeping in view situations where classical inconsistency conditions do not apply. For instance in [17], we find the base logic being taken as the *paraconsistent* 4-valued logic resulting from the notion of *first degree entailment*. We also find discussions of paraconsistent belief revision in [21,16,13].

Most of everyday reasoning appears to be conducted, even if a model of just the possible versions of our beliefs is available. In this context, the notion of rough truth and rough consistency seem particularly appropriate for creating a 'non-classical' belief change framework. Agents, for us, are interested in situations where the rough truth of propositions/observations/beliefs matters. Beliefs are represented by the wffs of $\mathcal{L}_\mathcal{R}$. So modal wffs $L\alpha$ and $M\alpha$ express 'certain'/'necessary' and 'uncertain'/(but) 'possible' beliefs respectively, depending on the (in)completeness of the available information. Deduction is carried out through the apparatus of $\mathcal{L}_\mathcal{R}$, as it captures 'precisely' the semantics of rough truth. During revision/contraction of beliefs, we seek the preservation of *rough consistency*, and deductive closure with respect to $\mathcal{L}_\mathcal{R}$.

In order to make a beginning, we have followed the AGM line of thought, and defined *rough belief change* functions through eight basic postulates (cf. Section 3). The present article is an expanded version of [2]. In particular, we have added to the core postulates presented in [2], those concerning composite revision and contraction, specifically pertaining to conjunctions. As we shall see, the classical conjunction rule 'if $\{\alpha, \beta\}$, then $\alpha \wedge \beta$' does not hold in $\mathcal{L}_\mathcal{R}$ – hence these added postulates differ from their classical versions. Interrelationships of the revision and contraction functions, resulting from the use of the Levi and Harper identities, are proved. As $\mathcal{L}_\mathcal{R}$ is paraconsistent, the proposed change functions provide an example of paraconsistent belief change. The last section concludes the article.

2   The System $\mathcal{L}_\mathcal{R}$

The language of $\mathcal{L}_\mathcal{R}$ is that of a normal modal propositional logic, wffs being given as: $p \in \mathcal{P}|\neg\alpha|\alpha \wedge \beta|L\alpha$, where $\mathcal{P}$ denotes the set of all propositional variables. The other connectives, viz. $\vee, \rightarrow, \leftrightarrow, M$, are defined as usual. In the following, Γ is any set of wffs, and α, β are any wffs of $\mathcal{L}_\mathcal{R}$.

2.1   The Semantics

An *S5-model*, as mentioned in the Introduction, is a triple (X, R, π), where $X \neq \emptyset$, R is an equivalence relation on X, and the function π associates with every propositional variable p, a subset $\pi(p)$ of X (i.e. $\pi : \mathcal{P} \rightarrow 2^X$). π is extended to the set of all wffs of the language in the standard way. (We use the same notation π for the extension.)

$$\pi(\neg\alpha) \equiv \pi(\alpha)^c; \ \pi(\alpha \wedge \beta) \equiv \pi(\alpha) \cap \pi(\beta);$$
$$\pi(L\alpha) \equiv \{x \in X : y \in \pi(\alpha) \text{ for all } y \text{ such that } xRy\} \equiv \underline{\pi(\alpha)}.$$

Thus π interprets any wff of $S5$ as a rough set in the approximation space (X, R).

A wff α is *true* in an $S5$-model (X, R, π), provided $\pi(\alpha) = X$. One writes $\Gamma \models \alpha$ to denote that, whenever each member γ of Γ is true in (X, R, π), α is true in it as well.

Definition 1. *An $S5$-model $\mathcal{M} \equiv (X, R, \pi)$ is a rough model of Γ, if and only if every member γ of Γ is roughly true in $\mathcal{M}$, i.e. $\overline{\pi(\gamma)} = X$.*

Definition 2. *α is a rough semantic consequence of Γ (denoted $\Gamma \approx \alpha$) if and only if every rough model of Γ is a rough model of α. If Γ is empty, α is said to be roughly valid, written $\approx \alpha$.*

Example 1. Let $\Gamma \equiv \{p\}$, $X \equiv \{a, b, c\}$, $R \equiv \{(a, a), (b, b), (c, c), (a, b), (b, a)\}$, and π be such that $\pi(p) \equiv \{a, c\}$. Then $\pi(Mp) = \overline{\pi(p)} = X$, and so (X, R, π) is a rough model of Γ. In fact, observe that every rough model of $\{p\}$ is a rough model of Mp, i.e. $\{p\} \approx Mp$.

Let $M\Gamma \equiv \{M\gamma : \gamma \in \Gamma\}$. It is clear that

Observation 1. *$\Gamma \approx \alpha$, if and only if $M\Gamma \models M\alpha$.*

So, semantically, $\mathcal{L}_\mathcal{R}$ is equivalent to the (pre-)discussive logic J of Jaśkowski. In the following, we present a proof system for $\mathcal{L}_\mathcal{R}$, and in the process, an axiomatization for J. Proofs of some of the stated results are given in the Appendix.

2.2   The Rough Deductive Apparatus

Let us recall, first, the deductive apparatus of $S5$ [9]. α, β, γ are any wffs.
Axioms:
  (i) $\alpha \rightarrow (\beta \rightarrow \alpha)$; $(\alpha \rightarrow (\beta \rightarrow \gamma)) \rightarrow ((\alpha \rightarrow \beta) \rightarrow (\alpha \rightarrow \gamma))$;
        $(\neg\beta \rightarrow \neg\alpha) \rightarrow (\alpha \rightarrow \beta)$ (axioms of Propositional Calculus).
  (ii) $L(\alpha \rightarrow \beta) \rightarrow (L\alpha \rightarrow L\beta)$ (K).
  (iii) $L\alpha \rightarrow \alpha$ (T).
  (iv) $M\alpha \rightarrow LM\alpha$ $(S5)$.
Rules of inference:
  (i)   *Modus Ponens.*
  (ii)   If α is a theorem, so is $L\alpha$ (*Necessitation*).

This gives an axiom schemata for $S5$, and the consequence relation, denoted $\vdash$, gets to be closed under uniform substitution.

Let us now consider two rules of inference, R_1, R_2:

$$R_1. \qquad \frac{\alpha}{\beta} \qquad\qquad R_2. \quad \frac{M\alpha}{M\beta}$$
$$\text{where } \vdash M\alpha \to M\beta \qquad \frac{}{M\alpha \wedge M\beta}$$

The consequence relation defining the system $\mathcal{L}_\mathcal{R}$ is given as follows.

Definition 3. α *is a* rough consequence *of* Γ *(denoted* $\Gamma \hspace{-0.5em}\mid\hspace{-0.3em}\sim \alpha$*) if and only if there is a sequence* $\alpha_1, ..., \alpha_n (\equiv \alpha)$ *such that each* $\alpha_i (i = 1, ..., n)$ *is either (i) a theorem of* $S5$*, or (ii) a member of* Γ*, or (iii) derived from some of* $\alpha_1, ..., \alpha_{i-1}$ *by* R_1 *or* R_2*.*

If Γ *is empty,* α *is said to be a* rough theorem*, written* $\hspace{-0.5em}\mid\hspace{-0.3em}\sim \alpha$*.*

$\mathcal{L}_\mathcal{R}$ is therefore structural. The rule R_1 is meant to yield from a wff α, any wff β that is logically equivalent in possibility. A special case would be when β is *roughly equal* [14] to α. This can be expressed using the *rough equality connective* $\approx$ [3]: $\alpha \approx \beta \equiv (L\alpha \leftrightarrow L\beta) \wedge (M\alpha \leftrightarrow M\beta)$. It must be noted here that a different, but equivalent version (called RMP_1) of R_1 was considered in [3,5], to accommodate this special case. Another rule (RMP_2) was considered in [3,5], but it can be derived in $\mathcal{L}_\mathcal{R}$. R_2 was not considered there. However, it appears to be essential now for proving strong completeness. It replaces the classical conjunction rule, which is not sound here (cf. Observation 3(c)).

Some derived rules of inference:

$$DR_1. \qquad \frac{\alpha}{\begin{array}{c}\vdash \alpha \to \beta\\ \hline \beta\end{array}} \qquad\qquad DR_2. \quad \frac{(M\alpha)L\alpha}{\begin{array}{c}(M\alpha)L\alpha \to \beta\\ \hline \beta\end{array}}$$

$$DR_3. \qquad \frac{\begin{array}{c}M\alpha\\ \neg M\alpha\end{array}}{\beta} \qquad\qquad DR_4. \quad \frac{\begin{array}{c}\alpha \to M\beta\\ \alpha \to \neg M\beta\end{array}}{\neg\alpha}$$

$$DR_5. \qquad \frac{M\alpha}{\alpha} \qquad\qquad DR_6. \quad \frac{\alpha}{M\alpha}$$

$$DR_7. \qquad \frac{\begin{array}{c}M\alpha \to \gamma\\ M\beta \to \gamma\end{array}}{M\alpha \vee M\beta \to M\gamma} \qquad\qquad DR_8. \quad \frac{\begin{array}{c}\vdash \alpha \to \beta\\ \beta \to \gamma\end{array}}{\alpha \to \gamma}$$

Observation 2

    (a)   The system is then strictly enhanced, in the sense that the set of rough theorems properly contains that of $S5$*-theorems. E.g.* $\hspace{-0.5em}\mid\hspace{-0.3em}\sim \alpha \to L\alpha$*, but* $\nvdash \alpha \to L\alpha$*. The latter is well-known, and we obtain the former from* $\hspace{-0.5em}\mid\hspace{-0.3em}\sim M(\alpha \to L\alpha)$ *(an* $S5$*-theorem) and* DR_5*.*

(b) $\mid\!\sim$ *satisfies cut and monotonicity.*

(c) $\mid\!\sim$ *is clearly compact, by its definition. In fact, one can find a single wff δ with $\{\delta\}\mid\!\sim\alpha$, whenever $\{\gamma_1, ..., \gamma_n\}\mid\!\sim\alpha$: we take δ such that $\vdash M\delta \leftrightarrow (M\gamma_1 \wedge ... \wedge M\gamma_n)$. (For any γ_1, γ_2, an instance of such a δ would be the wff $((\gamma_1 \wedge \gamma_2) \vee (\gamma_1 \wedge M\gamma_2 \wedge \neg M(\gamma_1 \wedge \gamma_2))).)$*

Note 1. The axiomatization $\mathcal{A}$ of J presented in [6], comprises the rules: (i) if α is an $S5$-axiom, then $L\alpha$; (ii) if $L\alpha, L(\alpha \rightarrow \beta)$, then $L\beta$; (iii) if $L\alpha$, then α; (iv) if $M\alpha$, then α; (v) if $L\alpha$, then $LL\alpha$. One can prove that $\mathcal{A}$ and $\mathcal{L_R}$ are equivalent, by showing that the rules of one system are derivable in the other. For instance, (iv) of $\mathcal{A}$ is just DR_5 in $\mathcal{L_R}$, and (i)-(v) are derivable here by using Definition 3(i), DR_1 and DR_2. Properties of $\mathcal{A}$ (cf. [6]) result in the converse.

Two of the immediate results are

Theorem 1. *(Deduction) For any Γ, α, β, if $\Gamma \cup \{\alpha\}\mid\!\sim\beta$, then $\Gamma\mid\!\sim\alpha \rightarrow \beta$.*

Theorem 2. *(Soundness) If $\Gamma\mid\!\sim\alpha$, then $\Gamma\models\!\!\approx\alpha$.*

Observation 3

(a) $\mid\!\sim\alpha$, *if and only if* $\vdash M\alpha$: $(\Rightarrow)$ $\mid\!\sim\alpha$ *implies* $\models\!\!\approx\alpha$ *(by soundness), and so* $\models M\alpha$, *by Observation 1. This implies* $\vdash M\alpha$, *by completeness of $S5$.*
$(\Leftarrow)$ *By DR_5.*

(b) *If* $\vdash \alpha$, *then* $\mid\!\sim L\alpha$.

(c) *The classical rules of Modus Ponens and Necessitation fail to be sound with respect to the rough truth semantics. The rule $\{\alpha, \beta\}\mid\!\sim\alpha\wedge\beta$ is not sound either, but it is so for* modal *wffs, i.e. for wffs of the form $M\gamma$ or $L\gamma$, for some wff γ (by R_2 and DR_1).*

(d) *Interestingly, the soundness result establishes that the converse of the deduction theorem is not true – e.g. $\mid\!\sim p \rightarrow Lp$, but $\{p\}$ $\not\mid\!\sim Lp$, p being any propositional variable. However, the converse does go through, i.e. $\Gamma\mid\!\sim\alpha \rightarrow \beta$ implies $\Gamma \cup \{\alpha\}\mid\!\sim\beta$, if α is a modal wff – this is because of DR_2.*

(e) $\mid\!\sim\alpha \leftrightarrow (M)L\alpha$. *So there is no difference between the modal and non-modal wffs in terms of the object-level implication $\rightarrow$. However, as just noted, $\{\alpha\}\not\mid\!\sim L\alpha$ in general – indicating that the meta-level implication $\mid\!\sim$ does make this distinction.*

It should be remarked that many of these points have been observed in the context of the logic J, e.g. in [6].

2.3   Rough Consistency

Definition 4. *A set Γ of wffs is* roughly consistent *if and only if the set $M\Gamma \equiv \{M\gamma : \gamma \in \Gamma\}$ is $S5$-consistent. Γ is* roughly inconsistent *if and only if $L\Gamma \equiv \{L\gamma : \gamma \in \Gamma\}$ is $S5$-inconsistent.*

Observation 4
   (a)   *Inconsistency implies rough inconsistency. The converse may not be true, taking e.g. the set* $\{p, \neg Lp\}$, *p being any propositional variable.*
   (b)   *Consistency implies rough consistency. Again, the converse may not be true – taking e.g. the set* $\{p, \neg p\}$.
   (c)   *There may be sets which are simultaneously roughly consistent and inconsistent. The example in (b) applies.*
   (d)   *If* Γ *is not roughly consistent, it is roughly inconsistent.*

The following shows that the *ECQ* principle (cf. Introduction) is not satisfied by $\mathcal{L}_{\mathcal{R}}$.

Theorem 3. $\mathcal{L}_{\mathcal{R}}$ *is paraconsistent.*

Proof. There are wffs α, β such that $\{\alpha, \neg\alpha\} \not\hspace{-0.3em}\sim\beta$: one can easily show that $\{p, \neg p\} \not\hspace{-0.3em}\approx Lp$, p being any propositional variable. Hence, by soundness, the result obtains.   □

However, we do have $\{\alpha \wedge \neg\alpha\}\hspace{-0.3em}\sim\beta$, for all α, β, so that $\mathcal{L}_{\mathcal{R}}$ is *weakly* paraconsistent.

Theorem 4. Γ *is roughly consistent, if and only if it has a rough model.*

Theorem 5. *If* Γ *is not roughly consistent, then* $\Gamma\hspace{-0.3em}\sim\alpha$ *for every wff* α.

The proof of Theorem 4 uses only *S*5 properties, and that of Theorem 5 uses the rule of inference R_2 (in fact, DR_3).

Theorem 6. *If* $\Gamma \cup \{\neg M\alpha\}\hspace{-0.3em}\sim\beta$ *for every wff* β, *then* $\Gamma\hspace{-0.3em}\sim\alpha$.

Theorem 7. *(Completeness) If* $\Gamma\hspace{-0.3em}\approx\alpha$, *then* $\Gamma\hspace{-0.3em}\sim\alpha$.

Proof. We suppose that $\Gamma \not\hspace{-0.3em}\sim\alpha$. By Theorem 6 there is β such that $\Gamma \cup \{\neg M\alpha\} \not\hspace{-0.3em}\sim\beta$. Thus $\Gamma \cup \{\neg M\alpha\}$ is roughly consistent, using Theorem 5. By Theorem 4, $\Gamma \cup \{\neg M\alpha\}$ has a rough model, which yields $\Gamma \not\hspace{-0.3em}\approx\alpha$.   □

3   Belief Change Based on $\mathcal{L}_{\mathcal{R}}$

In classical belief revision, the base language is assumed (cf. [7]) to be closed under the Boolean operators of negation, conjunction, disjunction, and implication. The underlying consequence relation is supraclassical (includes classical consequence) and satisfies cut, deduction theorem, monotonicity and compactness. Because of the assumption of supraclassicality, the *ECQ* principle is satisfied by the base consequence.
   Paraconsistent belief change has been discussed, for instance, by Restall and Slaney [17], Tanaka [21], Priest [16], and Mares [13]. One of the objections raised against this approach is that, since the *ECQ* principle is rejected, if a paraconsistent logic is used as an underlying system for belief change, one may not be

committed to revision at all. As an answer, we find a proposal in [16,13] of revision when the belief set becomes *incoherent*, not inconsistent. In [17], the notion of first degree entailment is taken as a basis for belief change, and so beliefs can be both true and false, as well as neither true nor false. Postulates defining revision and contraction are presented, based on the resulting paraconsistent 4-valued logic. This is followed by constructions of contraction functions through the standard methods of epistemic entrenchment, partial meet contractions and Grove's system of spheres, each generalized appropriately to accommodate this reasoning framework. On the other hand, Tanaka proposes a paraconsistent version of Grove's sphere semantics for the AGM theory, using the three kinds of paraconsistent logic, viz. the non-adjunctive, positive plus and relevant systems. He further studies soundness of this semantics with respect to the AGM postulates.

Here, we would like to present an example of paraconsistent belief change, that uses the modal language of $\mathcal{L}_\mathcal{R}$ as the base. It may be noticed that, though the corresponding consequence relation C_R is not supraclassical, it satisfies the other properties mentioned earlier (cf. Theorem 1, Observation 2(b),(c)). We follow the classical line for the definitions. A *belief set* is a set Γ of wffs such that $C_R(\Gamma) = \Gamma$. For a pair (Γ, α), there is a unique belief set $\Gamma_r^* \alpha$ ($\Gamma -_r \alpha$) representing *rough revision (contraction)* of Γ with respect to α. The new belief set is defined through a set of eight basic postulates (that follow). The *expansion* $\Gamma +_r \alpha$ of Γ by the wff α is the belief set $C_R(\Gamma \cup \{\alpha\})$. It is expected that rough contraction/revision by two roughly equal beliefs would lead to identical belief sets. To express this, we make use of the rough equality connective $\approx$ (cf. Section 2.2).

The postulates for rough revision

R_r1 For any wff α and belief set Γ, $\Gamma_r^* \alpha$ is a belief set.
R_r2 $\alpha \in \Gamma_r^* \alpha$.
R_r3 $\Gamma_r^* \alpha \subseteq \Gamma +_r \alpha$.
R_r4 If $\neg \alpha \notin \Gamma$, then $\Gamma +_r \alpha \subseteq \Gamma_r^* \alpha$.
R_r5 $\Gamma_r^* \alpha$ is not roughly consistent, only if $\vdash \neg \alpha$.
R_r5' $\Gamma_r^* \alpha$ is roughly inconsistent, only if $\vdash \neg \alpha$.
R_r6 If $\vdash \alpha \approx \beta$, then $\Gamma_r^* \alpha = \Gamma_r^* \beta$.
R_r7 $\Gamma_r^*(M\alpha \wedge M\beta) \subseteq \Gamma_r^* \alpha +_r \beta$.
R_r8 If $\neg \beta \notin \Gamma_r^* \alpha$, then $\Gamma_r^* \alpha +_r \beta \subseteq \Gamma_r^*(M\alpha \wedge M\beta)$.

The postulates for rough contraction

C_r1 For any wff α and belief set Γ, $\Gamma -_r \alpha$ is a belief set.
C_r2 $\Gamma -_r \alpha \subseteq \Gamma$.
C_r3 If $\alpha \notin \Gamma$, then $\Gamma -_r \alpha = \Gamma$.
C_r4 If $\nvdash \alpha$, then $\alpha \notin \Gamma -_r \alpha$.
C_r4' If $\nvdash \alpha$, then $L\alpha \notin \Gamma -_r \alpha$.
C_r5 $\Gamma \subseteq (\Gamma -_r \alpha) +_r \alpha$, if α is of the form $L\beta$ or $M\beta$ for some wff β.
C_r6 If $\vdash \alpha \approx \beta$, then $\Gamma -_r \alpha = \Gamma -_r \beta$.

C_r7 $\Gamma -_r \alpha \cap \Gamma -_r \beta \subseteq \Gamma -_r (M\alpha \wedge M\beta)$.
C_r8 If $\alpha \notin \Gamma -_r (M\alpha \wedge M\beta)$, then $\Gamma -_r (M\alpha \wedge M\beta) \subseteq \Gamma -_r \alpha$.

Before elaborating on these, let us point out the AGM postulates that have undergone a change here. $\Gamma^*\alpha$, $\Gamma - \alpha$ and $\Gamma + \alpha$ denote the (classically) revised, contracted and expanded belief sets respectively, and $\Gamma_\perp$ the set of all wffs of the language.

Revision postulates
R_5 $\Gamma^*\alpha = \Gamma_\perp$ only if $\vdash \neg\alpha$.
R_6 If $\vdash \alpha \leftrightarrow \beta$, then $\Gamma^*\alpha = \Gamma^*\beta$.
R_7 $\Gamma^*(\alpha \wedge \beta) \subseteq \Gamma^*\alpha + \beta$.
R_8 If $\neg\beta \notin \Gamma^*\alpha$, then $\Gamma^*\alpha + \beta \subseteq \Gamma^*(\alpha \wedge \beta)$.

Contraction postulates
C_4 If $\not\vdash \alpha$, then $\alpha \notin \Gamma - \alpha$.
C_5 $\Gamma \subseteq (\Gamma - \alpha) + \alpha$.
C_6 If $\vdash \alpha \leftrightarrow \beta$, then $\Gamma - \alpha = \Gamma - \beta$.
C_7 $\Gamma - \alpha \cap \Gamma - \beta \subseteq \Gamma - (\alpha \wedge \beta)$.
C_8 If $\alpha \notin \Gamma - (\alpha \wedge \beta)$, then $\Gamma - (\alpha \wedge \beta) \subseteq \Gamma - \alpha$.

The major consideration here is to preserve rough consistency during belief change. The idea, expectedly, is that if $\Gamma +_r \alpha$ is roughly consistent, it could itself serve as $\Gamma_r^*\alpha$. Let us notice the difference with the classical scenario: suppose $\Gamma \equiv C_R(\{p\})$, p being any propositional variable. Then $\Gamma +_r \neg p$ is roughly consistent, and so it is $\Gamma_r^*\neg p$ itself. But, classically, $\Gamma^*\neg p \subset \Gamma + \neg p$. Since we also have the notion of rough inconsistency, there is the option of avoiding such inconsistency during belief change. It is thus that there are two versions of postulates involving consistency preservation.

R_r1 and C_r1 express the constraint of deductive closure. R_r2 places a natural requirement on revision, and since the result of contraction is generally smaller than the original belief set, we have C_r2. The result of revision must lie within that of expansion of a belief set, and if $\alpha \notin \Gamma$, there would be no question of retracting it from Γ – giving R_r3, C_r3 respectively. $\neg\alpha \notin \Gamma$ implies consistency and hence rough consistency of $\Gamma +_r \alpha$, so that, in view of the previous remarks, R_r4 is justifiable in the rough context.

In R_r5, we stipulate that $\Gamma_r^*\alpha$ is generally roughly consistent, except in the case when $\neg\alpha$ is roughly valid, i.e. in no situation 'definitely' α holds (though 'possibly' α may hold). C_r4 again stipulates that, in general, $\alpha \notin \Gamma -_r \alpha$, except when α 'is possible' in all situations. C_r4' could appear more relevant: 'definitely' α may follow from our beliefs despite contraction by α only if α is, in every situation, possible. The controversial *recovery* postulate C_5 in [1] is admitted here as C_r5, only in the case of contraction with a *definable/describable* [14] belief, i.e. α such that $\vdash \alpha \leftrightarrow L\alpha$. If $\vdash\alpha \approx \beta$, then $C_R(\alpha) = C_R(\beta)$ – leaving the belief sets after any change with respect to α, β identical. This is stipulated in R_r6, C_r6. The last two axioms express the relationship of change functions with respect to beliefs and their conjunctions. The failure of soundness of the classical conjunction rule $\{\alpha, \beta\}\!\!\sim\!\alpha \wedge \beta$ here, necessitates a modification in the AGM R_7, R_8, C_7, C_8 giving $R_r7, 8,\ C_r7, 8$ respectively. Let us further elaborate on R_r8. If $\Gamma_r^*(M\alpha \wedge M\beta)$ is roughly consistent, it cannot contain $\neg M(M\alpha \wedge M\beta)$, i.e. neither $\neg M\alpha$, nor $\neg M\beta$ can belong to the set. Now, if $\Gamma_r^*\alpha$ were also roughly consistent, $\neg M\alpha$ would not be its member, and therefore it would not be a member of the expansion $\Gamma_r^*\alpha +_r \beta$ either. However, $\neg M\beta$ could well be in $\Gamma_r^*\alpha +_r \beta$. Thus R_r8 says that, if $\neg\beta \notin \Gamma_r^*\alpha$ – whence $\neg M\beta \notin \Gamma_r^*\alpha$ – there is no difference between the two belief sets $\Gamma_r^*(M\alpha \wedge M\beta)$ and $\Gamma_r^*\alpha +_r \beta$. One may give a similar argument for C_r8.

Observation 5. R_r5' *implies* R_r5 *and* C_r4 *implies* C_r4'.

The following interrelationships between rough contraction and revision are then observed, if the *Levi* and *Harper identities* [7] are used.

Theorem 8. *Let the Levi identity give* $*_r$, *i.e.* $\Gamma_r^*\alpha \equiv (\Gamma -_r \neg\alpha) +_r \alpha$, *where the contraction function* $-_r$ *satisfies* $C_r1 - 8$. *Then* $*_r$ *satisfies* $R_r1 - 8$.

Proof. $R_r1 - 4$ follow easily.
R_r5: Suppose $\Gamma_r^*\alpha$ is not roughly consistent. By Theorem 5, $\Gamma_r^*\alpha\!\!\sim\!\beta$, for any wff β in $\mathcal{L_R}$. In particular, $\Gamma_r^*\alpha\!\!\sim\!M\beta$, and $\Gamma_r^*\alpha\!\!\sim\!\neg M\beta$, for any wff β. By assumption, $\Gamma_r^*\alpha \equiv C_R((\Gamma -_r \neg\alpha)\cup\{\alpha\}$. So, using deduction theorem and DR_4, $\Gamma -_r \neg\alpha\!\!\sim\!\neg\alpha$. Hence by $C_r1, 4$, $\!\!\sim\!\neg\alpha$. R_r6 can be proved by observing that $\!\!\sim\!\alpha \approx \beta$ if and only if $\!\!\sim\!\neg\alpha \approx \neg\beta$, and by using C_r6.

For proving $R_r7, 8$, we first note that $\Gamma -_r \neg\alpha = \Gamma -_r \neg M\alpha$, by Observation 3(e) and C_r6 ($\leftrightarrow$ is a special case of $\approx$).

Thus $\Gamma -_r \neg\alpha = \Gamma -_r (\neg(M\alpha \wedge M\beta) \wedge (M\alpha \rightarrow M\beta))$.   (*)

Secondly, because of DR_5, DR_6 and R_2,
$\Gamma +_r M\alpha \wedge M\beta = C_R(\Gamma \cup \{\alpha, \beta\})$, for any Γ.   (**)
R_r7: Let $\gamma \in \Gamma_r^*(M\alpha \wedge M\beta)$, i.e. $\Gamma -_r \neg(M\alpha \wedge M\beta) \cup \{(M\alpha \wedge M\beta)\}\!\!\sim\!\gamma$. By deduction theorem, $\Gamma -_r \neg(M\alpha \wedge M\beta)\!\!\sim\!(M\alpha \wedge M\beta) \rightarrow \gamma$.

So $(M\alpha \wedge M\beta) \rightarrow \gamma \in \Gamma -_r \neg(M\alpha \wedge M\beta) \subseteq \Gamma \subseteq \Gamma -_r (M\alpha \rightarrow M\beta)\cup\{M\alpha \rightarrow M\beta\}$, the last by C_r5, as $M\alpha \rightarrow M\beta$ is equivalent to a modal wff. It is then easy to see that $\Gamma -_r (M\alpha \rightarrow M\beta)\!\!\sim\!(M\alpha \wedge M\beta) \rightarrow \gamma$. So $(M\alpha \wedge M\beta) \rightarrow \gamma \in \Gamma -_r \neg(M\alpha \wedge M\beta) \cap \Gamma -_r (M\alpha \rightarrow M\beta) \subseteq \Gamma -_r (\neg(M\alpha \wedge M\beta) \wedge (M\alpha \rightarrow M\beta))$, by C_r7. Therefore, using (*), $\Gamma -_r \neg\alpha\!\!\sim\!(M\alpha \wedge M\beta) \rightarrow \gamma$. As $(M\alpha \wedge M\beta)$ is also equivalent to a modal wff, by converse of deduction theorem (cf. Observation 3(d)), $\Gamma -_r \neg\alpha \cup \{M\alpha \wedge M\beta\}\!\!\sim\!\gamma$. By (**), $\Gamma -_r \neg\alpha \cup \{\alpha, \beta\}\!\!\sim\!\gamma$. As $\Gamma -_r \neg\alpha \cup \{\alpha, \beta\} = \Gamma_r^*\alpha +_r \beta$, we have R_r7.
R_r8: Suppose $\neg\beta \notin \Gamma_r^*\alpha \equiv (\Gamma -_r \neg\alpha) +_r \alpha$, i.e. $\Gamma -_r \neg\alpha \cup \{\alpha\} \not\!\!\sim\!\neg\beta$. Then it can be shown that $\Gamma -_r \neg\alpha \cup \{\alpha\} \not\!\!\sim\!M\alpha \rightarrow \neg M\beta$. Let $\gamma \in \Gamma_r^*\alpha +_r \beta = \Gamma -_r \neg\alpha +_r \{\alpha, \beta\}$. Using (*) and C_r8, we have $\Gamma -_r \neg\alpha = \Gamma -_r (\neg(M\alpha \wedge M\beta) \wedge (M\alpha \rightarrow M\beta)) = \Gamma -_r (M\alpha \rightarrow \neg M\beta) \wedge (M\alpha \rightarrow M\beta)) \subseteq \Gamma -_r (\neg(M\alpha \wedge M\beta))$. Thus, by (**), $\gamma \in C_R(\Gamma -_r (\neg(M\alpha \wedge M\beta)) \cup \{\alpha, \beta\}) = \Gamma -_r (\neg(M\alpha \wedge M\beta)) +_r M\alpha \wedge M\beta = \Gamma_r^*(M\alpha \wedge M\beta)$.   □

Theorem 9. *Let $-_r$ be given by the Harper identity, i.e. $\Gamma -_r \alpha \equiv \Gamma \cap \Gamma_r^* \neg \alpha$.*
(a) If the revision function $*_r$ satisfies $R_r 1 - 4$, $R_r 5'$ and $R_r 6 - 8$, then $-_r$ satisfies $C_r 1 - 8$.
(b) If the revision function $*_r$ satisfies $R_r 1 - 8$, then $-_r$ satisfies $C_r 1 - 3$, $C_r 4'$ and $C_r 5 - 8$.

Proof. Again $C_r 1 - 3, 6$ follow easily.
$C_r 5$: Let α be $L\beta$, for some wff β. If $\gamma \in \Gamma$, it can be shown that $\alpha \rightarrow \gamma \in \Gamma \cap \Gamma_r^* \neg \alpha$. By DR_2, $\gamma \in (\Gamma -_r \alpha) +_r \alpha$.

As in case of the previous theorem, let us note before proving $C_r 7, 8$ that, using $R_r 6$ we have
$$\Gamma_r^* \neg \alpha = \Gamma_r^* \neg M\alpha = \Gamma_r^*((\neg M\alpha \vee \neg M\beta) \wedge \neg M\alpha). \tag{*}$$
$C_r 7$: By (*) and $R_r 7$, $\Gamma_r^* \neg \alpha = \Gamma_r^*((\neg M\alpha \vee \neg M\beta) \wedge \neg M\alpha) \subseteq \Gamma_r^* \neg (M\alpha \wedge M\beta) +_r \neg M\alpha$. Now let $\gamma \in \Gamma_r^* \neg \alpha \cap \Gamma_r^* \neg \beta$. Using deduction theorem,
$\Gamma_r^* \neg (M\alpha \wedge M\beta) \hspace{-0.3em}\sim\hspace{-0.3em} \neg M\alpha \rightarrow \gamma$. Similarly, $\Gamma_r^* \neg (M\alpha \wedge M\beta) \hspace{-0.3em}\sim\hspace{-0.3em} \neg M\beta \rightarrow \gamma$. Observe that $\vdash M \neg M\delta \leftrightarrow \neg M\delta$. Thus by DR_7, DR_8, $\Gamma_r^* \neg (M\alpha \wedge M\beta) \hspace{-0.3em}\sim\hspace{-0.3em} \neg (M\alpha \wedge M\beta) \rightarrow M\gamma$. As $R_r 2$ holds and $\neg (M\alpha \wedge M\beta)$ is equivalent to a modal wff, $\Gamma_r^* \neg (M\alpha \wedge M\beta) \hspace{-0.3em}\sim\hspace{-0.3em} M\gamma$, by DR_2. Finally, by DR_5, we have $\gamma \in \Gamma_r^* \neg (M\alpha \wedge M\beta)$.
$C_r 8$: Suppose $\alpha \notin \Gamma -_r (M\alpha \wedge M\beta) \equiv \Gamma \cap \Gamma_r^* \neg (M\alpha \wedge M\beta)$. So $\alpha \notin \Gamma$, or $\alpha \notin \Gamma_r^* \neg (M\alpha \wedge M\beta)$. In the former case, as $C_r 2, 3$ already hold, we trivially obtain $\Gamma -_r (M\alpha \wedge M\beta) \subseteq \Gamma = \Gamma -_r \alpha$. Consider the latter case.
Using DR_5, $\neg (\neg M\alpha) \notin \Gamma_r^* \neg (M\alpha \wedge M\beta)$. By $R_r 8$,
$\Gamma_r^* \neg (M\alpha \wedge M\beta) +_r \neg M\alpha \subseteq \Gamma_r^* (\neg (M\alpha \wedge M\beta) \wedge \neg M\alpha) = \Gamma_r^* \neg M\alpha = \Gamma_r^* \neg \alpha$.
So $\Gamma_r^* \neg (M\alpha \wedge M\beta) \subseteq \Gamma_r^* \neg (M\alpha \wedge M\beta) +_r \neg M\alpha \subseteq \Gamma_r^* \neg \alpha$, and $C_r 8$ holds.
(a) $C_r 4$: As $\neg \alpha \in \Gamma_r^* \neg \alpha$ (by $R_r 2$) as well as $\alpha \in \Gamma_r^* \neg \alpha$ (assumption), $\Gamma_r^* \neg \alpha$ is inconsistent, and hence roughly so. Thus by $R_r 5'$, $\hspace{-0.3em}\sim\hspace{-0.3em} \neg \neg \alpha$. It follows that $\hspace{-0.3em}\sim\hspace{-0.3em} \alpha$.
(b) $C_r 4'$: $L\alpha, \neg \alpha \in \Gamma_r^* \neg \alpha$, by assumption and $R_r 2$. So, using $S5$ properties, $M\Gamma_r^* \neg \alpha \vdash L\alpha$ as well as $M\Gamma_r^* \neg \alpha \vdash \neg L\alpha$, implying that $\Gamma_r^* \neg \alpha$ is not roughly consistent. By $R_r 5$, $\hspace{-0.3em}\sim\hspace{-0.3em} \neg \neg \alpha$. Thus $\hspace{-0.3em}\sim\hspace{-0.3em} \alpha$.   □

Observation 6
   (a)   Let $\mathbf{R}(-_r)$, $\mathbf{C}(*_r)$ denote respectively, the revision and contraction functions obtained by using the Levi and Harper identities. Then $\mathbf{R}(\mathbf{C}(*_r)) = *_r$. However, in general, $\mathbf{C}(\mathbf{R}(-_r)) \neq -_r$.
   (b)   Rough belief revision and contraction coincide with the corresponding classical notions if $\vdash \alpha \leftrightarrow L\alpha$ for every wff α, i.e. all beliefs are definable/describable ($S5$ collapses into classical propositional logic).

4   Conclusions

Some demands of 'rough' reasoning seem to be met by the rough consequence logic $\mathcal{L}_{\mathcal{R}}$. In particular, roughly consistent sets of premises find rough models, and roughly true propositions can be derived from roughly true premises. This offers grounds for the use of $\mathcal{L}_{\mathcal{R}}$ in a proposal of rough belief change. Beliefs are propositions in $\mathcal{L}_{\mathcal{R}}$. The postulates defining revision and contraction express the constraints imposed on a set of beliefs during change – in particular, rough consistency is preserved, as is deductive closure with respect to $\mathcal{L}_\mathcal{R}$. We obtain expected consequences. (i) Classical belief change is a special case of rough belief change (cf. Observation 6(b)). (ii) Unlike the classical case, the definitions are not completely interchangeable (cf. Theorems 8,9 and Observation 6(a)). It is interesting to observe that $\mathcal{L}_\mathcal{R}$ is equivalent to the well-studied paraconsistent logic J of Jaśkowski, proposed in a different context altogether. The proposal here is thus an example of paraconsistent belief change. In view of (ii) above, it may be worthwhile to check if a different form of the Levi/Harper identity could lead to complete interchangeability of the definitions. For example, one could examine the suggestion given in [21] for a changed version of the Levi identity – the 'reverse' Levi identity, viz. $\Gamma^*\alpha \equiv (\Gamma + \alpha) - \neg\alpha$.

Two of the integrity constraints in [1] postulate that during revision or contraction, the change effected on the belief set is minimal. More explicitly,

(a) there is minimal information loss during change, and

(b) the least important beliefs are retracted, when necessary.

A construction of contraction functions preserving (a) with respect to set inclusion, is given by the method of *partial meet contraction*. We have, in fact, obtained a definition of rough contraction through this method – but that would be matter for a separate report. The more computationally tractable method of construction of contraction preserving (b), is based on the notion of *epistemic entrenchment* (cf. [7]) – that sets a 'priority' amongst the beliefs of the system. So during contraction/revision, the less entrenched (important) beliefs may be disposed of to get the new belief set. There is a to and fro passage between an entrenchment ordering and a change function satisfying the contraction postulates, given by representation theorems. Our next goal is to try to define an appropriate notion of entrenchment amongst rough beliefs. With the proposed set (or possibly a modified set) of postulates of rough belief change, one can then check for the conditions under which representation theorems may be obtained – thus giving an explicit modelling of rough contraction.

References

1. Alchourrón, C., Gärdenfors, P., Makinson, D.: On the logic of theory change: partial meet functions for contraction and revision. *J. Symb. Logic* **50** (1985) 510–530.
2. Banerjee, M.: Rough truth, consequence, consistency and belief revision. In: S. Tsumoto et al., editors, *LNAI 3066: Proc. 4th Int. Conf. On Rough Sets and Current Trends in Computing (RSCTC2004), Uppsala, Sweden, June 2004*, pages 95–102. Springer-Verlag, 2004.
3. Banerjee, M., Chakraborty, M.K.: Rough consequence and rough algebra. In: W.P. Ziarko, editor, *Rough Sets, Fuzzy Sets and Knowledge Discovery, Proc. Int. Workshop on Rough Sets and Knowledge Discovery (RSKD'93)*, pages 196–207. Springer-Verlag, 1994.
4. Banerjee, M., Chakraborty, M.K.: Rough logics: a survey with further directions. In: E. Orłowska, editor, *Incomplete Information: Rough Set Analysis*, pages 579–600. Springer-Verlag, 1998.
5. Chakraborty, M.K., Banerjee, M.: Rough consequence. *Bull. Polish Acad. Sc. (Math.)* **41(4)** (1993) 299–304.

6. da Costa, N.C.A., Doria, F.A.: On Jaśkowski's discussive logics. *Studia Logica* **54** (1995) 33–60.
7. Gärdenfors, P., Rott, H.: Belief revision. In: D.M. Gabbay, C.J. Hogger and J.A. Robinson, editors, *Handbook of Logic in AI and Logic Programming, Vol.4: Epistemic and Temporal Reasoning*, pages 35–132. Clarendon, 1995.
8. Gomolińska, A., Pearce, D.: Disbelief change. In: *Spinning Ideas: Electronic Essays Dedicated to Peter Gärdenfors on His Fiftieth Birthday.* S. Halldén et al., editors, http://www.lucs.lu.se/spinning/, 1999.
9. Hughes, G.E., Cresswell, M.J.: *A New Introduction to Modal Logic.* Routledge, 1996.
10. Jaśkowski, S.: Propositional calculus for contradictory deductive systems. *Studia Logica*, **24** (1969) 143–157.
11. Komorowski, J., Pawlak, Z., Polkowski, L., Skowron, A.: Rough sets: a tutorial. In: S.K. Pal and A. Skowron, editors, *Rough Fuzzy Hybridization: A New Trend in Decision-Making*, pages 3–98. Springer-Verlag, 1999.
12. Lepage, F., Lapierre, S.: Partial logic and the dynamics of epistemic states. In: *Spinning Ideas: Electronic Essays Dedicated to Peter Gärdenfors on His Fiftieth Birthday.* S. Halldén et al., editors, http://www.lucs.lu.se/spinning/, 1999.
13. Mares, E.D.: A paraconsistent theory of belief revision. *Erkenntnis* **56** (2002) 229–246.
14. Pawlak, Z.: Rough sets. *Int. J. Comp. Inf. Sci.* **11** (1982) 341–356.
15. Pawlak, Z.: Rough logic. *Bull. Polish Acad. Sc. (Tech. Sc.)* **35(5-6)** (1987) 253–258.
16. Priest, G.: Paraconsistent belief revision. *Theoria* **67** (2001) 214–228.
17. Restall, G., Slaney, J.: Realistic belief revision. In: *Proc. 1st World Congress in the Fundamentals of Artificial Intelligence, Paris, July 1995*, 367-378.
18. Rott, H.: *Change, Choice and Inference: A Study of Belief Revision and Nonmonotonic Reasoning.* Clarendon, 2001.
19. *Spinning Ideas: Electronic Essays Dedicated to Peter Gärdenfors on His Fiftieth Birthday.* S. Halldén et al., editors, http://www.lucs.lu.se/spinning/, 1999.
20. *Studia Logica, Special Issue on Belief Revision*, **73**, 2003.
21. Tanaka, K.: What does paraconsistency do ? The case of belief revision. In: *The Logical Yearbook 1997*, T. Childers, editor, Filosophia, Praha, 188–197.
22. Wassermann, R.: Generalized change and the meaning of rationality postulates. *Studia Logica* **73** (2003) 299–319.

Appendix

Proofs of some derived rules:

DR_1: $\vdash \alpha \to \beta$ implies $\vdash M\alpha \to M\beta$. Therefore, by R_1, we get the rule.

DR_3: $\vdash \neg M\alpha \leftrightarrow M\neg M\alpha$. So, assuming $\neg M\alpha$, using DR_1 we get $M\neg M\alpha$. Further assuming $M\alpha$ therefore gives, by R_2, $M\alpha \wedge M\neg M\alpha$. But $\vdash (M\alpha \wedge M\neg M\alpha) \to \beta$, for any β. Using DR_1 again, we get the rule.

DR_5: As $\vdash MM\alpha \to M\alpha$, assuming $M\alpha$ gives α by R_1.

DR_7: Assuming $M\alpha \to \gamma$ and $M\beta \to \gamma$ gives, by DR_6, $M(M\alpha \to \gamma)$ and $M(M\beta \to \gamma)$. Using R_2 and DR_1, we get $(M\alpha \to M\gamma) \wedge (M\beta \to M\gamma)$. But $\vdash (M\alpha \to M\gamma) \wedge (M\beta \to M\gamma) \to (M\alpha \vee M\beta \to M\gamma)$. Hence by DR_1 again, we get the rule.

DR_8: $\vdash \alpha \rightarrow \beta$ implies $\vdash (\beta \rightarrow \gamma) \rightarrow (\alpha \rightarrow \gamma)$. So assuming $\beta \rightarrow \gamma$ gives, by DR_1, $\alpha \rightarrow \gamma$. $\qquad \square$

Some more derived rules:

$$DR_9. \quad \dfrac{\alpha \rightarrow \beta \quad \vdash \beta \rightarrow \gamma}{\alpha \rightarrow \gamma} \qquad DR_{10}. \quad \dfrac{\alpha \rightarrow M\beta \quad \alpha \rightarrow M\gamma}{\alpha \rightarrow M\beta \wedge M\gamma}$$

$$DR_{11}. \quad \dfrac{\vdash \alpha \quad \alpha \rightarrow \beta}{\beta} \qquad DR_{12}. \quad \dfrac{\alpha \rightarrow L\beta \quad \alpha \rightarrow L\neg\beta}{\neg\alpha}$$

Proofs:

DR_9, DR_{10}: Similar to that of DR_8 and DR_7 respectively.

DR_{11}: $\vdash \alpha$ implies $\vdash (\alpha \rightarrow \beta) \rightarrow M\beta$. Assuming $\alpha \rightarrow \beta$ gives $M\beta$, by DR_1. Finally, by DR_5, we get β.

DR_{12}: $\vdash L\gamma \rightarrow ML\gamma$, for any wff γ. So, assuming $\alpha \rightarrow L\beta, \alpha \rightarrow L\neg\beta$ gives, by DR_9, $\alpha \rightarrow ML\beta, \alpha \rightarrow ML\neg\beta$. Then by DR_{10}, we get $\alpha \rightarrow ML\beta \wedge ML\neg\beta$. This yields, by DR_9 again, $\alpha \rightarrow L\beta \wedge L\neg\beta$, or $\alpha \rightarrow \neg(L\beta \rightarrow M\beta)$. Using DR_1, one can obtain $\neg\neg(L\beta \rightarrow M\beta) \rightarrow \neg\alpha$, and thus $(L\beta \rightarrow M\beta) \rightarrow \neg\alpha$ (by DR_8). But $\vdash L\beta \rightarrow M\beta$. So, $\neg\alpha$, using DR_{11}. $\qquad \square$

Proof of Deduction Theorem (Theorem 1):

By induction on the number of steps of derivation of β from $\Gamma \cup \{\alpha\}$. Cases corresponding to use of the rules R_1 and R_2 can be accounted for, by applying $DR_1, DR_5, DR_6, DR_9, DR_{10}$. $\qquad \square$

Proof of Soundness Theorem (Theorem 2):

One proves that R_1 and R_2 preserve rough truth. E.g., let us consider an arbitrary rough model (X, R, π) of the premises of R_1, viz. $\alpha, M\alpha \rightarrow M\beta$, where the latter is an $S5$ theorem. By soundness of $S5$, $M\alpha \rightarrow M\beta$ is actually true in (X, R, π), i.e. $\overline{\pi(\alpha)} \subseteq \overline{\pi(\beta)}$. Also $\overline{\pi(\alpha)} = X$, α being roughly true in (X, R, π). Thus $\overline{\pi(\beta)} = X$. Similarly, if $\overline{\pi(\alpha)} = \overline{\pi(\beta)} = X$, then $\pi(M(M\alpha \wedge M\beta)) = \overline{\overline{\pi(\alpha)} \cap \overline{\pi(\beta)}} = \overline{\overline{\pi(\alpha)} \cap \overline{\pi(\beta)}} = X$. But $\overline{\overline{\pi(\alpha)} \cap \overline{\pi(\beta)}} \equiv \pi(M\alpha \wedge M\beta)$. Hence R_2 preserves rough truth as well. $\qquad \square$

Proof of Theorem 4:

It is based on the $S5$ result that if any set of the form $M\Gamma(L\Gamma)$ is $S5$-consistent, it must have a model in which all its wffs are true. $\qquad \square$

Proof of Theorem 5:

Compactness of $\vdash$ is used to get a finite subset $\{\gamma_1, \gamma_2, ..., \gamma_n\}$ of Γ, for which $\vdash \neg(M\gamma_1 \wedge M\gamma_2 \wedge ... \wedge M\gamma_n)$. As noted in Observation 2 (c), there is a wff γ such that $\vdash M\gamma \leftrightarrow (M\gamma_1 \wedge M\gamma_2 \wedge ... \wedge M\gamma_n)$. So $\vdash \neg M\gamma$. Both $M\gamma$ and $\neg M\gamma$ are then derived from $\{\gamma_1, \gamma_2, ..., \gamma_n\}$ (and hence from Γ) by using DR_6, R_2, DR_1 and finally, DR_3.

Proof of Theorem 6:

By applying the Deduction Theorem, DR_{12}, DR_1 and DR_5. $\qquad \square$

Rough Sets and Vague Concept Approximation: From Sample Approximation to Adaptive Learning

Jan Bazan[1], Andrzej Skowron[2], and Roman Swiniarski[3,4]

[1] Institute of Mathematics, University of Rzeszów
Rejtana 16A, 35-959 Rzeszów, Poland
bazan@univ.rzeszow.pl
[2] Institute of Mathematics
Warsaw University
Banacha 2, 02-097 Warsaw, Poland
skowron@mimuw.edu.pl
[3] Institute of Computer Science, Polish Academy of Sciences
Ordona 21, 01-237 Warsaw, Poland
and
Department of Mathematical and Computer Sciences
San Diego State University
5500 Campanile Drive San Diego, CA 92182, USA
rswiniar@sciences.sdsu.edu

Abstract. We present a rough set approach to vague concept approximation. Approximation spaces used for concept approximation have been initially defined on samples of objects (decision tables) representing partial information about concepts. Such approximation spaces defined on samples are next inductively extended on the whole object universe. This makes it possible to define the concept approximation on extensions of samples. We discuss the role of inductive extensions of approximation spaces in searching for concept approximation. However, searching for relevant inductive extensions of approximation spaces defined on samples is infeasible for compound concepts. We outline an approach making this searching feasible by using a concept ontology specified by domain knowledge and its approximation. We also extend this approach to a framework for adaptive approximation of vague concepts by agents interacting with environments. This paper realizes a step toward approximate reasoning in multiagent systems (MAS), intelligent systems, and complex dynamic systems (CAS).

Keywords: Vagueness, rough sets, approximation space, higher order vagueness, adaptive learning, incremental learning, reinforcement learning, constraints, intelligent systems.

1 Introduction

In this paper, we discuss the rough set approach to vague concept approximation. There has been a long debate in philosophy about vague concepts [18].

J.F. Peters and A. Skowron (Eds.): Transactions on Rough Sets V, LNCS 4100, pp. 39–62, 2006.
© Springer-Verlag Berlin Heidelberg 2006

Nowadays, computer scientists are also interested in vague (imprecise) concepts, e.g, many intelligent systems should satisfy some constraints specified by vague concepts. Hence, the problem of vague concept approximation as well as preserving vague dependencies especially in dynamically changing environments is important for such systems. Lotfi Zadeh [66] introduced a very successful approach to vagueness. In this approach, sets are defined by partial membership in contrast to crisp membership used in the classical definition of a set. Rough set theory [32] expresses vagueness not by means of membership but by employing the boundary region of a set. If the boundary region of a set is empty it means that a particular set is crisp, otherwise the set is rough (inexact). The non-empty boundary region of the set means that our knowledge about the set is not sufficient to define the set precisely. In this paper, some consequences on understanding of vague concepts caused by inductive extensions of approximation spaces and adaptive concept learning are presented. A discussion on vagueness in the context of fuzzy sets and rough sets can be found in [40].

Initially, the approximation spaces were introduced for decision tables (samples of objects). The assumption was made that the partial information about objects is given by values of attributes and on the basis of such information about objects the approximations of subsets of objects form the universe restricted to sample have been defined [32]. Starting, at least, from the early 90s, many researchers have been using the rough set approach for constructing classification algorithms (classifiers) defined over extensions of samples. This is based on the assumption that available information about concepts is partial. In recent years, there have been attempts based on approximation spaces and operations on approximation spaces for developing an approach to approximation of concepts over the extensions of samples (see, e.g., [48,50,51,56]). In this paper, we follow this approach and we show that the basic operations related to approximation of concepts on extension of samples are inductive extensions of approximation spaces. For illustration of the approach we use approximation spaces defined in [47]. Among the basic components of approximation spaces are neighborhoods of objects defined by the available information about objects and rough inclusion functions between sets of objects. Observe that searching for relevant (for approximation of concepts) extensions of approximation spaces requires tuning many more parameters than in the case of approximation of concepts on samples. The important conclusion from our considerations is that the inductive extensions used in constructing of algorithms (classifiers) are defined by arguments "for" and "against" of concepts. Each argument is defined by a tuple consisting of a degree of inclusion of objects into a pattern and a degree of inclusion of the pattern into the concepts. Patterns in the case of rule-based classifiers can be interpreted as the left hand sides of decision rules. The arguments are discovered from available data and can be treated as the basic information granules used in the concept approximation process. For any new object, it is possible to check the satisfiability of arguments and select arguments satisfied to a satisfactory degree. Such selected arguments are fused by conflict resolution strategies for obtaining the classification decision. Searching for relevant approximation spaces in the case of approximations over extensions of samples requires discovery of many parameters and patterns including selection of relevant attributes defining information about objects, discovery of relevant patterns for approximated concepts, selection of measures (similarity or closeness) of objects into discovered patters for concepts, structure and parameters of conflict resolution strategy. This causes that in the case of more compound concepts the searching process becomes infeasible (see, e.g., [6,63]). We propose to use as hints in the searching for relevant approximation spaces for compound concepts an additional domain knowledge making it possible to approximate such concepts. This additional knowledge is represented by a concept ontology [3,4,5,26,27,28,45,46,48,49,57] including concepts expressed in natural language and some dependencies between them. We assume that the ontology of concept has a hierarchical structure. Moreover, we assume that for each concept from ontology there is given a labelled set of examples of objects. The labels show the membership for objects relative to the approximated concepts. The aim is to discover the relevant conditional attributes for concepts on different levels of a hierarchy. Such attributes can be constructed using the so-called production rules, productions, and approximate reasoning schemes (AR schemes, for short) discovered from data (see, e.g. [3,4,5,26,27,28,45,46,48,49,57])). The searching for relevant arguments "for" and "against" for more compound concepts can be simplified because using domain knowledge.

Notice, that the searching process for relevant approximation spaces is driven by some selected quality measures. While in some learning problems such measures can be selected in a relatively easy way and remain unchanged during learning in other learning processes they can be only approximated on the basis of a partial information about such measures, e.g., received as the result of interaction with the environment. This case concerns, e.g., adaptive learning. We discuss the process of searching for relevant approximation spaces in different tasks of adaptive learning [1,7,12,15,21,22,24,58]. In particular, we present illustrative examples of adaptation of observation to the agent's scheme, incremental learning, reinforcement learning, and adaptive planning. Our discussion is presented in the framework of multiagent systems (MAS). The main conclusion is that the approximation of concepts in adaptive learning requires much more advanced methods. We suggest that this approach can be also based on approximation of ontology. In adaptive learning, the approximation of concepts is constructed gradually and the temporary approximations are changing dynamically in the learning process in which we are trying to achieve the approximation of the relevant quality. This, in particular, causes, e.g., boundary regions to change dynamically during the learning process in which we are attempting to find the relevant approximation of the boundary regions of approximated vague concepts. This is consistent with the requirement of the higher order vagueness [18] stating that the borderline cases of vague concepts are not crisp sets. In Sect. 5, we point out some consequences of this fact for further research on the rough set logic.

This paper is an extension and continuation of several papers (see, e.g., [3,4,5,26,27,28,44,45,46,48,49,50,56]) on approximation spaces and vague concept approximation processes. In particular, we discuss here a problem of adaptive learning of concept approximation. In this case, we are also searching for relevant approximation of the quality approximation measure. In a given step of the learning process, we have only a partial information about such a measure. On the basis of this information we construct its approximation and we use it for inducing approximation spaces relevant for concept approximation. However, in the next stages of the learning process, it may happen that after receiving new information form the environment, it is necessary to reconstruct the approximation of the quality measure and in this way we obtain a new "driving force" in searching for relevant approximation spaces during the learning process.

This paper is organized as follows. In Section 2, we discuss inductive extensions of approximation spaces. We emphasize the role of discovery of special patterns and the so called arguments in inductive extensions. In Section 3, the role of approximation spaces in hierarchical learning is presented. Section 3, outlines and approach based on approximation spaces in adaptive learning. In Sect. 5 (Conclusions), we summarize the discussion presented in the paper and we present some further research directions based on approximation spaces to approximate reasoning in multiagent systems and complex adaptive systems.

2   Approximation Spaces and Their Inductive Extensions

In [32], any approximation space is defined as a pair (U, R), where U is a universe of objects and $R \subseteq U \times U$ is an indiscernibility relation defined by an attribute set.

The lower approximation, the upper approximation and the boundary region are defined as crisp sets. It means that the higher order vagueness condition is not satisfied [18]. We will return to this issue in Section 4.

We use the definition of approximation space introduced in [47]. Any approximation space is a tuple $AS = (U, I, \nu)$, where U is the universe of objects, I is an uncertainty function, and ν is a measure of inclusion called the inclusion function, generalized in rough mereology to the rough inclusion [47,51].

In this section, we consider the problem of approximation of concepts over a universe U^*, i.e., subsets of U^*. We assume that the concepts are perceived only through some subsets of U^*, called samples. This is a typical situation in machine learning, pattern recognition, or data mining [10]. In this section we explain the rough set approach to induction of concept approximations. The approach is based on inductive extension of approximation spaces.

Now we will discuss in more detail the approach presented in [50,51]. Let $U \subseteq U^*$ be a finite sample and let $C_U = C \cap U$ for any concept $C \subseteq U^*$. Let $AS = (U, I, \nu)$ be an approximation space over the sample U. The problem we consider is how to extend the approximations of C_U defined by AS to approximation of C over U^*. We show that the problem can be described as searching for an extension

$AS^* = (U^*, I^*, \nu^*)$ of the approximation space AS relevant for approximation of C. This requires showing how to induce values of the extended inclusion function to relevant subsets of U^* that are suitable for the approximation of C. Observe that for the approximation of C, it is enough to induce the necessary values of the inclusion function ν^* without knowing the exact value of $I^*(x) \subseteq U^*$ for $x \in U^*$.

We consider an example for rule-based classifiers[1]. However, the analogous considerations for k-NN classifiers, feed-forward neural networks, and hierarchical classifiers [10] show that their construction is based on the inductive extension of inclusion function [51,44].

Usually, neighborhoods of objects in approximation spaces are defined by some formulas called patterns. Let us consider an example. Let AS^* be a given approximation space over U^* and let us consider a language L of patterns, where x denotes an object from U^*. In the case of rule-based classifiers, patterns are defined by feature value vectors. More precisely, in this case any pattern $pat(x)$ is defined by a formula $\bigwedge\{(a, a(x)) : a \in A$ and $v_a \in V_a\}$, where A is a given set of condition attributes [32]. An object $u \in U^*$ is satisfying $\bigwedge\{(a, a(x)) : a \in A$ and $v \in V_a\}$ if $a(u) = a(x)$ for any $a \in A$, i.e., if and only if x, u are A-indiscernible [32]. The set of objects satisfying $pat(x)$ in U^*, i.e., the semantics of $pat(x)$ in U^*, is denoted by $\|pat(x)\|_{U^*}$. Hence, $\|pat(x)\|_{U^*} = [x]_A$ where $[x]_A$ is the A-indiscernibility class of $x \in U^*$ [32]. By $\|pat(x)\|_U$ we denote the restriction of $\|pat(x)\|$ to $U \subseteq U^*$, i.e., the set $\|pat(x)\| \cap U$. In the considered case, we assume that any neighborhood $I(x) \subseteq U$ in AS is expressible by a pattern $pat(x)$. It means that $I(x) = \|pat(x)\|_U \subseteq U$, where $\|pat(x)\|_U$ denotes the meaning of $pat(x)$ restricted to the sample U.

We assume that for any object $x \in U^*$, only partial information about x (resulting, e.g., from a sensor measurement) represented by a pattern $pat(x) \in L$ with semantics $\|pat(x)\|_{U^*} \subseteq U^*$ defining the neighborhood of x in U^* is available. Moreover, only partial information such as $\|pat(x)\|_U$ is available about this set. In particular, relationships between information granules over U^*, e.g., $\|pat(x)\|_{U^*}$ and $\|pat(y)\|_{U^*}$, for different $x, y \in U^*$, are known, in general, only to a degree estimated by using relationships between the restrictions of these sets to the sample U, i.e., between sets $\|pat(x)\|_{U^*} \cap U$ and $\|pat(y)\|_{U^*} \cap U$.

The set $\{pat(x) : x \in U\}$ of patterns (defined by the whole set of attributes A from from an approximation space AS) is usually not relevant for approximation of the concept $C \subseteq U^*$. Such patterns can be too specific or not general enough, and can directly be applied only to a very limited number of new sample elements. For example, for a new object $x \in U^* \setminus U$ the set $\|pat(x)\|_U$ can be empty.

However, by using some generalization strategies, one can induce from patterns belonging to $\{pat(x) : x \in U\}$ some new patterns that are relevant for approximation of concepts over U^*.

---

[1] For simplicity of reasoning we consider only binary classifiers, i.e. classifiers with two decision classes. One can easily extend the approach to the case of classifiers with more decision classes.

Usually, first we define a new set PAT of patterns, which are candidates for relevant approximation of a given concept C. A typical example of the set of such patterns used in the case of rule based classifiers can be defined by dropping some descriptors from patterns constructed over the whole set of attributes, i.e., $\{\bigwedge\{(a, a(x)) : a \in B \text{ and } v_a \in V_a\} : B \subseteq A \text{ and } x \in U\}$. Among such patterns we search for the left hand sides of decision rules.

The set $PATTERNS(AS, L, C)$ can be selected from PAT using some quality measures evaluated on meanings (semantics) of patterns from this set restricted to the sample U. Often such measures are based on the numbers of examples from the concept C_U and its complement that support (satisfy) a given pattern. For example, if the confidence coefficient

$$\frac{card(\|pat\|_U \cap C_U)}{card(\|pat\|_U)}, \tag{1}$$

where $pat \in PAT$, is at least equal to a given threshold and the support

$$\frac{card(\|pat\|_U \cap C_U)}{card(U)}, \tag{2}$$

is also at least equal to a given threshold than we select pat as a member of $PATTERNS(AS, L, C)$.

Next, on the basis of some properties of sets definable by patterns from $PATTERNS(AS, L, C)$ over U, we induce approximate values of the inclusion function $\nu^*(X, C)$ on subsets of $X \subseteq U^*$ definable by any such pattern and the concept C. For example, we assume that the value of the confidence coefficient is not changing significantly if we move from U to U^*, i.e.,

$$\frac{card(\|pat\|_U \cap C_U)}{card(\|pat\|_U)} \approx \frac{card(\|pat\|_{U^*} \cap C)}{card(\|pat\|_{U^*})}, \tag{3}$$

Next, we induce the value of ν^* on pairs (X, Y) where $X \subseteq U^*$ is definable by a pattern from $\{pat(x) : x \in U^*\}$ and $Y \subseteq U^*$ is definable by a pattern from $PATTERNS(AS, L, C)$. For example, if $pat(x) = \bigwedge\{(a, a(x)) : a \in A \text{ and } v_a \in V_a\}$ and pat is obtained from $pat(x)$ by dropping some conjuncts then $\nu^*(\|pat(x)\|_{U^*}, \|pat\|)_{U^*}) = 1$ because $\|pat(x)\|_{U^*} \subseteq \|pat\|_{U^*}$. In a more general case, one can estimate the degree of inclusion of $\|pat(x)\|_{U^*}$ into $\|pat\|_{U^*}$ using some similarity degrees defined between formulas from PAT and $PATTERNS(AS, L, C)$. For example, one can assume that the values of attributes on x which occur in pat are not necessarily the same but similar. Certainly, such a similarity should be also defined or learned from data.

Finally, for any object $x \in U^* \setminus U$ we induce the degree $\nu^*(\|pat(x)\|_{U^*}, C)$ applying a conflict resolution strategy $Conflict\_res$ (e.g., a voting strategy) to the family of tuples:

$$\{(\nu^*(\|pat(x)\|_{U^*}, \|pat\|_{U^*}), pat, \nu^*(\|pat\|_{U^*}, C)) : pat \in PATTERNS(AS, L, C)\}. \tag{4}$$

Let us observe that conflicts can occur due to inductive reasoning in estimation of values of ν^*. For some $x \in U^*$ and $pat, pat' \in PATTERNS(AS, L, C)$ the values $\nu^*(\|pat(x)\|_{U^*}, \|pat\|_{U^*})$, $\nu^*(\|pat(x)\|_{U^*}, \|pat'\|_{U^*})$ can be both large (i.e., close to 1) and at the same time the value $\nu^*(\|pat\|_{U^*}, C)$ can be small (i.e., close to 0) and the value of $\nu^*(\|pat'\|_{U^*}, C)$ can be large.

Values of the inclusion function for the remaining subsets of U^* can be chosen in any way – they do not have any impact on the approximations of C. Moreover, observe that for the approximation of C we do not need to know the exact values of uncertainty function I^* – it is enough to induce the values of the inclusion function ν^*. The defined extension ν^* of ν to some subsets of U^* makes it possible to define an approximation of the concept C in a new approximation space AS^*.

To reduce the number of conditions from (4) one can use the so called arguments "for" and "against" discussed, e.g., in [49].

Any C-argument, where $C \subseteq U^*$ is a concept is a triple

$$(\epsilon, pat, \epsilon') \tag{5}$$

where $\epsilon, \epsilon' \in [0, 1]$ are degrees and pat is a pattern from $PATTERNS(AS, L, C)$.

The argument $arg = (\varepsilon, pat, \varepsilon')$ is satisfied by a given object $x \in U^*$, in symbols $x \models_C arg$, if and only if the following conditions are satisfied:

$$\nu^*(\|pat(x)\|_{U^*}, \|pat\|_{U^*}) \geq \varepsilon; \tag{6}$$
$$\nu^*(\|pat\|_{U^*}, C) \geq \varepsilon'.$$

The idea of C-arguments is illustrated in Figure 1.

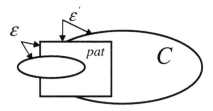

Fig. 1. C-argument

Instead of all conditions from (4) only some arguments "for" and "against" C are selected and the conflict resolution strategy is based on them. For any new object the strategy resolves conflicts between selected arguments "for" and "against" C which are satisfied by the object.

The very simple strategy for selection of arguments is the following one. The C-argument $arg = (\varepsilon, pat, \varepsilon')$ is called the argument "for" C if $\varepsilon \geq t$ and $\varepsilon' \geq t'$, where $t, t' > 0.5$ are given thresholds. The argument arg is "against" C, if this argument is the argument for the complement of C, i.e., for $U^* \setminus C$. However, in general, this may be not relevant method and the relevant arguments should be selected on the basis of more advanced quality measures. They can take into account, e.g., the support of patterns in arguments (see Section 4.1), their coverage, independence from other arguments, or relevance in searching for arguments used for approximation of more compound concepts in hierarchical learning.

One can define the lower approximation and upper approximation of the concept $C \subseteq U^*$ in the approximation space AS^* by

$$LOW(AS^*, C) = \{x \in U^* : \nu^*(I^*(x), C) = 1\}, \tag{7}$$
$$UPP(AS^*, C) = \{x \in U^* : \nu^*(I^*(x), C) > 0\}.$$

From the definition, in the case of standard rough inclusion [48], we have:

$$LOW(AS^*, C) \cap U \subseteq C \cap U \subseteq UPP(AS^*, C) \cap U. \tag{8}$$

However, in general the following equalities do not hold:

$$LOW(AS, C \cap U) = LOW(AS^*, C) \cap U, \tag{9}$$
$$UPP(AS, C \cap U) = UPP(AS^*, C) \cap U.$$

One can check that in the case of standard rough inclusion [48] we have:

$$LOW(AS, C \cap U) \supseteq LOW(AS^*, C) \cap U, \tag{10}$$
$$UPP(AS, C \cap U) \subseteq UPP(AS^*, C) \cap U.$$

Following the minimal length principle [41,42,52] some parameters of the induced approximation spaces are tuned to obtain a proper balance between the description length of the classifier and its consistency degree. The consistency degree on a given sample U of data can be represented by degrees to which the sets defined in equalities (9) are close. The description length is measured by description complexity of the classifier representation. Among parameters which are tuned are attribute sets used in the classifier construction, degrees of inclusion of patterns defined by objects to the left hand sides of decision rules, degrees of inclusion of patterns representing the left hand sides of decision rules in the decision classes, the specificity or support of these patterns, parameters of the conflict resolution strategy (e.g., set of arguments and parameters of arguments).

We can summarize our considerations in this section as follows. The inductive extensions of approximation spaces are basic operations on approximation spaces in searching for relevant approximation spaces for concept approximation. The approximation of concepts over U^* is based on searching for relevant approximation spaces AS^* in the set of approximation spaces defined by inductive extensions of a given approximation space AS. For any object $x \in U^* \setminus U$, the value $\nu^*(I^*(x), C)$ of the induced inclusion function ν^* is defined by conflict resolution strategy from collected arguments *for* classifying x to C and from collected arguments *against* classifying x to C.

3   Approximation Spaces in Hierarchical Learning

The methodology for approximation spaces extension presented in Section 2 is widely used for construction of rule based classifiers. However, this methodology cannot be directly used for concepts that are compound because of problems with inducing of the relevant set $PATTERNS(AS, L, C)$ of patterns. For such compound concepts, hierarchical learning methods have been developed (see, e.g., [2,3,4,5,26,27,28,45,46,48,49,57]).

We assume that domain knowledge is available about concepts. There is given a hierarchy of concepts and dependencies between them creating the so-called *concept ontology*. Only partial information is available about concepts in the hierarchy.

For concepts from the lowest level of hierarchy, decision tables with condition attributes representing sensory measurements are given. Classifiers for these concepts are induced (constructed) from such decision tables. Assuming that classifiers have been induced for concepts from some level l of the hierarchy, we are aiming at inducing classifiers for concepts on the next $l + 1$ level of the hierarchy. It is assumed that for concepts on higher levels there are given samples of objects with information about their membership values relative to the concepts. The relevant patterns for approximation of concepts from the $l + 1$ level are discovered using (i) these decision tables, (ii) information about dependencies linking concepts from the level $l + 1$ with concepts from the level l, and (iii) patterns discovered for approximation of concepts from the level l of the hierarchy. Such patterns define condition attributes (e.g., by the characteristic functions of patterns) in decision tables. Next, using the condition attributes approximation of concepts are induced. In this way, also, the neighborhoods for objects on the level $l + 1$ are defined. Observe also that the structure of objects on the higher level $l + 1$ is defined by means of their parts from the level l. In this section, for simplicity of reasoning, we assume that on each level the same objects are considered. To this end, we also assume that rough inclusion functions from approximation spaces are standard rough inclusion functions [48].

Now we outline a method of construction of patterns used for approximation of concepts from a given level of concept hierarchy by patterns used for approximation of concepts belonging to the lower level of the hierarchy.

This approach has been elaborated in a number of papers cited above, in particular in [49]. Assume that a concept C belongs to a level $l + 1$ of the hierarchy. We outline the idea of searching for sets $PATTERNS(AS, L, C)$ of patterns for a concept C, where AS is an approximation space discovered for approximation of the concept C and L is a language in which discovered patterns are expressed.

To illustrate this idea, let us consider and example of a dependency for a concept C from domain knwoledge:

$$\text{if } C_1 \text{ and } C_2 \text{ then } C, \tag{11}$$

where C_1, C_2, C are vague concepts. Analogously, let us consider a dependency for the complement of C:

$$\text{if } C_1' \text{ and } C_2' \text{ then } \neg C. \tag{12}$$

In general, we should consider a set with many dependencies with different concepts on the right hand sides of dependencies (creating, e.g., a partition of the universe) and in the process of generating arguments "for" and "against" a selected concept C are involved other vague dependencies from the given set. Let us recall that such a set of concepts and dependencies between them is specified in a given domain knowledge.

To approximate the target concept C, relevant patterns for C and $\neg C$ should be derived. The main idea is presented in Figure 2 and Figure 3. We assume that for any considered concept and for each pattern selected for this concept a degree of its inclusion into the concept can be estimated.

In Figure 2 it is shown that for patterns pat_1, pat_2 (e.g., left hand sides of decision rules in case of a rule based classifiers) for (or against) C_1 and C_2 and their inclusion degrees ϵ_1 and ϵ_2 into C_1 and C_2, respectively, it is constructed a pattern pat for (or against) C together with estimation of its inclusion degree ϵ to the concept C.

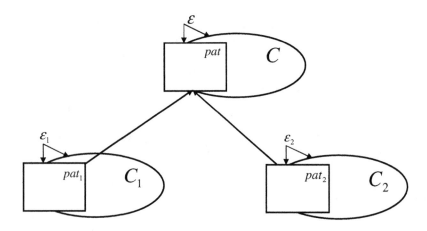

Fig. 2. An illustration of pattern construction

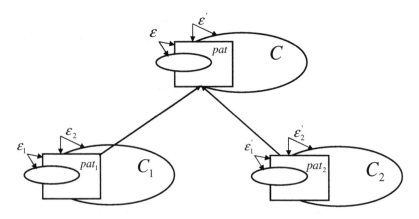

Fig. 3. An illustration of production rule

Figure 3 represents a construction of the target argument $(\epsilon, pat, \epsilon')$ for C from arguments $(\epsilon_1, pat_1, \epsilon_1')$ and $(\epsilon_2, pat_2, \epsilon_2')$ for C_1 and C_2, respectively. Such a construction

$$\text{if } (\epsilon_1, pat_1, \epsilon_1') \text{ and } (\epsilon_2, pat_2, \epsilon_2') \text{ then } (\epsilon, pat, \epsilon') \tag{13}$$

is called a *production rule* for the dependency (11). This production rule is true at a given object x if and only if the following implication holds:

$$\text{if } x \models_{C_1} (\epsilon_1, pat_1, \epsilon_1') \text{ and } x \models_{C_2} (\epsilon_2, pat_2, \epsilon_2') \text{ then } x \models_C (\epsilon, pat, \epsilon'). \tag{14}$$

Certainly, it is necessary to search for production rules of the high quality (with respect to some measures) making it possible to construct "strong" arguments in the conclusion of the production from "strong" arguments in the premises of the production rule. The quality of arguments is defined by means of relevant degrees of inclusion in these arguments and properties of patterns (such as support or description length).

The quality of arguments for concepts from the level $l + 1$ can be estimated on the basis properties of arguments for the concepts from the level l from which these arguments have been constructed. In this estimation are used decision tables delivered by domain experts. Such decision tables consist of objects with decision values equal to the membership degrees of objects relative to the concept or to its complement. In searching for productions of high quality, we use operations called *constrained sums* (see, e.g., [55]). Using these operations there are performed joins of information systems representing patterns appearing in arguments from the premise of production. The join is parameterized by constraints helping by tuning these parameters to filter the relevant objects from composition of patterns used for constructing a pattern for the concept C on the level $l + 1$ for the argument in the conclusion of the production rule. Moreover, the production rules may be composed into the so called approximation reasoning schemes (AR schemes, for short). This makes it possible to generate patterns for approximation of concepts on the higher level of the hierarchy (see, e.g., [2,3,4,5,26,27,28,46,49]). In this way one can induce gradually for any concept C in the hierarchy a relevant set of arguments (based on the relevant set of patterns $PATTERNS(AS, L, C)$ of patterns; see Section 3) for approximation of C.

We have recognized that for a given concept $C \subseteq U^*$ and any object $x \in U^*$, instead of crisp decision about the relationship of $I^*(x)$ and C, we can gather some arguments *for* and *against* it only. Next, it is necessary to induce from such arguments the value $\nu^*(I(x), C)$ using some strategies making it possible to resolve conflicts between those arguments [10,48]. Usually some general principles are used such as the minimal length principle [10] in searching for algorithms computing an extension $\nu^*(I(x), C)$. However, often the approximated concept over $U^* \setminus U$ is too compound to be induced directly from $\nu(I(x), C)$. This is the reason that the existing learning methods are not satisfactory for inducing high quality concept approximations in case of complex concepts [63]. There have been several attempts trying to omit this drawback. In this section we have discussed the approach based on hierarchical (layered) learning [57].

There are some other issues which should be discussed in approximation of compound vague concepts. Among them are issues related to adaptive learning and construction or reconstruction of approximation spaces in interaction with environments. In the following section, we consider an agent learning some concepts. This agent is learning the concepts in interaction with the environments. Different types of interaction are defining different types of adaptive learning processes. In particular one can distinguish incremental learning [13,23,61,65], reinforcement learning [9,14,17,34,39,60], competitive or cooperative learning [15]. There are several issues, important for adaptive learning that should be mentioned. For example, the compound target concept which we attempt to learn can gradually change over time and this concept drift is a natural extension for incremental learning systems toward adaptive systems. In adaptive learning it is important not only what we learn but also how we learn, how we measure changes in a distributed environment and induce from them adaptive changes of constructed concept approximations. The adaptive learning for autonomous systems became a challenge for machine learning, robotics, complex systems, and multiagent systems. It is becoming also a very attractive research area for the rough set approach. Some of these issues will be discussed in the following section.

4   Approximation Spaces in Adaptive Learning

There are different interpretations of the terms *adaptive learning* and *adaptive systems* (see, e.g., [1,7,12,15,21,22,24,58]). We mean by *adaptive* a system that learns to change with its environment. Our understanding is closest to the spirit of what appears in [7,12]. In complex adaptive systems (CAS), agents scan their environment and develop a schema for action. Such a schema defines interactions with agents surrounding it together with information and resources flow externally [7]. In this section, we concentrate only on some aspects of adaptive learning. The other issues of adaptive learning in MAS and CAS will be discussed elsewhere.

In particular, we would like to discuss the role of approximation spaces in adaptive learning.

In this paper, we consider the following exemplary situation. There is an agent ag interacting with another agent called the environment (ENV). Interactions are represented by actions [11,62] performed by agents. These actions are changing values of some sensory attributes of agents. The agent ag is equipped with ontology of vague concepts consisting of vague concepts and dependencies between them.

There are three main tasks of the agent ag: (i) adaptation of observation to the agent's scheme, (ii) adaptive learning of the approximations of vague concepts, and (iii) preserving constraints (e.g., expressed by dependencies between concepts).

Through adaptation of observation to the agent's scheme agent becomes more robust and can handle more variability [7].

Approximation of vague concepts by the agent ag requires development of searching methods for relevant approximation spaces which create the basis for approximation of concepts. Observe that the approximations of vague concepts are dynamically changing in adaptive learning when new knowledge about approximated concept is obtained by the agent ag. In particular, from this it follows that the boundary regions of approximated concepts are dynamically changing in adaptive learning. For each approximated concept we obtain a sequence of boundary regions rather than a single crisp boundary region. By generating this sequence we are attempting to approximate the set of borderline cases of a given vague concept. Hence, if the concept approximation problem is considered in adaptive framework the higher order postulate for vague concepts is satisfied (i.e., the set of borderline cases of any vague concept can not be crisp) [18,44,50].

The third task of the agent ag requires learning of a planning strategy. This is a strategy for predicting plans (i.e., sequences of actions) on the basis of observed changes in the satisfiability of the observed concepts from ontology. By executing plans the actual state of the system is transformed to a state satisfying the constraints. Changes in the environments can cause that the executed plans should be reconstructed dynamically by relevant adaptive strategies. In our example, actions performed by the agent ag are adjusting values of sensory attributes which are controllable by ag.

Before we will discuss the mentioned above tasks in more detail we would like to add some comments on interaction between agents.

The interactions among agents belong to the most important ingredients of computations realized by multiagent systems [21]. In particular, adaptive learning agents interact, in particular, with their environments. In this section, we will continue our discussion on adaptive learning by agents interacting with environment. Some illustrative examples of interactions which have influence on the learning process are presented.

Let us consider two agents ag and ENV representing the agent learning some concepts and the environment, respectively. By $ag_s(t)$ and $ENV_s(t)$ we denote (information about) the state of agents ag and ENV at the time t, respectively. Such an information can be represented, e.g., by a vector of attribute values A_{ag} and A_{ENV}, respectively [51]. The agent ag is computing the next state $ag_s(t+1)$ using his own transition relation $\longrightarrow_{ag}$ applied to the result of interaction of $ag_s(t)$ and $ENV_s(t)$. The result of such an interaction we denote by $ag_s(t) \oplus_{ENV} ENV_s(t)$ where $\oplus_{ENV}$ is an operation of interaction of ENV on the state of ag. Hence, the following condition holds:

$$ag_s(t) \oplus_{ENV} ENV_s(t) \longrightarrow_{ag} ag_s(t+1). \qquad (15)$$

Analogously, we obtain the following transition for environment states:

$$ag_s(t) \oplus_{ag} ENV_s(t) \longrightarrow_{ENV} ENV_s(t+1). \qquad (16)$$

In our examples, we will concentrate on two examples of interactions. In the first example related to incremental learning (see, e.g., [13,23,61,65]), we assume that $ag_s(t) \oplus_{ENV} ENV_s(t)$ is obtained by extending of $ag_s(t)$ by a new information about some new sample of objects labelled by decisions. The structure of $ag_s(t)$ is much more compound than in non-incremental learning. This will be discussed in one of the following section together with some aspects of adaptation in incremental learning. These aspects are related to searching for relevant approximation spaces. In the discussed case, we also assume that $ag_s(t) \oplus_{ag} ENV_s(t) = ENV_s(t)$, i.e., there is no interaction of the agent ag on the environment. In our second example, the agent ag can change the state of ENV by performing some actions or plans which change the state of the environment.

4.1   Adaptation of Observation to the Agent's Scheme

In this section, we present two illustrative examples of adaptation of observation to the agent's scheme. In the consequence of such an adaptation, the agent's scheme becomes more robust relative to observations.

In the first example, we consider instead of patterns $pat(x)$ (see Section 2) mode general patterns which are obtained by granulation of such patterns using a similarity relation τ. Assuming that the object description $pat(x)$ is defined by $\bigwedge\{(a, a(x)) : a \in A \text{ and } v_a \in V_a\}$ one can define such a similarity τ on description of objects, e.g., by a composition of similarity relations on attribute value sets (see, e.g., [20,25,47])[2]. Then instead of patterns $pat(x)$ we obtain patterns $pat_\tau(x)$ with the semantics defined by $\|pat_\tau(x)\|_{U^*} = \{y \in U^* : pat(x)\tau pat(y)\}$. The definition of satisfiability of arguments (6) changes as follows

$$\nu^*(\|pat_\tau(x)\|_{U^*}, \|pat\|_{U^*}) \geq \varepsilon; \tag{17}$$
$$\nu^*(\|pat\|_{U^*}, C) \geq \varepsilon'.$$

Observe, that $\|pat_\tau(x)\|_{U^*}$ is usually supported by many more objects than $\|pat(x)\|_{U^*}$. Hence, if it is possible to tune the parameters of τ in such a way that the first condition in (17) is satisfied for sufficiently large ε than the obtained argument is much more robust than the previous one, i.e., it is satisfied by much more objects than the previous one $pat(x)$ and at the same time the requirement related to the degrees of inclusion is preserved.

Our second example concerns construction of more robust production rules and productions (sets of production rules corresponding to the same dependency between vague concepts) (see Figure 4). Patterns in such productions represent different layers of vague concepts and are determined by the linguistic values of membership such as *small, medium, high* (see, e.g., [5]). These more general patterns are constructed using information granulation [51]. Let us consider a simple example of information granulation. Observe that the definition of the satisfiability of arguments given by (6) is not unique. One can consider the decision table (U, A, d), where A is a given set of condition attributes [32] and the decision d is the characteristic function of the set $Y_\varepsilon(pat) = \{y \in U : \nu(\|pat(y)\|_U, \|pat\|_U)) \geq \varepsilon\}$. From this decision table can be induced the classifier $Class(pat)$ for the concept $Y_\varepsilon^*(pat) = \{y \in U^* : \nu^*(\|pat(y)\|_{U^*}, \|pat\|_{U^*})) \geq \varepsilon\}$.

---

[2] Note, that the similarity relation τ has usually many parameters which should be tuned in searching for relevant similarity relations.

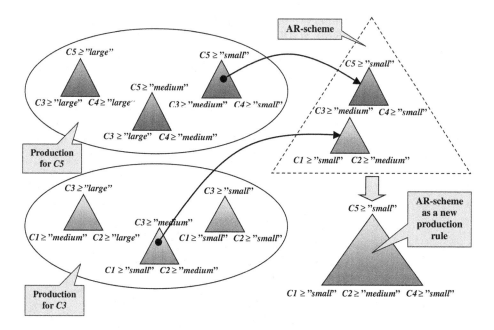

Fig. 4. An illustration of production and AR scheme

Any object $x \in U^*$ is satisfying the C-argument (5) if and only if the following condition is satisfied:

$$\nu^*(Y_\varepsilon^*(pat), C) \geq \varepsilon'. \tag{18}$$

The satisfiability of (18) is estimated by checking if the following condition holds on the sample U:

$$\nu(Y_\varepsilon(pat), C \cap U) \geq \varepsilon'. \tag{19}$$

We select only the arguments $(\varepsilon, pat, \varepsilon')$ with the maximal ε' satisfying (19) for given ε and pat.

Assume that $0 = \varepsilon_0 < \ldots < \varepsilon_{i-1} < \varepsilon_i < \ldots < \varepsilon_n = 1$. For any $i = 1, \ldots, n$ we granulate a family of sets

$$\{Y_\varepsilon^*(pat) : pat \in PATTERNS(AS, L, C) \text{ and } \nu^*(Y_\varepsilon^*(pat), C) \in [\varepsilon_{i-1}, \varepsilon_i)\} \tag{20}$$

into one set $Y_\varepsilon^*(\varepsilon_{i-1}, \varepsilon_i)$. Each set $Y_\varepsilon^*(\varepsilon_{i-1}, \varepsilon_i)$ is defined by an induced classifier $Class_\varepsilon(\varepsilon_{i-1}, \varepsilon_i)$. The classifiers are induced, in an analogous way as before, by constructing a decision table over a sample $U \subseteq U^*$. In this way we obtain a family of classifiers $\{Class_\varepsilon(\varepsilon_{i-1}, \varepsilon_i)\}_{i=1,\ldots,n}$.

The sequence $0 = \varepsilon_0 < \ldots < \varepsilon_{i-1} < \varepsilon_i < \ldots < \varepsilon_n = 1$ should be discovered in such a way that the classifiers $Class_\varepsilon(\varepsilon_{i-1}, \varepsilon_i)$ correspond to different layers of the concept C with linguistic values of membership. One of the method in searching for such sequence can be based on analysis of a histogram. This histogram represents a function $f(I)$ where $I \in \mathcal{J}$, $\mathcal{J}$ is a given uniform partition of the interval $[0,1]$, and $f(I)$ is the number of patterns from $\{Y_\varepsilon^*(pat) : pat \in PATTERNS(AS, L, C)\}$ with the inclusion degree into C from $I \subseteq [0,1]$.

4.2   Adaptation and Incremental Learning

In this section, we outline a searching process for relevant approximation spaces in incremental learning. Let us consider an example of incremental concept approximation scheme Sch (see Figure 5). By $Inf(C)$ and $Inf'(C)$ we denote

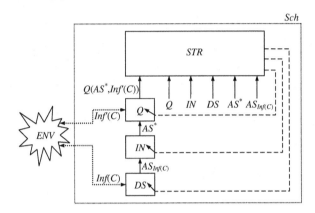

Fig. 5. An example of incremental concept approximation scheme

a partial information about the approximated concept (e.g., decision table for C or training sample) in different moments t and $t+1$ of time, respectively[3]. ENV denotes an environment, DS is an operation constructing an approximation space $AS_{Inf(C)}$ from a given sample $Inf(C)$, i.e., a decision table. IN is an inductive extension operation (see Section 2) transforming the approximation space $AS_{Inf(C)}$ into an approximation space AS^* used for approximation of the concept C; Q denotes a quality measure for the induced approximation space AS^* on a new sample $Inf'(C)$, i.e., an extension of the decision table $Inf(C)$. For example, the value $Q(AS^*, Inf'(C))$ can be taken as a ratio of the number of objects from $Inf'(C)$ that are classified correctly (relative to the decision values from $Inf'(C)$) by the classification algorithm (classifier) for C defined by AS^* (see Section 2) to the number of all objects in $Inf'(C)$.

The double-ended arrows leading into and out of ENV illustrate an interaction between agent ag and the environment ENV. In the case of a simple incremental learning strategy only samples of C are communicated by ENV to ag. More compound interactions between ag and ENV will be discuss later. They can be related to reaction from ENV on predicted by ag decisions (actions, plans) (see, e.g., award and penalty policies in reinforcement strategies [9,17,14,34,39,60]).

---

[3] For simplicity, in Figure 5 we do not present time constraints.

STR is a strategy that adaptively changes the approximation of C by modifying the quality measure Q, the operation of inductive extension IN, and the operation DS of constructing the approximation space $AS_{Inf(C)}$ from the sample $Inf(C)$. Dotted lines outgoing from the box labelled by SRT in Figure 5 are illustrating that the strategy STR after receiving the actual values of input its parameters is changing them (e.g., in the next moment of time). To make Figure 5 more readable the dotted lines are pointing to only one occurrence of each parameter of STR but we assume that its occurrences on the input for STR are modified too.

In the simple incremental learning strategy the quality measure is fixed. The aim of the strategy STR is to optimize the value of Q in the learning process. This means that in the learning process we would like to reach as soon as possible an approximation space which will guarantee the quality of classification measured by Q to be almost optimal. Still, we do not know how to control by STR this optimization. For example, should this strategy be more like the annealing strategy [19], then it is possible to perform more random choices at the beginning of the learning process and next be more "frozen" to guarantee the high convergence speed of the learning process to (semi-)optimal approximation space. In the case of more compound interactions between ag and ENV, e.g., in reinforcement learning, the quality measure Q should be learned using, e.g., awards or penalties received as the results of such interactions. This means that together with searching for an approximation space for the concept it is necessary to search for an approximation space over which the relevant quality measure can be approximated with high quality.

The scheme Sch describes an adaptive strategy ST modifying the induced approximation space AS^* with respect to the changing information about the concept C. To explain this in more detail, let us first assume that a procedure $new_C(ENV, u)$ is given returning from the environment ENV and current information u about the concept C a new piece of information about this concept (e.g., an extension of a sample u of C). In particular, $Inf^{(0)}(C) = new_C(ENV, \emptyset)$ and $Inf^{(k+1)}(C) = new_C(ENV, Inf^{(k)}(C))$ for $k = 0, \ldots$. In Figure 5 $Inf'(C) = Inf^{(1)}(C)$. Next, assuming that operations $Q^{(0)} = Q$, $DS^{(0)} = DS$, $IN^{(0)} = IN$ are given, we define $Q^{(k+1)}$, $DS^{(k+1)}$, $IN^{(k+1)}$, $AS^{(k+1)}_{(Inf^{(k+1)}(C)}$, and $AS^{*(k+1)}$ for $k = 0, \ldots$, by

$$(Q^{(k+1)}, DS^{(k+1)}, IN^{(k+1)}) = \qquad\qquad\qquad\qquad (21)$$
$$= STR(Q^{(k)}(AS^{*(k)}, Inf^{(k+1)}(C)), Q^{(k)}, IN^{(k)}, DS^{(k)}, AS^{*(k)}, AS^{(k)}_{Inf^{(k)}(C)})$$
$$AS^{(k+1)}_{Inf^{(k+1)}(C)} = DS^{(k+1)}(Inf^{(k+1)}(C)); \quad AS^{*(k+1)} = IN^{(k+1)}(AS^{(k+1)}_{Inf^{(k+1)}(C)}).$$

One can see that the concept of approximation space considered so far should be substituted by a more complex one represented by the scheme Sch making it possible to generate a sequence of approximation spaces $AS^{*(k)}$ for $k = 1, \ldots$ derived in an adaptive process of approximation of the concept C. One can also treat the scheme Sch as a complex information granule [48].

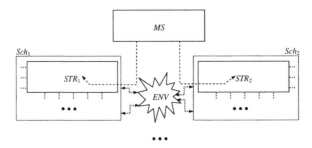

Fig. 6. An example of metastrategy in adaptive concept approximation

One can easily derive more complex adaptive schemes with metastrategies that make it possible to modify also strategies. In Figure 6 there is presented an idea of a scheme where a metastrategy MS can change adaptively also strategies STR_i in schemes Sch_i for $i = 1, \ldots, n$ where n is the number of schemes. The metastrategy MS can be, e.g., a fusion strategy for classifiers corresponding to different regions of the concept C.

4.3 Adaptation in Reinforcement Learning

In reinforcement learning [9,14,17,34,39,56,60], the main task is to learn the approximation of the function $Q(s, a)$, where s, a denotes a global state of the system and an action performed by an agent ag and, respectively and the real value of $Q(s, a)$ describes the reward for executing the action a in the state s. In approximation of the function $Q(s, a)$ probabilistic models are used. However, for compound real-life problems it may be hard to build such models for such a compound concept as $Q(s, a)$ [63]. In this section, we would like to suggest another approach to approximation of $Q(s, a)$ based on ontology approximation. The approach is based on the assumption that in a dialog with experts an additional knowledge can be acquired making it possible to create a ranking of values $Q(s, a)$ for different actions a in a given state s. We expect that in the explanation given by expert about possible values of $Q(s, a)$ are used concepts from a special ontology of concepts. Next, using this ontology one can follow hierarchical learning methods (see Section 3 and [2,3,4,5,26,27,28,45,46,48,49,57])) to learn approximations of concepts from ontology. Such concepts can have temporal character too. This means that the ranking of actions may depend not only on the actual action and the state but also on actions performed in the past and changes caused by these actions.

4.4 Adaptation and Planning

A more compound scheme than what was considered in the previous section can be obtained by considering strategies based on cooperation among the schemes for obtaining concept approximations of high quality. In Figure 7 an adaptive scheme for plan modification is presented. $PLAN$ is modified by a metastrategy MS that adaptively changes strategies in schemes Sch_i where $i = 1, \ldots, n$. This

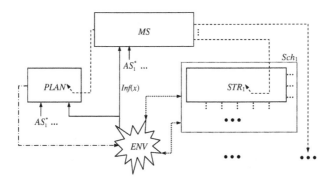

Fig. 7. An example of adaptive plan scheme

is performed on the basis of the derived approximation spaces AS_i^* induced for concepts that are guards (preconditions) of actions in plans and on the basis of information $Inf(x)$ about the state x of the environment ENV. The generated approximation spaces together with the plan structure are adaptively adjusted to make it possible to achieve plan goals.

The discussed example is showing that the context in which sequences of approximation spaces are generated can have complex structure represented by relevant adaptive schemes. The main goal of the agent ag in adaptive planning is to search for approximation of the optimal trajectory of states making it possible for the agent ag to achieve the goal, e.g., to keep as invariants some dependencies between vague concepts. Observe, that searching in adaptive learning for such a trajectory approximation should be performed together with adaptive learning of many other vague concepts which should be approximated, e.g., preconditions for actions, meta actions or plans.

One of the very important issue in adaptive learning is approximation of compound concepts used in reasoning about changes observed in the environment. The agent ag interacting with the environment ENV is recording changes in the satisfiability of concepts from her/his ontology. These changes should be expressed by relevant concepts (features) which are next used for construction of preconditions of actions (or plans) performed by the agent ag. In real-life problems these preconditions are compound concepts. Hence, to approximate such concepts we suggest to use an additional ontology of changes which can be acquired in a dialog with experts. All concepts from the ontology create a hierarchical structure. In this ontology relevant concepts characterizing changes in the satisfiability of concepts from the original ontology are included together with other simpler concepts from which they can be derived. We assume that such an ontology can be acquired in a dialog with experts. Concepts from this ontology are included in the expert explanations consisting of justifications why in some exemplary situations it is necessary to perform some particular actions in a particular order. Next, by approximation of the new ontology (see Section 3 and [2,3,4,5,26,27,28,45,46,48,49,57]) we obtain the approximation of the mentioned above compound concepts relevant for describing changes. This methodology can be used not only for predicting the relevant actions, meta actions or plans but also for the plan reconstruction. In our current projects we are developing the methodology for adaptive planning based on ontology approximation.

5   Conclusions

In the paper, we have discussed some problems of adaptive approximation of concepts by agents interacting with environments. These are the fundamental problems in synthesis of intelligent systems. Along this line important research directions perspective arise.

In particular, this paper realizes a step toward developing methods for adaptive maintenance of constraints specified by vague dependencies. Notice that there is a very important problem related to such a maintenance which should be investigated further, i.e., approximation of vague dependencies. The approach to this problem based on construction of arguments "for" and "against" for concepts from conclusions of dependencies on the basis of such arguments from premisses of dependencies will be presented in one of our next paper.

Among interesting topics for further research are also strategies for modeling of networks supporting approximate reasoning in adaptive learning. For example, AR schemes and AR networks (see, e.g., [48]) can be considered as a step toward developing such strategies. Strategies for adaptive revision of such networks and foundations for autonomous systems based on vague concepts are other examples of important issues.

In this paper also some consequences on understanding of vague concepts caused by inductive extensions of approximation spaces and adaptive concept learning have been presented. They are showing that in the learning process each temporary approximations, in particular boundary regions are crisp but they are only temporary approximations of the set of borderline cases of the vague concept. Hence, the approach we propose is consistent with the higher order vagueness principle [18].

There are some important consequences of our considerations for research on approximate reasoning about vague concepts. It is not possible to base such reasoning only on *static* models of vague concepts (i.e., approximations of given concepts [32] or membership functions [66] induced from a sample available at a given moment) and on multi-valued logics widely used for reasoning about rough sets or fuzzy sets (see, e.g., [31,36,66,69]). Instead of this there is a need for developing evolving systems of logics which in open and changing environments will make it possible to gradually acquire knowledge about approximated concepts and reason about them.

Acknowledgment

Many thanks to Professor James Peters for his incisive comments and for suggesting many helpful ways to improve this article.

The research has been supported by the grant 3 T11C 002 26 from Ministry of Scientific Research and Information Technology of the Republic of Poland.

References

1. Axaelrod, R.M.: *The Complexity of Cooperation.* Princeton University Press, Princeton, NJ (1997).
2. Bazan, J.G., Nguyen, H.S., Peters, J.F., Skowron, A., Szczuka, M.: Rough set approach to pattern extraction from classifiers. In: A. Skowron and M. Szczuka (Eds.), *Proceedings of the Workshop on Rough Sets in Knowledge Discovery and Soft Computing at ETAPS 2003* (RSKD 2003), Electronic Notes in Computer Science, **82**(4), Elsevier, Amsterdam, Netherlands, www.elsevier.nl/locate/entcs/volume82.html, April 12-13 (2003) 20–29.
3. Bazan, J.G., Peters, J.F., Skowron, A.: Behavioral pattern identification through rough set modelling. In [54] (2005) 688–697.
4. Bazan, J.G., Skowron, A.: On-line elimination of non-relevant parts of complex objects in behavioral pattern identification. In [30] (2005) 720–725.
5. Bazan, J.G., Skowron, A.: Classifiers based on approximate reasoning schemes. In B. Dunin-Keplicz, A. Jankowski A. Skowron, M. Szczuka (Eds.), *Monitoring, Security, and Rescue Tasks in Multiagent Systems MSRAS*, Advances in Soft Computing. Springer, Heidelberg (2005) 191–202.
6. Breiman, L.: Statistical modeling: The two cultures. *Statistical Science* **16**(3) (2001) 199–231.
7. Desai, A.: *Adaptive Complex Enterprices.* Communications ACM **48**(5) (2005) 32–35.
8. Dieterich, T.G.: Machine Learning Research: Four Current Directions. *AI Magazine* **18**(4) (1997) 97–136.
9. Dieterich, T.G.: Hierarchical reinforcement learning with the MAXQ value function decomposition. *Artificial Intelligence* **13** (2000) 227–303.
10. Friedman, J., Hastie, T., Tibshirani, R.: *The Elements of Statistical Learning: Data Mining, Inference, and Prediction.* Springer-Verlag, Heidelberg (2001).
11. Ghallab, M., Nau, D., Traverso, P.: *Automated Planning: Theory and Practice.* Elsevier, Morgan Kaufmann, CA (2004).
12. Gell-Mann, M.: *The Quark and the Jaguar.* Freeman and Co., NY (1994).
13. Giraud-Carrier, Ch.: A note on the utility of incremental learning. *AI Communications* **13**(4) (2000) 215–223.
14. McGovern, A.: *Autonomous Discovery of Temporal Abstractions from Interaction with an Environment.* PhD thesis, University of Massachusetts Amherst (2002).
15. Hoen, P.J., Tuyls, K., Panait, L., Luke, S., La Poutré, J.A.: An Overview of cooperative and competitive multiagent learning. In K. Tuyls, P. J. Hoen, K. Verbeeck, S. Sen (Eds.), *Learning and Adaption in Multi-Agent Systems: First International Workshop* (LAMAS 2005), Utrecht, The Netherlands, July 25 (2005) 1–46.
16. Kautz, H., Selman, B., Jiang, Y.: A General Stochastic Approach to Solving Problems with Hard and Soft Constraints. In D. Gu, J. Du, P. Pardalos (Eds.), *The Satisfiability Problem: Theory and Applications*, DIMACS Series in Discrete Mathematics and Theoretical Computer Science **35**, American Mathematical Society (1997) 573–586.
17. Kaelbling, L.P., Littman, M.L., Moore, A.W.: Reinforcement learning: A survey. *Journal of Artificial Intelligence Research* **4** (1996) 237–285.

18. R. Keefe, R.: *Theories of Vagueness*. Cambridge Studies in Philosophy, Cambridge, UK (2000).
19. Kirkpatrick, S., Gelatt, C.D., Vecchi, M.P.: Optimization by simulated annealing. *Science* **220**(4598) (1983) 671–680.
20. Krawiec, K., Słowiński, R., Vanderpooten, D.: Lerning decision rules from similarity based rough approximations. In L. Polkowski, A. Skowron (Eds.), *Rough Sets in Knowledge Discovery 2: Applications, Case Studies and Software Systems*, Physica-Verlag, Heidelberg (1998) 37–54.
21. Luck, M., McBurney, P., Shehory, O., Willmott, S.: *Agent Technology: Computing as Interaction. A Roadmap for Agent-Based Computing*. Agentlink III, the European Coordination Action for Agent-Based Computing, University of Southampton, UK (2005).
22. Marx, L.M.: Adaptive learning and iterated weak dominance. *Games and Economic Behavior* **26** (1999) 253–278.
23. Michalski, R.S., Mozetic, I., Hong, J., Lavrac, N.: The multi-purpose incremental learning system AQ15 and its testing application to three medical domains. In *Proceedings of the 5th National Conference on Artificial Intelligence*, Philadelphia, PA. Morgan Kaufmann (1986) 1041–1047.
24. Milgrom, P., Roberts, J.: Adaptive and sophisticated learning in normal form games. *Games Economic Behavior* **3**, (1991) 82–100.
25. Nguyen, S.H.: Regularity analysis and its applications in Data Mining. In L. Polkowski, T. Y. Lin, S. Tsumoto (Eds.), *Rough Set Methods and Applications: New Developments in Knowledge Discovery in Information Systems*, Physica-Verlag, Heidelberg, Germany (2000) 289–378.
26. Nguyen, S.H., Bazan, J., Skowron, A., Nguyen, H.S.: Layered learning for concept synthesis. In J. F. Peters, A. Skowron (Eds.), *Transactions on Rough Sets I: Journal Subline, Lecture Notes in Computer Science* **3100**, Springer, Heidelberg, Germany (2004) 187–208.
27. Nguyen, T.T., Skowron, A.: Rough set approach to domain knowledge approximation. In [64] (2003) 221–228.
28. Nguyen, T.T.: Eliciting domain knowledge in handwritten digit recognition. In [30] (2005) 762–767.
29. Pal, S.K, Polkowski, L., Skowron, A. (Eds.): *Rough-Neural Computing: Techniques for Computing with Words*, Cognitive Technologies, Springer-Verlag, Heidelberg (2004).
30. Pal, S.K, Bandoyopadhay, S., Biswas, S. (Eds.). *Proceedings of the First International Conference on Pattern Recognition and Machine Intelligence* (PReMI'05), December 18-22, 2005, Indian Statistical Institute, Kolkata, Lecture Notes in Computer SCience **3776**. Springer-Verlag, Heidelberg, Germany (2005).
31. Pavelka, J.: On Fuzzy Logic I-III. *Zeit. Math Logik Grund. Math.* **25** (1979) 45–52, 119-134, 447-464.
32. Pawlak, Z.: *Rough Sets: Theoretical Aspects of Reasoning about Data*. System Theory, Knowledge Engineering and Problem Solving **9**, Kluwer Academic Publishers, Dordrecht (1991).
33. Pawlak, Z., Skowron, A.: Rough Membership Functions. In R. R. Yager, M. Fedrizzi, J. Kacprzyk (eds.), *Advances in the Dempster-Schafer Theory of Evidence*, John Wiley and Sons, New York (1994) 251–271.
34. Peters, J.F., Henry, C.: Reinforcement Learning with Approximation Spaces. *Fundamenta Informaticae* **71**(2-3) (2006) 323–349.

35. Peters, J.F., Skowron, A. (Eds.): Transactions on Rough Sets III: Journal Subline, Lecture Notes in Computer Science, vol. 3400. Springer, Heidelberg, Germany (2005).
36. Polkowski, L.: *Rough Sets: Mathematical Foundations*. Physica-Verlag, Heidelberg (2002).
37. Polkowski, L., Skowron, A.: Rough Mereology: A New Paradigm for Approximate Reasoning. *International Journal of Approximate Reasoning* **15**, (1996) 333–365.
38. Polkowski, L., Skowron, A.: Rough Mereological Calculi of Granules: A Rough Set Approach to Computation. *Computational Intelligence* **17**, (2001) 472–492.
39. Randlov, J.: *Solving Complex Problems with Reinforcement Learning*. Ph.D. Thesis, University of Copenhagen (2001).
40. Read, S.: *Thinking about Logic. An Introduction to the Philosophy of Logic*. Oxford University Press, Oxford, New York (1995).
41. Rissanen, J.: Modeling by shortes data description. Automatica **14** (1978) 465–471.
42. Rissanen, J.: Minimum-description-length principle. In: S. Kotz, N. Johnson (Eds.), Encyclopedia of Statistical Sciences, John Wiley & Sons, New York, NY (1985) 523–527.
43. Skowron, A.: Rough Sets in KDD. *16-th World Computer Congress* (IFIP'2000), Beijing, August 19-25, 2000, In Z. Shi, B. Faltings, M. Musen (Eds.), *Proceedings of Conference on Intelligent Information Processing* (IIP2000), Publishing House of Electronic Industry, Beijing (2000) 1–17.
44. Skowron, A.: Rough Sets and Vague Concepts. *Fundamenta Informaticae* **4(1-4)** (2005) 417–431.
45. Skowron, A.: Approximate reasoning in distributed environments. In N. Zhong, J. Liu (Eds.), *Intelligent Technologies for Information Analysis*, Springer, Heidelberg (2204) 433–474.
46. Skowron, A., Peters, J.F.: Rough sets: Trends and challenges. In [64] (2003) 25–34.
47. Skowron, A., J. Stepaniuk, S.: Tolerance Approximation Spaces. *Fundamenta Informaticae* **27** (1996) 245–253.
48. Skowron, A., Stepaniuk, J.: Information Granules and Rough-Neural Computing. In [29] (2004) 43–84.
49. Skowron, A., Stepaniuk, J.: Ontological framework for approximation. In [53] (2005) 718–727.
50. Skowron, A., Swiniarski, R.: Rough sets and higher order vagueness. In [53] (2005) 33–42.
51. Skowron, A., Swiniarski, R., Synak, P.: Approximation spaces and information granulation. In [35] (2004) 175–189.
52. Ślęzak, D.: Approximate entropy reducts. Fundamenta Informaticae **53** (2002) 365–387.
53. Ślęzak, D., Wang, G., Szczuka, M., Düntsch, I., Yao, Y. (Eds.): *Proceedings of the 10th International Conference on Rough Sets, Fuzzy Sets, Data Mining, and Granular Computing* (RSFDGrC'2005), Regina, Canada, August 31-September 3, 2005, Part I, Lecture Notes in Artificial Intelligence, vol. 3641. Springer-Verlag, Heidelberg, Germany (2005).
54. Ślęzak, D., Yao, J.T., Peters, J.F., Ziarko, W., Hu, X. (Eds.): *Proceedings of the 10th International Conference on Rough Sets, Fuzzy Sets, Data Mining, and Granular Computing* (RSFDGrC'2005), Regina, Canada, August 31-September 3, 2005, Part II, Lecture Notes in Artificial Intelligence, vol. 3642. Springer-Verlag, Heidelberg, Germany (2005).
55. Stepaniuk, J., Bazan, J., Skowron, A.: Modelling complex patterns by information systems. *Fundamenta Informaticae* **67**(1-3) (2005) 203–217.

56. Stepaniuk, J., Skowron, A., Peters, J.F., Swiniarski, R.: Calculi of approximation spaces. *Fundamenta Informaticae* **72**(1-3) (2206) (to appear).
57. Stone, P.: *Layered Learning in Multi-Agent Systems: A Winning Approach to Robotic Soccer.* The MIT Press, Cambridge, MA (2000).
58. Sun, R. (Ed.): *Cognition and Multi-Agent Interaction. From Cognitive Modeling to Social Simulation.* Cambridge University Press, New York, NY (2006).
59. Suraj, Z.: The synthesis problem of concurrent systems specified by dynamic information systems. In L. Polkowski, A. Skowron (Eds.), *Rough Sets in Knowledge Discovery 2: Applications, Case Studies and Software Systems*, Physica-Verlag, Heidelberg (1998) 418–448.
60. Sutton, R.S., Barto, A.G.: *Reinforcement Learning: An Introduction.* MIT Press, Cambridge, MA (1998).
61. Utgoff, P.: An incremental ID3. In *Proceedings of the Fifth International Workshop on Machine Learning*, Ann Arbor, Michigan, June 12-14, 1988, Morgan Kaufmann (1988) 107–120.
62. Van Wezel, W., Jorna, R., Meystel, A.: *Planning in Intelligent Systems: Aspects, Motivations, and Methods.* John Wiley & Sons, Hoboken, New Jersey (2006).
63. Vapnik, V.: *Statistical Learning Theory.* John Wiley & Sons, New York, NY (1998).
64. Wang, G., Liu, Q., Yao, Y., Skowron, A. (Eds.): *Proceedings of the 9-th International Conference on Rough Sets, Fuzzy Sets, Data Mining, and Granular Computing* (RSFDGrC'2003), Chongqing, China, May 26-29, 2003, Lecture Notes in Artificial Intelligence **2639**, Springer-Verlag, Heidelberg, Germany (2003).
65. Wojna, A.: Constraint Based Incremental Learning of Classification Rules. In [71] (2001) 428-435.
66. Zadeh, L.A.: Fuzzy sets. *Information and Control* **8** (1965) 333–353.
67. Zadeh, L.A.: Outline of a new approach to the analysis of complex systems and decision processes. *Trans. on Systems, Man and Cybernetics* **SMC-3** (1973) 28-44.
68. Zadeh, L.A.: The concept of a linguistic variable and its application to approximate reasoning I, II, III, *Information Sciences* **8** (1975) 199-257, 301-357; **9** 43-80.
69. Zadeh, L.A.: Fuzzy Logic = Computing with Words. *IEEE Transactions on Fuzzy Systems* **2** (1996) 103–111.
70. Ziarko, W.: Variable Precision Rough Set Model. *Journal of Computer and System Sciences* **46** (1993) 39–59.
71. Ziarko, W., Yao, Y. (Eds.): Proceedings of the 2nd International Conference on Rough Sets and Current Trends in Computing (RSCTC'2000), Banff, Canada, October 16-19, 2000, Lecture Notes in Artificial Intelligence, vol. 2005. Springer-Verlag, Heidelberg, Germany, 2001.

Matching 2D Image Segments with Genetic Algorithms and Approximation Spaces

Maciej Borkowski and James F. Peters

Department of Electrical and Computer Engineering,
University of Manitoba
Winnipeg, Manitoba R3T 5V6 Canada
{maciey, jfpeters}@ee.umanitoba.ca

Abstract. [1]This article introduces an approach to matching 2D image segments using approximation spaces. The rough set approach introduced by Zdzisław Pawlak provides a ground for concluding to what degree a particular set of similar image segments is a part of a set of image segments representing a norm or standard. The number of features (color difference and overlap between segments) typically used to solve the image segment matching problem is small. This means that there is not enough information to permit image segment matching with high accuracy. By contrast, many more features can be used in solving the image segment matching problem using a combination of evolutionary and rough set methods. Several different uses of a Darwinian form of a genetic algorithm (GA) are introduced as a means to partition large collections of image segments into blocks of similar image segments. After filtering, the output of a GA provides a basis for finding matching segments in the context of an approximation space. A coverage form of approximation space is presented in this article. Such an approximation space makes it possible to measure the the extent that a set of image segments representing a standard covers GA-produced blocks. The contribution of this article is the introduction of an approach to matching image segments in the context of an approximation space.

Keywords: Approximation space, coverage, genetic algorithm, image, 2D matching, rough sets, image segment.

1 Introduction

Considerable work on the application of rough set methods in image processing has been reported (see, e.g., [37,2,18,51,52]). This paper introduces an approach to matching image segments in the context of approximation spaces. The basic model for an approximation space was introduced by Pawlak in 1981 [30], elaborated in [28,32], generalized in [46,47,50], and applied in a number of ways (see, e.g., [36,38,39,48,11]). An approximation space serves as a formal counterpart of perception or observation [28], and provides a framework for approximate reasoning about vague concepts. Image segmentation (see,

---

[1] Transactions on Rough Sets V, 2006, to appear.

J.F. Peters and A. Skowron (Eds.): Transactions on Rough Sets V, LNCS 4100, pp. 63–101, 2006.

e.g.,[43,4,8,13,14,23,29,56,54]), and the image segment matching problem (see, e.g.,[12,55,42,53]) have been widely studied . The goal of an image-matching system is to match the segments from the two given images. Color and over-lap are the two features of image segments that are commonly used to solve the matching problem. To achieve more accuracy in matching image segments, a combination of an evolutionary approach to finding sets of similar segments and approximation spaces are used. The evolutionar approach is realized with a genetic algorithm (GA) that partitions collections of image segments into sets of similar image segments. Filtering out GA-produced sets of image segments with the best match is carried out in the context of an approximation space. This approach makes it possible to solve the image segment matching problem with larger sets of features that yield more information about segments. This approach also results in more accurate matching of image segments. An overview of the 2D image segment matching method presented in this article is shown in Fig. 1.

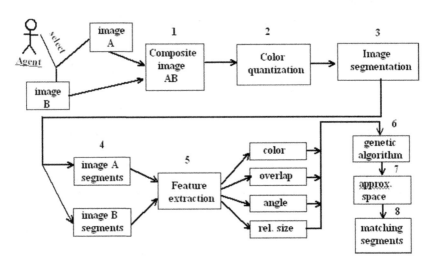

Fig. 1. 2D Image Segment Matching Steps

The matching process begins by forming a composite of a pair of images, then carrying out color quantization (step 2 in Fig. 1). After that, the quantized image is segmented, which results in a pair of segmented images. Next, feature values of image segment pairs are obtained in step 5 in Fig. 1. Then a GA is applied to a collection of image segment pairs, which are partitioned into sets. After eliminating non-disjoint sets of segment pairs, the coverage of the remaining sets of segment pairs is measured relative to a standard (norm), which is a set of image segment pairs that represent certain knowledge. The end result in step 8 of Fig. 1 is a collection of best matching pairs of image segments. This is in keeping with the original view of approximation spaces as counterparts of perception (in this case, approximations provide a framework for visual perception). The contribution of this paper is the use of approximation spaces to solve the image segment matching problem.

This paper is organized as follows. A brief introduction to rough set theory is given in Sect. 2. Set approximation is presented in Sect. 2.1, and the structure of generalized approximation spaces is given in Sect. 2.2. The basic structure of a Darwinian form of genetic algorithm is presented in Sect. 3. Fundamental terminology for 2D digital images and classical 2D image processing techniques are presented in Sect. 4 and Sect. 5, respectively. Upper and lower approximation of sets of image segment pairs is described in Sect. 6 and Sect. 7, respectively. A detailed presentation of GAs for image processing is given in Sect. 8. An approach to matching image segments is presented in Sect. 9.

2   Basic Concepts About Rough Sets

This section briefly presents some fundamental concepts in rough set theory that provide a foundation for the image processing described in this article. In addition, a brief introduction to approximation spaces is also given, since approximation spaces are used to solve the 2D matching problem.

2.1   Rough Set Theory

The rough set approach introduced by Zdzisław Pawlak [31,32,33] provides a ground for concluding to what degree a set image segment pairs representing a standard cover a set of similar image segment pairs. The term "coverage" is used relative to the extent that a given set is contained in standard set. An overview of rough set theory and applications is given in [40,21]. For computational reasons, a syntactic representation of knowledge is provided by rough sets in the form of data tables. A data (information) table IS is represented by a pair (U, A), where U is a non-empty, finite set of elements and A is a non-empty, finite set of attributes (features), where $a : U \longrightarrow V_a$ for every $a \in A$. For each $B \subseteq A$, there is associated an equivalence relation $Ind_{IS}(B)$ such that $Ind_{IS}(B) = \{(x, x') \in U^2 | \forall a \in B, a(x) = a(x')\}$. Let $U/Ind_{IS}(B)$ denote a partition of U determined by B (i.e., $U/Ind_{IS}(B)$ denotes the family of all equivalence classes of $Ind_{IS}(B)$), and let $B(x)$ denote a set of B-indiscernible elements containing x. $B(x)$ is called a block, which is in the partition $U/Ind_{IS}(B)$. For $X \subseteq U$, the sample X can be approximated from information contained in B by constructing a B-lower and B-upper approximation denoted by B_*X and B^*X, respectively, where $B_*X = \cup \{B(x)|B(x) \subseteq X\}$ and $B^*X = \cup\{B(x)|B(x) \cap X \neq \emptyset\}$. The B-lower approximation B_*X is a collection of blocks of sample elements that can be classified with full certainty as members of X using the knowledge represented by attributes in B. By contrast, the B-upper approximation B^*X is a collection of blocks of sample elements representing both certain and possibly uncertain knowledge about X. Whenever B_*X is a proper subset of B^*X, i.e., $B_*X \subset B^*X$, the sample X has been classified imperfectly, and is considered a rough set.

2.2 Approximation Spaces

This section gives a brief introduction to approximation spaces. The basic model for an approximation space was introduced by Pawlak in 1981 [30], elaborated in [28,32], generalized in [46,47,50], and applied in a number of ways (see, e.g., [36,38,48,11]). An approximation space serves as a formal counterpart of perception or observation [28], and provides a framework for approximate reasoning about vague concepts.

A very detailed introduction to approximation spaces considered in the context of rough sets is presented in [40]. The classical definition of an approximation space given by Zdzisław Pawlak in [30,32] is represented as a pair (U, Ind), where the indiscernibility relation Ind is defined on a universe of objects U (see, e.g., [44]). As a result, any subset X of U has an approximate characterization in an approximation space. A generalized approximation space was introduced by Skowron and Stepaniuk in [46,47,50]. A *generalized approximation space* is a system $GAS = (\ U, N, \nu\ )$ where

- U is a non-empty set of objects, and $\mathcal{P}(U)$ is the powerset of U,
- $N : U \rightarrow \mathcal{P}(U)$ is a neighborhood function,
- $\nu : \mathcal{P}(U) \text{ x } \mathcal{P}(U) \rightarrow [0, 1]$ is an overlap function.

A set $X \subseteq U$ in a GAS if, and only if X is the union of some values of the neighborhood function. In effect, the uncertainty function N defines for every object x a set of similarly defined objects [45]. That is, N defines a neighborhood of every sample element x belonging to the universe U (see, e.g., [35]). Generally, N can be created by placing constraints on the value sets of attributes (see, e.g., [40]) as in (1).

$$y \in N(x) \Leftrightarrow \max_a \{ dist_a(a(x), a(y)) \} \le \epsilon. \tag{1}$$

where $dist_a$ is a metric on the value set of a and ϵ represents a threshold [40]. Specifically, any information system $IS = (U, A)$ defines for any $B \subseteq A$ a parameterized approximation space $AS_B = (U, N_B, \nu)$, where $N_B = B(x)$, a B-indiscernibility class in the partition of U [45]. The rough inclusion function ν computes the degree of overlap between two subsets of U. Let $\mathcal{P}(U)$ denote the powerset of U. The overlap function ν is commonly defined as standard rough inclusion (SRI) $\nu : \mathcal{P}(U) \text{ x } \mathcal{P}(U) \rightarrow [0, 1]$ as defined in (2).

$$\nu_{SRI}(X, Y) = \begin{cases} \frac{|X \cap Y|}{|X|}, & \text{if } X \ne \emptyset, \\ 1, & \text{if } X = \emptyset. \end{cases} \tag{2}$$

for any $X, Y \subseteq U$, where it is understood that the first term is the smaller of the two sets. The result is that $\nu_{SRI}(X, Y)$ represents the proportion of X that is "included" in Y. However, we are interested in the larger of the two sets (assume that the $card(Y) \ge card(X)$) because we want to see how well Y "covers" X, where Y represents a standard for evaluating sets of similar image segments. Standard rough coverage (SRC) ν_{SRC} can be defined as in (3).

$$\nu_{SRC}(X, Y) = \begin{cases} \frac{|X \cap Y|}{|Y|}, & \text{if } Y \ne \emptyset, \\ 1, & \text{if } Y = \emptyset. \end{cases} \tag{3}$$

In other words, $\nu_{SRC}(X, Y)$ returns the degree that Y covers X. In the case where $X = Y$, then $\nu_{SRC}(X, Y) = 1$. The minimum coverage value $\nu_{SRC}(X, Y) = 0$ is obtained when $X \cap Y = \emptyset$ (i.e., X and Y have no elements in common).

3   Genetic Algorithms

Evolution has been characterized as an optimization process [9,19,25]. Darwin observed "organs of extreme perfection" that have evolved [5]. Genetic algorithms (GAs) belong to a class of evolutionary algorithms introduced by John Holland in 1975 [19] as a means of studying evolving populations. A GA has three basic features:

- **Representation:** each population member has a representation,
- **Method of Selection:** fitness of each population member is evaluated,
- **Method of Variation (Crossover):** create new population member by combining the best features from pairs of highly fit individuals.

Crossover is the fundamental operation used in classical genetic algorithms. Mutation is another method used in GAs to induce variations in the genes of a chromosome representing a population member. The basic steps in a genetic algorithm are described as follows. Let $P(t)$ denote an initial population of individual structures, each with an initial fitness at time t. Then an iteration begins. Individuals in $P(t)$ are selected for mating and copied to a mating buffer $C(t)$ at time step t. Combine individuals in $C(t)$ to form a new mating buffer $C'(t)$. Construct a new population P_{t+1} from P_t and $C'(t)$. A desired fitness is used as a stopping criterion for the iteration in a GA. A representation of a very basic GA that uses only the crossover operation is given in Alg. 1.

Algorithm 1. Basic GA

Input   : population P_t, mating pool C_t
Output: evolved population P_T at time T
$t = 0$;
Initialize fitness of members of P_t;
while (*Termination condition not satisfied*) **do**
    $t = t + 1$;
    Construct mating pool C_t from P_{t-1};
    Crossover structures in C_t to construct new mating pool C'_t;
    Evaluate fitness of individuals in C'_t;
    Construct new population P_t from P_{t-1} and C'_t;
end

GAs have proven to be useful in searching for matching segments in pairs of images (see Sect. 8). In preparation for the GA approach to matching images, some basic terminology (see Sect. 4), rudiments of classical 2D image processing (see Sect. 5), and image matching using rough set methods (e.g., upper approximation approach in Sect. 6 and lower approximation approach in Sect. 7) are presented.

4   Classical 2D Matching Terminology

This section gives an introduction to the basic definitions of technical terms associated with the classical approach to 2D matching images.

Definition 1. ([12,24,17]) **Pixel.** *A pixel (also referred to as a* image element, picture element, *or pel) is an element of a digital image.*

Definition 2. ([12]) **Color.** *A color of a pixel is a mapping from a space of all colors perceived by humans into a finite set of integer numbers grouped into three components. Each component Red, Green and Blue is represented by a number from 0 to 255. The total number of different colors represented by a pixel is 16,777,216.*

Definition 3. ([12]) **Grayscale.** *Grayscale represents a subset of RGB space, where all components have equal values, eg. Red = Green = Blue. There are only 256 such combinations and therefore the grayscale values can be represented only by one number from the range 0 to 255.*

In a digital image, a pixel (short for *picture element*) is an element that has a numerical value that represents a grayscale or RGB intensity value. Pixels are part of what are known as 4-neighborhoods, which provide a basis for identifying image segments.

Definition 4. ([12,24]) **4-Neighborhood.** *A pixel p with coordinates (x,y) has 4 neighbors (2 vertical and 2 horizontal neighbors) at coordinates*

$$(x + 1, y), (x - 1, y), (x, y + 1), (x, y - 1)$$

Definition 5. ([12]) **Pixel Membership.** *Pixel $p_1 = (x_1, y_1)$ belongs to a 4-neighborhood of pixel $p_2 = (x_2, y_2)$ if and only if exactly one coordinate of p_2 differs from the corresponding coordinate of p_1. This difference must be equal to 1.*

Definition 6. Segment. *A segment is a collection of 4-neighborhood connected pixels, which have the same color.*

The process of matching segments described in this article is based on four parameters described in the section 5.3. These parameters are *degree of overlap* between segments, *angle of rotation* between segments, *distance between mean colors* of segments and *ratio of cardinalities* of both segments. A combination of a genetic algorithm and rough set-based post processing is used to combine the information from all four parameters to find the best matches between the segments.

The problem of finding the match between image segments is not trivial. The four parameters required for matching image segments sometimes contain contradictory information about the quality of match. Thus it is impossible to find proper matches using only one or two of these parameters. The simplest approach is to find the matches with the smallest distance in the space defined by the four parameters. This space is denoted by Ω and consists of vectors where each coordinate value is the difference of some parameter values.

Definition 7. Image Segment Parameter Space. *Define space Ω to be a subspace of $\mathbb{R}^4$ such that*

$$\Omega = \mathbb{C} \times \mathbb{O} \times \mathbb{A} \times \mathbb{RC} \subseteq \mathbb{R}^4.$$

where $\mathbb{C}$, $\mathbb{O}$, $\mathbb{A}$ and $\mathbb{RC}$ denote domains of the four parameters' values, i.e. the distance between mean colors of segments, degree of overlap between segments, the angle of rotation between segments and ratio of cardinalities of both segments, respectively.

The match between two points s, $t \in \Omega$ can be calculated as a weighted distance between their parameters' values.

Definition 8. Distance in Image Segment Parameter Space. *The distance between two vectors s and t such that s, $t \in \Omega$ is defined to be a distance between these points in the space Ω weighted by the vector $w = (C, O, A, Rc)$.*

$$\|s - t\|_w = \sqrt{C \cdot (s_1 - t_1)^2 + O \cdot (s_2 - t_2)^2 + A \cdot (s_3 - t_3)^2 + Rc \cdot (s_4 - t_4)^2}.$$

where $s = (s_1, s_2, s_3, s_4)$ and $t = (t_1, t_2, t_3, t_4)$.

Here, the weight vector (C, O, A, Rc) denotes the importance of each parameter. Each such vector and an ideal vector ζ define a measure of the quality of a match.

Definition 9. Measure of Quality. *A measure of quality of a match between two segments s parametrized by the vector $w = (C, O, A, Rc)$ and the ideal solution ξ is given by the distance between points s and ξ in Ω space.*

$$Q_{w,\xi}(s) = \|s - \xi\|_w .$$

The problem with the Def. 9 is with defining the ideal solution ξ. The first two parameters in w can be defined, where the difference in color $C = 0$ and the overlap between two segments $O = 1$. The remaining two parameters (A and Rc) in w can be defined only with respect to some set of matches. It does not make sense to define the ideal angle of rotation between segments, since it depends on the images and can be different for any pair of images. Therefore, the ideal solution can be defined only in the first two positions. In order to make the remaining two parameters not influential, the w vector must contain zeros in the third and fourth position. As a result, the ideal vector is defined as

$$\zeta = (0, 1, 0, 0) , w = (x, y, 0, 0) .$$

where $x, y \in \mathbb{R}$. Unfortunately, this solution uses only two parameters instead of four. This can lead to wrong classification as shown in the Fig. 9 or Fig. 12 ($\circ$ denotes the correct match and $+$ denotes the closest match using $Q_{w,\xi}$ measure).

An algorithm which uses all four image segment features should generate a set of possible good matches. A genetic algorithm (GA) is the example of such an algorithm. It is possible to design a genetic algorithm (see, e.g., GA Alg. 6 and Alg. 7) which orders image segments. This form of GA can be considered an image segment matching algorithm, which uses all four features in the image segment feature space. The basis for this form of image segment matching algorithm is explained in Sect. 5.

5   2D Image Processing

This section introduces the basic concepts that will be used in a GA-based image segment matching algorithm. At the 2D image processing level, the information available about digital images comes from the locations of pixels and their RGB values. The main goal of 2D image matching is to match all pixels from one image with corresponding pixels from a second image. This operation is known as *image registration* [7], [57], [3].

5.1   Image Segmentation

Quantization has been defined as a process of converting analog signal to digital signal [10]. A quantizer is defined as a mapping from an uncountably infinite space of values into a finite set of *output levels*. In proposed system the source signal is digital image. Its domain is a finite set (pixels) of integer numbers (colors). Since colors are represented by three components, namely *Red*, *Green* and *Bule*, and each component is described by one byte, the input signal is already finite. Thus, the term quantization is rather used as mapping from a finite set of numbers to another finite set of numbers, where the cardinality of the destination set is smaller than the source set. In what follows, the Lloyd quantization algorithm [10] has been used, see Alg. 2.

Algorithm 2. The Lloyd Algorithm [10] (alg. Q_n)

Input: image I, required number of colors n
Output: optimal codebook with n entires C_{opt}

Initialize codebook C_1 with n entries randomly, set $m = 1$
repeat
  Based on codebook C_m and using nearest neighbour condition partition the image I into the quantization cells R_m
  Using centroid condition find optimal codebook C_{m+1} for cells R_m
  Set $m = m + 1$
until *distortion caused by C_m small enough*
Set $C_{opt} = C_m$

A quantization mapping is usually expressed by a codebook. A codebook is a set of n colors which are used to represent the original image. The mapping is performed by replacing the original color with the closest color from the codebook. The optimal codebook of size n is the set of colors which minimizes the distortion caused by the codebook. Here, the distortion is calculated as the squared difference between all components of the original color and its nearest neighbor from the codebook.

The Lloyd algorithm consists of main two steps, which are repeated until the distortion caused by the codebook is small enough. The first step is the partitioning of the input image based on the current codebook. The partitioning is performed using nearest neighbour condition, e.g. each pixel is assigned to the cell closest to the color of given pixel. In the second step, a new codebook is created based on the partitioning from the first step. Each codebook entry is replaced by a centroid of all colors of pixels from the corresponding cell.

Color quantization is used as an aid in image segmentation. It works only in the color space. The actual segmentation needs to take into account also a spacial information, namely the position of pixels. Only the combination of color and spatial information leads to identification of image segments. The averaging step fills the gap regarding the use of spatial information. It's only purpose is to average information carried out by pixels representing similar colors. The term *similar* is in this context precisely defined. Assume, that an original image denoted by I_o is given. First, quantization reducing number of colors to n_1 is performed. This step is denoted by formula 4 to obtain a quantized image denoted by $I_{q_{n_1}}$ (symbol Q represents the algorithm 2, where I_o is the input image I and n_1 is the required number of colors n).

$$I_o \xrightarrow{Q_{n_1}} I_{q_{n_1}} \tag{4}$$

As a result of quantization, the quantized image $I_{q_{n_1}}$ contains only n_1 colors. In the next step, the information from $I_{q_{n_1}}$ image is used to average the colors among all pixels, which are *connected*.

In quantized image, regions of pixels of the same color can be identified. These regions create segments. To each such segment is assigned a color, which is an average of all original colors from pixels belonging to this region. This step is denoted by the formula in (5), see also Alg. 3.

$$I_{q_{n_1}} \xrightarrow{Av_{I_o}} I_{Av_{n_1}} \tag{5}$$

The image $I_{Av_{n_1}}$ resulting from (5) has more than n_1 colors, where pixels are grouped into segments. This procedure, namely steps defined in (4) and (5), is

---

Algorithm 3. The Spacial Color Averaging (alg. $Av_{I_{s_n}}$)

Input: image I
Output: averaged image I_A

Mark all pixels from I as *not processed*
foreach *not processed pixel p in I* **do**
  Find segment $S(p)$ in I containing pixel p
  Assign to each corresponding pixel in I_A from $S(p)$ an average color of all pixels from $S(p)$
  Set all pixels from $S(p)$ as *processed*
end

---

repeated. The number of colors gradually decreases in consecutive iterations so that the creation of segments can be observed.

The unwanted effect of the algorithm defined this way is that if a segment is created at some step, there are no chances to change it in consecutive steps. In other words, the first quantization plays a crucial role in entire process. In addition, the resulting image still contains a lot of details (even though the number of colors was reduced). An example of such image processed using seven iterations described by the succession of mappings in (6), where numbers n_i for $i = 1, 2, ..., 6^2$ are 256, 64, 32, 16, 12, 8 and 4 are shown in left side of figure 2.

$$I_{s_{n_{i-1}}} \xrightarrow{Q_{n_i}} I_{q_{n_i}} \xrightarrow{Av_{I_{s_{n_{i-1}}}}} I_{Av_{n_i}} \tag{6}$$

Fig. 2. Hydrant image after 7 iterations of (6) (left) and (7) (right)

To make the entire segmentation process more robust and force the creation of bigger segments, one extra step for each stage defined by (6) is added. That is, after the colors are recreated from the original image, a 3 by 3 median filter is used. This causes almost uniform areas to blur even more and allows edges of neighboring segments to overlap. As a result, all small details from the image are lost, and big uniform segments are formed instead. The final formula describing one step of this iterated algorithm is shown in (7).

$$I_{s_{n_{i-1}}} \xrightarrow{Q_{n_i}} I_{q_{n_i}} \xrightarrow{Av_{I_{s_{n_{i-1}}}}} I_{Av_{n_i}} \xrightarrow{M_{3x3}} I_{s_{n_i}} \tag{7}$$

The M_{3x3} symbol denotes the median filter which is applied to each pixel from an input image. The median filter is applied to 3 by 3 neighborhood of given pixel $p(x, y)$.

$$M_{3x3}(p) = median\{p(x-1, y-1), p(x, y-1), p(x+1, y-1), p(x-1, y),$$

$$p(x, y), p(x+1, y), p(x-1, y+1), p(x, y+1), p(x+1, y+1)\} \tag{8}$$

---

[2] For $i = 0$ it is assumed that $I_{s_{n_0}} = I_o$, and after each iteration $I_{s_{n_{i-1}}} = I_{Av_{n_{i-1}}}$.

In order to find the median, all pixels are sorted by their color value and the one in the middle (e.g. at the 5-th place) is chosen.

The right side of figure 2 shows the result of applying seven-step iterative algorithm (with the same values as in previous example), where each step is described by (7). There are still many small segments, but comparing with the corresponding image, where the median filter was not used, their number was greatly reduced.

Figure 3 shows all steps of applying formula (7). The image in the first row and leftmost column is the original image. The second image in the first row, is a result of 8-bit quantization. The third image shows the result of applying 3 by 3 median filter. In the second row, the second iteration is shown. Leftmost image shows the result of averaging colors in segments from previous step. Middle image shows result of 5-bit quantization and rightmost image shows result of applying 3 by 3 median filter. The remaining five rows are organized the same way as the second row.

5.2   Segment Selection

At this stage it is assumed that a digital image is divided into segments. To increase the chances of identifying the same segments in both images, image quantization is performed on one large image, which is a composite of two individual images placed next to the other. After segmentation of the composite image, the two images in the composite are extracted and the analysis continues on the separate images. In this step, only some segments from all of the segments created so far are selected. The reason for this is the high number of segments and their shape. During a procedure to match shapes (described in section 8), all segments from both images are matched using a GA search for segments, which satisfy specific criteria. A GA is used because there is a need to work with the smallest number of segments possible. In addition, the matching algorithm requires that each segment satisfy some additional properties.

- **Lower bound on segment size:** avoid too few segments,
- **Upper bound on segment size:** avoid too many segments,
- **Convexity factor:** avoid perspective distortion.

Segment size is measured by the number of pixels belonging to a given segment. A lower bound on segment size is need for the following reasons. First, if there are not too many pixels, for example less than 10, the pixels can describe only a small number of distinctive shapes. Matching of such shapes is very difficult, since such a small number of pixels does not have enough power to uniquely represent fairly distinctive shape. Second, if all tiny segments are considered, the search space for matching segments becomes too large. There is a small chance that these tiny shapes can be uniquely matched.

The explanation for upper bound of segment size is motivated by characteristics of most images. Usually, images contain large areas of a solid color. For outdoor images it can be the sky, for indoor images it can be the walls of the room the image is shot in. These solid areas function as a background for the

Fig. 3. Quantized images obtained by iterating (7)

given scene. The shape of the background is not unique and it changes due to perspective transformation. By setting upper bound for segment size all segments, which can be part of the background are filtered out. For this research, this limit is set to be 30% of the entire image area.

The last constraint in matching image segments is the convexity factor, which deals with perspective distortion and filters out shapes, which are difficult to match. To get a deeper insight into this problem, consider what detected segments represent and how they differ from image to image. Each image is a 2D representation of a 3D scene. Similarly, segments which are flat represent 3D objects. The transformation from 3D space into 2D images flattens objects in a sense that the information from different parts of an object is represented in a small area. For example, consider the silhouette of a tripod. Given one segment representing an entire tripod, each leg is separated from the other legs by some background pixels (at least at the bottom of the tripod). Depending on the angle of the camera, some legs can be quite close to each other. The shape of a tripod changes dramatically with the change of view angle. Attempts to match such shapes should be avoided. This example shows that objects which are spread in all three dimensions are separated by some pixels not belonging to the object. This condition is expressed for flat images in terms of convexity. A segment S is *convex* if each point from a straight line connecting any two points in S also belongs to S [12].

Definition 10. Convexity factor. *The convexity factor for a segment S is a number $C_f(S)$ between 0 and 1 specifying how many lines between all combinations of points from segment S lie entirely inside the segment S.*

$$C_f(S) = \frac{\# \text{ lines entirely inside } S}{\# \text{ all possible lines}}.$$

To filter out segments which are potentially difficult to match, a threshold for a convexity factor is set and only segments greater than the threshold are selected. Based on experiments, a threshold of 0.5 has worked well.

Implementation of an algorithm used to calculate a convexity factor from the definition requires n^2 lines to be tested, where n is the number of segment pixels. In order to speed up the calculations the estimation is performed. The estimation process is applied on two levels. First, not all combinations of points are checked. Instead, randomly selected $50 \cdot n$ pairs of points are chosen. Second, instead of checking if an entire line is contained within a segment, only checks for 7 points are performed: middle point of a line, one fourth, three fourths and remaining multiples of $1/8$, namely, $1/8^{th}$, $3/8^{th}$, $5/8^{th}$ and $7/8^{th}$ of the line. In order to calculate each of these points only as few as two additions and two divisions are required, which makes this algorithm very fast with a complexity of $O(n)$.

5.3   Feature Generation

In this section, the following four features used for matching segments are elaborated.

- **degree of overlap** between segments,
- **angle of rotation** between segments,
- **distance between mean colors** of segments,
- **ratio of cardinalities** of segment pairs.

This section describes how these features are extracted from two sets of segments (one set of segments for each image). Sect. 8 elaborates about how the actual matching is performed using these features. First, recall that segments are only two dimensional representations of three dimensional objects. Due to the change of view angle, segments undergo transformation, which alters their shape. Therefore, simple comparison of shapes is not enough to pair segments.

Before the matching can start, values for the four features for all combinations of segments from both sets are generated. First two features are generated by an algorithm which tries to find the biggest overlap between two segments.

Overlap. This parameter measures the overlap between two segments. To calculate the overlap, two segments are plotted in one image using the same color (one_seg denotes the number of pixels belong to one segment). Pixels which belong to both segments are denoted by a second color (two_seg denotes the number of pixels belonging to both segments). A measure of the overlap between a pair of image segments is computed using Eq. 9.

$$overlap = e^{-\frac{|P_{one\_seg}|}{|P_{two\_seg}|}}. \tag{9}$$

where P_i denotes pixels of i-*th* color. For $|P_{one\_seg}| \neq 0$ and $|P_{two\_seg}| = 0$ it is assumed that $\frac{|P_{one\_seg}|}{|P_{two\_seg}|} = \infty$. In other words, *overlap* measures how well one segment matches the other. The minimum value for *overlap* is zero. In this case, the number of pixels belonging to both segments is equal to zero, which means that the segments do not intersect. A maximum *overlap* = 1 occurs when both segments have the same same shape and are located at the same position. In the case where one_seg = 0, $e^0 = 1$. For all other cases, *overlap* $\epsilon(0, 1)$.

The formula 9 was chosen for two reasons. First, it rescales the range of overlap values from $(0, \infty)$ to $(0, 1)$ interval. A finite interval is easier to handle than the infinite one. Second, the exponential function compresses the output of the original $\frac{|P_{one\_seg}|}{|P_{two\_seg}|}$ function in the range where P_{one_seg} is much greater than P_{two_seg} (for example, where $\frac{|P_{two\_seg}|}{|P_{one\_seg}|} < 2.5$, see figure 4). The absolute value of the slope of the overlap function from Eq. 9 is much smaller than the slope of the $\frac{|P_{one\_seg}|}{|P_{two\_seg}|}$ function. This allows for easier comparison of overlap values in the last stage of overlapping, e.g. when there is much more common pixels than not matched ones. The smaller slope means that small changes in the ratios of common/not matched pixels will not cause huge changes of the overlap function.

Fig. 5 illustrates best overlap. For better visualization, the two segments are plotted using different colors. The intersection is denoted by the brightest shade of gray. The lefthand side of Fig. 5 shows both segments with their original rotation, scale and position. The righthand side of Fig. 5 shows the two segments with maximum *overlap* = 0.85. Observe that the area occupied by only

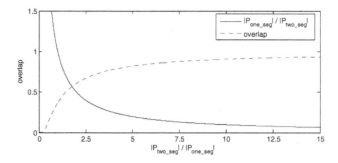

Fig. 4. The result of applying exponential into overlap function

Fig. 5. Preliminary overlap of 2 segments(left), and best overlap (right)

one segment has been significantly decreased in comparison with the original configuration.

Example 1. The figure 6 shows three sample steps out of many steps performed during segment matching of the figure 2. These three steps explain the idea behind the overlap formula introduced in the Eq. 9.

The first image in the figure 6 shows the first stage when the two segments do not have any pixels in common. The area of the first image is 41083 pixels and the area of the second one is 36447 pixels. Since there are no common pixels $|P_{one\_seg}| = 41083 + 36447 = 77530$ and $|P_{two\_seg}| = 0$. From the assumption for $|P_{one\_seg}| \neq 0$ and $|P_{two\_seg}| = 0$ the fraction $\frac{|P_{one\_seg}|}{|P_{two\_seg}|} = \infty$. Because of the formula 9 the overlap is not equal to infinity but a finite number $e^{-\infty} = 0$. It is easier to deal with finite numbers than infinite.

The second image shows one of the intermediate steps, where the two segment have a lot of pixels in common, but also a lot of non overlapping pixels. The area of the first segment, which does not intersect with the second one is 18219 pixels. The area of the second segment, which dos not intersect with the first one is 13583 pixels. The area of overlap between these segments is 22864 pixels. Therefore, $|P_{one\_seg}| = 18219 + 13583 = 31802$ and $|P_{two\_seg}| = 22864$. The overlap is equal to $overlap = e^{-\frac{|P_{one\_seg}|}{|P_{two\_seg}|}} = e^{-\frac{31802}{22864}} = e^{-1.391} = 0.248$.

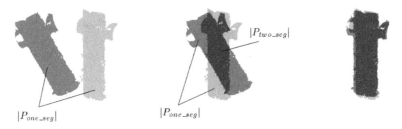

Fig. 6. Three sample steps of segment matching

The third image shows the final result of search for the best overlap. The first segment, was being translated, rotated and rescaled to maximize the overlap function. In this position the overlap is maximum. Here, the $|P_{one\_seg}| = 5540 +$ $1097 = 6637$ and $|P_{two\_seg}| = 35350$. Thus, $overlap = e^{-\frac{6637}{35350}} = e^{-0.187} = 0.828$.

The figure 7 shows the entire process of finding the best overlap for segments from figure 2. The horizontal axis denotes the iteration number. In each iteration the position, rotation and scale for the first segment is altered to minimize the overlap function. In the left part of the figure 7 a ratio $\frac{|P_{one\_seg}|}{|P_{two\_seg}|}$ is plotted. For the first several iterations it takes on high values compared to the end of matching process. In fact, the ending of the matching process is more important, since it can detect small differences in segments' shapes. Therefore, the overlap function, showed in the right part of the figure 7, is more sensitive to changes in the second half of the matching process. When two segments do not overlap significantly, the overlap function is close to zero. Only after there is a lot of overlap between segments, see the middle image from the figure 6, the overlap function changes more rapidly to emphasis the change in overlap.

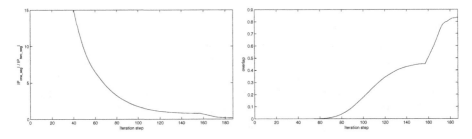

Fig. 7. The process of finding the best overlap for $\frac{|P_{one\_seg}|}{|P_{two\_seg}|}$ and $overlap$ parameter

Angle. The angle of rotation is the relative angle which one segment must be rotated to maximize the overlap between two segments.

Color. The previous two parameters (*overlap* and *angle*) dealt with geometrical properties of segments. The *color* parameter takes into account the color of a segment. Recall that all pixels from one segment are assigned the mean value of the colors from the original image. $C_{diff}(i,j)$ denotes the distance between the

RGB vectors of colors for a pair of segments. If the i-th segment's color is denoted by $C_i = (R_i, G_i, B_i)$ and j-th segment's color is denoted by $C_j = (R_j, G_j, B_j)$, then $C_{diff}(i, j)$ is defined by Eq. 10.

$$C_{diff}(i, j) = |C_i - C_j| = \sqrt{(R_i - R_j)^2 + (G_i - G_j)^2 + (B_i - B_j)^2}. \qquad (10)$$

Ratio of Cardinalities. The $Ratio of Cardinalities$ parameter is a measure of the relative size of a pair of segments. Let S_i, S_j denote sets of 4-neighborhood connected pixels for image segment i and j, respectively. Further, let RC denote a measure of the $Ratio\ of\ Cardinalities$, which is defined by Eq. 11.

$$RC(i, j) = \frac{|S_i|}{|S_j|}. \qquad (11)$$

5.4   Exhaustive Feature Matching

The goal of the matching algorithm is to produce a set of segment pairs so that each segment of a pair belongs to a different images. Given n_1 segments identified in the first image and n_2 segments identified in the second image, the total number of possible pairs is $n_1 \cdot n_2$. From $n_1 \cdot n_2$ matches only small number corresponds to the correct matches. In order to allow for grouping of several image segment matches a hypothesis is introduced.

The central notion of the searching algorithm is a *hypothesis*. A hypothesis it is is a set of image segment matches. A hypothesis is created by assuming that all four parameters for correctly matched segments are in the same range of values. In other words, a hypothesis identifies a set of paired segments.

The algorithm searches through the space of matches using hypotheses to validate each pair. It can be characterized by the average rotation angle between segments and average ratio of cardinalities of both segments, where the average is taken with respect to all pairs in the hypothesis. The rotation angle between segments corresponds to the rotation between images and the ratio of cardinalities corresponds to the difference in distances between object and the camera for the two views. Thus, a hypothesis contains only pairs of segments, which are similar to each other with respect to these two conditions. If the difference in a segment's shape is not caused by the change of view point, for example, the difference comes from the fact that non-matching segments are being considered. In that case, the values for relative rotation and ratio of cardinalities are random for different pairs. When there is a big difference in these two parameters, it is not possible to extract segments. On the other hand, if the difference in these parameters is caused by the change of the view point, it is the same for any two correctly paired segments. This allows for creation of bigger hypotheses with higher probability that each contains only correct matches.

The four parameters are denoted by the following tables:

$C(i, j)$: color difference,
$O(i, j)$: overlap,

$A(i, j)$: angle of relative rotation,
$RC(i, j)$: ratio of cardinalities.

where (i, j) denotes i-th segment from the first image and j-th segment from the second image, respectively. Next, a brief description of how these parameters are used to evaluate hypotheses, is given.

Ratio of cardinalities. This condition uses the ratio of cardinalities parameter $RC(i, j)$. If all the pairs from a given hypothesis are correct matches, then the value of this parameter for each pair should be in the same range. The minimal and maximal values of $RC(i, j)$ for all pairs from a given hypothesis are found. The minimum and maximum values should be in a $\pm RC_{th}$ range from the mean value of all rations of cardinality for given hypothesis. If any $RC(i, j)$ value from given hypothesis is outside the interval $[(1 - RC_{th}) * \overline{RC}, (1 + RC_{th}) * \overline{RC}]$ then a given hypothesis is not valid and is discarded. Otherwise, the next check is performed.

Angle of rotation. This check utilizes the assumption that for correct matches the angles of rotation $A(i, j)$ should be similar to each other for all pairs from a given hypothesis. First, the average angle of all angles is calculated (except for the pair added last). Then the rotation angle from the pair added last, is compared to the average angle. If the absolute value of the difference is greater than some threshold A_{th}, then the given hypothesis fails the check and is removed from the system. Otherwise, the next check is performed.

Triangle property. After passing the *Ratio of cardinalities* and *Angle of rotation* checks, a newly added pair in a given hypothesis is checked against the triangle property. This property assures that a newly added pair preserves the order in which any three segments are arranged in a triangle. Given three segments in one image, one can connect the centroids of these segments creating a triangle. The vertices of this triangle can be ordered in clockwise or counterclockwise order. After repeating the same procedure for corresponding segments in a second image, a second triangle is formed. By checking the order of the vertices in the second triangle, one can validate the correctness of matches. If the order of vertices is not the same, this does *not* mean that the matching in not correct. The order is preserved between two different views if the triangle of interest is face up on the same side. The centroids of segments need not lay on the plane in the real 3D space. This means that while moving from one view to the second one, the triangle formed by these segments is flipped to the other side, which reverses the order of the vertices. Nevertheless, this effect is very hard to obtain. Notice, that the identified segments would have to look the same from both sides. In most cases, the change in position between the two views is too small to cause this to happen. Hence, despite this special case, the power of discriminating bad matches is very useful for this application and is utilized in this check to decrease the number of hypotheses.

In Alg. 4, the centroid of a segment from a new pair is used to build triangles with all combinations of centroids from the hypothesis. The corresponding triangles for segments from a second image are built as well. If the order of

---

Algorithm 4. Matching Segments

Input: tables $C(i,j), O(i,j), A(i,j), RC(i,j)$, centroids of all segments
Output: hypothesis with highest score, sets of matches M

Set the set of all hypotheses $M = \emptyset$, and $N_M = |M|$;
for *all segments s_i in the first image* **do**

> Create pairs $P_i = \bigcup_j P_{ij}$ with all segments S_j from second image;
> Remove from P_i all pairs P_{ij} such that
> $C(i,j) > C_{th}$ or $O(i,j) < O_{th}$;
> Add $N_P = |P_i|$ pairs to N_M existing hypotheses
> producing total of $N_M + N_M \cdot N_P + N_P$ hypotheses;
> **foreach** *hypothesis $M_k \in M$* **do**
>
> > Set $\overline{RC} = \sum_{i,j} \frac{RC(i,j)}{|M_k|}$
> > **if** $\exists_{RC(i,j)}.RC(i,j) \notin [(1 - RC_{th}) \cdot \overline{RC}, (1 + RC_{th}) \cdot \overline{RC}]$ *or*
> > $|A(i_{new}, j_{new}) - mean_{P(i,j) \in M_k \setminus P(i_{new}, j_{new})} A(i,j)| > A_{th}$ *or*
> > $P(i_{new}, j_{new})$ *changes triangle order* **then**
> > > | Remove M_k from M;
> >
> > **end**
>
> **end**

end

---

vertices for any of these corresponding triangles do not match, the hypothesis fails the check and is removed from the set M. Otherwise, the algorithm finishes the pruning part and moves to the growing step.

The last part of the matching Alg. 4 identifies the hypothesis, which is the most likely to contain only correct matches. After applying algorithm 4, the set M consists of many hypotheses, which satisfy all conditions. From them, only one hypothesis is selected. The measure of correctness is the number of hypotheses the pair associated with a pair of image segments. Each pair is assigned a number based on its hypothesis count. Then each hypothesis is assigned a score, which is the sum of all measures of correctness of all pairs belonging to a given hypothesis. The hypothesis with the highest score is selected as the output of Alg. 4. The list of pairs from the selected hypothesis consists of correctly paired segments from both images.

6 Single Point Standard (Upper Approximation)

The goal of an image-matching system is to match segments from two given images. Consider, for example, Fig. 8 shows generated segments for the Wearever®[3] box scene. The left image in Fig. 8 contains 68 segments and the right image contains 51 segments.

---

[3] Trademark of the WearEver Company, `http://www.wearever.com`

Fig. 8. Generated segments for the Wearever box scene

Let $IS = (U, A)$ be an information system, where U is a set of pairs of image segments, and A is a set of image segment attributes. The attributes in A are defined relative to two segments, namely, *degree of overlap, angle of rotation, distance between mean colors* and *ratio of cardinalities*. Hence, each attribute value is indexed by two numbers which are the indices of the segments in a pair $x \in U$. For example, let S_i, S_j be sets of image segments for image i and image j, respectively. Then the subscripting for the image segment pair (s_i, s_j) specifies that $s_i \in S_i$ and $s_j \in S_j$. Most ranges of values for the segment attributes have been adjusted so that they are in the interval $[0, 1]$ or $[-1, 1]$. A summary of the segment attributes is given in the table 1.

Table 1. Attributes' ranges

color	range $[0, 1]$. 0 means identical colors; 1, all channels differ by the maximal value, e.g. the color value is 1 if one segments has color $(0, 0, 0)$ in RGB space and the second segment has color $(255, 255, 255)$ in RGB space (where for each channel the range of values is from 0 to 255).
overlap	range $[0, 1]$. 0 means no overlap between segments; 1, identical segments (after translation, rotation and scaling).
angle of rotation	range $[-1, 1]$. 0 means no rotation between segments; 1, rotation by 180 degrees, where the sign denotes the direction of rotation.
RC	range $[0, \infty]$. 1 means that both segments have the same area. For $RC \in [0, 1]$ the second segment is greater than the first one. For $RC > 1$ the first segment is greater than the second one.

The angle of rotation for a proper match is unknown. Hence, use of this attribute does not introduce any new information and is not considered in what follows, since the standard for this attribute is unknown.

The rough matching is performed relative to the standard set $B^*\mathcal{Z}$, which is an ideal match of two image segments. In other words, $B^*\mathcal{Z}$ is the optimal case for matching two identical image segments and such case may, but does not have to exist in the real data.

All segment pairs are ranked based on the information represented by $B^*\mathcal{Z}$. Different upper approximations can be constructed by changing the equivalence relation and subsets of attributes used to obtain $B^*\mathcal{Z}$. In the original K-means clustering algorithm [27], data points are arranged so that they are clustered around K centers. In this work, an equivalence relation based on the K-means clustering algorithm has been introduced (see, e.g., [37]), and which we summarize in this section. Briefly, two segments s_i and s_j are in relation $Ind_K(B)$ if and only if the values of all attributes for s_i and s_j are associated with the same cluster. $Ind_K(B)$ is formally defined in Eq. 12.

$$Ind_K(B) = \left\{ \begin{array}{l} (s_i, s_j) \in U^2 | \ \forall a \in B, \\ \exists l.1 \leq l \leq K, a(s_i) \in C_l \wedge a(s_j) \in C_l \end{array} \right\}. \tag{12}$$

where C_l denotes the l-th cluster from the set of K clusters. Let the set $\mathcal{Z}$ be defined as in 13.

$$\mathcal{Z} = \left\{ \begin{array}{l} x \in U \times U | \\ color(x) = 0, \\ overlap(x) = 1. \end{array} \right\} \tag{13}$$

The set $\mathcal{Z}$ consists of matched pairs of segments with attribute values specified in 13. Let $B(x)$ be a block in the partition of U, which is a set of B-indiscernible pairs of image segments containing x. At this point, there is interest in finding the upper approximation of $\mathcal{Z}$, which is described in (14).

$$B^*(\mathcal{Z}) = \{x \mid B(x) \cap \mathcal{Z} \neq \emptyset\}. \tag{14}$$

Alg. 5 gives the steps for ranking segment matches using the upper approximation. To each vote is assigned the same unit weight. Because cases where 2 attributes are used include cases where 1 attribute is used, the effective weights are greater for cases with multiple attributes used. The table 2 show the effective voting weights for the algorithm 5.

Figures 9 and 10 show sample voting results for two segments. A circle ∘ denotes the good match made by visual inspection of the two images, and a cross + denotes the segment which is the closest to the standard $\mathcal{Z}$. That is, for a given segment i from the first image, a + denotes the segment j_{min} from the second image such that

$$j_{min} = \min_j |\mathcal{Z} - \{C(i,j), O(i,j)\}|.$$

Fig. 9 shows how the information is extracted from the generated attributes using the upper approximation $B^*(\mathcal{Z})$ of the set $\mathcal{Z}$. The cross + shows that

---

Algorithm 5. Matching Segments Using Upper Approximation

Input: set of attributes $\mathcal{A} = \{C(i,j), O(i,j)\}$
Output: ranking of all segment pairs s_{ij}

for *(all segments s_i in the first image)* **do**
  for *(K=2 to (# of segments in the second image)/2)* **do**
    Perform K-means clustering for each attrib. separately
    for *(each subset B of the set of all attributes $\mathcal{A}$)* **do**
      Find $B^*(\mathcal{Z})$
      for *(each segment s_j from the second image)* **do**
        if *($s_j \in B^*(\mathcal{Z})$)* **then**
          | vote for pair s_{ij}
        end
      end
    end
  end
end

Table 2. Voting table $\mathcal{T}$ for algorithm 5

# of attributes in B	# of votes	effective # of votes
1	1	1
2	1	3

the best match using the distance between the given three parameters is with segment number 44. However, the correct match is with segment number 18. The number of votes for the segment number 18 is higher than the number of votes for the segment 44. This means that using this algorithm, segment 18 is more likely to be chosen as the match than segment 44.

The problem which is still to be solved is the high number of segments with high votes. For example, in Fig. 9, segments 18, 20 and 33 have high votes and it is not possible to select the best match. Hence, there is interest in considering the lower approximation $B_*(\mathcal{Z})$ of the set $\mathcal{Z}$.

7   Interval Standard (Lower Approximation)

This section presents an extension of the method described in Sect. 6. The lower approximation $B_*(\mathcal{Z})$ is derived relative to $\mathcal{Z}$, which is defined as the approximation of a set of image segment pairs that constitute a perfect match. However, this is a bit unrealistic and not flexible. To allow for some tolerance in concluding that an image segment pair constitutes a match, an interval interpretation of the attribute values of image segment pairs is introduced. That is, the attribute values associated with image segment pairs in the set $\mathcal{Z}$ are parametrized by a

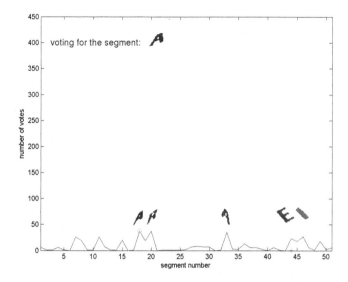

Fig. 9. Voting results: ○ good match, + the closest match

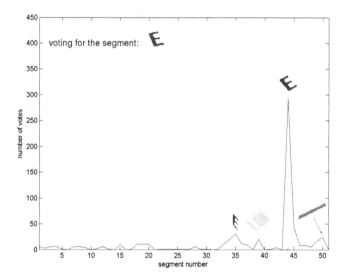

Fig. 10. Voting results: ○ good match, + the closest match

parameter δ, which denotes the optimal value for each attribute. In effect, each attribute value of each image segment pair $x \in \mathcal{Z}$ belongs to a small interval containing δ. Using this approach, $\mathcal{Z}$ is defined as in Eq. 15.

$$\mathcal{Z} = \left\{ \begin{array}{l} x \in U \times U| \\ color(x) \in (0, \delta), \\ overlap(x) \in (1 - \delta, 1). \end{array} \right\}. \tag{15}$$

where the attribute values for each image segment pair x in $\mathcal{Z}$ belong to intervals for color and overlap specified in Eq. 15.

For experiments, the parameter δ was set to 0.1. The results for different δ values did not differ significantly from the ones shown here. The formula for calculating the lower approximation is given in 16.

$$B_*(\mathcal{Z}) = \{x \mid B(x) \subseteq \mathcal{Z}\}. \tag{16}$$

The new matching algorithm is essentially the same as Alg. 5, except that a *Find $B_*(\mathcal{Z})$* operation has been added.

As can be seen from Fig. 12, the best results are obtained for the 'single point standard'. The notation circle $\circ$ in Fig. 12 denotes a good match made by visual inspection of the two images, and a cross $+$ indicates a segment pair which is the closest to $\mathcal{Z}$. The 'interval standard' method fails to yield one segment pair as a good match. Instead, it yields several segments with equally high votes. This means that this method cannot be used by itself as the deciding method for solving the matching problem. However, the interval standard method can be used as an aid, since the correct solution is usually among the segment pairs with the highest votes.

Example 2. The figure 11 shows two images used to explain in more detail the idea of the standard $\mathcal{Z}$. Left part of the figure 11 shows the first image. It consists of twelve segments created from the letters of a word "Matching". Notice, there are only eight letters in a word "Matching". The remaining four segments are: white area in the letter 'a' (denoted by a.), a dot in 'i' letter (denoted by i˙), upper white area in the letter 'g' (denoted by g˙) and lower white area in the letter 'g' (denoted by g.).

The right part of the image 11 shows the same letters as the first image. Only, for the second image, they underwent geometrical transformations: image warping, rescaling and rotation. In addition, brightness of each letter from the second image was randomly altered.

In order to construct the standard $\mathcal{Z}$ the color difference and overlap parameters were calculated. The color difference values are shown in the table 3 and overlap values are shown in the table 4. Values corresponding to proper matches are denoted by bold face font in both tables. For example, the pair of segments 'M' from both images is characterized by the pair (0.070,0.614), where the first number denotes the color difference and the second number denotes the overlap value for these two segments.

The creation of standard $\mathcal{Z}$ for given parameter δ is straight forward. From the equation 15, standard is a set of all segment pairs for which the color difference and the overlap values are in some interval, e.g. color difference is less than δ and overlap is greater than $1 - \delta$. For example, for $\delta = 0.1$ there are only two segments satisfying the above requirements. These pairs are (i˙,i˙) for which color(i˙,i˙)= 0.075 < 0.1, overlap(i˙,i˙)= 0.913 > 0.9, and (g˙,g˙) for which color(g˙,g˙)= 0.017 < 0.1, overlap(g˙,g˙)= 0.931 > 0.9, see tables 3 and 4. Therefore, for $\delta = 0.1$ the standard $\mathcal{Z} = \{(i˙,i˙), (g˙,g˙)\}$. This means that the segments

Fig. 11. Segments for example 2

Table 3. Color table

	M	a	t	c	h	i	i·	n	g	a.	g·	g.
M	**0.070**	0.574	0.671	0.066	0.572	0.532	0.532	0.639	0.068	0.784	0.784	0.784
a	0.579	**0.064**	0.707	0.549	0.606	0.492	0.492	0.213	0.647	0.778	0.778	0.778
t	0.654	0.693	**0.048**	0.595	0.091	0.875	0.875	0.528	0.657	0.773	0.773	0.773
c	0.118	0.539	0.628	**0.011**	0.532	0.545	0.545	0.596	0.126	0.792	0.792	0.792
h	0.571	0.608	0.179	0.531	**0.045**	0.753	0.753	0.449	0.585	0.647	0.647	0.647
i	0.481	0.452	0.898	0.564	0.760	**0.069**	0.069	0.554	0.558	0.483	0.483	0.483
i·	0.482	0.454	0.894	0.565	0.756	0.075	**0.075**	0.552	0.559	0.477	0.477	0.477
n	0.649	0.305	0.556	0.636	0.449	0.551	0.551	**0.108**	0.710	0.579	0.579	0.579
g.	0.082	0.617	0.679	0.171	0.573	0.507	0.507	0.666	**0.078**	0.705	0.705	0.705
a.	0.719	0.727	0.800	0.790	0.679	0.542	0.542	0.657	0.768	**0.014**	0.014	0.014
g·	0.715	0.726	0.799	0.786	0.677	0.540	0.540	0.657	0.764	0.017	**0.017**	0.017
g.	0.718	0.728	0.800	0.790	0.679	0.542	0.542	0.659	0.767	0.014	0.014	**0.014**

Table 4. Overlap table

	M	a	t	c	h	i	i·	n	g	a.	g·	g.
M	**0.614**	0.084	0.019	0.058	0.372	0.040	0.001	0.242	0.210	0.001	0.001	0.002
a	0.367	**0.383**	0.143	0.206	0.221	0.165	0.010	0.497	0.276	0.003	0.005	0.010
t	0.141	0.330	**0.832**	0.292	0.249	0.001	0.001	0.151	0.153	0.016	0.142	0.203
c	0.181	0.266	0.202	**0.722**	0.228	0.250	0.043	0.316	0.220	0.069	0.081	0.121
h	0.052	0.411	0.331	0.139	**0.714**	0.247	0.002	0.549	0.251	0.008	0.004	0.021
i	0.301	0.275	0.001	0.306	0.134	**0.716**	0.133	0.230	0.166	0.267	0.508	0.423
i·	0.003	0.256	0.001	0.113	0.032	0.216	**0.913**	0.033	0.018	0.421	0.675	0.807
n	0.333	0.390	0.116	0.368	0.239	0.149	0.012	**0.847**	0.280	0.001	0.002	0.024
g.	0.234	0.268	0.118	0.109	0.155	0.115	0.001	0.091	**0.713**	0.001	0.001	0.010
a.	0.001	0.045	0.320	0.102	0.028	0.185	0.726	0.001	0.007	**0.861**	0.818	0.721
g·	0.001	0.060	0.152	0.054	0.015	0.274	0.730	0.003	0.002	0.675	**0.931**	0.873
g.	0.001	0.108	0.251	0.130	0.042	0.425	0.618	0.337	0.039	0.726	0.865	**0.860**

i· and g· are the most similar segments in both images. The matching of the remaining segments is performed relative to this match.

The table 5 shows several sets $\mathcal{Z}$ for different values of parameter δ. The bigger the parameter δ the more matches are included in the standard $\mathcal{Z}$. This

Table 5. $\mathcal{Z}$ table vs. δ parameter

δ	$\mathcal{Z}$	(color, overlap)
0.05	$\emptyset$	
0.1	$\{(i´,i´),(g´,g´)\}$	$\{(0.075,0.913),(0.017,0.931)\}$
0.15	$\{(i´,i´),(a.,a.),(g´,g´),$ $(g.,g´),(g´,g.),(g.,g.)\}$	$\{(0.075,0.913),(0.014,0.861),(0.017,0.931),$ $(0.014,0.865),(0.017,0.873),(0.014,0.860)\}$
0.2	$\{(t,t),(i´,i´),(n,n),$ $(a.,a.),(a.,g´),(g´,g´),$ $(g.,g´),(g´,g.),(g.,g.)\}$	$\{(0.048,0.832),(0.075,0.913),(0.108,0.847),$ $(0.014,0.861),(0.014,0.818),(0.017,0.931),$ $(0.014,0.865),(0.017,0.873),(0.014,0.860)\}$
0.3	$\{(t,t),(c,c),(h,h),$ $(i,i),(i´,i´),(n,n),$ $(g,g),(a.,a.),(g.,a.),$ $(a.,g´),(g´,g´),(g.,g´),$ $(a.,g.),(g´,g.),(g.,g.)\}$	$\{(0.048,0.832),(0.011,0.722),(0.045,0.714),$ $(0.069,0.716),(0.075,0.913),(0.108,0.847),$ $(0.078,0.713),(0.014,0.861),(0.014,0.726),$ $(0.014,0.818),(0.017,0.931),(0.014,0.865),$ $(0.014,0.721),(0.017,0.873),(0.014,0.860)\}$

is the core of the rough set approach, where the definition of an 'ideal match' is not fixed, but can be adjusted based on available information and data. Notice, for smaller δ values the standard $\mathcal{Z}$ is an empty set, which means that the identical segments are not present in both images. On the other hand, bigger δ values produce a standard, which contains incorrect matches. Nevertheless, at this stage, the matching correctness is not crucial, in fact, segments $(g.,g´)$ form better match than $(g.,g.)$, since (color, overlap) values for the first pair are $(0.014, 0.865)$ and for the second pair $(0.014, 0.860)$. The correctness of this match cannot be determined using only color difference and overlap values, but other parameters must be used as well.

As shown in this example, the δ parameter allows for adjusting how strict the definition of an ideal match is. This was not possible in the Upper Approximation based approach, see equation 13.

8   Genetic Approach for Matching

The genetic approach for segment matching creates a framework for the search based on any set of features extracted from images. Features can be extracted using segments, lines or Harris corners [15]. In addition, because the genetic algorithm is used, the search space can be very large. This allows for selection of matches from large set of pixels from both images.

The central notion of a classical, Darwinian form of genetic algorithm is the chromosome. It consists of genes. A *gene* represents a pair of matched features for two images. There are many types of features that can be used in the algorithm. More abstract forms (called also *shapes*) are source of features. Three methods of deriving features have been considered so far, and are summarized in the table 6. One of the functions of the gene is to hide the differences between the features for the genetic algorithm. The genetic algorithm does not discern between the genes and processes them the same way. The set of features creates a *chromosome* (or

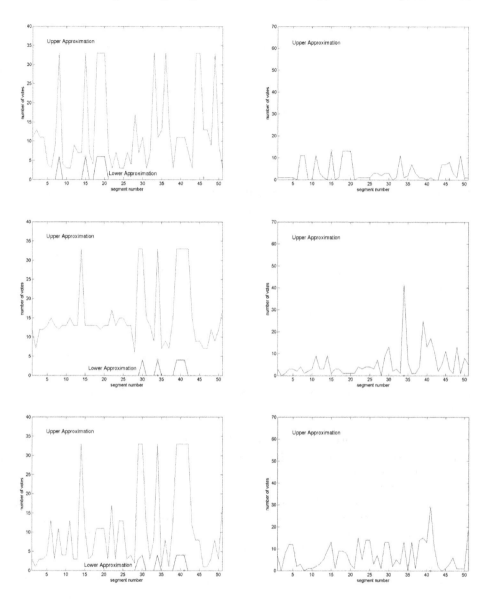

Fig. 12. Voting results: 'interval standard' (left) and 'single point standard' (right)

hypothesis as described in Sect. 5.4). A genetic algorithm tries to select the best hypothesis, which consists of only correct matches. The term *hypothesis* is used in the context of matching features from the images, since a hypothesis identifies possible matches of features. The term *chromosome* is used in the context of the genetic algorithm, since the chromosome is the member of the population.

The overview of the structures used in the algorithm is given in the figure 13. Three kinds of feature generators are denoted by three paths at the bottom of

Table 6. Features and their corresponding abstract forms

Abstract Form	Complexity	Features
point	simple	cross-correlation
line	moderate	cross-correlation, angle, RC
segment	complex	color, overlap, angle, RC

the image. The tasks of image processing blocks are generation of points, lines and segments. After this step, the identified shapes are passed to the feature extraction blocks. These blocks use selected shapes to generate features. The term 'feature' needs more explanation. Usually, the term *feature* means a mapping of observed object in the universe to an attribute value. In this case, a feature is not an attribute or aspect of a single image segment or pixel, but a result of a comparison of a pair of image segments or pixels. In the case of the color of a pair of image segments, a feature is the difference in the average colors of the two segments. In general, a feature $\mathcal{F}(x, y)$ for a pair of observed objects x and y is a scalar from some pre-defined range $[a, b]$ as defined in (17).

$$\mathcal{F} : (x, y) \longrightarrow c \in \mathbb{R} \quad a \leq c \leq b. \tag{17}$$

In addition, there is one point $\kappa \in [a, b]$, which denotes the value for which a pair of objects are not discernible with respect to the given feature. For example, for colors of image segment pairs, the possible range of color differences can be defined as $[0, 1]$, where 0 denotes two identical colors and 1 denotes the maximum difference of colors allowed by an image's color depth. In this case, $\kappa = 0$.

The fact that the features are calculated as the difference between attributes for pairs of digital images is denoted by the 'fusion' box in the figure 13. The procedure represented by the fusion box combines the information from pairs of images to generate feature values.

Next, the description of the chromosome is given (see figure 14). A gene represents a match between two shapes from a pair of images. This is indicated by a pair of indices of two corresponding shapes. Genes in each chromosome are divided into three blocks: point block, line block and segment block. Each block contains indices of matched shapes of a given type, namely point, line or segment.

The number of genes in each group can be zero. The genetic algorithm does not discern between different types of genes as long as both halves of the gene are of the same type. The order of the chromosome is the sum of lengths of all blocks, e.g. $np + nl + ns$.

The current version of the genetic algorithm has only image segment genes implemented. Therefore, point and line blocks are always empty. The creation of the genes is constrained by the rules, which assure that only reasonable matches are considered. These rules control the color difference, overlap and the ratio of cardinalities. Let *color_th*, *overlap_th*, *rc_th* denote the maximum values allowed for the color difference, overlap and the ratio of cardinalities for a pair

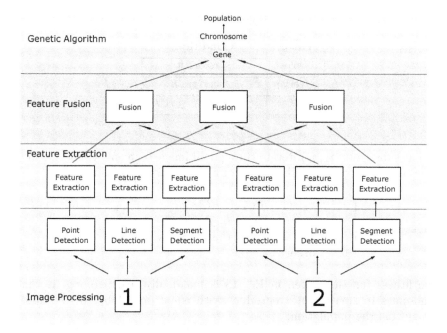

Fig. 13. The overview of matching algorithm

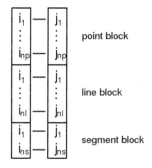

Fig. 14. The chromosome. Each block contains indices of matched shapes.

of image segments, respectively. For the ratio of cardinalities parameter the threshold denotes the the maximum difference of the areas of a pair of segments, e.g. for $rc\_th = 2$ it means that $\frac{1}{2} \leq RC \leq 2$. Table 7 gives constraints for feature values during the creation of genes.

Let Ch denote a chromosome from a population evaluated by a genetic algorithm. Further, let ang, $\overline{ang}$, rc_th, ang_th denote angle of rotation, average angle of rotation, ratio of cardianlities threshold, and angle threshold, respectively. The $RC(i_k, j_k)$ parameter was defined in Eq. 11 and the $\overline{RC}$ symbol denotes the mean value of ratios of cardinalities for all genes from given chromosome. The subscript k iterates from 1 to the number of genes in given chromosome

Table 7. Rules for creating genes

Feature	Condition
Color	$\leq color\_th$
Overlap	$\geq overlap\_th$
RC	$\geq 1/rc\_th \wedge \ \leq rc\_th$

such that subscripts (i_k, j_k) iterate through all image segments contained in the chromosome, see figure 14. The fitness of Ch is determined using Eq. 18.

$$Fitness(Ch) = \begin{cases} 1 & \forall_k \ (1 - RC_{th}) \cdot \overline{RC} \leq RC(i_k, j_k) \leq (1 + RC_{th}) \cdot \overline{RC} \\ & \wedge \ max_k \ |ang(i_k, j_k) - \overline{ang}| < ang\_th \\ & \wedge \ \text{passes the triangle check,} \\ 0 & \text{otherwise.} \end{cases}$$

(18)

The fitness function given in Eq. 18 is maximally selective, e.g. it causes chromosomes to survive and reproduce with equal probability or die and be removed from the population.

Alg. 6 is based on the standard procedure for genetic algorithms described in [6]. The only genetic operator implemented so far is the crossover operation. New genes are not created by Alg. 6. All unique genes appearing in the population are created before evolutionary iteration starts. The crossover operation cannot split halves of existing genes. Two chromosomes of lengths n_1 and n_2 can only be concatenated to form a new chromosome of a length $n_1 + n_2$, which contains all genes from the original two chromosomes. The repetition of left and right handed parts of the gene within a chromosome is not allowed either. This means that only two chromosomes with different sets of left and right handed parts can mate and create an offspring.

After implementing Alg. 6, all genes are scored using the Alg 7. The symbol γ denotes partitioning introduced by the chromosomes. Each chromosome forms a block of genes. Genes within a block (a chromosome) are considered to

---

Algorithm 6. Genetic Algorithm

Input: tables $C(i, j), O(i, j), A(i, j), RC(i, j)$, centroids of all segments
Output: ordered set of matches $\mathcal{O}$

Create the set of genes S using rules from table 7
while *(stop condition is not true)* **do**
  Apply genetic operator: crossover
  Evaluate fitness function
  Remove chromosomes with fitness function below some threshold
end
Order all genes into set $\mathcal{O}$ using *GA based ordering* algorithm, see alg. 7

Algorithm 7. GA based ordering

Input: set of genes G, partitioning of this set γ
Output: ordered set of matches $\mathcal{O}$

Create the set of counters for all genes in the set G
Set initial values of these counters to 0
for *(all blocks from γ)* **do**
    for *(all genes s_i from given block)* **do**
       | Increase counter of gene s_i by 1
    end
end
Return sorted in descending order list of genes $\mathcal{O} = sort(G)$

---

be indiscernible. Since, different chromosomes can contain the same genes this partitioning forms a tolerance relation.

The genes are sorted by the number of chromosomes they appear into. The chromosome which contains the most common genes is selected as the output of the simulation. All sorted genes are returned in the set $\mathcal{O}$.

9 2D Matching with Approximation Spaces

This section considers an approach to processing the output from the genetic algorithm 6 within the context of an approximation space. Let S be a set of n best genes returned by Alg. 6. Let $B(x)$ be a block of genes equivalent to x, and let B_*S be the lower approximation of the set S. There is advantage in using B_*S as a standard, and measuring how well B_*S "covers" each block. In this way, it is possible to select a block covered to the greatest extent by the standard, and which represents the set of best image segment matches. The steps of this approach to finding the set of best matches are given in Alg. 8.

The results are shown in the plot in the figure 15. In the plot from the figure 15, $n = 114$ is called the pool of genes. The most important thing to observe in this plot is that rough coverage does better between genes numbers 37 and 49. This means Alg. 8 sorts the genes better than the pure GA represented in Alg. 6.

9.1 Tolerance Relation vs. Equivalence Relation

The output of the GA in Alg. 6 is a set of hypotheses. In other words, Alg. 6 produces sets containing pairs of image segments. Each such set (hypothesis) corresponds to one chromosome. The crossover operation in Alg. 6 produces a chromosome, which is a copy of two input chromosomes. Therefore, the resulting partitioning is a tolerance relation (see property 1). Alg. 9 converts the tolerance relation induced by Alg. 6 into an equivalence relation. This is done by removing the overlapping sets in the partition created by Alg. 6.

Algorithm 8. The algorithm for selecting best matches using rough coverage

---

Input: set of all possible matches
Output: ordered set of matches

Run the GA to get the partitioning of the genes
Create and initialize to 0 the rough coverage weights for each gene
Create a set S of top n genes
for *(each block $B(x_i)$)* **do**
⎢   Evaluate rough inclusion value

$$Rcover(B(x_i), S) = \frac{|B(x_i) \cap B_*S|}{|B_*S|}$$

⎢   Increase weights of genes from $B(x_i)$ by $Rcover(B(x_i), S)$
end

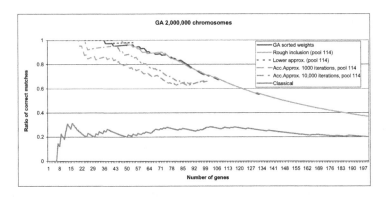

Fig. 15. Rough coverage vs. % correct matches with 2,000,000 chromosomes

Observe that Alg. 9 searches for the chromosomes with the highest weight (starting from the longest chromosomes) and removes all chromosomes which have non-empty intersection with a given chromosome. The resulting partitioning of all genes C forms an equivalence relation (see theorem 2).

The down side to converting the tolerance relation induced by Alg. 6 into an equivalence relation is the reduction of the number of blocks. For example, in case of the system consisting of ≈818,000 chromosomes with 4920 genes where the longest chromosome has length 29, the conversion to equivalence relation decreases the number of blocks with cardinality greater than one to 1229. This means that the number of blocks after the conversion is less than 0.16% of the number of tolerance relation blocks.

Experimental results show that the equivalence relation does not have enough power to improve the ordering produced by the GA in Alg. 6. This is due to the

---

Algorithm 9. Conversion of tolerance rel. to equivalence relation

---

Input: set of genes G, set of chromosomes Ch, partitioning γ of set G
    expressed by sets $Ch_I \in Ch$
Output: partitioning Ind of set G, which is an equivalence relation

1 Order all genes using *GA based ordering*, see alg. 7
2 Set Ind to $\emptyset$

   /* starting from the longest chromosomes   */
4 **for** *(all chromosomes' lengths I)* **do**
5 | **for** *(all chromosomes Ch_I of the length I)* **do**
6 | | Find the chromosome $ch_{max} \in Ch_I$ with the highest score and move
   | | it to Ind
7 | | **for** *(all chromosomes ch_k in Ch)* **do**
8 | | | **if** *($ch_k \cap ch_{max} \neq \emptyset$)* **then**
10 | | | | Remove ch_k from Ch
11 | | | **end**
12 | | **end**
13 | **end**
14 **end**

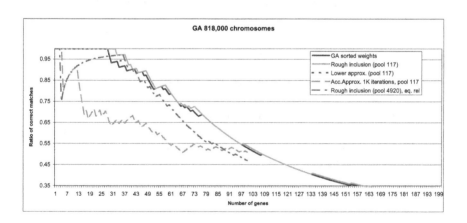

Fig. 16. Ratio of correct matches for tolerance and equivalence relation

small number of blocks in the equivalence relation. Fig. 16 shows sample results
for 818,000 chromosomes.

Definition 11. *Tolerance Relation* (from [21])
A binary relation $\tau \subseteq X \times X$ is called a **tolerance relation** if and only if τ is

1. reflexive, an object is in relation with itself $x\tau x$,
2. symmetric, if $x\tau y$ then $y\tau x$.

Definition 12. Equivalence Relation *(from [21])*
A binary relation $R \subseteq X \times X$ is called an **equivalence relation** if and only if R is a tolerance relation and is

1. transitive, if xRy and yRz then xRz.

Property 1. The crossover operator in the GA in Alg 6 produces a partitioning of the set of genes G, such that created blocks have non-empty intersection.

Proof. Let Ch denote the set of all chromosomes ch_k such that $Ch = \bigcup_k ch_k$. If L denotes the longest chromosome in Ch, then all chromosomes can be grouped into subsets of Ch, namely $Ch_L, Ch_{L-1}, \ldots, Ch_I, \ldots, Ch_1$, where $Ch_I \subseteq Ch$ and index I denotes the length of the chromosome[4]. The crossover operator *crossov* can be defined as follows:

$$crossov : Ch_I \times Ch_J \to Ch_K, \quad ch_k = crossov(ch_i, ch_j)$$

where $ch_i \in Ch_I$, $ch_j \in Ch_J$, $ch_k \in Ch_K$, $K = I + J$, $K \geq 2$ and $I, J \geq 1$.

After applying the crossover operator, the chromosomes ch_i and ch_j are not removed from the set of all chromosomes Ch, i.e. $ch_i, ch_j, ch_k \in Ch$. This means that a gene g_t which belongs originally to ch_i belongs also to the chromosome ch_k.

Now, for any chromosome ch_k of order greater than 1.

$$\forall_{g_t \in ch_k} \exists l \neq k \mid g_t \in ch_l$$

From the fact that the order of a chromosome ch_k is greater than 1, we have $ch_k = crossov(ch_i, ch_j)$. What follows is that $l = i$ or $l = j$.   ∎

The above proof shows that for each chromosome ch_k of order $K > 1$ there exists a chromosome of order less than K, which is a part of ch_k. Therefore, for each such chromosome ch_k there exist at least two chromosomes that have non empty intersection with it.

Theorem 1. The GA in Alg 6 produces a partitioning of the set of all genes, which corresponds to the tolerance relation γ (and not an equivalence relation).

Proof. Chromosomes ch_k produced by the GA consist of the genes from the set G. Each chromosome can be considered as a block of indistinguishable genes. Thus, they create a partitioning of the set G. This partitioning corresponds to the tolerance relation if the relation determined by this partitioning is reflexive and symmetric. The relation γ is based on the fact that two elements belong to the same chromosome, i.e.

$$x\gamma y \quad \text{iff} \quad \exists_i \mid x \in ch_i \text{ and } y \in ch_i$$

---

[4] The index written with a small letter by a chromosome ch does not indicate the length of the chromosome.

where $ch_i \in Ch$. From the definition, belonging to a set is symmetric and reflexive.

The partitioning Ch covers all genes G because Ch includes the set of chromosomes of order one Ch_1, which is identical with the set of all genes G, i.e. $Ch_1 = G$, $Ch_1 \in Ch$ therefore $G \subseteq Ch$.

The relation γ is not an equivalence relation because chromosomes ch may have non-empty intersections (from property 1). As the result, the transitivity constraint is not satisfied. If for $k \neq l$ holds $x \in ch_k$, $y \in ch_k$, $y \in ch_l$, $z \in ch_l$ and $x \notin ch_l$ and $z \notin ch_k$ then $x\gamma y$, $y\gamma z$ and x is not in relation with z.   ■

Theorem 2. Alg. 9 converts the partition γ produced by the GA in Alg. 6 into a partition Ind which is an equivalence relation.

Proof. The partitioning Ind is a subset of the partition γ. Thus, there are two conditions that must be satisfied for the partition Ind to be an equivalence relation:

– all blocks from Ind must have empty intersection:

  The step 10 from algorithm 9 assures that all subsets from Ind have empty intersection.
– all blocks from Ind must cover the entire space of genes G:

  The step 4 from algorithm 9 starts with the longest chromosomes and ends with the shortest $Ch_L, Ch_{L-1}, \ldots, Ch_I, \ldots, Ch_1$, where L is the length of the longest chromosome in the system. The shortest chromosome is of length one, i.e. it is a gene. Notice, none of the genes which are not included in the set Ind will be removed from the set Ch_1 because their intersection with ch_{max} is empty. This means that in the last iteration of loop 4 all missing genes will be added to the set Ind.   ■

9.2   Classical vs. Rough Matching Methods

Classical 2D image segment matching method is usually limited to two features, namely, color difference and the overlap between two segments (see, e.g., [12,55,24]). This is a severe limitation because these two features do not yield enough information to permit accurate image segment matching. By contrast, in designing genes in chromosomes used in evolutionary 2D segment matching, the number of features associated with a gene can be quite large.

In this study, four parameters for each gene and 2,000,000 chromosomes have been used. Similarly, using the rough coverage methods, the number of features (parameters) associated with an image segment can be large. In this study, four features are used, namely, the distance between mean colors of segments, degree of overlap between segments, the angle of rotation between segments and ratio of cardinalities of both segments. In addition, rough coverage values computed within the context of an approximation space, represent a comparison between each of the blocks containing similar pairs of image segments and a set representing a norm (e.g., $B_* S$). In effect, the rough coverage matching method

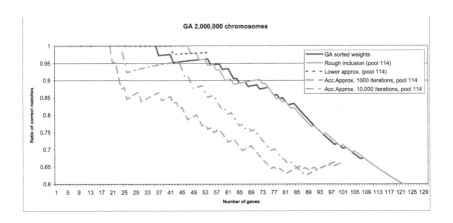

Fig. 17. Rough coverage vs. % of correct matches (zoomed Fig. 15)

yields better results because it uses more information about the image segments being compared. This is one way to explain the plots in Fig. 15.

In Fig. 17, the left upper corner of the plot from Fig. 15 is shown. Fig. 17 illustrates the advantage of the rough coverage approach compared to the other methods. Recall, that the problem of segment matching is considered in the context of 2D to 3D conversion. The 2D to 3D conversion algorithm takes as an input paired pixels, which are generated from paired segments. Any mismatch at the segment matching stage propagates to the pixel matching stage and finally into 2D to 3D conversion. Therefore, a crucial requirement for image segment matching is to generate as little wrong matches as possible. Fig. 17 shows that the rough coverage approach yields the biggest number of correct matches, i.e. the first wrong match occurs after finding 47 good matches. For the weights generated by the GA used in in Alg. 8, the first mismatch occurs after only 36 correct matches. Hence, rough coverage greatly reduces the number of mismatches, which improves the robustness of the overall 2D to 3D conversion process.

10   Conclusion

An approach to using a combination of genetic algorithms and approximation spaces in solving the image segment matching problem is given in this paper. Approximation spaces are used as a form of visual perception of pairs of images, which is step towards 2D image classification in the case where one of the paired images plays the role of a reference image for matching purposes.

Acknowledgements

The authors would like to thank Andrzej Skowron for his observations concerning approximation spaces and his helpful suggestions concerning this article.

This research has been supported by Natural Sciences and Engineering Research Council of Canada (NSERC) grant 185986 and grant T247 from Manitoba Hydro.

References

1. Bishop, C.M.:*Neural Networks For Pattern Recognition*. Oxford Univ. Press, UK (1995).
2. Borkowski, M.: Digital Image Processing in Measurement of Ice Thickness on Power Transmission Lines: A Rough Set Approach, M.Sc. Thesis, Supervisor: J.F. Peters, Department of Electrical and Computer Engineering, University of Manitoba (2002).
3. Gottesfeld Brown,L.: *A Survey Of Image Registration Techniques*. ACM Computing Surveys, 24, No. 4, December (1992).
4. Caelli,T., Reye,D.: On the classification of image regions by color, texture and shape, Pattern Recognition, 26 (1993) 461-470.
5. Darwin, C.: On the Origin of the Species by Means of Natural Selection or the Preservation of Favoured Races in the Struggle for Life. Murray, London (1859).
6. Duda,R.O., Hart,P.E., Stork, D.G: *Pattern Classification*, 2nd Edition. John Wiley & Sons, Chichester (2001) 373-377.
7. Fitzpatrick, J.M, Hill, D.L.G., Calvin,R.M. Jr.: *Chapter 8: Image Registration*. In: Sonka, M., Fitzpatrick, J.M., *Handbook of Medical Imaging, Vol. 2: Medical Image Processing and Analysis*. SPIE PRESS Vol. PM80, Bellingham, USA, (2000).
8. Fu, K.S., Mui, J.K.: A survey of image segmentation, Pattern Recognition, 13 (1981) 3-16.
9. Eshelman, L.J.: Genetic algorithms. In: Back, T., Fogel, D.B., Michalewicz, Z. (Eds.), Handbook of Evolutionary Computation. Oxford University Press, Oxford, UK (1997).
10. Gersho,A., Gray,R.M.: *Vector Quantization And Signal Compression*. Kluwer Academic Publishers,Dordrecht,The Netherlands (1992).
11. Gomolinska, A.: Rough validity, confidence, and coverage of rules in approximation spaces, Transactions on Rough Sets III (2005) 57-81.
12. Gonzalez,R.C., Woods,R.E.: *Digital Image Processing*, 2nd Edition. Prentice Hall, NJ (2002).
13. Gray, R.M., Neuhoff, D.L.: Quantization, IEEE Transactions on Information Theory, 44 (1998) 1-63.
14. Haralick, R.M., Shapiro, L.G.: Image segmentation techniques, Computer Vision, Graphics, and Image Processing, 29 (1985) 100-132.
15. Harris,C., Stephens,M.: *A Combined Corner And Edge Detector*. In: Proceedings of the 4th Alvey Vision Conference, Manchester (1988) 189192.
16. Hartigan, J.A., Wong, M.A.: A K-means clustering algorithm, Applied Statistics, 28 (1979) 100-108.
17. Heaton, K.G.: Physical Pixels, M.Sc. Thesis, MIT (1994).
18. Hirano, H., Tsumoto, S.: Segmentation of medical images based on approximation in rough set theory. In: J.J. Alpigini, J.J., Peters, J.F., Skowron, A., Zhong, N. (Eds.), LNAI 2475. Springer-Verlag, Berlin (2002) 554-563.
19. Holland, J.H.: Adaptation in Natural and Artificial Systems. University of Michigan Press, Ann Arbor, MI (1975).

20. Paton, R.C.: Principles of genetics. In: T. Back, T., Fogel, D.B., Michalewicz, Z. (Eds.), Handbook of Evolutionary Computation. Oxford University Press, Oxford, UK (1997).
21. Komorowski,J., Pawlak,Z., Polkowski,L., Skowron,A.: *Rough Sets: A Tutorial.* In: Pal, S.K., Skowron, A. (Eds.), Rough Fuzzy Hybridization. A New Trend in Decision-Making. Springer-Verlag Singapore Pte. Ltd., Singapore (1999) 1-14.
22. Konrad, E., Orlowska, E., Pawlak, Z.: Knowledge Representation Systems. Definability of Informations. Institute for Computer Science, Polish Academy of Sciences Report 433, April (1981).
23. Lee, S.Uk, Chung, S.Y., Park, R.H.: A comparative performance study of several global thresholding techniques for segmentation, Computer Graphics and Image Processing, vol. 52 (1990) 171-190.
24. M. Loog, B.v. Ginneken, R.P.W. Duin: Dimensionality reduction by canonical contextual correlation projections. In: Pajdla, T., Matas, J. (Eds.), Lecture Notes in Computer Science 3021. Springer, Berlin (2004) 562-573.
25. Mayr, E.: Toward a New Philosophy of Biology: Observations of an Evolutionist. Belknap, Cambridge, MA (1988).
26. Mees, C.E.K., James, T.H.: The Theory of the Photograpahics Process. Macmillan, UK (1966).
27. Moody, J., Darken, C.J.: Fast learning in networks of locally-tuned processing units, Neural Computation 1/2 (1989) 281-294.
28. Orłowska, E.: Semantics of Vague Concepts. Applications of Rough Sets. Institute for Computer Science, Polish Academy of Sciences Report 469, March (1982).
29. Pal, N.R., Pal, S.K.: A review of image segmentation techniques, Pattern Recognition, 26/9 (1993).
30. Pawlak, Z.: Classification of Objects by Means of Attributes. Institute for Computer Science, Polish Academy of Sciences Report 429, March (1981).
31. Pawlak, Z.: Rough Sets. Institute for Computer Science, Polish Academy of Sciences Report 431, March (1981)
32. Pawlak, Z.: Rough sets, *International J. Comp. Inform. Science*, 11 (1982) 341-356.
33. Pawlak, Z.: *Rough Sets. Theoretical Reasoning about Data*, Theory and Decision Library, Series D: System Theory, Knowledge Engineering and Problem Solving, vol. 9. Kluwer Academic Pub., Dordrecht (1991).
34. Peters, J.F.: Approximation space for intelligent system design patterns. *Engineering Applications of Artificial Intelligence*, 17(4), 2004, 1-8.
35. Peters, J.F., Skowron, A., Synak, P., Ramanna, S.: Rough sets and information granulation. In: Bilgic, T., Baets, D., Kaynak, O. (Eds.), Tenth Int. Fuzzy Systems Assoc. World Congress IFSA, Instanbul, Turkey, *Lecture Notes in Artificial Intelligence* 2715. Springer-Verlag, Heidelberg (2003) 370-377.
36. J.F. Peters, Approximation spaces for hierarchical intelligent behavioral system models. In: Keplicz, B.D., Jankowski, A., Skowron, A., Szczuka, M. (Eds.), Monitoring, Security and Rescue Techniques in Multiagent Systems, *Advances in Soft Computing*. Springer-Verlag, Heidelberg (2004) 13-30.
37. Peters, J. F., Borkowski, M.: K-means indiscernibility relation over pixels. In: Tsumoto, S., Slowinski, R., Komorowski, J., Grzymala-Busse, J.W., *Lecture Notes in Artificial Intelligence* (LNAI), 3066. Springer-Verlag, Berlin (2004), 580-535
38. Peters, J.F.: Rough ethology, Transactions on Rough Sets III (2005) 153-174.
39. Peters, J.F., Henry, C.: Reinforcement learning with approximation spaces. Fundamenta Informaticae 71 (2006) 1-27.
40. L. Polkowski, *Rough Sets. Mathematical Foundations.* Springer-Verlag, Heidelberg (2002).

41. Polkowski, L., Skowron, A. (Eds.), Rough Sets in Knowledge Discovery 2, *Studies in Fuzziness and Soft Computing* 19. Springer-Verlag, Heidelberg (1998).
42. Rosenfeld, A.: Image pattern recognition, Proc. IEEE, 69 (1981).
43. Seemann, T.: Digital Image Processing Using Local Segmentation, Ph.D. Thesis, supervisor: Peter Tischer, Faculty of Information Technology, Monash University, Australia, April (2002).
44. Skowron, A.: Rough sets and vague concepts. *Fundamenta Informaticae*, XX (2004) 1-15.
45. Skowron, A., Stepaniuk, J.: Modelling complex patterns by information systems. *Fundamenta Informaticae*, XXI (2005) 1001-1013.
46. Skowron, A., Stepaniuk, J.: Information granules and approximation spaces, in Proc. of the 7th Int. Conf. on Information Processing and Management of Uncertainty in Knowledge-based Systems (IPMU98), Paris, (1998) 1354–1361.
47. Skowron, A., Stepaniuk, J.: Generalized approximation spaces. In: Lin, T.Y.,Wildberger, A.M. (Eds.), Soft Computing, Simulation Councils, San Diego (1995) 18-21.
48. Skowron, A., Swiniarski, R., Synak, P.: Approximation spaces and information granulation, Transactions on Rough Sets III (2005) 175-189.
49. Sahoo, P.K., Soltani, S., Wong, A.K.C.: A survey of thresholding techniques, Computer Vision, Graphics and Image Processing, 41 (1988) 233-260.
50. Stepaniuk, J.: Approximation spaces, reducts and representatives, in [41], 109–126.
51. Szczuka, M.S., Son, N.H.: Analysis of image sequences for unmanned aerial vehicles. In: Inuiguchi, M., Hirano, S., Tsumoto, S. (Eds.), Rough Set Theory and Granular Computing. Springer-Verlag, Berlin (2003) 291-300.
52. WITAS project homepage: `http://www.ida.liu.se/ext/witas/`
53. Yang, G.Z., Gillies, D.F.: Computer Vision Lecture Notes, Department of Computing, Imperial College, UK, 2001. See `http://www.doc.ic.ac.uk/ gzy`.
54. Zhang, Y.J.: Evaluation and comparison of different segmentation algorithms, Pattern Recognition Letters, 18 (1997) 963-974.
55. Zhang, C., Fraser, C.S.: Automated registration of high resolution satellite imagery for change detection, Research Report, Department of Geomatics, University of Melbourne (2003).
56. Zhang, Y.J., Gerbrands, J.J.: Objective and quantitative segmentation evaluation and comparison, Signal Processing, 39 (1994) 43-54.
57. Zitová, B., Flusser,J.: *Image Registration Methods: A Survey*, Image and Vision Computing 21 (2003).

An Efficient Algorithm for Inference in Rough Set Flow Graphs

C.J. Butz, W. Yan, and B. Yang

Department of Computer Science, University of Regina,
Regina, Canada, S4S 0A2
{butz, yanwe111, boting}@cs.uregina.ca

Abstract. Pawlak recently introduced *rough set flow graphs* (RSFGs) as a graphical framework for reasoning from data. No study, however, has yet investigated the complexity of the accompanying inference algorithm, nor the complexity of inference in RSFGs. In this paper, we show that the traditional RSFG inference algorithm has exponential time complexity. We then propose a new RSFG inference algorithm that exploits the factorization in a RSFG. We prove its correctness and establish its polynomial time complexity. In addition, we show that our inference algorithm never does more work than the traditional algorithm. Our discussion also reveals that, unlike traditional rough set research, RSFGs make implicit independency assumptions regarding the problem domain.

Keywords: Reasoning under uncertainty, rough set flow graphs.

1 Introduction

Very recently, Pawlak [7,8] introduced *rough set flow graphs* (RSFGs) as a graphical framework for uncertainty management. RSFGs extend traditional rough set research [9,10] by organizing the rules obtained from decision tables as a *directed acyclic graph* (DAG). Each rule is associated with three coefficients, namely, *strength*, *certainty* and *coverage*, which have been shown to satisfy Bayes' theorem [7,8]. Pawlak also provided an algorithm to answer queries in a RSFG and stated that RSFGs are a new perspective on Bayesian inference [7]. No study, however, has yet investigated the complexity of Pawlak's inference algorithm, nor the complexity of inference in RSFGs.

In this paper, our analysis of the traditional RSFG inference algorithm [7,8] establishes that its time complexity is exponential with respect to the number of nodes in a RSFG. We then propose a new inference algorithm that exploits the factorization in a RSFG. We prove the correctness of our algorithm and establish its polynomial time complexity. In addition, we show that our algorithm never does more work than the traditional algorithm, where work is the number of additions and multiplications needed to answer a query. The analysis in this manuscript also reveals that RSFGs make implicit assumptions regarding the problem domain. More specifically, we show that the *flow conservation assumption* [7] is in fact a *probabilistic conditional independency* [13] assumption.

J.F. Peters and A. Skowron (Eds.): Transactions on Rough Sets V, LNCS 4100, pp. 102–122, 2006.
© Springer-Verlag Berlin Heidelberg 2006

It should be noted that the work here is different from our earlier work [2] in several important ways. In this manuscript, we propose a new algorithm for RSFG inference and establish its polynomial time complexity. On the contrary, we established the polynomial complexity of RSFG inference in [2] by utilizing the relationship between RSFGs and *Bayesian networks* [11]. Another difference is that here we show that RSFG inference algorithm in [7,8] has exponential time complexity, an important result not discussed in [2].

This paper is organized as follows. Section 2 reviews probability theory, RS-FGs and a traditional RSFG inference algorithm [7,8]. That the traditional inference algorithm has exponential time complexity is shown in Section 3. In Section 4, we propose a new RSFG inference algorithm. We prove the correctness of this new algorithm and establish its polynomial time complexity in Section 5. Section 6 shows that it never does more work than the traditional algorithm. In Section 7, we observe that RSFGs make independence assumptions. The conclusion is presented in Section 8.

2   Definitions

In this section, we review probability theory and RSFGs.

2.1   Probability Theory

Let $U = \{v_1, v_2, \ldots, v_m\}$ be a finite set of variables. Each variable v_i has a finite domain, denoted $dom(v_i)$, representing the values that v_i can take on. For a subset $X = \{v_i, \ldots, v_j\}$ of U, we write $dom(X)$ for the Cartesian product of the domains of the individual variables in X, namely, $dom(X) = dom(v_i) \times \ldots \times dom(v_j)$. Each element $c \in dom(X)$ is called a *configuration* of X. If c is a configuration on X and $Y \subseteq X$, then by c_Y we denote the configuration on Y by dropping from c the values of those variables not in Y.

A *potential* [12] on $dom(U)$ is a function ϕ on $dom(U)$ such that the following two conditions both hold: (i) $\phi(u) \geq 0$, for each configuration $u \in dom(U)$, and (ii) $\phi(u) > 0$, for at least one configuration $u \in dom(U)$. For brevity, we refer to ϕ as a potential on U rather than $dom(U)$, and we call U, not $dom(U)$, its domain [12]. By XY, we denote $X \cup Y$.

A *joint probability distribution* (jpd) [12] on U is a function p on U such that the following two conditions both hold: (i) $0 \leq \phi(u) \leq 1$, for each configuration $u \in U$, and (ii) $\sum_{u \in U} \phi(u) = 1.0$.

Example 1. Consider five attributes Manufacturer (M), Dealership (D), Age (A), Salary (S), Position (P). One jpd $p(U)$ on $U = \{M, D, A, S, P\}$ is depicted in Appendix I.

We say X and Z are *conditionally independent* [13] given Y, denoted $I(X, Y, Z)$, in a joint distribution $p(X, Y, Z, W)$, if

$$p(X, Y, Z) = \frac{p(X, Y) \cdot p(Y, Z)}{p(Y)},$$

where $p(V)$ denotes the marginal [12] distribution of a jpd $p(U)$ onto $V \subseteq U$ and $p(Y) > 0$.

The following theorem provides a necessary and sufficient condition for determining when a conditional independence holds in a problem domain.

Theorem 1. *[5] $I(X, Y, Z)$ iff there exist potentials ϕ_1 and ϕ_2 such that for each configuration c on XYZ with $p(c_Y) > 0$, $p(c) = \phi_1(c_{XY}) \cdot \phi_2(c_{YZ})$.*

Example 2. Recall the jpd $p(U)$ in Example 1. The marginal $p(M, D, A)$ of $p(U)$ and two potentials $\phi(M, D)$, $\phi(D, A)$ are depicted in Table 1. By definition, conditional independence $I(M, D, A)$ holds in $p(U)$ as $p(M, D, A) = \phi(M, D) \cdot \phi(D, A)$.

Table 1. The marginal $p(M, D, A)$ of $p(U)$ in Example 1 and potentials $\phi(M, D)$ and $\phi(D, A)$

M	D	A	$p(M, D, A)$	M	D	$\phi(M, D)$	D	A	$\phi(D, A)$
Toyota	Alice	Old	0.036	Toyota	Alice	0.120	Alice	Old	0.300
Toyota	Alice	Middle	0.072	Toyota	Bob	0.060	Alice	Middle	0.600
Toyota	Alice	Young	0.012	Toyota	Dave	0.020	Alice	Young	0.100
Toyota	Bob	Old	0.024	Honda	Bob	0.150	Bob	Old	0.400
Toyota	Bob	Middle	0.036	Honda	Carol	0.150	Bob	Middle	0.600
Toyota	Dave	Old	0.002	Ford	Alice	0.050	Carol	Middle	0.600
Toyota	Dave	Middle	0.006	Ford	Bob	0.150	Carol	Young	0.400
Toyota	Dave	Young	0.012	Ford	Carol	0.050	Dave	Old	0.100
Honda	Bob	Old	0.060	Ford	Dave	0.250	Dave	Middle	0.300
Honda	Bob	Middle	0.090				Dave	Young	0.600
Honda	Carol	Middle	0.090						
Honda	Carol	Young	0.060						
Ford	Alice	Old	0.015						
Ford	Alice	Middle	0.030						
Ford	Alice	Young	0.005						
Ford	Bob	Old	0.060						
Ford	Bob	Middle	0.090						
Ford	Carol	Middle	0.030						
Ford	Carol	Young	0.020						
Ford	Dave	Old	0.025						
Ford	Dave	Middle	0.075						
Ford	Dave	Young	0.150						

2.2   Rough Set Flow Graphs

Rough set flow graphs are built from decision tables. A *decision table* [10] represents a potential $\phi(C, D)$, where C is a set of conditioning attributes and D is a decision attribute.

Example 3. Recall the five attributes $\{M, D, A, S, P\}$ from Example 1. Consider the set $C = \{M\}$ of conditioning attributes and the decision attribute D. Then one decision table $\phi(M, D)$ is shown in Table 2. Similarly, decision tables $\phi(D, A)$, $\phi(A, S)$ and $\phi(S, P)$ are also depicted in Table 2.

Table 2. *Decision tables* $\phi(M, D)$, $\phi(D, A)$, $\phi(A, S)$ *and* $\phi(S, P)$

M	D	$\phi(M, D)$
Toyota	Alice	120
Toyota	Bob	60
Toyota	Dave	20
Honda	Bob	150
Honda	Carol	150
Ford	Alice	50
Ford	Bob	150
Ford	Carol	50
Ford	Dave	250

D	A	$\phi(D, A)$
Alice	Old	51
Alice	Middle	102
Alice	Young	17
Bob	Old	144
Bob	Middle	216
Carol	Middle	120
Carol	Young	80
Dave	Old	27
Dave	Middle	81
Dave	Young	162

A	S	$\phi(A, S)$
Old	High	133
Old	Medium	67
Old	Low	22
Middle	High	104
Middle	Medium	311
Middle	Low	104
Young	High	26
Young	Medium	52
Young	Low	181

S	P	$\phi(S, P)$
High	Executive	210
High	Staff	45
High	Manager	8
Medium	Executive	13
Medium	Staff	387
Medium	Manager	30
Low	Executive	3
Low	Staff	12
Low	Manager	292

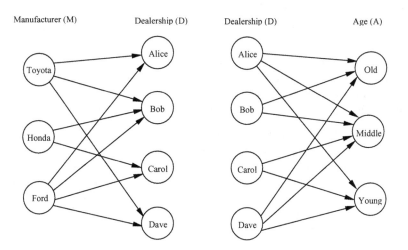

Fig. 1. The DAGs of the binary RSFGs for the decision tables $\phi(M, D)$ and $\phi(D, A)$ in Table 2, respectively. The coefficients are given in part of Table 3.

Each decision table defines a binary RSFG. The set of nodes in the flow graph are $\{c_1, c_2, \ldots, c_k\} \cup \{d_1, d_2, \ldots, d_l\}$, where $c_1, c_2, \ldots, c_k$ and $d_1, d_2, \ldots, d_l$ are the values of C and D appearing in the decision table, respectively. For each row

Table 3. The top two tables are the *strength* $\phi(a_i, a_j)$, *certainty* $\phi(a_j|a_i)$ and *coverage* $\phi(a_i|a_j)$ coefficients for the edges (a_i, a_j) in Fig. 1. These two tables together with the bottom two tables are the coefficients for the edges in Fig. 2.

M	D	$\phi_1(M,D)$	$\phi_1(D\|M)$	$\phi_1(M\|D)$
Toyota	Alice	0.120	0.600	0.710
Toyota	Bob	0.060	0.300	0.160
Toyota	Dave	0.020	0.100	0.070
Honda	Bob	0.150	0.500	0.420
Honda	Carol	0.150	0.500	0.750
Ford	Alice	0.050	0.100	0.290
Ford	Bob	0.150	0.300	0.420
Ford	Carol	0.050	0.100	0.250
Ford	Dave	0.250	0.500	0.930

D	A	$\phi_2(D,A)$	$\phi_2(A\|D)$	$\phi_2(D\|A)$
Alice	Old	0.050	0.300	0.230
Alice	Middle	0.100	0.600	0.190
Alice	Young	0.020	0.100	0.080
Bob	Old	0.140	0.400	0.630
Bob	Middle	0.220	0.600	0.420
Carol	Middle	0.120	0.600	0.230
Carol	Young	0.080	0.400	0.310
Dave	Old	0.030	0.100	0.140
Dave	Middle	0.080	0.300	0.150
Dave	Young	0.160	0.600	0.620

A	S	$\phi_3(A,S)$	$\phi_3(S\|A)$	$\phi_3(A\|S)$
Old	High	0.133	0.600	0.506
Old	Medium	0.067	0.300	0.156
Old	Low	0.022	0.100	0.072
Middle	High	0.104	0.200	0.395
Middle	Medium	0.311	0.600	0.723
Middle	Low	0.104	0.200	0.339
Young	High	0.026	0.100	0.099
Young	Medium	0.052	0.200	0.121
Young	Low	0.181	0.700	0.589

S	P	$\phi_4(S,P)$	$\phi_4(P\|S)$	$\phi_4(S\|P)$
High	Executive	0.210	0.800	0.929
High	Staff	0.045	0.170	0.101
High	Manager	0.008	0.030	0.024
Medium	Executive	0.013	0.030	0.058
Medium	Staff	0.387	0.900	0.872
Medium	Manager	0.030	0.070	0.091
Low	Executive	0.003	0.010	0.013
Low	Staff	0.012	0.040	0.027
Low	Manager	0.292	0.950	0.885

in the decision table, there is a directed edge (c_i, d_j) in the flow graph, where c_i is the value of C and d_j is the value of D. Clearly, the defined graphical structure is a *directed acyclic graph* (DAG). Each edge (c_i, d_j) is labelled with three coefficients. The *strength* of (c_i, d_j) is $\phi(c_i, d_j)$ obtained from the decision table. From $\phi(c_i, d_j)$, we can compute the *certainty* $\phi(d_j|c_i)$ and the *coverage* $\phi(c_i|d_j)$.

Example 4. Consider the decision tables $\phi(M, D)$ and $\phi(D, A)$ in Table 2. The DAGs of the binary RSFGs are illustrated in Fig. 1, respectively. The strength, certainty and coverage of the edges of the flow graphs in Fig. 1 are shown in the top two tables of Table 3.

In order to combine the collection of binary flow graphs into a general flow graph, Pawlak makes the *flow conservation* assumption [7]. This means that, for an attribute A appearing as a decision attribute in one decision table $\phi_1(C_1, A)$ and also as a conditioning attribute in another decision table $\phi_2(A, D_2)$, we have

$$\sum_{C_1} \phi_1(C_1, A) = \sum_{D_2} \phi_2(A, D_2).$$

Example 5. The two binary RSFGs in Example 4 satisfy the flow conservation assumption, since in Table 3, $\phi_1(D) = \phi_2(D)$. For instance, $\phi_1(D = \text{"Alice"}) = 0.170 = \phi_2(D = \text{"Alice"})$.

A *rough set flow graph* (RSFG) [7,8] is a DAG, where each edge is associated with the strength, certainty and coverage coefficients from a collection of decision tables satisfying the flow conservation assumption.

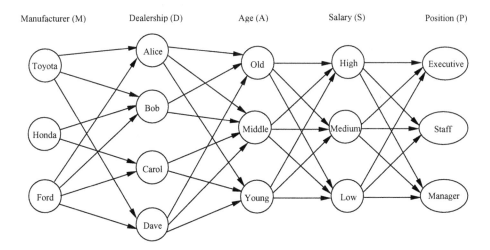

Fig. 2. The *rough set flow graph* (RSFG) for $\{M, D, A, S, P\}$, where the strength, certainty and coverage coefficient are given in Table 3

Example 6. The RSFG for the decision tables in Table 2 is the DAG in Fig. 2 together with the strength, certainty and coverage coefficients in Table 3.

The task of RSFG inference is to compute a binary RSFG on $\{A_i, A_j\}$, namely, a DAG on $\{A_i, A_j\}$ and the coefficient table, denoted $Ans(A_i, A_j)$, which is a table with strength, certainty and coverage columns. We use the term *query* to refer to any request involving strength, certainty or coverage.

Example 7. Consider a query on $\{M, P\}$ posed to the RSFG in Example 6. The answer to this query is the binary RSFG defined by Table 4 and Fig. 3.

Pawlak proposed Algorithm 1 to answer queries in a RSFG.

Algorithm 1. Algorithm 1. [7,8]

---

    input : A RSFG and a query on $\{A_i, A_j\}$, $i < j$.
    output: The coefficient table $Ans(A_i, A_j)$ of the binary RSFG on $\{A_i, A_j\}$.
    $\phi(A_j|A_i) = \sum_{A_{i+1},\ldots,A_{j-1}} \phi(A_{i+1}|A_i) \cdot \phi(A_{i+2}|A_{i+1}) \cdot \ldots \cdot \phi(A_j|A_{j-1})$;
    $\phi(A_i|A_j) = \sum_{A_{i+1},\ldots,A_{j-1}} \phi(A_i|A_{i+1}) \cdot \phi(A_{i+1}|A_{i+2}) \cdot \ldots \cdot \phi(A_{j-1}|A_j)$;
    $\phi(A_i, A_j) = \phi(A_i) \cdot \phi(A_j|A_i)$;
    return($Ans(A_i, A_j)$);

Algorithm 1 is used to compute the coefficient table of the binary RSFG on $\{A_i, A_j\}$. The DAG of this binary RSFG has an edge (a_i, a_j) provided that $\phi(a_i, a_j) > 0$ in $Ans(A_i, A_j)$. We illustrate Algorithm 1 with Example 8.

Table 4. Answering a query on $\{M, P\}$ posed to the RSFG in Fig. 2 consists of this coefficient table $Ans(M, P)$ and the DAG in Fig. 3

M	P	$\phi(M, P)$	$\phi(P|M)$	$\phi(M|P)$
Toyota	Executive	0.053132	0.265660	0.234799
Toyota	Staff	0.095060	0.475300	0.214193
Toyota	Manager	0.051808	0.259040	0.157038
Honda	Executive	0.067380	0.224600	0.297764
Honda	Staff	0.140820	0.469400	0.317302
Honda	Manager	0.091800	0.306000	0.278259
Ford	Executive	0.105775	0.211550	0.467437
Ford	Staff	0.207925	0.415850	0.468505
Ford	Manager	0.186300	0.372600	0.564703

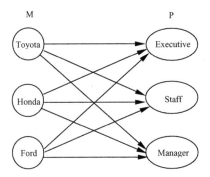

Fig. 3. Answering a query on $\{M, P\}$ posed to the RSFG in Fig. 2 consists of the coefficient table $Ans(M, P)$ in Table 4 and this DAG on $\{M, P\}$

Example 8. Given a query on $\{M, P\}$ posed to the RSFG in Fig. 2. Let us focus on $M = $ "*Ford* " and $P = $ "*Staff* ", which we succinctly write as "*Ford* " and "*Staff* ", respectively. The certainty $\phi($"*Staff* "|"*Ford* ") is computed as:

$$\phi(\text{``Staff''}|\text{``Ford''}) = \sum_{D,A,S} \phi(D|\text{``Ford''}) \cdot \phi(A|D) \cdot \phi(S|A) \cdot \phi(\text{``Staff''}|S).$$

The coverage $\phi($"*Ford* "|"*Staff* ") is computed as:

$$\phi(\text{``Ford''}|\text{``Staff''}) = \sum_{D,A,S} \phi(\text{``Ford''}|D) \cdot \phi(D|A) \cdot \phi(A|S) \cdot \phi(S|\text{``Staff''}).$$

The strength $\phi($"*Ford* ", "*Staff* ") is computed as:

$$\phi(\text{``Ford''}, \text{``Staff''}) = \phi(\text{``Ford''}) \cdot \phi(\text{``Staff''}|\text{``Ford''}).$$

The DAG of this binary RSFG on $\{M, P\}$ is depicted in Fig. 3.

In Example 8, computing coefficients $\phi($"*Ford* ", "*Staff* "), $\phi($"*Staff* "|"*Ford* ") and $\phi($"*Ford* "|"*Staff* ") in $Ans(M, P)$ in Table 4 required 181 multiplications and 58 additions. No study, however, has formalized the time complexity of Algorithm 1.

3   Complexity of Traditional Algorithm in RSFG

In this section, we establish the time complexity of Algorithm 1.

Theorem 2. *Consider a RSFG on m variables $U = \{A_1, A_2, \ldots, A_m\}$. Let $|dom(A_i)| = n$, for $i = 1, \ldots, m$. Let (a_i, a_{i+1}) be an edge in the RSFG, where $a_i \in dom(A_i)$, $a_{i+1} \in dom(A_{i+1})$ and $i = 1, \ldots, m-1$. To answer a query on $\{A_i, A_j\}$, the time complexity of Algorithm 1 is $O(ln^l)$, where $l = j - i + 1$.*

Proof. To compute the certainty $\phi(A_j|A_i)$, let

$$\psi_1(A_i, A_{i+1}, \ldots, A_j) = \phi(A_{i+1}|A_i) \cdot \phi(A_{i+2}|A_{i+1}) \cdot \ldots \cdot \phi(A_j|A_{j-1}). \qquad (1)$$

The potential $\psi_1(A_i, A_{i+1}, \ldots, A_j)$ has n^l rows, since $|dom(A_i)| = n$ for each variable. By Equation (1), computing the certainty for one row requires $l - 2$ multiplications. Therefore, $\psi_1(A_i, A_{i+1}, \ldots, A_j)$ is constructed by $(l-2)(n^l)$ multiplications. The second step is to determine

$$\phi(A_j|A_i) = \sum_{A_{i+1}, \ldots, A_{j-1}} \psi_1(A_i, A_{i+1}, \ldots, A_j). \qquad (2)$$

There are exactly n^{l-2} rows in $\psi_1(A_i, A_{i+1}, \ldots, A_j)$ with $A_i = a_i$ and $A_j = a_j$. Thus, computing $\phi(A_j = a_j|A_i = a_i)$ requires $n^{l-2} - 1$ additions. Since there are n^2 configurations in $\phi(A_j|A_i)$, to compute $\phi(A_j|A_i)$ requires $(n^2)(n^{l-2} - 1)$ additions. That is, $n^l - n^2$ additions are required for Equation (2). As shown above, the complexity to compute Equation (1) is $O(ln^l)$ and that to compute Equation (2) is $O(n^l)$. Therefore, computing the certainty $\phi(A_j|A_i)$ has time complexity $O(ln^l)$. It is easily seen that computing the coverage $\phi(A_i|A_j)$ requires exactly the same amount of work as required for computing the certainty $\phi(A_j|A_i)$. Thus, computing the coverage $\phi(A_i|A_j)$ has time complexity $O(ln^l)$. The strength $\phi(A_i, A_j)$ is defined as the product $\phi(A_i) \cdot \phi(A_j|A_i)$, which involves n^2 multiplications. Since the computation of Algorithm 1 is dominated by that for certainty (coverage), the time complexity is $O(ln^l)$.   □

The exponential time complexity of Algorithm 1 lies in the fact that it does not exploit the factorization during inference. However, this does not mean that Algorithm 1 is always inefficient in all practical situations.

4   An Efficient Algorithm for RSFG Inference

In this section, we will introduce an efficient algorithm to answer queries in a RSFG and establish its complexity.

   The main idea is to exploit the factorization to eliminate variables one by one, instead of all at once as Algorithm 1 does. We focus on computing the coefficient table $Ans(A_i, A_j)$ with the DAG of the output RSFG understood.

Algorithm 2. Algorithm 2

---

input : A RSFG and a query on $\{A_i, A_j\}$, $i < j$.
output: The coefficient table $Ans(A_i, A_j)$ of the binary RSFG on $\{A_i, A_j\}$.
for $k = (i+1)$ *to* $(j-1)$ **do**
$\quad\left|\begin{array}{l} \phi(A_{k+1}|A_i) = \sum_{A_k} \phi(A_k|A_i) \cdot \phi(A_{k+1}|A_k); \\ \phi(A_i|A_{k+1}) = \sum_{A_k} \phi(A_i|A_k) \cdot \phi(A_k|A_{k+1}); \end{array}\right.$
end
$\phi(A_i, A_j) = \phi(A_i) \cdot \phi(A_j|A_i);$
return($Ans(A_i, A_j)$);

We illustrate Algorithm 2 with the following example.

Example 9. Recall Example 8. Again, we focus on the edge (*"Ford "*, *"Staff "*) in the DAG in Fig. 3. According to Algorithm 2, variables $\{D, A, S\}$ need be eliminated. Consider variable D. The certainty $\phi(A|\text{"Ford "})$ is

$$\phi(A|\text{"Ford "}) = \sum_D \phi(D|\text{"Ford "}) \cdot \phi(A|D),$$

while the coverage $\phi(\text{"Ford "}|A)$ is

$$\phi(\text{"Ford "}|A) = \sum_D \phi(\text{"Ford "}|D) \cdot \phi(D|A).$$

The consequence is that variable D has been eliminated, while variables M and A have been linked via the certainty $\phi(A|\text{"Ford "})$ and coverage $\phi(\text{"Ford "}|A)$. Similarly, eliminating A yields $\phi(S|\text{"Ford "})$ and $\phi(\text{"Ford "}|S)$. Finally, consider eliminating variable S. The certainty $\phi(\text{"Staff "}|\text{"Ford "})$ is

$$\phi(\text{"Staff "}|\text{"Ford "}) = \sum_S \phi(S|\text{"Ford "}) \cdot \phi(\text{"Staff "}|S),$$

while the coverage $\phi(\text{"Ford "}|\text{"Staff "})$ is

$$\phi(\text{"Ford "}|\text{"Staff "}) = \sum_S \phi(\text{"Ford "}|S) \cdot \phi(S|\text{"Staff "}).$$

The strength $\phi(\text{"Ford "}, \text{"Staff "})$ is determined as

$$\phi(\text{"Ford "}, \text{"Staff "}) = \phi(\text{"Ford "}) \cdot \phi(\text{"Staff "}|\text{"Ford "}).$$

In Example 9, computing $\phi(\text{"Ford "}, \text{"Staff "})$, $\phi(\text{"Staff "}|\text{"Ford "})$ and $\phi(\text{"Ford "}|\text{"Staff "})$ in $Ans(M, P)$ in Table 4 only required 45 multiplications and 30 additions. Recall that Algorithm 1 required 181 multiplications and 58 additions.

5   Theoretical Foundation

In this section, we show correctness of Algorithm 2 and prove Algorithm 2 is efficient by analyzing its time complexity in the worst case.

5.1 Correctness of the New RSFG Inference Algorithm

Here we prove that Algorithm 2 is correct. Let us first review two well known results.

Lemma 1. *[12] If ϕ is a potential on U, and $X \subseteq Y \subseteq U$, then marginalizing ϕ onto Y and subsequently onto X is the same as marginalizing ϕ onto X.*

Lemma 1 indicates that a marginal can be obtained by a series of marginalizations in any order. For example,

$$\sum_{A,B} \phi(A,B,C) \;=\; \sum_A (\sum_B \phi(A,B,C)) \;=\; \sum_B (\sum_A \phi(A,B,C)).$$

Lemma 2. *[12] If ϕ is a potential on X and ψ is a potential on Y, then the marginalization of $\phi \cdot \psi$ onto X is the same as ϕ multiplied with the marginalization of ψ onto $X \cap Y$.*

For instance,

$$\sum_C \phi(A,B) \cdot \phi(B,C) \;=\; \phi(A,B) \cdot \sum_C \phi(B,C).$$

Now let us turn to the correctness of Algorithm 2.

Theorem 3. *Given a query on $\{A_i, A_j\}$ posed to a RSFG on $U = \{A_1, A_2, \ldots, A_m\}$, where $1 \leq i < j \leq m$. The answer produced by Algorithm 2 is correct.*

Proof. We show the claim by proving that the answer table $Ans(A_i, A_j)$ produced by Algorithm 2 contains the strength $\phi(A_i, A_j)$, the certainty $\phi(A_j|A_i)$ and the coverage $\phi(A_i|A_j)$ computed by Algorithm 1. To answer the certainty $\phi(A_j|A_i)$, Algorithm 1 is expressed by Equation (3),

$$\phi(A_j|A_i) = \sum_{A_{i+1}, A_{i+2}, \ldots, A_{j-1}} \phi(A_{i+1}|A_i) \cdot \phi(A_{i+2}|A_{i+1}) \cdot \ldots \cdot \phi(A_j|A_{j-1}). \quad (3)$$

By Lemma 1 and Equation (3), $\phi(A_j|A_i)$ is equal to

$$\sum_{A_{i+1}} \sum_{A_{i+2}} \cdots \sum_{A_{j-2}} \sum_{A_{j-1}} \phi(A_{i+1}|A_i) \cdot \phi(A_{i+2}|A_{i+1}) \cdot \ldots \cdot \phi(A_j|A_{j-1}). \quad (4)$$

By Lemma 2 and Equation (4), $\phi(A_j|A_i)$ is equal to

$$\sum_{A_{i+1}} \sum_{A_{i+2}} \cdots \sum_{A_{j-2}} \phi(A_{i+1}|A_i) \cdot \ldots \cdot \phi(A_{j-2}|A_{j-3}) \cdot \sum_{A_{j-1}} \phi(A_{j-1}|A_{j-2}) \cdot \phi(A_j|A_{j-1}). \quad (5)$$

By recursively using Lemma 2, Equation (5) can be rewritten as,

$$\sum_{A_{i+1}} \phi(A_{i+1}|A_i) \cdot \sum_{A_{i+2}} \phi(A_{i+2}|A_{i+1}) \cdot \ldots \cdot \sum_{A_{j-1}} \phi(A_{j-1}|A_{j-2}) \cdot \phi(A_j|A_{j-1}). \quad (6)$$

By Equations (3) - (6), the computation of the certainty $\phi(A_j|A_i)$ by Algorithm 1 is expressed as,

$$\phi(A_j|A_i) = \sum_{A_{i+1}} \phi(A_{i+1}|A_i) \cdot \ldots \cdot \sum_{A_{j-1}} \phi(A_{j-1}|A_{j-2}) \cdot \phi(A_j|A_{j-1}). \quad (7)$$

Equation (7) is the construction of the certainty $\phi(A_j|A_i)$ in Algorithm 2. It can be similarly shown that the strength $\phi(A_i, A_j)$ and coverage $\phi(A_i|A_j)$ produced by Algorithms 1 and 2 are the same. □

5.2   Complexity of the New RSFG Inference Algorithm

In this subsection, we establish the computational complexity of Algorithm 2.

Theorem 4. *Consider a RSFG on m variables $U = \{A_1, A_2, \ldots, A_m\}$. Let $|dom(A_i)| = n$, for $i = 1, \ldots, m$. Let (a_i, a_{i+1}) be an edge in the RSFG, where $a_i \in dom(A_i)$, $a_{i+1} \in dom(A_{i+1})$ and $i = 1, \ldots, m-1$. To answer a query on $\{A_i, A_j\}$, the time complexity of Algorithm 2 is $O(ln^3)$, where $l = j - i + 1$.*

Proof. The certainty $\phi(A_j|A_i)$ is computed by eliminating each variable A_k between A_i and A_j in the RSFG. For a variable A_k, Algorithm 2 first computes

$$\psi_2(A_{k-1}, A_k, A_{k+1}) = \phi(A_k|A_{k-1}) \cdot \phi(A_{k+1}|A_k). \quad (8)$$

The potential $\psi_2(A_{k-1}, A_k, A_{k+1})$ has n^3 rows, since $|dom(A_i)| = n$ for each variable. Computing the certainty for one row requires 1 multiplication. Therefore, potential $\psi_2(A_{k-1}, A_k, A_{k+1})$ is constructed by n^3 multiplications. The second step is to determine

$$\phi(A_{k+1}|A_{k-1}) = \sum_{A_k} \psi_2(A_{k-1}, A_k, A_{k+1}). \quad (9)$$

There are n rows in $\psi_2(A_{k-1}, A_k, A_{k+1})$ with $A_{k-1} = a_{k-1}$ and $A_{k+1} = a_{k+1}$. Thus, computing $\phi(A_{k+1} = a_{k+1}|A_{k-1} = a_{k-1})$ requires $n-1$ additions. Since there are n^2 configurations in $\phi(A_{k+1}|A_{k-1})$, $(n^2)(n-1)$ additions are required to compute $\phi(A_{k+1}|A_{k-1})$ in Equation (9). Therefore, the time complexity to compute the certainty $\phi(A_{k+1}|A_{k-1})$ is $O(n^3)$. Since there are $l-2$ variables between A_i and A_j, the time complexity to compute the desired certainty $\phi(A_j|A_i)$ has time complexity $O(ln^3)$. Similar to the proof of Theorem 2, it follows that the time complexity of Algorithm 2 is $O(ln^3)$. □

Theorem 4 shows that Algorithm 2 has polynomial time complexity in the worst case. Therefore, Algorithm 2 is an efficient algorithm for RSFG inference in all practical situations.

6   Related Work

In this section, we show Algorithm 2 never performs more work than Algorithm 1. To show this claim let us first characterize the computation performed by Algorithm 1 and Algorithm 2 when answering a query.

We need only focus on how the certainty $\phi(A_j|A_i)$ is computed from a RSFG on $U = \{A_1, A_2, \ldots, A_m\}$ with certainties $\phi(A_2|A_1), \phi(A_3|A_2), \ldots, \phi(A_m| A_{m-1})$. For simplicity, we eliminate variables in the following order: $A_{j-1}, A_{j-2}, \ldots, A_{i+1}$.

Algorithm 1 computes the following product $\psi_1(A_i, A_{i+1}, \ldots, A_j)$:

$$\psi_1(A_i, A_{i+1}, \ldots, A_j)$$
$$= \phi(A_{i+1}|A_i) \cdot \ldots \cdot \phi(A_{j-2}|A_{j-3}) \cdot \phi(A_{j-1}|A_{j-2}) \cdot \phi(A_j|A_{j-1})$$

via a series of binary multiplications, namely,

$$\psi_1(A_i, A_{i+1}, \ldots, A_j)$$
$$= \phi(A_{i+1}|A_i) \cdot [\ldots \cdot [\phi(A_{j-2}|A_{j-3}) \cdot [\phi(A_{j-1}|A_{j-2}) \cdot \phi(A_j|A_{j-1})]] \ldots]. \quad (10)$$

According to Equation (10), the first multiplication is as follows,

$$\psi_1(A_{j-2}, A_{j-1}, A_j) = \phi(A_{j-1}|A_{j-2}) \cdot \phi(A_j|A_{j-1}). \quad (11)$$

The intermediate multiplications are performed as follows,

$$\psi_1(A_{k-1}, A_k, \ldots, A_j) = \phi(A_k|A_{k-1}) \cdot \psi_1(A_k, A_{k+1}, \ldots, A_j), \quad (12)$$

where $k = (j-2), \ldots, (i+1)$.

After computing $\psi_1(A_i, A_{i+1}, \ldots, A_j)$, Algorithm 1 eliminates variables $A_{i+1}, A_{i+2}, \ldots, A_{j-1}$ via a series of marginalizations, namely,

$$\sum_{A_{i+1}} \sum_{A_{i+2}} \cdots \sum_{A_{j-1}} \psi_1(A_i, A_{i+1}, \ldots, A_j).$$

An intermediate marginalization takes the form,

$$\psi_1(A_i, \ldots, A_{l-1}, A_j) = \sum_{A_l} \psi_1(A_i, \ldots, A_{l-1}, A_l, A_j), \quad (13)$$

where $l = (j-1), \ldots, (i+2)$. The final marginalization yields

$$\phi(A_j|A_i) = \sum_{A_{i+1}} \psi_1(A_i, A_{i+1}, A_j). \quad (14)$$

Now consider how Algorithm 2 computes the certainty $\phi(A_j|A_i)$. As previously mentioned, Algorithm 2 eliminates variables $A_{j-1}, \ldots, A_{i+1}$ one by one. Algorithm 2 computes,

$$\phi(A_j|A_i)$$
$$= \sum_{A_{i+1}} \phi(A_{i+1}|A_i) \cdot \ldots \cdot \sum_{A_{j-2}} \phi(A_{j-2}|A_{j-3}) \cdot \sum_{A_{j-1}} \phi(A_{j-1}|A_{j-2}) \cdot \phi(A_j|A_{j-1}).$$
$$(15)$$

According to Equation (15), the first multiplication in Algorithm 2 is,

$$\psi_2(A_{j-2}, A_{j-1}, A_j) = \phi(A_{j-1}|A_{j-2}) \cdot \phi(A_j|A_{j-1}). \tag{16}$$

Algorithm 2 then performs intermediate additions and multiplications, iteratively,

$$\phi(A_j|A_{j-2}) = \sum_{A_{j-1}} \psi_2(A_{j-2}, A_{j-1}, A_j),$$

$$\psi_2(A_{j-3}, A_{j-2}, A_j) = \phi(A_{j-2}|A_{j-3}) \cdot \phi(A_j|A_{j-2}),$$

$$\phi(A_j|A_{j-3}) = \sum_{A_{j-2}} \psi_2(A_{j-3}, A_{j-2}, A_j),$$

$$\vdots$$

$$\psi_2(A_i, A_{i+1}, A_j) = \phi(A_{i+1}|A_i) \cdot \phi(A_j|A_{i+1}).$$

Therefore, an intermediate marginalization takes the form,

$$\phi(A_j|A_{l-1}) = \sum_{A_l} \psi_2(A_{l-1}, A_l, A_j), \tag{17}$$

where $l = (j-1), \ldots, (i+2)$. An intermediate multiplication takes the form,

$$\psi_2(A_{k-1}, A_k, A_j) = \phi(A_k|A_{k-1}) \cdot \phi(A_j|A_k), \tag{18}$$

where $k = (j-2), \ldots, (i+1)$. After these intermediate additions and multiplications, the final marginalization yields the desired certainty $\phi(A_j|A_i)$:

$$\phi(A_j|A_i) = \sum_{A_{i+1}} \psi_2(A_i, A_{i+1}, A_j). \tag{19}$$

Lemma 3 shows that the intermediate potentials computed in the multiplication process of Algorithm 2 are marginalizations of the larger potentials computed in Algorithm 1. Lemma 4 shows that the intermediate potentials computed in the marginalization process of Algorithm 2 have no more rows than the marginalizations of the larger potentials computed in Algorithm 1.

Lemma 3. *To answer a query on* $\{A_i, A_j\}$ *posed to a RSFG on* $U = \{A_1, A_2, \ldots, A_m\}$, $\phi(A_j|A_k)$ *in Equation (18) of Algorithm 2 is a marginal of* $\psi_1(A_k, A_{k+1}, \ldots, A_j)$ *in Equation (12) of Algorithm 1.*

Proof. By definition, the marginal of $\psi_1(A_k, A_{k+1}, \ldots, A_j)$ onto $\{A_k, A_j\}$ is:

$$\sum_{A_{k+1}, \ldots, A_{j-1}} \psi_1(A_k, A_{k+1}, \ldots, A_j). \tag{20}$$

By Algorithm 1, Equation (20) is equal to,

$$\sum_{A_{k+1}, \ldots, A_{j-1}} \phi(A_{k+1}|A_k) \cdot \ldots \cdot \phi(A_{j-1}|A_{j-2}) \cdot \phi(A_j|A_{j-1}). \tag{21}$$

By Lemmas 1 and 2, Equation (21) can be rewritten as:

$$\sum_{A_{k+1}} \phi(A_{k+1}|A_k) \cdot \ldots \cdot \sum_{A_{j-1}} \phi(A_{j-1}|A_{j-2}) \cdot \phi(A_j|A_{j-1}). \qquad (22)$$

By Equation (7),

$$\phi(A_j|A_k) = \sum_{A_{k+1}} \phi(A_{k+1}|A_k) \cdot \ldots \cdot \sum_{A_{j-1}} \phi(A_{j-1}|A_{j-2}) \cdot \phi(A_j|A_{j-1}). \qquad (23)$$

By Equations (20) - (23),

$$\phi(A_j|A_k) = \sum_{A_{k+1},\ldots,A_{j-1}} \psi_1(A_k, A_{k+1}, \ldots, A_j).$$

Therefore, $\phi(A_j|A_k)$ is the marginal of $\psi_1(A_k, A_{k+1}, \ldots, A_j)$ onto variables $\{A_k, A_j\}$. $\qquad \square$

Lemma 4. *To answer a query on $\{A_i, A_j\}$ posed to a RSFG on $U = \{A_1, A_2, \ldots, A_m\}$, $\psi_2(A_{l-1}, A_l, A_j)$ in Equation (17) of Algorithm 2 has no more rows than the marginal of $\psi_1(A_i, \ldots, A_{l-1}, A_l, A_j)$ in Equation (13) of Algorithm 1 onto variables $\{A_{l-1}, A_l, A_j\}$.*

Proof. By definition, the marginal of $\psi_1(A_i, \ldots, A_{l-1}, A_l, A_j)$ onto variables $\{A_{l-1}, A_l, A_j\}$ is:

$$\sum_{A_i,\ldots,A_{l-2}} \psi_1(A_i, \ldots, A_{l-1}, A_l, A_j). \qquad (24)$$

By Algorithm 1, Equation (24) is equal to,

$$\sum_{A_i,\ldots,A_{l-2}} \phi(A_{i+1}|A_i) \cdot \ldots \cdot \phi(A_{l-2}|A_{l-3}) \cdot \phi(A_{l-1}|A_{l-2}) \cdot \phi(A_l|A_{l-1}) \cdot \phi(A_j|A_l). \qquad (25)$$

By Lemma 2, Equation (25) is equal to,

$$\phi(A_l|A_{l-1}) \cdot \phi(A_j|A_l) \cdot \sum_{A_i,\ldots,A_{l-2}} \phi(A_{i+1}|A_i) \cdot \ldots \cdot \phi(A_{l-2}|A_{l-3}) \cdot \phi(A_{l-1}|A_{l-2}). \qquad (26)$$

By Lemmas 1 and 2, Equation (26) can be rewritten as:

$$\phi(A_l|A_{l-1}) \cdot \phi(A_j|A_l) \cdot \sum_{A_i} (\sum_{A_{i+1}} \phi(A_{i+1}|A_i) \cdot \ldots \cdot \sum_{A_{l-2}} \phi(A_{l-2}|A_{l-3}) \cdot \phi(A_{l-1}|A_{l-2})). \qquad (27)$$

By Equation (7), $\sum_{A_{i+1}} \phi(A_{i+1}|A_i) \cdot \ldots \cdot \sum_{A_{l-2}} \phi(A_{l-2}|A_{l-3}) \cdot \phi(A_{l-1}|A_{l-2})$ yields $\phi(A_{l-1}|A_i)$. Thus, Equation (27) can be rewritten as:

$$\phi(A_l|A_{l-1}) \cdot \phi(A_j|A_l) \cdot \sum_{A_i} \phi(A_{l-1}|A_i). \qquad (28)$$

By Equation (18),

$$\psi_2(A_{l-1}, A_l, A_j) = \phi(A_l|A_{l-1}) \cdot \phi(A_j|A_l).$$ (29)

Substituting Equation (29) into Equation (28), we obtain:

$$\psi_2(A_{l-1}, A_l, A_j) \cdot \sum_{A_i} \phi(A_{l-1}|A_i).)$$ (30)

By Equations (24) - (30),

$$\sum_{A_i,\ldots,A_{l-2}} \psi_1(A_i, \ldots, A_{l-1}, A_l, A_j) = \psi_2(A_{l-1}, A_l, A_j) \cdot \sum_{A_i} \phi(A_{l-1}|A_i).$$

Therefore, $\psi_2(A_{l-1}, A_l, A_j)$ has no more rows than the marginal of $\psi_1(A_i, \ldots, A_{l-1}, A_l, A_j)$ onto variables $\{A_{l-1}, A_l, A_j\}$. □

We use the above analysis to show the following two results. Lemma 5 says that Algorithm 2 never performs more multiplications than Algorithm 1 when answering a query. Lemma 6 says the same except for additions.

Lemma 5. *Given a query on $\{A_i, A_j\}$ posed to a RSFG on $U = \{A_1, A_2, \ldots, A_m\}$, Algorithm 2 never performs more multiplications than Algorithm 1.*

Proof. It can be seen from Equations (11) and (16) that Algorithms 1 and 2 use the same number of multiplications to compute the first potential $\psi_1(A_{j-2}, A_{j-1}, A_j)$ and $\psi_2(A_{j-2}, A_{j-1}, A_j)$. Therefore, Algorithm 1 and Algorithm 2 perform the same number of multiplications provided that precisely two potentials need be multiplied to answer a query. On the other hand, Algorithm 2 never performs more multiplications than Algorithm 1 provided that there are at least three potentials to be multiplied. By Lemma 3, $\phi(A_j|A_k)$ is the marginal of $\psi_1(A_k, A_{k+1}, \ldots, A_j)$ onto $\{A_k, A_j\}$. Therefore, all multiplications in Equation (18) performed by Algorithm 2 for computing the certainty $\phi(A_j|A_i)$ must necessarily be performed in Equation (12) by Algorithm 1. It can be similarly shown that Algorithm 2 never performs more multiplications than Algorithm 1 when computing the strength $\phi(A_i, A_j)$ or coverage $\phi(A_i|A_j)$. Therefore, Algorithm 2 never performs more multiplications than Algorithm 1 when answering a query. □

Lemma 6. *Given a query on $\{A_i, A_j\}$ posed to a RSFG on $U = \{A_1, A_2, \ldots, A_m\}$, Algorithm 2 never performs more additions than Algorithm 1.*

Proof. It can be seen from Equations (14) and (19) that Algorithms 1 and 2 use the same number of additions to eliminate the last variable A_{i+1} from the potential $\psi_1(A_i, A_{i+1}, A_j)$ and $\psi_2(A_i, A_{i+1}, A_j)$. Therefore, Algorithm 1 and Algorithm 2 perform the same number of additions provided that precisely one variable need be eliminated to answer a query. On the other hand, Algorithm 2 never performs more additions than Algorithm 1, provided that there are at least two variables to be eliminated. By Lemma 4, $\psi_2(A_{l-1}, A_l, A_j)$ has no more rows than the marginal of $\psi_1(A_i, \ldots, A_{l-1}, A_l, A_j)$ onto $\{A_{l-1}, A_l, A_j\}$. Therefore, summing out A_l from $\psi_2(A_{l-1}, A_l, A_j)$ combines no more rows than needed from $\psi_1(A_i, \ldots, A_{l-1}, A_l, A_j)$. Since combining n rows requires $n - 1$ additions, Algorithm 2 never performs more additions than Algorithm 1 for computing the certainty $\phi(A_j|A_i)$. That Algorithm 2 never performs more additions than Algorithm 1 when computing the strength $\phi(A_i, A_j)$ or coverage $\phi(A_i|A_j)$ follows in a similar fashion. Therefore, Algorithm 2 never performs more additions than Algorithm 1 when answering a query.   □

Lemmas 5 and 6 indicate that Algorithm 2 never performs more work than Algorithm 1.

7   Other Remarks on Rough Set Flow Graphs

One salient feature of rough sets is that they serve as a tool for uncertainty management without making assumptions regarding the problem domain. On the contrary, we establish in this section that RSFGs, in fact, make implicit independency assumptions regarding the problem domain.

Two tables $\phi_1(A_i, A_j)$ and $\phi_2(A_j, A_k)$ are *pairwise consistent* [3,13], if

$$\phi_1(A_j) = \phi_2(A_j). \tag{31}$$

Example 10. In Table 3, $\phi_1(M, D)$ and $\phi_2(D, A)$ are pairwise consistent. For instance, $\phi_1(D = \text{``Alice''}) = 0.170 = \phi_2(D = \text{``Alice''})$.

Consider $m - 1$ potentials $\phi_1(A_1, A_2), \phi_2(A_2, A_3), \ldots, \phi_{m-1}(A_{m-1}, A_m)$, such that each consecutive pair is pairwise consistent, namely,

$$\phi_i(A_{i+1}) = \phi_{i+1}(A_{i+1}), \tag{32}$$

for $i = 1, 2, \ldots, m - 2$. Observe that the schemas of these decision tables form an *acyclic hypergraph* [1]. Dawid and Lauritzen [3] have shown that if a given set of potentials satisfies Equation (32) and are defined over an acyclic hypergraph, then the potentials are marginals of a unique potential $\phi(A_1, A_2, \ldots, A_m)$, defined as:

$$\phi(A_1, A_2, \ldots, A_m) = \frac{\phi_1(A_1, A_2) \cdot \phi_2(A_2, A_3) \cdot \ldots \cdot \phi_{m-1}(A_{m-1}, A_m)}{\phi_1(A_2) \cdot \ldots \cdot \phi_{m-2}(A_{m-1})}. \tag{33}$$

In [7,8], the flow conservation assumption is made. This means that a given set of $m - 1$ decision tables $\phi_1(A_1, A_2), \phi_2(A_2, A_3), \ldots, \phi_{m-1}(A_{m-1}, A_m)$ satisfies Equation (32). By [3], these potentials are marginals of a unique potential $\phi(A_1, A_2, \ldots, A_m)$ defined by Equation (33), which we will call the *collective potential*. The collective potential $\phi(A_1, A_2, \ldots, A_m)$ represents the problem domain from a rough set perspective.

In order to test whether independencies are assumed to hold, it is necessary to normalize $\phi(A_1, A_2, \ldots, A_m)$. (Note that the normalization process has been used in [7,8].) Normalizing $\phi(A_1, A_2, \ldots, A_m)$ yields a jpd $p(A_1, A_2, \ldots, A_m)$ by multiplying $1/N$, where N denotes the number of all cases. It follows from Equation (33) that

$$
\begin{aligned}
p(A_1, A_2, \ldots, A_m) &= \frac{1}{N} \cdot \phi(A_1, A_2, \ldots, A_m) \\
&= \frac{1}{N} \cdot \frac{\phi_1(A_1, A_2) \cdot \phi_2(A_2, A_3) \cdot \ldots \cdot \phi_{m-1}(A_{m-1}, A_m)}{\phi_1(A_2) \cdot \ldots \cdot \phi_{m-2}(A_{m-1})}. \quad (34)
\end{aligned}
$$

We now show that RSFGs make implicit independency assumptions regarding the problem domain.

Theorem 5. *Consider a RSFG defined by $m - 1$ decision tables $\phi_1(A_1, A_2)$, $\phi_2(A_2, A_3), \ldots, \phi_{m-1}(A_{m-1}, A_m)$. Then $m - 2$ probabilistic independencies $I(A_1, A_2, A_3 \ldots A_m), I(A_1 A_2, A_3, A_4 \ldots A_m), \ldots, I(A_1 \ldots A_{m-2}, A_{m-1}, A_m)$ are satisfied by the jpd $p(A_1, A_2, \ldots, A_m)$, where $p(A_1, A_2, \ldots, A_m)$ is the normalization of collective potential $\phi(A_1, A_2, \ldots, A_m)$ representing the problem domain.*

Proof. Consider $I(A_1, A_2, A_3 \ldots A_m)$. By Equation (34), let

$$
\phi'(A_1, A_2) = \phi_1(A_1, A_2) \quad (35)
$$

and

$$
\phi''(A_2, A_3, \ldots, A_m) = \frac{1}{N} \cdot \frac{\phi_2(A_2, A_3) \cdot \ldots \cdot \phi_{m-1}(A_{m-1}, A_m)}{\phi_1(A_2) \cdot \ldots \cdot \phi_{m-2}(A_{m-1})}. \quad (36)
$$

By substituting Equations (35) and (36) into Equation (34),

$$
p(A_1, A_2, \ldots, A_m) = \phi'(A_1, A_2) \cdot \phi''(A_2, A_3, \ldots, A_m). \quad (37)
$$

By Theorem 1, Equation (37) indicates that $I(A_1, A_2, A_3 \ldots A_m)$ holds. It can be similarly shown that $I(A_1 A_2, A_3, A_4 \ldots A_m), \ldots, I(A_1 \ldots A_{m-2}, A_{m-1}, A_m)$ are also satisfied by the jpd $p(A_1, A_2, \ldots, A_m)$. □

Example 11. Decision tables $\phi(M, D), \phi(D, A), \phi(A, S)$ and $\phi(S, P)$ in Table 2 satisfy Equation (32) and are defined over an acyclic hypergraph $\{MD, DA, AS, SP\}$. This means they are marginals of a unique collective potential,

$$
\phi(M, D, A, S, P) = \frac{\phi(M, D) \cdot \phi(D, A) \cdot \phi(A, S) \cdot \phi(S, P)}{\phi(D) \cdot \phi(A) \cdot \phi(S)}. \quad (38)
$$

The normalization of $\phi(M, D, A, S, P)$ is a jpd $p(M, D, A, S, P)$,

$$
p(M, D, A, S, P) = \frac{1}{1000} \cdot \frac{\phi(M, D) \cdot \phi(D, A) \cdot \phi(A, S) \cdot \phi(S, P)}{\phi(D) \cdot \phi(A) \cdot \phi(S)}, \quad (39)
$$

where the number of all cases $N = 1000$. To show $I(M, D, ASP)$ holds, let

$$
\phi'(M, D) = \phi(M, D) \quad (40)
$$

and

$$\phi''(D, A, S, P) = \frac{1}{1000} \cdot \frac{\phi(D, A) \cdot \phi(A, S) \cdot \phi(S, P)}{\phi(D) \cdot \phi(A) \cdot \phi(S)}. \tag{41}$$

Substituting Equations (40) and (41) into Equation (39),

$$p(M, D, A, S, P) = \phi'(M, D) \cdot \phi''(D, A, S, P). \tag{42}$$

By Theorem 1, the independence $I(M, D, ASP)$ holds in $p(M, D, A, S, P)$. It can be similarly shown that $I(MD, A, SP)$ and $I(MDA, S, P)$ are also satisfied by $p(M, D, A, S, P)$.

The important point is that the flow conservation assumption [7] used in the construction of RSFGs implicitly implies probabilistic conditional independencies holding in the problem domain.

8   Conclusion

Pawlak [7,8] recently introduced the notion of rough set flow graph (RSFGs) as a graphical framework for reasoning from data. In this paper, we established that the RSFG inference algorithm suggested in [7,8] has exponential time complexity. The root cause of the computational explosion is a failure to exploit the factorization defined by a RSFG during inference. We proposed a new RSFG algorithm exploiting the factorization. We showed its correctness and established its time complexity is polynomial with respect to number of nodes in a RSFG. In addition, we showed that it never performs more work than the traditional algorithm [7,8]. These are important results, since they indicate that RSFGs are an efficient framework for uncertainty management. Finally, our study has revealed that RSFGs, unlike previous rough set research, make implicit independency assumptions regarding the problem domain. Future work will report on the complexity of the inference in generalized RSFGs [4]. As the order in which variables are eliminated affects the amount of computation performed [6], we will also investigate this issue in RSFGs.

Acknowledgments

The authors would like to thank A. Skowron for constructive comments and suggestions.

References

1. Beeri, C., Fagin, R., Maier, D. and Yannakakis, M.: On The Desirability of Acyclic Database Schemes. Journal of the ACM, 30(3) (1983) 479-513
2. Butz, C.J., Yan, W. and Yang, B.: The Computational Complexity of Inference Using Rough Set Flow Graphs. The Tenth International Conference on Rough Sets, Fuzzy Sets, Data Mining, and Granular Computing, Vol. 1 (2005) 335-344

3. Dawid, A.P. and Lauritzen, S.L.: Hyper Markov Laws in The Statistical Analysis of Decomposable Graphical Models. The Annals of Satistics, Vol. 21 (1993) 1272-1317
4. Greco, S., Pawlak, Z. and Slowinski, R.: Generalized Decision Algorithms, Rough Inference Rules and Flow Graphs. The Third International Conference on Rough Sets, and Current Trends in Computing (2002) 93-104
5. Hajek, P., Havranek T. and Jirousek R.: Uncertain Information Processing in Expert System. (1992)
6. Madson, A.L. and Jensen, F.V.: Lazy Propagation: A Junction Tree Inference Algorithm based on Lazy Evaluation, Artificial Intelligence, 113 (1-2) (1999) 203-245.
7. Pawlak, Z.: Flow Graphs and Decision Algorithms. The Ninth International Conference on Rough Sets, Fuzzy Sets, Data Mining, and Granular Computing (2003) 1-10
8. Pawlak, Z.: In Pursuit of Patterns in Data Reasoning from Data - The Rough Set Way. The Third International Conference on Rough Sets, and Current Trends in Computing (2002) 1-9
9. Pawlak, Z.: Rough Sets. International Journal of Computer and Information Sciences, Vol. 11, Issue 5 (1982) 341-356
10. Pawlak, Z.: Rough Sets: Theoretical Aspects of Reasoning about Data. Kluwer Academic (1991)
11. Pearl, J.: Probabilistic Reasoning in Intelligent Systems: Networks of Plausible Inference. Morgan Kaufmann, San Francisco, California (1988)
12. Shafer, G.: Probabilistic Expert Systems. Society for the Institute and Applied Mathematics, Philadelphia (1996)
13. Wong, S.K.M., Butz, C.J. and Wu, D.: On the Implication Problem for Probabilistic Conditional Independency, IEEE Transactions on Systems, Man, and Cybernetics, Part A: Systems and Humans, Vol. 30, Issue 6. (2000) 785-805

Appendix I

Table 5. A jpd $p(U)$ is shown in Tables 5 and 6, where $U = \{M, D, A, S, P\}$

M	D	A	S	P	p(U)	M	D	A	S	P	p(U)
Toyota	Alice	Old	High	Executive	0.017280	Toyota	Dave	Old	Medium	Staff	0.000540
Toyota	Alice	Old	High	Staff	0.003672	Toyota	Dave	Old	Medium	Manager	0.000042
Toyota	Alice	Old	High	Manager	0.000648	Toyota	Dave	Old	Low	Executive	0.000002
Toyota	Alice	Old	Medium	Executive	0.000324	Toyota	Dave	Old	Low	Staff	0.000008
Toyota	Alice	Old	Medium	Staff	0.009720	Toyota	Dave	Old	Low	Manager	0.000190
Toyota	Alice	Old	Medium	Manager	0.000756	Toyota	Dave	Middle	High	Executive	0.000960
Toyota	Alice	Old	Low	Executive	0.000036	Toyota	Dave	Middle	High	Staff	0.000204
Toyota	Alice	Old	Low	Staff	0.000144	Toyota	Dave	Middle	High	Manager	0.000036
Toyota	Alice	Old	Low	Manager	0.003420	Toyota	Dave	Middle	Medium	Executive	0.000108
Toyota	Alice	Middle	High	Executive	0.011520	Toyota	Dave	Middle	Medium	Staff	0.003240
Toyota	Alice	Middle	High	Staff	0.002448	Toyota	Dave	Middle	Medium	Manager	0.000252
Toyota	Alice	Middle	High	Manager	0.000432	Toyota	Dave	Middle	Low	Executive	0.000012
Toyota	Alice	Middle	Medium	Executive	0.001296	Toyota	Dave	Middle	Low	Staff	0.000048
Toyota	Alice	Middle	Medium	Staff	0.003888	Toyota	Dave	Middle	Low	Manager	0.001140
Toyota	Alice	Middle	Medium	Manager	0.003024	Toyota	Dave	Young	High	Executive	0.000960
Toyota	Alice	Middle	Low	Executive	0.000144	Toyota	Dave	Young	High	Staff	0.000204
Toyota	Alice	Middle	Low	Staff	0.000576	Toyota	Dave	Young	High	Manager	0.000036
Toyota	Alice	Middle	Low	Manager	0.013680	Toyota	Dave	Young	Medium	Executive	0.000072
Toyota	Alice	Young	High	Executive	0.000960	Toyota	Dave	Young	Medium	Staff	0.002160
Toyota	Alice	Young	High	Staff	0.000204	Toyota	Dave	Young	Medium	Manager	0.000168
Toyota	Alice	Young	High	Manager	0.000036	Toyota	Dave	Young	Low	Executive	0.000084
Toyota	Alice	Young	Medium	Executive	0.000072	Toyota	Dave	Young	Low	Staff	0.000336
Toyota	Alice	Young	Medium	Staff	0.002160	Toyota	Dave	Young	Low	Manager	0.007980
Toyota	Alice	Young	Medium	Manager	0.000168	Honda	Bob	Old	High	Executive	0.028800
Toyota	Alice	Young	Low	Executive	0.000084	Honda	Bob	Old	High	Staff	0.006120
Toyota	Alice	Young	Low	Staff	0.000336	Honda	Bob	Old	High	Manager	0.001080
Toyota	Alice	Young	Low	Manager	0.007980	Honda	Bob	Old	Medium	Executive	0.000540
Toyota	Bob	Old	High	Executive	0.011520	Honda	Bob	Old	Medium	Staff	0.016200
Toyota	Bob	Old	High	Staff	0.002448	Honda	Bob	Old	Medium	Manager	0.001260
Toyota	Bob	Old	High	Manager	0.000432	Honda	Bob	Old	Low	Executive	0.000060
Toyota	Bob	Old	Medium	Executive	0.000216	Honda	Bob	Old	Low	Staff	0.000240
Toyota	Bob	Old	Medium	Staff	0.006480	Honda	Bob	Old	Low	Manager	0.005700
Toyota	Bob	Old	Medium	Manager	0.000504	Honda	Bob	Middle	High	Executive	0.014400
Toyota	Bob	Old	Low	Executive	0.000024	Honda	Bob	Middle	High	Staff	0.003060
Toyota	Bob	Old	Low	Staff	0.000096	Honda	Bob	Middle	High	Manager	0.000540
Toyota	Bob	Old	Low	Manager	0.002280	Honda	Bob	Middle	Medium	Executive	0.001620
Toyota	Bob	Middle	High	Executive	0.005760	Honda	Bob	Middle	Medium	Staff	0.048600
Toyota	Bob	Middle	High	Staff	0.001224	Honda	Bob	Middle	Medium	Manager	0.003780
Toyota	Bob	Middle	High	Manager	0.000216	Honda	Bob	Middle	Low	Executive	0.000180
Toyota	Bob	Middle	Medium	Executive	0.000648	Honda	Bob	Middle	Low	Staff	0.000720
Toyota	Bob	Middle	Medium	Staff	0.019440	Honda	Bob	Middle	Low	Manager	0.017100
Toyota	Bob	Middle	Medium	Manager	0.001512	Honda	Carol	Middle	High	Executive	0.014400
Toyota	Bob	Middle	Low	Executive	0.000072	Honda	Carol	Middle	High	Staff	0.003060
Toyota	Bob	Middle	Low	Staff	0.000288	Honda	Carol	Middle	High	Manager	0.000540
Toyota	Bob	Middle	Low	Manager	0.006840	Honda	Carol	Middle	Medium	Executive	0.001620
Toyota	Dave	Old	High	Executive	0.000960	Honda	Carol	Middle	Medium	Staff	0.048600
Toyota	Dave	Old	High	Staff	0.000204	Honda	Carol	Middle	Medium	Manager	0.003780
Toyota	Dave	Old	High	Manager	0.000036	Honda	Carol	Middle	Low	Executive	0.000180
Toyota	Dave	Old	Medium	Executive	0.000018	Honda	Carol	Middle	Low	Staff	0.000720

Table 6. A jpd $p(U)$ is shown in Tables 5 and 6, where $U = \{M, D, A, S, P\}$

M	D	A	S	P	$p(U)$	M	D	A	S	P	$p(U)$
Honda	Carol	Middle	Low	Manager	0.017100	Ford	Bob	Middle	Medium	Staff	0.048600
Honda	Carol	Young	High	Executive	0.004800	Ford	Bob	Middle	Medium	Manager	0.003780
Honda	Carol	Young	High	Staff	0.001020	Ford	Bob	Middle	Low	Executive	0.000180
Honda	Carol	Young	High	Manager	0.000180	Ford	Bob	Middle	Low	Staff	0.000720
Honda	Carol	Young	Medium	Executive	0.000360	Ford	Bob	Middle	Low	Manager	0.017100
Honda	Carol	Young	Medium	Staff	0.010800	Ford	Carol	Middle	High	Executive	0.004800
Honda	Carol	Young	Medium	Manager	0.000840	Ford	Carol	Middle	High	Staff	0.001020
Honda	Carol	Young	Low	Executive	0.000420	Ford	Carol	Middle	High	Manager	0.000180
Honda	Carol	Young	Low	Staff	0.001680	Ford	Carol	Middle	Medium	Executive	0.000540
Honda	Carol	Young	Low	Manager	0.039900	Ford	Carol	Middle	Medium	Staff	0.016200
Ford	Alice	Old	High	Executive	0.007200	Ford	Carol	Middle	Medium	Manager	0.001260
Ford	Alice	Old	High	Staff	0.001530	Ford	Carol	Middle	Low	Executive	0.000060
Ford	Alice	Old	High	Manager	0.000270	Ford	Carol	Middle	Low	Staff	0.000240
Ford	Alice	Old	Medium	Executive	0.000135	Ford	Carol	Middle	Low	Manager	0.005700
Ford	Alice	Old	Medium	Staff	0.004050	Ford	Carol	Young	High	Executive	0.001600
Ford	Alice	Old	Medium	Manager	0.000315	Ford	Carol	Young	High	Staff	0.000340
Ford	Alice	Old	Low	Executive	0.000015	Ford	Carol	Young	High	Manager	0.000060
Ford	Alice	Old	Low	Staff	0.000060	Ford	Carol	Young	Medium	Executive	0.000120
Ford	Alice	Old	Low	Manager	0.001425	Ford	Carol	Young	Medium	Staff	0.003600
Ford	Alice	Middle	High	Executive	0.004800	Ford	Carol	Young	Medium	Manager	0.000280
Ford	Alice	Middle	High	Staff	0.001020	Ford	Carol	Young	Low	Executive	0.000140
Ford	Alice	Middle	High	Manager	0.000180	Ford	Carol	Young	Low	Staff	0.000560
Ford	Alice	Middle	Medium	Executive	0.000540	Ford	Carol	Young	Low	Manager	0.013300
Ford	Alice	Middle	Medium	Staff	0.016200	Ford	Dave	Old	High	Executive	0.012000
Ford	Alice	Middle	Medium	Manager	0.001260	Ford	Dave	Old	High	Staff	0.002550
Ford	Alice	Middle	Low	Executive	0.000060	Ford	Dave	Old	High	Manager	0.000450
Ford	Alice	Middle	Low	Staff	0.000240	Ford	Dave	Old	Medium	Executive	0.000225
Ford	Alice	Middle	Low	Manager	0.005700	Ford	Dave	Old	Medium	Staff	0.006750
Ford	Alice	Young	High	Executive	0.000400	Ford	Dave	Old	Medium	Manager	0.000525
Ford	Alice	Young	High	Staff	0.000085	Ford	Dave	Old	Low	Executive	0.000025
Ford	Alice	Young	High	Manager	0.000015	Ford	Dave	Old	Low	Staff	0.000100
Ford	Alice	Young	Medium	Executive	0.000030	Ford	Dave	Old	Low	Manager	0.002375
Ford	Alice	Young	Medium	Staff	0.000900	Ford	Dave	Middle	High	Executive	0.012000
Ford	Alice	Young	Medium	Manager	0.000070	Ford	Dave	Middle	High	Staff	0.002550
Ford	Alice	Young	Low	Executive	0.000035	Ford	Dave	Middle	High	Manager	0.000450
Ford	Alice	Young	Low	Staff	0.000140	Ford	Dave	Middle	Medium	Executive	0.001350
Ford	Alice	Young	Low	Manager	0.003325	Ford	Dave	Middle	Medium	Staff	0.040500
Ford	Bob	Old	High	Executive	0.028800	Ford	Dave	Middle	Medium	Manager	0.003150
Ford	Bob	Old	High	Staff	0.006120	Ford	Dave	Middle	Low	Executive	0.000150
Ford	Bob	Old	High	Manager	0.001080	Ford	Dave	Middle	Low	Staff	0.000600
Ford	Bob	Old	Medium	Executive	0.000540	Ford	Dave	Middle	Low	Manager	0.014250
Ford	Bob	Old	Medium	Staff	0.016200	Ford	Dave	Young	High	Executive	0.012000
Ford	Bob	Old	Medium	Manager	0.001260	Ford	Dave	Young	High	Staff	0.002550
Ford	Bob	Old	Low	Executive	0.000060	Ford	Dave	Young	High	Manager	0.000450
Ford	Bob	Old	Low	Staff	0.000240	Ford	Dave	Young	Medium	Executive	0.000900
Ford	Bob	Old	Low	Manager	0.005700	Ford	Dave	Young	Medium	Staff	0.027000
Ford	Bob	Middle	High	Executive	0.014400	Ford	Dave	Young	Medium	Manager	0.002100
Ford	Bob	Middle	High	Staff	0.003060	Ford	Dave	Young	Low	Executive	0.001050
Ford	Bob	Middle	High	Manager	0.000540	Ford	Dave	Young	Low	Staff	0.004200
Ford	Bob	Middle	Medium	Executive	0.001620	Ford	Dave	Young	Low	Manager	0.099750

Intelligent Algorithms for Movie Sound Tracks Restoration

Andrzej Czyżewski, Marek Dziubiński, Łukasz Litwic,
and Przemysław Maziewski

Multimedia Systems Department
Gdansk University of Technology
ul. Narutowicza 11/12, 80-952 Gdańsk, Poland
{andcz, mdziubin, llitwic, przemas}@sound.eti.pg.gda.pl

Abstract. Two algorithms for movie sound tracks restoration are discussed in the paper. The first algorithm is the unpredictability measure computation applied to the psychoacoustic model-based broadband noise attenuation. A learning decision algorithm, based on a neural network, is employed for determining useful audio signal components acting as maskers of the noisy spectral parts. An application of the rough set decision system to this task is also considered. An iterative method for calculating the sound masking pattern is presented. The second of presented algorithms is the routine for precise evaluation of parasite frequency modulations (wow) utilizing sinusoidal components extracted from the sound spectrum. The results obtained employing proposed intelligent signal processing algorithms, as well as the relationship between both routines, will be presented and discussed in the paper.

Keywords: Audio restoration, noise reduction, wow evaluation.

1 Introduction

Noise is a common disturbance in archival recordings and a suitable solution is presented in this paper. Acoustic noise reduction is a subject of extensive research, carried out in the last decades. Several approaches during this time were studied, such as adaptive filtering [1,2,3], autocorrelation [4,5] and statistical methods [6,7,8], parametric models for spectrum estimation [9,10], and some techniques based on intelligent algorithms (including rough set - based audio signal enhancement approach) have been investigated in recent years [11,12]. In addition, multi-channel representation of signals was considered, with regard to microphone matrices [13,14,15]. The main stream approaches, which are based on assumption that undistorted signal is not correlated with parasite noise and the noise is stationary, of additive type are: Kalman [10][16][17] and Wiener filtration [18,19], stochastic modeling of the signals [20,21] and spectral subtraction [22,23]. These methods however, do not utilize concepts of perceptual filtration, thus they do not take into account some subjective properties of the human auditory system [24].

J.F. Peters and A. Skowron (Eds.): Transactions on Rough Sets V, LNCS 4100, pp. 123–145, 2006.

The engineered algorithm, described in this paper, utilizes phenomena related to sound perception and is based on perceptual filtering. In addition, the restoration process is controlled by an intelligent algorithm. The intelligent reasoning, based on a neural network, is the core decision unit responsible for classifying noisy patterns. The approach has been presented in authors earlier paper [25]. Application of perceptual filtering has been exploited by several researchers [26], for various audio processing applications, such as audio coding standards [27,28]. As it was demonstrated in our earlier work [29], utilizing concepts of perceptual filtering for noise removal may be very effective. However such an approach requires employing a precise sound perception modeling [30], rather than implementation of the simplified one exploited in the MPEG coding standard [27].

Another problem related to archive audio recorded in movie sound tracks is parasitic frequency modulation (FM) originated from motor speed fluctuations, tape damages and inappropriate editing techniques. This kind of distortion is usually defined as wow or flutter or modulation noise, depending on the frequency range of the parasitic modulation frequency. Wow defect is typically defined as frequency modulation in the range up to 6Hz, flutter is the frequency modulation between 6-15Hz and modulation noise (or scrape flutter)describes 96Hz frequency modulation. In this paper we will focus on the wow defect.

As particularly wow leads to undesirable changes of all of the sound frequency components, sinusoidal sound analysis originally proposed by McAulay and Quatieri [31] was found to be very useful in the defects evaluation. In such an approach tracks depicting tonal components changes (MQ tracks) are determined to obtain a precise wow characteristic [32][33]. The statistical methods for post-processing of MQ tracks for monophonic audio signals were introduced by Godsill and Rayner [34][35][36]. Their approach is build on three processing steps. Firstly, a DFT magnitude-based peak tracking algorithm is used for tonal component estimation. It provides a set of harmonic components trajectories. This set constitutes data for the FM estimation which leads to the second processing step. Each trajectory is denoted by its center frequency being unknown and considered as varying one. The variations are attributed both to the investigated FM (tracking misleading) and unknown noisy-like components (e.g. genuine musical pitch deviations). The noise-like components are assumed to be independent, identically distributed Gaussian processes. Thereby, the likelihood function of the unknown center frequencies and the parasite FM are obtained. Further the maximum likelihood solution (ML) is used for wow characteristic evaluation. Details of this method can be found in the papers [34][35][36]. Additionally a priori information on wow distortion can be introduced through Bayesian probability framework. The maximum a posteriori estimator (MAP) was proposed by Godsill [33]. In the last processing step, the estimated wow characteristic enables signal restoration. The sinc based non-uniform resampling was used for this purpose. For details on the non-uniform resampling in wow restoration the literature can be inquired [38].

A method for statistical processing of MQ trajectories for polyphonic audio signals was introduced by Walmsley, Godsill and Rayner. The method is an addition to the Bayesian approach to the monophonic signals. The parasite FM are modeled using latent variables estimated jointly across numerous neighboring frames. At the end the Markov chain the Monte Carlo methods are used for parameters estimation [34].

The MQ-based approach to wow characteristic evaluation was also studied by authors of this paper. However, because of the observation that wow can be a non-periodic distortion [37], in authors' approach the emphasis was put not only on the statistically-based post-processing of the MQ tracks, but also on the precise estimation of tonal components variations. Regardless of the MQ-based routines also some new algorithms for wow evaluation, e.g., the novel time-based autocorrelation analysis, were proposed by authors of this paper. More details are described in other papers [38,39,40,41,42,43].

Algorithms for wow evaluation were presented also by Nichols [44]. The methods strengthened by the non-uniform resampler are available in the application called JPITCH. Two of the included automatic wow evaluation methods are devoted to short-term pitch defects (i.e. with time duration less than 1 second). The first method for short-term defects processing is similar to the concept of the MQ tracks analysis utilized by Godsill and Rayner. Nichols however, introduces an iterative procedure for accurate peak frequencies estimation. The second method for short-term FM processing is based on a novel concept of graphical processing of a spectrogram. The spectrogram is searched for peaks which, after excluding some false elements, are joined to form trajectories. Then the set of tracks is post-processed to obtain the wow characteristic. The third method proposed by Nichols concentrates on the long-term distortions (much longer than 1 second, i.e., drift). The algorithm is frame-based. In each one-second segment, the detune estimate is calculated as weighted product of the highest magnitude peaks. The results are filtered to remove false oscillations. The results, obtained using one the 3 algorithms, allows for signal restoration performed by means of incommensurate sampling rate conversion.

The recording media features were utilized in wow evaluation. Wolfe and Howarth presented a methodology for parasite FM cancelation in analogue tape recordings [45,46]. The proposed procedures utilize information about the carrier speed variation extracted from the magnetic bias signal.

The magnetic bias as well as the power line hum, which is also suitable for wow extraction, were studied by authors of this paper [39,40]. In authors' approach both artefacts are read synchronously with the audio content and are post-processed to obtain the wow characteristic. Based on this information the restoration is performed using non-uniform resampling [38].

Notwithstanding all of the cited proposals there is still a need for further algorithmic approach to the wow restoration as it can be very complex sharing periodic or accidental nature. Therefore this paper, addresses the problem of wow extraction, in this case also employing soft computing.

Although both distortions (wow and noise), seem to be significantly different in terms of algorithmic solutions used for restoration, and also in terms of perception,they both must be considered simultaneously in the signal reconstruction process. It is necessary, due to the fact that wow (if introduced first), may disturb stationary character of noise, and may affect noise reduction algorithm performance. In case of such a scenario, it is necessary to carry out wow compensation as the first step, and noise reduction as the next one, assuming that the wow compensation algorithm will restore the stationary character of noise. Alternatively, if the additive and stationary noise have been introduced before the wow modulation occurred, the processing stages must be reversed (i.e. noise should be removed first), because demodulation process influences stationary character of noise. It must be stressed that only non-white noise is sensitive to the parasite modulation disturbance (simply because its frequency characteristic is not flat, thus shifting spectral components by frequency modulation locally affects noise spectrum shape.This will be demonstrated in the experiments section. It is also important to mention that for the case of archival recordings, acoustic noise almost always has the non-white character. The unpredictability measure which is an important factor allowing noise recognition and reduction can be also interpreted in terms of rough set theory. This observation is bridging the gap between the meaning of uncertainty in signal processing and the notion of uncertainty in some soft computing methods.

2   Noise Removal Algorithm

Modeling of the basiliar membrane behavior, especially masking phenomena occurring on this membrane, is one of the most important feature of the contemporary audio coding standards [27,28], although it may be applied also for the noise reduction purposes [29,30]. More detailed information on psychoacoustics principles of signal processing can be found in abundant literature [24][47][48][49] and also in our papers [25][30].

The tonality descriptors play a very significant role in perceptual filtering. Application of the precise perceptual model [30], requires calculation of the Unpredictability Measure parameter [49] for each spectrum bin, in each processing frame, based on which it is then possible to calculate the masking offset. The masking offset for the excitation of b_x Barks at frequency of b_x Barks is given by the formula:

$$O_{k,x} = \alpha_k^t \cdot (14.5 + bark(x)) + (1 - \alpha_k^t) \cdot 5.5. \tag{1}$$

The tonality index α_k^t of the excitation of b_x Barks is assumed to be related directly to the *Unpredictability Measure* parameter ($\alpha_k^t = c_k^t$), where c_k^t is calculated in the following way:

$$c_k^t = \frac{\sqrt{(r_k^t \cdot \cos \Phi_k^t - \hat{r}_k^t \cdot \cos \hat{\Phi}_k^t)^2 + (r_k^t \cdot \sin \Phi_k^t - \hat{r}_k^t \cdot \sin \hat{\Phi}_k^t)^2}}{r_k^t + |\hat{r}_k^t|}. \tag{2}$$

for r_k^t denoting spectral magnitude and Φ_k^t denoting phase, both at time t, while $\hat{r}_k^t$ and $\hat{\Phi}_k^t$ represent the predicted values of Φ_k^t, and are referred to the past information (calculated for two previous signal sample frames):

$$
\begin{cases} \hat{r}_k^t = r_k^{t-1} + (r_k^{t-1} - r_k^{t-2}) \\ \hat{\Phi}_k^t = \Phi_k^{t-1} + (\Phi_k^{t-1} - \Phi_k^{t-2}) \end{cases} \Rightarrow \begin{cases} \hat{r}_k^t = 2r_k^{t-1} - r_k^{t-2}, \\ \hat{\Phi}_k^t = 2\Phi_k^{t-1} - \Phi_k^{t-2}. \end{cases} \tag{3}
$$

Thus, based on the literature [47], the masking threshold of the Basilar membrane T, stimulated by the single excitation of b_x Barks and of magnitude equal to S_x is calculated with regard to:

$$
\begin{cases} T_{i,x} = S_i \cdot 10^{-s_1 \cdot (b_x - b_i)/10 - O_{i,x}}, & b_x \leq b_i, \\ T_{j,x} = S_j \cdot 10^{-s_2 \cdot (b_j - b_x)/10 - O_{j,x}}, & b_x > b_j. \end{cases} \tag{4}
$$

where S_i, S_j are magnitudes related to excitations b_i, b_j and global masking threshold is obtained by summing up all of individual excitations.

2.1   Perceptual Noise Reduction System

The perceptual noise reduction system [25], requires an assumption that the acoustic noise is of additive type. The spectral representation of the disturbance is obtained with the use of spectral subtraction techniques [50]. Because noise suppression in this approach is based on masking some spectral components of the disturbing noise, it is necessary to determine which components should be masked and which should act as maskers. For this reason, so called rough estimate $\hat{X}^{ref}(j\omega)$ of the clean signal's spectrum is obtained with accordance to spectral subtraction method [50] based on the iterative algorithm represented by the Noise Masking block in Fig. 1.

The algorithm was proposed earlier[29,30], however it was recently improved and extended with a learning decision algorithm. The new *Decision System* module [25] containing a neural network is responsible for determining which components are going to be treated as maskers U(*useful* components), and which represent distortions and are going to be masked D (*useless* components). The basic classification (without neural network application described in Sect. 4) can be carried out on the basis of the following expressions:

$$
U = \{\hat{X}_i^{ref}; \quad |\hat{X}_i^{ref}| > T_i^{ref} \wedge |Y_i| > T_i^Y, 1 \leq i \leq N/2\}. \tag{5}
$$

$$
D = \{Y_i; \quad |\hat{X}_i^{ref}| \leq T_i^{ref} \vee |Y_i| \leq T_i^Y, 1 \leq i \leq N/2\}. \tag{6}
$$

where i denotes spectrum component indexes, U and D are sets containing useful and useless information. T^{ref} is the masking threshold caused by the presence of $\hat{X}^{ref}(j\omega)$, and T^Y is the masking threshold of the input signal: $Y^{ref}(j\omega)$.

Lowering of the masking threshold preserves more noise of the input signal, so the influence of the reconstruction filter is significantly smaller than it is in

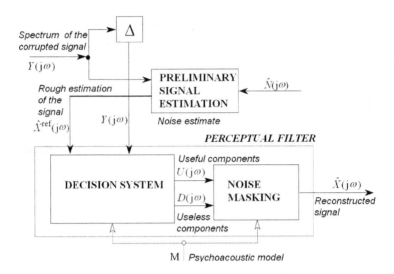

Fig. 1. General lay-out of noise reduction system

case of the uplifting method, giving less distorted output signal. Modified global masking threshold T_x^β at barks can be calculated with regard to formula:

$$T_x^\beta = \sum_{j \in U_L(x)} T_{j,x} + \sum_{j \in D_L(x)} T_{j,x}^\beta + \sum_{i \in U_H(x)} T_{i,x} + \sum_{i \in D_H(x)} T_{i,x}^\beta. \tag{7}$$

where $T_{i,x}^\beta$ and $T_{j,x}^\beta$ represent new masking thresholds, caused by reduced single excitations and β is vector containing reduction factor values for the noisy components. $U_L(x)$ and $U_H(x)$ (similarly $D_L(x)$ and $D_H(x)$) denote subset of U (or subset of D) containing elements with frequencies lower or equal (L) to b_x barks, and frequencies higher than b_x barks (H).

Since values of β may differ for the elements of D, and changing each value affects T_x^β, thus it is impractical to calculate all reducing factor values directly. For this reason sub-optimal iterative algorithm was implemented [11].

The value of *StopThreshold* should be larger or equal to 0. In practical experiments choosing *StopThreshold* = 0.01 compromises both noise reduction quality and computational efficiency.

2.2   Unpredictability Measure Application

Calculation of the masking offset, described by Eq. 1 plays a significant role in the masking threshold calculation. In noisy signals, tonal components that are occurring just above the noise floor, may be not very well represented by the *Unpredictability Measure* (*UM*) parameter due to the strong influence of the noisy content. A practical solution to this problem is extending time domain resolution, by increasing overlap of the frames used only for unpredictability calculation. Standard *Unpredictability Measure* (2-3) refers to the fragment of

---

input: Magnitude spectrum of signal S, a set of noisy components D
output: Masking threshold T,Reducing factor values β

Calculating initial masking threshold;
$T_x^\beta \leftarrow$ `CalculateThreshold` ;
$\beta_x \leftarrow \frac{T_x^\beta}{S_x^D}$;
Stop condition;
while $\forall x \in D : \min(\beta_x) < 1 \wedge \max(\beta_x) - \min(\beta_x) <$ StopThreshold **do**
    Reducing distance between noisy components and masking threshold;
    foreach $x \in D$ **do**
      | $S_x^D \leftarrow S_x^D \cdot \beta_x$;
    end
    Calculating current masking threshold and current reducing factor values;
    $T_x^\beta \leftarrow$ `CalculateThreshold` ;
    $\beta_x \leftarrow \frac{T_x^\beta}{S_x^D}$;
end

Algorithm 1. Algorithm for Reducing Factor Values Computation

the signal represented by 3 consecutive frames, i.e. beginning of this fragment (T_{start}) is at the beginning of the frame with $t - 2$ index and the end of the fragment (T_{start}) is at the end of frame with t index, with accordance to (3). Consequently, the same fragment is divided into N equally spaced frames, so that the improved UM can be expressed as:

$$\bar{c}_k^t = \frac{1}{N-2} \sum_{n=1}^{N-2} c_k^{t^n} . \tag{8}$$

where

$$c_k^{t^n} = \frac{dist\left((\hat{r}_k^{t^n}, \hat{\Phi}_k^{t^n}), (r_k^{t^n}, \Phi_k^{t^n})\right)}{r_k^{t^n} + |\hat{r}_k^{t^n}|} . \tag{9}$$

and

$$\begin{cases} \hat{r}_k^{t^n} = r_k^{t^n-1} + (r_k^{t^n-1} - r_k^{t^n-2}) \\ \hat{\Phi}_k^{t^n} = \Phi_k^{t^n-1} + (\Phi_k^{t^n-1} - \Phi_k^{t^n t-2}) \end{cases} \Rightarrow \begin{cases} \hat{r}_k^{t^n} = 2r_k^{t^n-1} - r_k^{t^n-2}, \\ \hat{\Phi}_k^{t^n} = 2\Phi_k^{t^n-1} - \Phi_k^{t^n-2}. \end{cases} \tag{10}$$

while $T_{start} \leq t^n - 2 < t^n - 1 < t^n \leq T_{stop}$ and $c_k^t = \bar{c}_k^t$.

Additionally, classification of the spectrum components in non-linear spectral subtraction, can be extended by some psychoacoustic parameters, i.e. the tonality description values. By analyzing time-frequency domain behavior of the UM vectors calculated for each frame, it is easy to spot tracks representing harmonic content of the signal, even though the simply averaged sequence may not result in very high $\bar{c}_k^t$. Basing on this observation, artificial neural network was deployed as the decision system for classifying, $c_k^{t^n}$ patterns. A set of training data was obtained from the noise fragment and from the noisy signal - c_k^{tn} vectors of the noise represented useless components, while those obtained from the noisy input signal, classified as useful components with standard spectral subtraction algorithm, represented patterns of the useful signal. The neural network used in the training was a feed-forward, back-propagation structure with three layers. The hyperbolic tangent sigmoid transfer function was chosen to activate first two layers, while hard limit transfer function was employed to activate the output layer. The weights and biases, were updated during the training process, according to Levenberg-Marquardt optimization method. A method of controlling the generalization process was also used. Such an approach is very effective for recovering sinusoidal components, however it does not significantly improve recovery of non-tonal components. It allows to increase efficiency of the tonal components detection by 60%, which results in decreased amount of artefacts in processed signals, but does not reflects *Signal-to-Noise-Ratio* (SNR) significantly. As was said, the UM vectors calculated for each frame may represent useful or useless components, depending on the current content of the processed frame. Instead of the neural network also the rough set inference system can be applied to provide the decision about the kind of the frame content. Since the basic rough operators (the partition of a universe into classes of equivalence, C-lower approximation of a set X and calculation of a positive region) can be performed more efficiently when objects are ordered, the applied algorithm often executes sorting of all objects with respect to a set of attributes [11]. The vectors $\bar{c}_k^t$ recorded for consecutive signal frames can be gathered in the form of a decision table T. The values of $c_k^{t^n}$ are conditional attributes and binary values associated to the vectors $\bar{c}_k^t$ provide decision attribute d_k. Therefore the object d_k in the table is described by the following relation:

$$c_k^{t^1}, c_k^{t^2}, ..., c_k^{t^{N-2}} => d_k \qquad (11)$$

The rule discovery procedure based on the rough set principles assumes that only conditional attributes require quantization. The uniform quantization is proposed to that end in the presented concept. In the execution mode, the input vector of noisy audio parameters $\bar{c}_k^t$ is quantized, and then processed by the set of generated rules. Subsequently, the algorithm splits the decision table T into two tables: consisting of only certain rules and of only uncertain rules. For them both, there is additional information associated with every object in them concerning the minimal set of indispensable attributes and the rough measure. The rough measure of the decision rule provides one of the most basic concepts related to rough set theory.

There is an interpretation dependency between the rough measure and the unpredictability measure $\bar{c}_k^t$ as in eq. (8). The notions that can be found in the literature, such as: Measurement Uncertainty, Sampling Uncertainty, Mathematical Modelling Uncertainty, Causal Uncertainty are all related to the problem of making uncertain decisions. The noisy data processing is an evident example of making uncertain decisions, because UM (Unpredictability Measure) represents the margin of uncertainty while interpreting spectrum shape in terms of useful or useless components representation. Similar approach was tried by one of authors in the past [11] [29] bringing evidence that rough set decision system provides a valid alternative to the neural networks applied to this kind of experiments. A fundamental notion of the rough set-based learning system is the need to discover dependencies between given features of a problem to be classified.

The concepts related to rough measure of the rule μ_{RS} are well covered in the literature. This measure associated with each rule is defined as follows:

$$\mu_{RS} = \frac{|X \cap Y|}{|Y|} \qquad (12)$$

where: X- is the concept, and Y- the set of examples described by the rule.

A parameter was defined also allowing one to optimize the rule generation process, e.g. in pattern recognition tasks [10]. This parameter was called the rule strength r and is defined as follows:

$$r = c(\mu_{RS} - n_\mu), \qquad (13)$$

$$n_\mu \in< 0, 1) \qquad (14)$$

where: c - number of cases supporting the condition part of the rule,
n_μ- neutral point of the rough measure.

The neutral point n_μ of the rough measure μ_{RS} is one of the parameters of the rule generation system to be set experimentally by its operator during the testing of the system. This parameter allows the regulation of the influence of possible rules on the process of decision making. The value of n_μ is selected experimentally after building the knowledge base of the system, so that it becomes possible for an operator to adjust the amount of noise to be suppressed through influencing the margin of uncertainty related to qualifying components to disjoint sets of useful and useless ones [11].

3   Parasitic Modulation Compensating Algorithm

The wow defect distorts the tonal structure of contaminated audio signal thus tonal component analysis was found to be very useful for wow defect evaluation [32]. The analysis of tonal components was proposed to be performed by means of sinusoidal modeling approach. This approach expresses the audio signal as a sum of sinusoidal components having slowly-varying frequencies and amplitudes. For the audio signal $x(t)$ the following relation can be shown:

$$x(t) = \sum_{p=1}^{P} a_p(t)cos\big(\phi_p(t)\big). \qquad (15)$$

$$\phi_p(t) = \phi_p(0) + 2\pi \int_0^t f_p(u)du. \qquad (16)$$

where P corresponds to number of sinusoidal components (partials). The parameters a_p and f_p correspond to amplitude and frequency values of partial. The successive values of f_p ,which create the frequency track, also called MQ track, are processed to obtain the wow modulation pattern called *Pitch*Variation Curve (*PVC*) [37]. The block-diagram depicting main stages of sinusoidal modeling approach is presented in Fig. 2. In the first stage (Time-Frequency Analysis),

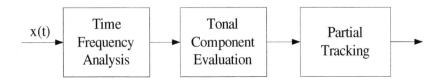

Fig. 2. Block diagram of sinusoidal modeling approach

joint time-frequency analysis is performed. In most applications the Short-time Fourier Transform (*STFT*) is used. Since frame-based processing is usually applied, *DFT* of every time (analysis) frame is computed. The second stage (Tonal Component Evaluation) determines the tonal components in every frame, which are matched to existing tracks or create new ones in the last stage (Partial Tracking).

Sinusoidal modeling is very powerful tool used mainly in additive synthesis, however it must be noticed that sound modeling and analysis is much more straightforward in case of monophonic than in polyphonic sources. Moreover, sounds from archival recordings are likely to be contaminated by several distortions e.g. noises, clicks, which makes the sinusoidal analysis much harder to perform. Finally, the wow distortion itself introduces frequency tracks modulation, which if strong makes tracking process complicated.

For the mentioned reasons, a somewhat different approach to sinusoidal analysis than in the case of additive synthesis applications is proposed. Since wow can be effectively evaluated on the basis of a single component (bias, hum [39,40]) it assumed that also individual tonal components can sufficiently contribute to wow evaluation. Therefore only the most salient tonal components are in the interest of the presented algorithm.

The essential part of the sinusoidal analysis algorithm is the partial tracking stage. The original approach for partial tracking, proposed by McAulay and Quatieri [31], was based on frequency matching:

$$\left| f_k^{i-1} - f_l^i \right| < \Delta_f. \tag{17}$$

where f_k^{i-1} is the frequency of the processed track in frame $i - 1$ and f_l^i is the frequency of matched peak in frame i. The parameter Δ_f (frequency deviation) is the maximum frequency distance between track and its continuation. This approach has been continuously developed including such enhancements like Hidden Markov Models (HMM) [51], or linear prediction (LPC)[52], however in most applications only magnitude spectrum information is utilized for frequency tracking. The approach presented in this paper utilizes both magnitude and phase spectra for partial tracking.

3.1 Frequency Track Evaluation Algorithm

The block-diagram of frequency track evaluation algorithm is presented in Fig. 3. The consecutive analysis frames of input signal are obtained by windowing. The Hamming is used for good main-lobe to side-lobe rejection ratio. Zero-phase windowing is performed to remove linear trend from phase spectrum [53]. The signal can be optionally zeropadded to improve frequency resolution. DFT of every analysis frame is computed for magnitude and phase spectra.

In the next step, which is depicted as Peak Picking in Fig.3, candidates for tonal components are evaluated as the meaningful peaks of magnitude spectrum due to the following formula:

$$X_m(k-1) < X_m(k) \qquad \wedge \qquad X_m(k+1) < X_m(k). \tag{18}$$

where $X_m(k)$ is value of magnitude spectrum X_m in kth bin.

In authors' previous papers [41,42], a special processing was employed for reliable determination of tonal components using some known methods like Sinusoidal Likeness Measure [54]. In further work a different approach was taken. It is assumed that invalid components are rejected during tracking stage [55]. This kind of tonal component validation over several frames appears to be more appropriate than single frame validation. The peaks evaluated by (14) are sorted

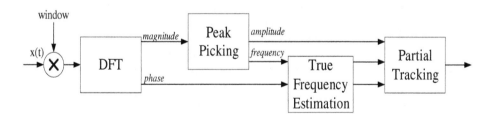

Fig. 3. Block-diagram of frequency track evaluation algorithm

due to their magnitude in order to indicate the most salient components, which are assumed to depict wow defect most reliably. After the peaks are evaluated, their *true* (instantaneous) frequencies are computed using frequency reassignment method (15)[56]. This is essential (see Sect. 4) since finite frequency resolution introduces significant error to frequency estimation. The frequency reassignment method assigns the value of frequency to center of gravity of each bin instead of geometrical center [57].

$$\hat{f}_0 = k\frac{F_s}{N} + \Im\left\{\frac{X_{w'}(k)}{X_w(k)}\right\}\frac{Fs}{2\pi}. \tag{19}$$

where $X_w(k)$ and $X_{w'}(k)$ are *DFT* spectra using window and its derivative.

The last stage of presented processing is partial tracking. The novel approach presented in this paper employs both magnitude and phase spectrum information for partial tracking. The original frequency criterion for partial tracking (13) is found very useful in track creation process however it has very significant drawback. The criterion indicates the peaks which are closest to the processed track, but it does not provide any further information whether the continuation is valid or not. The frequency distance cannot be considered as the measure because tracks variations are increasing with growth. Also track's amplitude, though useful during partial tracking process, does not provide such information. It is assumed that phase spectrum analysis can supply the algorithm with this kind of information.

For the analyzed partial having phase value ϕ_k^{i-1} in frame $i - 1$, the phase value predicted in frame i is equal to:

$$\hat{\phi}_k^i = \phi_k^{i-1} + \frac{R}{N_{DFT}}. \tag{20}$$

where N_{DFT} is the length of the zeropadded *DFT* and R is the frame hop distance. The error of phase prediction can be evaluated as:

$$\phi_{err} = \left|\hat{\phi}_k^i - \phi_l^i\right|. \tag{21}$$

The prediction error ϕ_{err} is in the range $[0; \pi)$. The value near 0 suggests phase coherence in two adjacent frames. Otherwise, if value of error is near π, it may indicate phase incoherence and invalid track continuation (see Sect. 4).

The presented algorithm for partial tracking employs frequency matching criterion (13) for continuity selection and phase prediction error (17) for validation of that continuity. The track termination is associated with a high value of phase prediction error, being different approach from the original frequency deviation condition (13).

3.2 *PVC* Generation

PVC (Pitch Variation Curve) controls the non-uniform resampler during wow compensation process [38,39]. The *PVC* can be obtained from a tonal component by normalization of its values. The normalization should be performed in such way, which ensures the value of 1 for the parts of signal which are not distorted. In the presented approach it is assumed that a first few frames are not contaminated with the distortion. Thus the normalization of tracks values to relative values is performed in the following manner:

$$RF_k(i) = \frac{f_k^i}{f_k^1}. \tag{22}$$

where f_k^i is the value of frequency track in the frame i ,and f_k^1 is the value of frequency track in the first frame.

It is sufficient to utilize only one frequency track for PVC generation, when there is a frequency track which has valid values throughout the whole selected region. More often, however, PVC must be evaluated on the basis of a few tracks. The median was found to provide a satisfactory PVC evaluation for this purpose. The median-based PVC is evaluated as follows:

$$PVC_{median}(i) = median\big(RF_k(i)\big). \tag{23}$$

The accuracy of PVC computation according to the presented manner depends strongly on the accuracy of tonal component evaluation. Thus the main effort in the presented approach is put on validity of tonal component tracking.

4   Experiments and Results

The presented experiments concern the presented issues of noise reduction, wow evaluation and wow-noise dependencies respectively. The sound examples of presented experiments are available at the web site [58].

4.1   Experiments Concerning Noise Reduction

It is important to notice, that for the comparison purposes in the informal subjective tests the same spectral subtraction algorithm was used to calculate the rough estimate $\hat{X}^{ref}$ as for the perceptual reconstruction. Figure 4 presents time-domain changes of the masked noise for a female singing recorded with 44100 Hz sampling rate. The second part of the experiments was devoted to analysis of the performance of the intelligent unpredictability measure pattern classification employed to spectral subtraction. Spectrograms in Fig. 5 present signal recovered with the standard linear spectral subtraction method, and with spectral subtraction improved by the UM vector classification system (as described in Sect. 2.2).

4.2   Experiments Concerning Wow Compensation

For the experiments some archival recordings from the Polish National Film Library and from the Documentary and Feature Film Studio were chosen. The sound examples, taken from magnetic sound tracks, were mainly contaminated with accidental wow. Figures 6 and 7 reveal the motivation for the use of phase prediction error. In both figures the fragments of frequency tracks are displayed simultaneously with the value of the phase prediction error. In Fig. 6 the low value of the phase prediction error corresponds to continuous frequency track (solid white line in the spectrogram). Figure 7, on the other hand, presents the track which is terminated due to the high value of the phase detection error. This example shows that the track's phase analysis may effectively contribute to track evaluation stage.

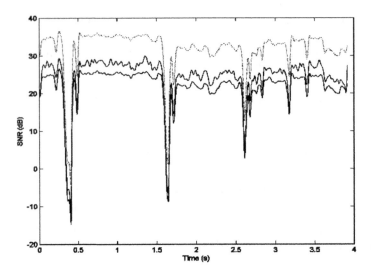

Fig. 4. Time-varying SNR for 24 dB noise attenuation, calculated for each processed frame, for input signal (the lowest curve), for perceptually reconstructed signal (the middle curve) and for signal restored with spectral subtraction method (the highest curve), which was used as the rough estimate of the restored signal

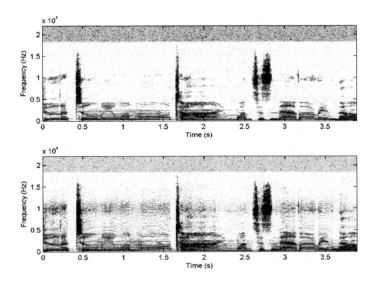

Fig. 5. Spectrograms of signal restored with spectral subtraction (upper plot), and with spectral subtraction enhanced by intelligent pattern recognition system (lower plot)

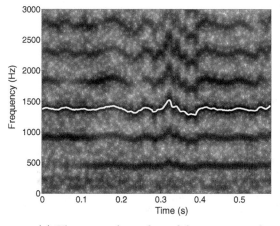

(a) The correctly evaluated frequency track

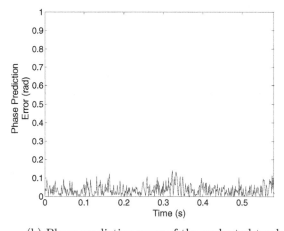

(b) Phase prediction error of the evaluated track

Fig. 6. Evaluation of correctly frequency track and corresponding phase prediction error

The next plot (Fig. 8) shows the outcome of the true frequency estimation of tracks values. It can be noticed that the slow modulation of the track (thin line) was not detected due to a finite frequency resolution of DFT. However when a frequency reassignment is applied the true shape of wow modulation is extracted (thick line).

The algorithm applied to the fragment of archival magnetic soundtrack resulted in tonal component detection showed in Fig. 9. In the presented example the additional parameter, which controlled the maximum number of tracks was set to 8.

Figure 10 shows two $PVCs$. The first one (thin line) was evaluated according to (19), i.e., the median-based estimate, utilizing the tracks displayed in Fig. 9.

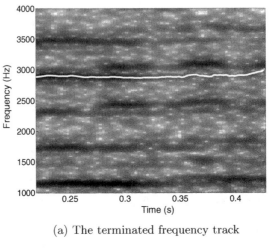

(a) The terminated frequency track

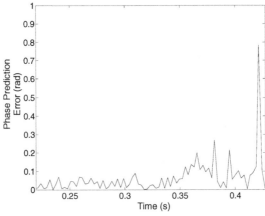

(b) High values of phase prediction error indicating track termination

Fig. 7. Evaluation of terminated frequency track and corresponding phase prediction error

The other PVC (thick line) was computed on the basis of only one track from the same sound excerpt. The latter enabled a nearly transparent reconstruction since the tracks variations corresponded to the distortion's variations. The PVC was based on estimation from the set of tracks also resulted in a satisfactory reconstruction, however, it was possible to hear a difference between these two reconstructions. This implies that in cases where wow can be evaluated from an individual component, this kind of a processing should be performed. However, more common situation is that PVC has to be computed using a set of evaluated tracks thus the median-based PVC estimation should be used (see (19)).

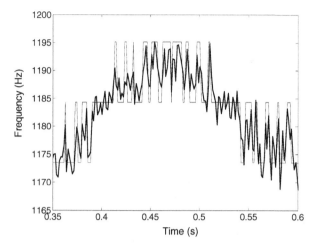

Fig. 8. Improvement of frequency resolution by means of frequency reassignment

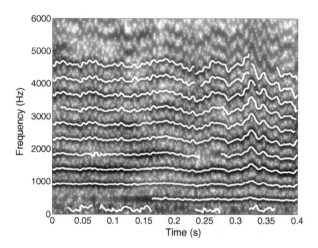

Fig. 9. Spectrogram of analyzed sound and evaluated frequency tracks

4.3   Experiments Concerning Wow and Noise Dependencies

The most common situation in case of archival recordings is the simultaneous existence of various distortions, e.g.: noise, wow, hiss, clicks etc. Therefore it is very essential to apply the algorithms in the right order to get the best subjective quality of the restored sound. The aim of this section is to show some dependencies between the two distortions: wow and noise.

The performed experiments utilized the two algorithms presented in this paper. The algorithms operated on a clarinet sound with the simulated noise and the wow distortion respectively. The sound spectrogram is presented in Fig. 11.

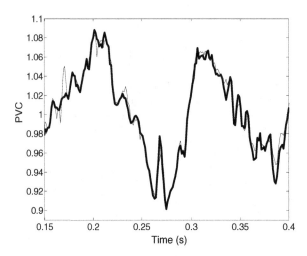

Fig. 10. Fragment of PVC courses. Thick line corresponds to PVC evaluated from single track. Thin line corresponds to PVC evaluated from a set of tracks.

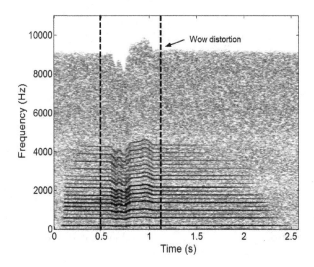

Fig. 11. Spectrogram of clarinet sound with noise and wow distortion

It can be noticed that the noise is modulated, i.e., its stationarity is distorted, due to subsequently added wow. In such situation two restoration scenarios are available. The first scenario involves the noise reduction before the wow compensation. The result of this operation is showed in Fig. 12. It can be noticed in the spectrogram that this processing introduced some parasite artifacts, i.e., noise modulation, to the restored signal. The result of the second restoration scenario

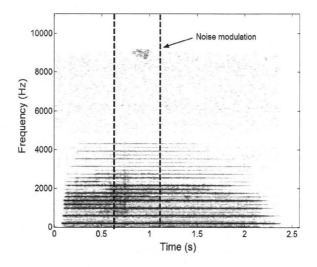

Fig. 12. Spectrogram of reconstructed signal. Wow removal was performed before noise removal which resulted in parasitic artefacts.

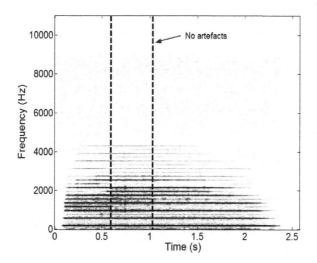

Fig. 13. Spectrogram of restored sound. Noise removal was performed before wow which resulted in valid restoration of the sound.

which involved firstly the wow compensation and the noise reduction afterwards is given in Fig. 13. Contrarily to the first restoration, the restored sound does not contain any audible defects. Also no visible artifacts can be noticed in the spectrogram.

5 Conclusions

As an extension of the spectral subtraction, the intelligent-pattern-recognition system involving the *Unpredictability Measure* for spectrum components classification was presented. Applying some properties of the human auditory system to noise reduction allowed preserving much more of the input signal's energy and consequently enabled decreasing unfavorable influence of the reconstruction filter. An artificial neural network was successfully applied as the decision system for classifying uncertain noisy spectral patterns. An application of the rough set decision system to this task was also discussed. Thus the elaborated method allowed more effective noise reduction characterized by a small number of the parasite artifacts.

The utilization of frequency tracks enabled compensation of wow distortion. It was found that the employment of the magnitude and phase spectrum in partial tracking process improved its tracking abilities comparing to the earlier algorithms [42]. The introduction of the phase prediction error criterion enabled the validation of the tracking procedure. Thereby the tracking algorithm was found to be more suitable for distorted polyphonic sounds comparing to the original magnitude-based approach.

The experiments concerning wow and noise dependencies showed that during the restoration process much attention must be paid to applying the restoration algorithms in the correct order. The experiments also showed that wow influences other distortion characteristics, e.g., the noise stationarity, thus it can seriously affect the quality of the restored sound.

Acknowledgments

Research funded by the Commission of the European Communities, Directorate-General of the Information Society within the Integrated Project No. FP6-507336 entitled:"PRESTOSPACE - Preservation towards storage and access. Standardized Practices for Audio-visual Contents Archiving in Europe". The research was also subsidized by the Foundation for Polish Science, Warsaw and by the Ministry of Science and and Information Society Technologies - Dec. No. 155/E-359/SPB/6.PR UE/DIE 207/04-05.

References

1. Vaseghi, S.: Advanced Signal Processing and Noise Reduction. Wiley & Teubner, New York (1997)
2. Welch, G., Bishop, G.: An Introduction to the Kalman Filter. Technical Report of The University of North Carolina in Chapel Hill, USA, No. 95-041
3. Widrow, B., Stearns, S.: Adaptive Signal Processing.Prentice-Hall Intl. Inc., New Yersey (1985)
4. Kunieda, N., Shimamura, T., Suzuki, J., Yashima, H.: Reduction of Noise Level by SPAD (Speech Processing System by Use of Auto-Difference Function). In: International Conference on Spoken Language Processing, Yokohama (1994)

5. Yoshiya, K., Suzuki, J.: Improvement in Signal-to-Noise Ratio by SPAC (Speech Processing System Using Autocorrelation Function). In: Electronics and Communications in Japan, vol.61-A, No. 3 (1978) 18-24
6. Eprahim,Y.: A Bayesian Estimation for Speech Ebhacement Using Hidden Markov Models. In: IEEE Transactions on Signal Processing, vol. 40, No.4 (1992) 725-735
7. Eprahim, Y.: Statistical-Model-Based Speech Enhacement Systems. In: Proceedings of the IEEE, vol. 80,No. 10 (1992) 1526-1555
8. Feder, M., Oppenheim, A., Weinstein E.: Maximum Likelihood Noise Cancellation Using the EM Algorithm. In: IEEE Transactions on Acoustics Speech and Signal Processing, vol. 37, No. 2 (1989) 204-216
9. Lim, J., Oppenheim, A.: Enhancement and Bandwidth Compression of Noisy Speech. In: Proceedings of the IEEE, vol. 67, No. 12 (1979) 1586-1604
10. Czyzewski A., Kaczmarek A., Speaker-independent recognition of isolated words using rough sets. Proc. Second Annual Joint Conference on Information Sciences, 28. Sept. - 01 Oct. 1995, North Carolina, USA, pp. 397-400, (1995).
11. Czyzewski, A., Krolikowski, R.: Neuro-Rough Control of Masking Tresholds for Audio Signal Enhancements. In: Neuro Computing, vol. 36, No. 1-4 (2001) 5-27
12. Knecht, W., Schenkel, M., Moschytz, G., Neural Network Filters for Speech Enhancement. In: IEEE Transactions on Speech and Audio Processing, vol.3,No.6 November (1995) 433-438
13. Asano, F., Hayamizu, S., Yamada, T., Nakamura, S.: Speech Enhacement Based on the Subspace Method. In: IEEE Transactions on Speech and Audio Processing, vol. 8, No. 5 Spetember (2000) 497-507
14. Elko, G.: Adaptive Noise Cancellation with Directional Microphones. In: Proceedings of the IEEE Workshop on Applications of Signal Processing to Audio and Acoustics, New Paltz (1997)
15. Wallace, G.: The JPEG: Still Picture Compression Standard. In: Communication of the ACM, vol. 34, No. 4 April (1991) 31-44
16. Gibson, J., Koo, B.: Filtering of Colored Noise for Speech Enhancement and Coding. In: IEEE Transactions on Signal Processing, vol. 39, No. 8 August (1991) 1732-1742
17. Lee, K., Jung, S.: Time-Domain Approach Using Multiple Kalman filters and EM Algorithm to Speech Enhacement with Stationary Noise. IEEE Transaction on Signal Processing, vol. 44, No. 3 May (2000) 282-291
18. Ikeda, S.,Sugiyama, A.: An Adaptive Noise Canceller with Low Signal Distortion for Speech Codecs. In: IEEE Transactions on Signal Processing, vol. 47, No. 3 March (1999) 665-674
19. Sambur, M.: Adaptive Noise Cancelling for Speech Signals. In: IEEE Transactions on Acoustics Speech and Signal Processing, vol. ASSP-26, No. 5 October (1978) 419-423
20. Eprahim, Y., Malah, D., Juang, B. :On the Application of Hidden Markov Models for Enhancing Noisy Speech. In: IEEE Transactions on Acoustics Speech and Signal Processing, vol. 37, No. 12 December (1989) 1846-1856
21. Sameti, H.,Sheikhzadeh, H., Brennan, R.: HMM-Based Strategies for Enhacement of Speech Signals Embeeded in Nonstationary Noise. In: IEEE Transactions on Speech and Audio Processing, vol. 6, No. 5 September (1998) 445-455
22. Sim, B., Tong, Y., Chang, J., Tan, C.: A Parametric Formulation of the Generalized Spectral Subtraction Method. In: IEEE Transactions on Speech and Audio Processing, vol. 6, No. 4 July (1998) 328-337
23. Vaseghi, S., Frayling-Cork, R.: Restoration of Old Gramophonic Recordings. In: Journal of Audio Engineering Society, vol. 40,No.10 October (1997) 791-800

144   A. Czyżewski et al.

24. Zwicker, E, Zwicker, T.: Audio Engineering and Psychoacoustics: Matching Signals to the Final Receiver,the Human Auditory System. In: Journal of Audio Engineering Society, Vol. 39, No. 3 (1991) 115-126
25. Czyzewski, A., Dziubinski, M.: Noise Reduction in Audio Employing Spectral Unpredictability Measure and Neural Net. Knowledge-Based Intelligent Information and Engineering Systems: 8th International Conference, KES'04, LNAI [Lecture Notes in Artificial Intelligence] 3213, Springer - Verlag, Berlin, Heidelberg, Wellington, New Zealand, September (2004), Part I: 743-750
26. Tsoukalas, D., et al.: Perceptual Filters for Audio Signal Enhacement. In: Journal of Audio Engineering Society, Vol. 45, No. 1/2 (1997) 22-36
27. MPEG-4,International Standard ISO/IEC FCD 14496-3, Subpart 4 (1998)
28. Shlien, S.: Guide to MPEG-1 Audio Standard. In: IEEE Transactions on Broadcasting, Vol.40,(1994) 206-218
29. Czyzewski, A., Krolikowski, R.: Noise Reduction in Audio Signals Based on the Perceptual Coding Approach. Proceedings of the IEEE Workshop on Applications of Signal Processing to Audio and Acoustics, October, New Paltz (1999) 147-150
30. Krolikowski, R., Czyzewski, A.: Noise Reduction in Acoustic Signals Using the Perceptual Coding. 137th Meeting of Acoustical Society of America, Berlin, (1998) CD-Preprint
31. McAulay, J., Quatieri, T.F.: Speech Analysis/Synthesis Based on a Sinusoidal Representation. In: IEEE Transactions on Acoustics, Speech, and Signal Processing, Vol. 34, No. 4 August (1986) 744-754
32. Godsill, J. S., Rayner, J. W.: The Restoration of Pitch Variation Defects in Gramophone Recordings. In: Proceedings of the IEEE Workshop on Applications of Signal Processing to Audio and Acoustics, October, New Paltz (1993)
33. Godsill, J. S.: Recursive Restoration of Pitch Variation Defects in Musical Recordings. In: Proceedings of International Conference on Acoustics, Speech, and Signal Processing, Vol. 2, April Adelaide, (1994) 233-236
34. Walmsley, P. J., Godsill, S. J., Rayner P. J. W.: Polyphonic Pitch Tracking Using Joint Bayesian Estimation of Multiple Frame parameters. In: Proceedings of 1999 IEEE Workshop on Applications of Signal Processing to Audio and Acoustics, October New Paltz (1999)
35. Godsill, J. S., Rayner, P. J. W.: Digital Audio Restoration. In: Kahrs M., Brandenburg K.(Eds.), Applications of Digital Signal Processing to Audio and Acoustics, Kluwer Academic Publishers, (1998) 41-46
36. Godsill, J. S., Rayner, P. J. W.: Digital Audio Restoration - A Statistical Model-Based Approach, Springer-Verlag, London (1998)
37. Czyzewski, A., Maziewski, P., Dziubinski, M., Kaczmarek, A., Kostek, B.: Wow Detection and Compensation Employing Spectral Processing of Audio. 117 Audio Engineering Society Convention, Convention Paper 6212, October San Francisco (2004)
38. Maziewski, P.: Wow Defect Reduction Based on Interpolation Techniques, In Proceedings of 4th Polish National Electronic Conference, vol. 1/2, June (2005) 481-486.
39. Czyzewski, A., Dziubinski, M., Ciarkowski, A., Kulesza, M., Maziewski, P., Kotus, J.: New Algorithms for Wow and Flutter Detection and Compensation in Audio. 118th Audio Engineering Society Convention, Convention Paper No. 6212, May Barcelona (2005)
40. Czyzewski, A., Maziewski, P., Dziubinski, M., Kaczmarek, A., Kulesza, M., Ciarkowski, A.: Methods for Detection and Removal of Parasitic Frequency Modulation in Audio Recordings. AES 26th International Conference, Denver, July (2005)

41. Litwic, L., Maziewski, P.: Evaluation of Wow Defects Based on Tonal Components Detection and Tracking. In: Proceeding of 11th International AES Symposium, Krakow June (2005) 145 -150
42. Czyzewski, A., Dziubinski, M., Litwic, L., Maziewski, P.: Intelligent Algorithms for Optical Tracks Restoration. In: Lecture Notes in Artifictial Intelligence Vol. 3642 August/September (2005) 283-293
43. Ciarkowski, A., Czyzewski, A., Kulesza, M., Maziewski, P.: DSP Techniques in Wow Defect Evaluation. In: Proceedings of Signal Processing 2005 Workshop September (2005) 103-108
44. Nichols, J.: An Interactive Pitch Defect Correction System for Archival Audio. AES 20th International Conference, October, Budapest (2001)
45. Howarth, J., Wolfe, P.: Correction of Wow and Flutter Effects in Analog Tape Transfers. 117 Audio Engineering Society Convention, Convention Paper 6213, October San Francisco (2004)
46. Wolfe, P., Howarth, J.: Nonuniform Sampling Theory in Audio Signal Processing. 116 Audio Engineering Society Convention, Convention Paper 6123, May Berlin (2004)
47. Beerends, J., Stemerdink, J.: A Perceptual Audio Quality Measure Based on a Psychoacoustic Sound Representation. In: Journal of Audio Engineering Society, Vol. 40, No. 12 (1992) 963-978
48. Humes, L.: Models of the Additivity of Masking. In: Journal of Acoustical Society of America, vol. 85 (1989) 1285-1294
49. Brandenburg, K.: Second Generation Perceptual Audio Coding: The Hybrid Coder. In: Proceedings of the 90th Audio Engineering Society Convention, Convetion Paper 2937 Montreux (1990)
50. Vaseghi, S.: Advanced Signal Processing and Digital Noise Reduction, Wiley&Teubner, New York (1997)
51. Depalle, P., Garcia, G., Rodet, X.: Analysis of Sound for Additive Synthesis: Tracking of Partials Using Hidden Markov Models. In: Proceedings of IEEE International Conference on Speech and Signal Processing (ICASSP' 93) , 1993
52. Lagrange, M., Marchand,S., Rault, J.B.: Tracking Partials for Sinusoidal Modeling of Polyphonic Sounds. In: Proceedings of IEEE International Conference on Speech and Signal Processing (ICASSP'05) Philadelphia March 2005
53. Serra, X.: Musical Sound Modeling with Sinusoids plus Noise. In: Pope, S., Picalli, A., De Poli, G., Roads, C. (eds.): Musical Signal Processing, Swets & Zeitlinger Publishers (1997)
54. Rodet, X.: Musical Sound Signal Analysis/Synthesis: Sinusoidal + Residual and Elementary Waveform Models. In: Proceedings of IEEE Symposium on Time-Frequency and Time-Scale Analysis (1997)
55. Lagrange, M.:, Marchand, S., Rault, J.B.: Sinusoidal Parameter Extraction and Component Selection in a Non-stationary Model. Proc. of the 5th Int. Conference on Digital Audio Effects, September, Hamburg (2002)
56. Auger, F., Flandrin, P.: Improving the Readability of Time-frequency and Time-scale Representations by the Reassignment Method. In: IEEE Transactions on Signal Processing, vol. 43, No. 5 May (1995) 1068-1089
57. Keiler, F., Marchand, S.:Survey on Extraction of Sinusoids in Stationary Sounds. In: Proceedings of the 5th International Conference on Digital Audio Effects, Hamburg September (2002)
58. Sound examples: http://sound.eti.pg.gda.pl/~llitwic/SoundRest/

Rough Set-Based Application to Recognition of Emotionally-Charged Animated Character's Gestures

Bożena Kostek and Piotr Szczuko

Multimedia Systems Department, Gdańsk University of Technology
{bozenka, szczuko}@sound.eti.pg.gda.pl

Abstract. This research study is intended to analyze emotionally-charged animated character's gestures. Animation methods and rules are first shortly reviewed in this paper. Then the experiment layout is presented. For the purpose of the experiment, the keyframe method is used to create animated objects characterized by differentiating emotions. The method comprised the creation of an animation achieved by changing the properties of a temporal structure of an animated sequence. The sequence is then analyzed in terms of identifying the locations and spacing of keyframes, as well as the features that could be related to emotions present in the animation. On the basis of this analysis several parameters contained in feature vectors describing each object emotions at key moments are derived. The labels are assigned to particular sequences by viewers participating in subjective tests. This served as a decision attribute. The rough set system is used to process the data. Rules related to various categories of emotions are derived. They are then compared with the ones used in traditional animation. Also, the most significant parameters are identified. The second part of the experiment is aimed at checking the viewers' ability to discern less dominant emotional charge in gestures. A time-mixing method is proposed and utilized for the generation of new gestures emotionally-charged with differentiated intensity. Viewers' assessment of the animations quality is presented and analyzed. Conclusions and future experiments are shortly outlined.

Keywords: animation, rough set analysis, subjective tests, emotion, non-verbal modality.

1 Introduction

At the beginning of XX century, simultaneously with the invention of the movie, the animation was born. At the very early stage of animation, to create the animated motion, a small black-and-white picture was drawn on several layers of celluloid. Continuity of motion was achieved by introducing very small changes between each frame (cel), drawn one after one. Later the color drawing on papers and on foils attached to the peg was invented. Therefore it was possible to create frames in varied order – first most important and powerful poses were

J.F. Peters and A. Skowron (Eds.): Transactions on Rough Sets V, LNCS 4100, pp. 146–166, 2006.
© Springer-Verlag Berlin Heidelberg 2006

designed, then transitional frames were filled in, called also in-betweens [3]. The same approach is utilized in the computer animation systems with keyframes [22]. Both traditional and computer keyframe animation aim at animating a character with a highly human appearance, personality and emotions though using different artistic and technical means [20,22].

Correct utilization of poses and transitions between them can implicate different emotional features of the character's motion. As a result of experience derived from the first years of traditional animation the animation rules were created in 1910 by the animators from the Walt Disney studio. These rules state how to achieve specified features, utilizing posing, keyframes, and phases of motion [20,22]. The know-how of a hand-made animation and the rules are being passed through generations of animators. They have a subjective nature, and were never analyzed by scientific means. That is why the starting point of this study is to analyze animations by means of the classification system based on rough sets which is a very suitable method while dealing with uncertain, subjective data [15,16].

We start our work with these traditional animation rules and our aim is to check whether it is possible to generate adequate rules based on motion parametrization and rough set analysis. Further, this is to see whether it is possible to generate automatically animation sequences with a desired emotional features based on the knowledge base and rules generated by the rough set method, without an animator interference.

Communication between humans extends beyond verbal techniques. Gestures comprising body and head movements, and also postures are often employed to convey information in human-to-human communication. Moreover, in some situations they can deliver or enhance the unspoken or obscured by noise speech message. Perception of gestures is also very rarely misinterpreted. This non-verbal modality is therefore of a great importance [6,9], and can be easily associated with emotions. In the paper of Mehrabian [12], and also Hatice et al. [8] it was stated that 93 percent of human communication is nonverbal and the most expressive way humans display emotions is through facial expressions and body gestures.

Therefore, the scenario for this study is as follows. One can analyze and parameterize features of a hand-made animation and then correlate them with the description of emotions. These data can be utilized for the creation of a knowledge base containing feature vectors derived from the analysis of the animated character's emotions. In that way the animator's tasks would be limited to designing a simple animated sequence, and delivering a description of the desired emotions to the expert system. These data should next be processed in the system, in which the animation parameters are modified, and as a result an emotionally-featured animation is generated. The next phase is the assessment of the conveyed emotion rendered into the animation to confirm its quality. The second part of the experiment, presented in this paper, concerns checking the viewers' ability to discern less dominant emotional charge in the gestures of the character. A time-mixing method is proposed and utilized for the generation of new emotionally-charged gestures with graded intensity. The viewers' assessment of the created animations quality is presented and analyzed. The second, and the superior aim of this research is however to design a system designated for the computer-based animation that would be capable to create more realistic animations according to the animator's requirements. Elements of such a system are shortly described.

This paper presents an extension of the research study carried out by the authors in the domain of animation [11]. The first part (Sections 1-3) presents a revised version of the original paper, and the second part (Section 4) describes new work, i.e. animation generation experiments focused on of emotional features contained in motion. In addition future work is outlined.

The paper is organized as follows: Section 2 briefly reviews research on application of computer methods in animation domain, and problems related to acquisition of emotional features contained in motion. Section 3 is related to the parametrization of animation and to the analysis of experimental data collected during subjective tests. Section 4 presents an animation generation method and then focuses on subjective verification of quality of generated animations. Subjective scores gathered from the experts are next analyzed and an attempt to automatically classify emotional features is made. Section 5 presents conclusions, and in Section 6 future experiments are outlined.

2   Computer-Based Animation Methods

The simplest definition of the animation process is the creation of motion by changing the properties of objects over time. A keyframe is a time point when a property has been set or changed. In traditional animation master animators draw the keyframes and assistants do all the in-between frames. In computer animation in-between frames are calculated or interpolated by the computer. The animator first sets up main poses in time and space then the system fills in the transitional frames by interpolating locations, rotations and torque of objects and characters' skeletons. The interpolation process, i.e. the acceleration, slow-in and slow-out phases can be changed, and keyframes may be inserted or transposed. Although the computer animation process is much faster than the traditional one, the quality of computer animations still lacks naturalness or individual style.

As was mentioned in Section 1, a set of rules was proposed by traditional animators, describing technical means to achieve realistic and emotionally featured motion. One of the basic rules is anticipation. This refers to the preparation for an action. Before the main action starts there should always be a preparation for it, which is a slight movement in a direction opposite to the direction of the main action. If the preparation phase is long, the performing character will be perceived as weak, or hesitating. Short anticipation gives effect of a self-confident, strong character. Follow-through is a rule related to the physics of the moving body. The motion always starts near the torso: first the arm moves, then the forearm, and the last is the hand. Therefore keyframes for bones in the forearm and hand should be delayed comparing to the arm bone. The delay adds a whip-like effect to the motion, and the feeling of the elasticity and flexibility. Another rule connected to the physiology of motion is overshoot. The last bone in the chain, for example the hand, cannot stop instantly. It should overpass the target position, go back to it, and finally slowly stop. Stops are never complete. A rule called 'moving hold' is related to keeping a character in a pose for some time. Very small movements of the head, eyes, and limbs should be introduced to maintain the realistic quality of the motion. Natural motion almost always goes along arcs. Only a vertically falling object moves absolutely straight. Rotation of joints in human body implicate curve moves, but also head movement from left to right if straight, will not appear to be natural. Other rules, not mentioned here, are related to staging, i.e. posing a character in front of the camera, exaggeration of motion, squashing and stretching for achieving cartoon-like effects, and so on. These rules do not lie within the scope of the objectives of this research study.

Numerous research studies have been devoted to achieving a realistic motion with computer methods. Physical simulations of a human body motion were created, resulting in a realistic motion during jump, flight, free fall, etc. [7]. In these cases it is always necessary to assign boundary conditions such as how the motion starts, when and where it should stop, and on this basis the transitional phase is calculated. Such a method makes the animation realistic but neglects emotional features. An attempt was also made to connect emotions and energy corresponding to the motion, therefore a highly energy-charged motion was assumed as happy and lively, and a low energy consuming motion as tired and sad. In self-training algorithms the energy consumption was utilized as the target function. The method was tested on a 4-legged robot creature, and is still in the research phase [14].

Genetic algorithms and neural networks were applied to create a model of a human, and to teach it to move and react in a human-like manner [21]. Data related to the physiology of a human body were gathered and the target functions were composed related to keeping vertical position, not falling over, and reaching a desired location on foot. A system developed for this purpose was trained to accomplish the mentioned tasks. Various behaviors were implemented, such as walking, falling, running, limping, jumping, reacting to being hit by different forces, but the developed method actually lacks in emotional acting. There were also some efforts to create new controllers for the animation process. For example a recorded motion of the pen drawing on the computer tablet is mapped to some motion parameters, like changes of location, rotation and speed of the selected character's bone, or the movement of the character's eyes. It gives the animator new technical means to intuitively act with the controller, and map that action onto the character [18]. A very similar and well-known method is the motion capture [4], consisting of the registration of sensor motions, which are attached to the key parts of the performer's body. It is possible to use a single sensor as a controller for the selected motion parameter [17]. Unfortunately this method is expensive, moreover processing of the captured data is very unintuitive and complex, and in this case the editing by hand is nearly impossible. A more advanced version of motion capture, called performance capture, introduces the recording of facial expressions with similar equipment (sensors are attached in crucial points of the face). The results can be seen in a computer-animated movie "The Polar Express", with Tom Hanks playing five roles – a main child character and some adult characters [23].

Advanced computer-based animation methods for motion generation may result in a psychological effect called "Uncanny Valley" [13]. One of the results of increasing anthropomorphism of animated characters is the growth in the viewers' positive emotional response. At some point, however, when a character very closely resembling a human being still misses some human nuances, viewers suddenly tend to express disquiet and negative response. Developing further anthropomorphism increases a positive response, until full human features are reached (Figure 1). As Bryant stated, judging from the relative depth of the curves Mori apparently considers motion more important than simple appearance, though he stresses that both are affected at least as much by subtle nuances as by more striking factors [5]. Examples of *uncanny valley* effect are present in some recent computer-animated movies like almost-photorealistic "Final Fantasy: Spirits Within", "The Polar Express" mentioned above, and "Final Flight of the Osiris" (part of the Animatrix movie). Based on the reaction of the audience, animators cannot agree whether a human-like motion in animations should be kept below uncanny valley to avoid the sense of wrongness, or should it be increased to achieve full resemblance to a human being.

Methods mentioned above either generate motion neglecting its emotional features, or allow live motion recording and processing but have no means to record features related to emotion. Our work is focused on adapting animation

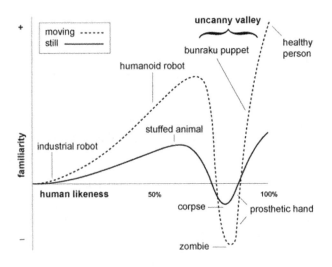

Fig. 1. Illustration of *uncanny valley* effect [13]

rules to computer generated animation. Parametrization of animated motion and analysis of its features perception along with guidelines derived from the traditional animation should enable to generate motion with explicit emotional features as required by an animator.

3 Experiment Layout

This Section presents the experiment framework. It is assumed that the animated motion could be described in a similar way as the animator describes it. This means that using the data of the character's bones in keyframes along with the interpolation data is sufficient to comprehensively analyze and generate realistic and emotionally featured motion. It is assumed that the animation rules presented earlier are related to these data, thus it is desirable to treat them as a knowledge base, and to evaluate both the correspondence to emotional features of a motion, and the parameterization effectiveness performed in terms of categorization.

The first stage of our work involves an analysis of a created animation. It determines the important parameters of the animation in terms of automatic classification, such as temporal information, an example of which are the time-related parameters describing the positions of keyframes in the animation. These parameters are analyzed in terms of usefulness for the recognition of the emotion. The next stage is to remove redundant and unimportant parameters. Then the assumption is made and verified that the parameters classified as important for automatic classification are also important for human viewers. Next, based on the above information, more advanced animations are prepared by changing the temporal structure of the given animation sequence resulting in gradual changes of emotion intensity. The source animations expressing two different emotions are combined together to create new sequences. Their quality and emotional features are subjectively assessed. This serves to answer the question whether the exaggeration seen in animated characters is not only accepted but also preferred by the viewers.

3.1 Assessment of Emotion Expressed by Gestures

For the purpose of this study one of the authors, a semi-professional animator, basing on animation rules described in the Introduction, created a series of animations. They present two arm gestures (picking up a hypothetical thing and pointing at a hypothetical thing) expressing different emotions: fear, anger, sadness, happiness, love, disgust and surprise. Animations consist of the rotations of joints as shown in Figure 2. In such a case none of other bone parameters are changed. In total 36 test animations were prepared. It is assumed that each animation should clearly reproduce the emotion prescribed. However, taking the subjective character of emotions into account reveals that additional assessment of perceived emotions is needed. This will help verifying the emotional definition of each sequence.

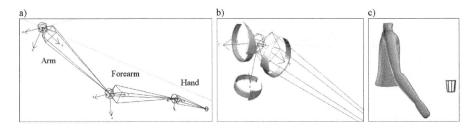

Fig. 2. Presentation of a chain of bones (a), possible bone rotations (b), torso with animated arm (c)

The following subjective test was conducted. For each animation presented in a random order, viewers were asked to classify the type of the presented emotion, by picking only one emotion from the list, and, choosing one of the ranks from a 3-point scale (1 - just noticeably, 2 - noticeably, 3 - definitely) in order to denote how much the emotion was clear to them. Additional features (i.e. strength naturalness, smoothness, and lightness of the emotion) were evaluated within a 5-point scale (e.g. 1 - unnatural, 3 - neutral, 5 - natural). Twelve non-specialists took part in this test. 17 of 36 animations were described by the test participants as having different emotions than the prescribed ones. Assuming that a subjective assessment is a very important factor while describing motion or gestures of an animated character, the answers gathered from the viewers were taken into account in the next stage of the analysis instead of prior assumptions about emotions contained in the animation sequences. The results of matching emotions with the animation sequences revealed difficulties in selecting only one of the descriptions. The participants reported equivocal interpretations of gestures which showed similarities e.g. between "negative" emotions like fear, surprise, and anger. Therefore, to deal with such uncertain data, the rough set-based analysis [15,16] was applied for further processing. 27.9% of the acquired data described the emotion of anger, 19.4% of fear, 16.5% of happiness, 13.3% of love, and below 10% of disgust, sadness and surprise.

3.2 Parametrization of Animation Data

For the purpose of the analysis of the animation data, various parameters of the keyframed motion are proposed. Parameters are based on the creation process of the animation. In that process, first a simple animation version is prepared, with a character's key poses spaced in time with interpolated motion between them. Later the animator adds other important elements of motion such as for example anticipation and overshoot, by inserting new keyframes (Figure 3). Animations for the test were prepared based only on the rotation of bones, therefore there are separate keyframes for the rotations along X-, Y-, Z-axes, and in such a case the amplitude literally means the rotation angle (see Figure 4).

Animation data are taken directly from the animation application, therefore keyframes and parameters values are available directly. Long animated sequences

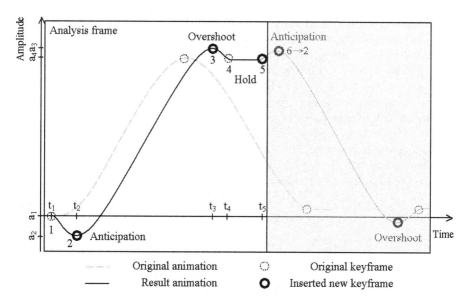

Fig. 3. Presentation of animation parameters. Original animation is without important elements such as anticipation and overshoot. The final animation is a variation of the original one, with anticipation and overshoot inserted. For that purpose it is necessary to add new keyframes, which change the curve of the motion amplitude. Keyframe times are marked as t_i, and the values of amplitude for them as a_i. White range represents analysis frame with keyframes from 1 to 5. Gray range is next analysis frame, where keyframe 6 corresponds to frame 2, and is also responsible for anticipation phase.

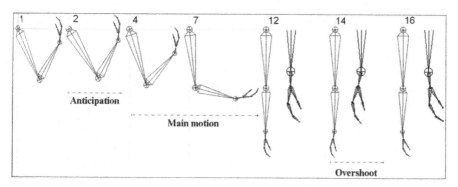

Fig. 4. An animation utilizing bone rotations with keyframe numbers and motion phases marked. Anticipation is very short (2 frames long) and subtle motion in the direction opposite to the main motion. The main phase usually extends across many frames and changes of rotation for bones are distinct. Overshoot is a short phase before the complete stop, when the last bones in a chain overpass the target position (hand bones are magnified to visualize the overpass in frame 14).

are segmented into parts with one main motion phase and one hold phase (see Figure 3, "Analysis frame"). Each segment is analyzed and included as a pattern in a decision table (Table 1). For each animation segment, values of animation parameters related to the amplitudes of particular phases ($A_a = a_2 - a_1$, $A_m = a_3 - a_2$, $A_o = a_4 - a_3$), their lengths ($t_m = t_3 - t_2, t_a = t_2 - t_1$, $t_o = t_4 - t_3, t_h = t_5 - t_4$), and speeds ($V_m = A_m/t_m, V_a = A_a/t_a, V_o = A_o/t_o$) are calculated. Variables a_i and t_i are depicted in Figure 3. The decision parameter for the animation segment is an emotion selected most often by the viewers while rating the animation. This table serves as an input to the rough set system. The system task is to evaluate rules describing interrelations between the calculated parameters and the features of the motion.

Table 1. Decision table

Bone	Rot	$t_1 ... t_5$	$a_1 ... a_5$	A_m	t_m	V_m	A_a	t_a	V_a	A_o	t_o	V_o	t_h	A_a/A_m	t_a/t_m	Decision
Arm	Rot_X	1 ... 6	1.3 ... 1	1.2	2	0.6	1	1	1	0.5	1	0.5	2	0.83	0.5	Surprise
Hand	Rot_Y	1 ... 8	1 ... 1.4	1	2	0.5	2	2	1	2	1	2	3	2	1	Fear
...	...	...	...	...	...	...	...	...	...	...	...	...	...	...		
Hand	Rot_Z	1 ... 7	2 ... 0.5	2	2	1	1	2	0.5	0.2	1	0.2	2	0.5	1	Happiness

3.3   Data Processing

For the generation of rules based on the decision table, the Rough Set Exploration System was used [1,2,19]. During the processing, the automatic discretization of parameter values was performed. Local discretization method was utilized. In general all parameter values (attributes) are continuous, and for purpose of generalization, and processing of previously unseen object a discretization is usually performed. Local discretization method uses Maximal Discernibility (MD) Heuristics [2]. An attribute A is selected. The algorithm searches for a cut value $c \in A$, which discern a largest number of pair of objects. Then all pairs discerned by selected c are removed, and new maximal discernibility cut value is searched for those left. The procedure is repeated until no objects are left. Then the algorithm is repeated for next attribute A. An attribute can be excluded from further processing, as no important for classification, if no cut can be found.

As a result of discretization some parameters are automatically excluded at this level as not important for defining the relations searched. 12 parameters were left in the decision table: 'Bone', 'Rotation axis', the amplitude for the first keyframe a_1, the length and amplitude of the anticipation phase (t_a, A_a respectively), the amplitude for the anticipation keyframe (a_2), the length and speed of the main motion phase (t_m, V_m respectively), the time for the overshoot keyframe (t_3), length of a hold phase (t_h), speed of an overshoot phase (V_o), time for an ending keyframe (t_5) (see Figure 3 for reference). There were 1354 rules containing the above parameters generated by a genetic algorithm available in the Rough Set Exploration System [19]. Total coverage was 1.0, and total accuracy of the object classification from the decision table was 0.9. Shortening of the rules resulted in a set of 1059 rules, giving total coverage of 0.905, and the accuracy of the classification equaled 0.909.

Further steps aimed at decreasing the number of parameters. It was assumed that parameters with 1, 2 or 3 ranges of discretization could be removed, because they are used in discerning the lowest number of object classes. As a result classification accuracy loss should not be significant, but the number of rules would be lower. Therefore from the mentioned 12 parameters, only 6 were left: 'Bone', and all the others having more than 3 discretization ranges, i.e. a_1 parameter with 10 discretization ranges, t_a - 12 ranges, t_m - 6 ranges, t_3 - 7 ranges, and t_h - 5 ranges. This resulted in the generation of 455 rules with total coverage of 1.0 and the accuracy of 0.871 (compare to 0.9 for all parameters). After shortening, 370 rules were left, giving the coverage of 0.828 and the accuracy of 0.93.

After discarding the last amplitude parameter left at this stage i.e. a_1, when the 'Bone' and time parameters were used, the results did not change much. There were 197 rules, with the coverage of 1.0 and the accuracy of 0.871 - the same as before. Also after shortening, when only 160 rules were left, the coverage was the same as before (0.828), and the accuracy equaled 0.933. The confusion matrix for the classification using the parameters 'Bone', t_a, t_m, t_3, t_h is presented in Table 2. Removing by hand some additional parameters caused a great loss in accuracy. This seems a very satisfying result, showing that time parameters are especially important for identifying emotions in motion. In the derived 160 rules generated for five parameters, 100 use only time parameters without the 'Bone' parameter. 102 rules have the size of 2, and 58 rules have the size of 3. Maximal support for the rule is 29, minimal is 3, and the mean is 5.7.

For rules, related to each class, parameter values were analyzed, which resulted in the creation of representative sets. For example, for a surprise emotion, most objects in the decision table have t_a=(7.5,8.5), t_m=(17.5,Inf), t_3=(45.5,51.5), t_h=(18.5,19.5) for all bones, and for 'love' - t_a=(16.5,19.5), t_m= (17.5,Inf), t_3= (36.0,45.5), t_h= (19.5,21.0). That information will be utilized later to generate variations of animations introducing desired emotional features.

Table 2. Results of classification of objects with the derived set of 160 rules

	Predicted:						
	Anger	Sadness	Love	Happiness	Fear	Surprise	Disgust
Actual: Anger	103	0	0	0	0	0	0
Sadness	4	31	0	0	0	0	0
Love	0	0	37	0	0	0	0
Happiness	0	0	0	50	3	0	0
Fear	0	0	8	0	57	0	4
Surprise	0	0	0	0	0	16	0
Disgust	0	4	0	0	0	0	24

Examples of the derived rules are: IF (Bone='Forearm') AND (t_m="(-Inf,5.5)") THEN (Decision=Anger), and IF (t_a="(16.5,19.5)") AND (t_m="(-Inf,5.5)") THEN (Decision=Anger), which can be rewritten in a natural language as: *"if the forearm main motion phase is very short then the emotion is translated as anger"*, and *"if the anticipation in motion is long and the main motion phase is very short then emotion is translated as anger"*.

4   Animation Generation

This section presents an experiment related to combining two source animations and assessing emotional feature of the generated result. It is verified whether the animations generated are clearly readable to the observer, and whether their subjective emotional features can be recognized and correctly understood by the experts. Next an attempt is made to correctly classify all animations in domain of their time parameters, utilizing the rough set analysis. Significant parameters are found, and rules are generated that may enable to create new animation sequences with adequate emotional features assigned to motion. The starting point is however generalization of the gathered data by the rough set method.

Ten new animations were created according to the rules of traditional animation as well as to the rules derived in the previous section. They present a cartoon-like adolescent character expressing different emotions which can be analyzed in pairs:

- 1st pair - Clasping hands together with determination or calm (occurring often during speech)
- 2nd pair - Throwing arms into the air and dropping them back down to its sides with anger or happiness
- 3rd pair - Reacting with fear or interest to an object appearing suddenly within the field of view
- 4th pair - Folding arms with pensiveness or stubbornness
- 5th pair - Throwing arms down with strong disappointment (almost anger) or fatigue

Animations consist mostly of hands gesticulation (Figure 5). An additional motion is limited to some subtle legs and torso movements. Facial expression is not present in this study. This is because the purpose of this experiment is to assess emotions only by viewing gestures, thus the limitations applied should guarantee that only gesticulation is taken into account by the test participants. In each pair of the keyframes important poses of a body are the same, but temporal structure varies to achieve different emotions. As stated in the previous section time parameters are more important in conveying emotions than amplitudes. This means that utilizing the same keyframes but changing timing can result in producing different emotions.

In the experiment, an assumption was made that for each pair the sequences prepared represent the extreme intensities of emotions, and a continuous transition from one emotion to another is possible. Animations originating from those

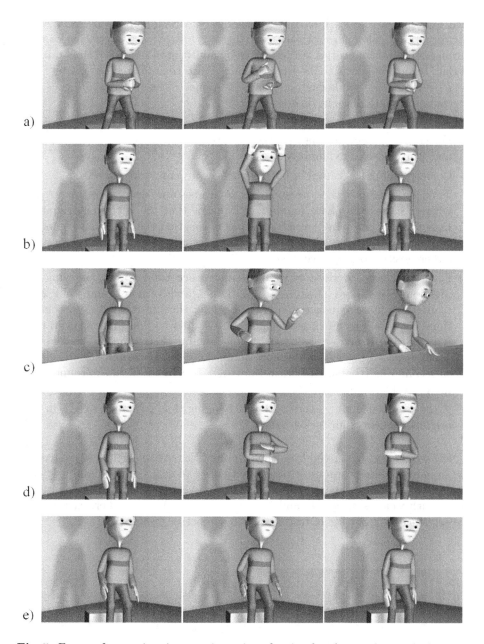

Fig. 5. Frames from animations: a. 1st pair - clasping hands together with determination or calm (occurring often during speech), b. 2nd pair - throwing arms into the air and dropping them back down to its sides with anger or happiness, c. 3rd pair - reacting with fear or interest to an object appearing suddenly within the field of view, d. 4th pair - folding arms with pensiveness or stubbornness, e. 5th pair - throwing arms down with strong disappointment (almost anger) or fatigue

with extreme intensities but conveying emotion with different intensity were then generated. Therefore for each pair three new sequences presenting combination of extreme intensities are generated by applying the so-called time-mixing method of interpolation in the time domain. Since time-related parameters are the most significant while characterizing emotional features of motion, thus the interpolation takes place in time domain.

- For each sequence all phases such as: anticipation, main motion, overshoot and stop phase were numbered.
- Corresponding keyframes in both sequences in pairs were found.
- Time parameters of the corresponding keyframes were denoted as t_{ineg}, t_{ipos}, where i states the phase number, neg means that the keyframe originates from the first emotion in a pair, and pos means that the keyframe originates from another one, the latter does not necessarily mean that emotions are either positive or negative.
- For each i new values of t_{ij} were calculated according to the following expression: $t_{ij} = t_{ineg} + (t_{ipos} - t_{ineg}) \cdot j/(n+1)$, where $n = 3$, and $j = 1, 2...n$ (generating more sequences requires higher n); j denotes the intensity of emotions, i.e. $j = 1$ result in animation quite similar to first animation in pair (that with keyframes t_{ineg}), $j = 2$ result in animation with time parameters equally spaced between two source animations, for $j = 3$ animation is closer to second one (that with keyframes t_{ipos}). The values of t_{ij} are rounded to integer value.
- j^{th} new sequence was generated by exchanging time parameters t_{ineg} in the original sequences for t_{ij} for each i.

New sequences are derived from the combinations of two original sequences in the domain of time positions of their keyframes (we call it time-mixing). It can also be interpreted as linear combination of two vectors of animation keyframe times, with weights $j/(n+1)$ and $(n+1-j)/(n+1)$. Altogether, each of the five sequences forms a set of test signals in subjective tests.

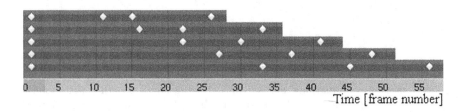

Fig. 6. Time positioning of the animation keyframes in one set. From top to the bottom: the original first sequence ('*determination*'), transitions generated between two original sequences resulting in j_1, j_2, j_3 sequences, and the second original sequence ('*calm*').

4.1 Subjective Test

The aim of the subjective test was to verify if a linear combination (time-mixing) of animations representing two different emotions can result in a realistic, readable animation with an emotion of a lessened intensity, but still a clear one. 24 persons took part in the test. Their task was to assign an appropriate label to the emotions presented in the sequences. From each set, two sequences were randomly chosen and displayed three times. If the rendered emotions were the same, the assignment should go to the one with higher intensity. Any uncertainty was to be marked with a question mark. Table 3 shows the percentage of selected options for each test sequence in five sets.

Table 3. Results of emotion assessment

1st set:

Signals:	Emotion perceived [%]:		
	calm	determ.	uncertain
calm	93.8	6.3	0.0
j1	95.8	2.1	2.1
j2	85.4	6.3	8.3
j3	43.8	50.0	6.3
determination	33.3	62.5	4.2

2nd set:

Signals:	Emotion	perceived	[%]:
	anger	happiness	uncertain
anger	56.5	32.6	10.9
j1	50.0	41.3	8.7
j2	45.7	47.8	6.5
j3	28.3	50.0	21.7
happiness	0.0	100.0	0.0

3rd set:

Signals:	Emotion perceived [%]:		
	fear	interest	uncertain
fear	89.1	6.5	4.3
j1	87.0	10.9	2.2
j2	73.9	17.4	8.7
j3	8.7	82.6	8.7
interest	4.3	95.7	0.0

4th set:

Signals:	Emotion	perceived	[%]:
	pensiveness	stub.	uncertain
pensiveness	93.8	0.0	6.3
j1	27.1	58.3	14.6
j2	29.2	58.3	12.5
j3	4.2	87.5	8.3
stubbornness	10.4	83.3	6.3

5th set:

Signals:	Emotion	perceived	[%]:
	disappointment	fatigue	uncertain
disappointment	95.8	2.1	2.1
j1	62.5	27.1	10.4
j2	58.3	29.2	12.5
j3	20.8	68.8	10.4
fatigue	52.1	43.8	4.2

In some cases (e.g. 'pensiveness' and 'j_1' in the 4th set) weaker intensity of the emotion time-mixed with another animation imposed the decrease of readability of this emotion. In many cases a time-mixing animation with the same weights applied (each 'j_2' in Table 3) creates readable animation ('calm' in the 1st set). In some cases (e.g. 'determination' in the 1st set, 'anger' in the 2nd set, 'stubbornness' in the 4th set) one emotion from the pair is mistakenly taken for another. This indicates that even slight changes in the temporal structure of the sequences result in a decreased readability of the given emotion. This could be solved by adding other emotional elements to an animation (e.g., facial movement, voice, or introducing exaggerated gestures whenever possible).

The results of choosing the emotion with a higher intensity, when both presented emotions seem to be the same are shown in Table 4. The gradation of the emotion is verified in the following manner. In the generation stage, sequences were marked as *neg*, j_1, j_2, j_3, *pos*. If two sequences are for example from the 1st set, then *neg* means '*calm*' and *pos* - '*determination*'. Let us assume that the test participant viewed *neg* and j_2, interpreted both as '*calm*', and chose the second one as less intense. In such a case the answer is correct, because j_2 is a result of time-mixing of temporal structures of '*calm*' with '*determination*' sequences, and therefore its intensity is weaker. When j_1 and j_2 are compared, and then named by the test participant as '*determination*', j_2 should be marked as more intense, because its distance to the sequence (*pos* - '*determination*') in the domain of the keyframe time positions is smaller then the distance of j_1.

Table 4. Results of emotion gradation in a subjective test

[%]	correct	incorrect	uncertain
1st set:	81.3	15.6	3.1
2nd set:	61.7	24.3	13.9
3rd set:	77.5	14.5	8.0
4th set:	94.8	2.1	3.1
5th set:	74.0	16.7	9.4

4.2   Data Mining

Although the number of generated sequences is very small, a knowledge base is created and the rough set analysis is performed for its knowledge discovery ability and rule calculation for further processing, especially for generation/classification of new animation sequences.

Analysis sets are presented in Table 5. Values in 'Case 1, 2, and 3' are utilized in three separate classification processes. The sequences were taken from all five sets described in the previous sections: two sequences of each pair and all transitions between them. Sequences j_1 and j_3 were labeled with a corresponding emotion name (Case 1) and a positive/negative quality (Case 3). For 'pensiveness', 'stubbornness', and 'disappointment' an alternative emotion name was also assigned, similar in expression (Case 2). All j_2 sequences were marked as neutral, because the results of the subjective test show that viewers, as Table 3 presents, in many cases are not able to classify them to either class. Moreover, the j_2 sequences originating from an equal combination (time-mixing) of two emotions are supposed to be neutral.

Table 5. Analysis sets

	Signal:	Case 1 decision	Case 2 decision	Case 3 decision
1st set:	calm	calm	calm	positive
	j1	calm	calm	positive
	j3	determination	determination	negative
	determination	determination	determination	negative
2nd set:	anger	anger	anger	negative
	j1	anger	anger	negative
	j3	happiness	happiness	positive
	happiness	happiness	happiness	positive
3rd set:	fear	fear	fear	negative
	j1	fear	fear	negative
	j3	interest	interest	positive
	interest	interest	interest	positive
4th set:	pensiveness	pensiveness	calm	positive
	j1	pensiveness	calm	positive
	j3	stubbornness	anger	negative
	stubbornness	stubbornness	anger	negative
5th set:	disappointment	disappointment	anger	negative
	j1	disappointment	anger	negative
	j3	fatigue	fatigue	negative
	fatigue	fatigue	fatigue	negative
All sets:	j2	neutral	neutral	neutral

Sequences were analyzed in terms of keyframes timing, as was described in Section 3.2. Lengths of anticipation, main motion, overshoot, hold, and stop phases were calculated, and fed along with 'Decision' data to the rough set analysis system.

For 'Case 1', each of the 'Decision' parameters is related only to two objects, therefore classification seems to be irrelevant (Figure 7). For a proper classification almost all parameters are required (anticipation, motion, overshoot, stop). 30 rules were generated and the classification achieved accuracy was 1.0. After filtering the rules, i.e. removing ones with support equal 1, 10 rules have been left, the coverage for 'calm' and 'fatigue' decreased to 0, but for other objects the classification accuracy remained at 1.0 level.

For 'Case 2' the 'pensiveness' objects are treated as 'calm', and the 'stubbornness' and 'disappointment' – as 'anger', because they are similar in expression. The set of 28 rules was generated, with the classification accuracy of 1.0. After filtration a set of 12 rules was left, the coverage for the 'fatigue' objects decreased to 0, but for all other objects the classification accuracy remained at 1.0. Then, for 'Case 3' all objects were partitioned into three classes - with a positive, negative, and neutral emotional expression (Figure 9). 'Case 3' is related to the meaning of gestures as identified by human. This means that each decision from 'Case 1' is now replaced with a label 'negative', 'positive' or 'neutral', e.g. both disappointment and fatigue are negative emotions, therefore label

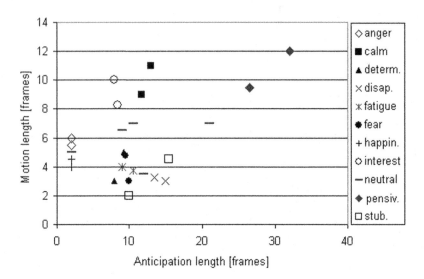

Fig. 7. Parameters of objects in 'Case 1'. Objects of classes are situated too close to each other, and too many rules are needed to classify all of them, resulting in irrelevant generalization.

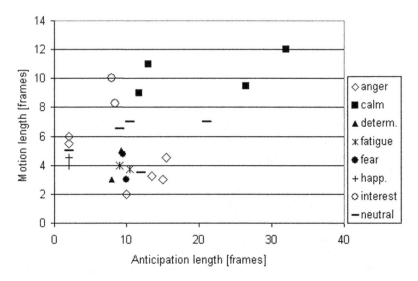

Fig. 8. Parameters of objects in 'Case 2'. 'Pensiveness' objects are now labeled as 'calm', 'stubbornness' and 'disappointment' – as 'anger', because they are similar in expression. 28 rules allow for correct classification of all objects. 12 rules can classify all except 'fatigue'.

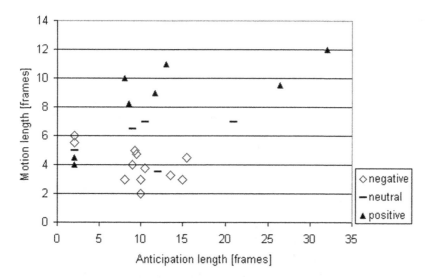

Fig. 9. Parameters of objects in 'Case 3'. All objects were labeled as 'negative', 'positive', and 'neutral', referring to their more abstract expression. 13 rules are enough to correctly classify all objects.

Table 6. Confusion matrix for 'Case 3'

	Predicted:					
	negative	positive	neutral	No.of obj.	Accuracy	Coverage
Actual: negative	11	0	0	12	1	0.917
positive	0	8	0	8	1	1
neutral	0	0	2	5	1	0.4

"negative" is assigned to them. A set of 13 rules was generated, with the classification accuracy and coverage equal 1.0. In that case 3 of 4 rules related to 'neutral' class have support equal 1. Objects of that class are isolated from each other, as can be seen in Figure 6, therefore generalization possibility is limited. After the filtration, when rules with low support are removed, 9 rules have been left, 4 describing a 'negative' class, 4 describing a 'positive' class, and 1 describing 'neutral' class. The results of classification are presented in Table 6. Accuracy is 1.0 for each class, but one 'negative' object and 3 'neutral' were omitted. These rules, even with cover not equal 1.0, may be used for generalization purposes and new object generation.

5   Conclusions

The results show that generating new animations by combining two animations with different emotions assigned to them does not result in a readable animation. A change in an original animation often creates confusion about the rendered emotion. It seems that there exists a narrow margin for the animation timing modifications that assure no change in emotional features, yet it is still advisable to create animations with as highest emotion intensity as possible. It should be noted that a good readability of a motion achieved by its exaggeration is always present in silent movies and animated movies made by hand. In computer animations this fact is often forgotten. Moreover, animating non-realistic characters requires that the exaggerated motion is supplemented by a non-realistic facial animation and a cartoonish-like style.

Presented results verify very important facts already known to animators. Firstly, the results show that a temporal structure of an animated sequence and the timing of its keyframes are important for expressing the emotion in an animation. Secondly, motion is unreadable when its expression is too far from the extreme intensity. On the other hand, a clear expressive gesticulation could be achieved by a proper timing of keyframes. Other animation means such as for example full body motion, facial expression, and modification of key poses are also helpful when presenting similar emotions like fatigue and calm. These additional cues enhance distinction between various emotional states, therefore utilizing features related to gesticulation only, may not be sufficient.

Lack of the relation between amplitudes and emotions, presented in Section 3, gives possibilities to introduce many constrains to motion, like the exact specification of the target position for a grabbing, or walking sequence. In that case amplitudes remain unchanged to satisfy constrains, but timing could vary, resulting in different emotional expressions.

6   Future Experiments

As seen from the presented experiments most of the rules generated in the rough set-based analysis are closely congruent to the traditional animation rules. Thus, in future experiments these rules will be used in the expert system. The input parameters fed to the system will consist of data obtained from the keyframes of a simple animation created for an easy manipulation along with the description of the desired emotional feature. Output parameters are the ones needed to create a realistic motion sequence, i.e. the positions of keyframes and the lengths of anticipation and overshoot phases.

For any simple animation, a set of modifications could be created, utilizing differentiated values of emotional descriptions. This can be done using the "computing with words" approach and the fuzzy logic processing [24]. Rules generated by the rough set analysis will be the basis of the system. In addition, the rough set measure is to be applied as the weight in the fuzzy processing [10]. With the output parameters from the fuzzy logic module, changes will be introduced to animation, resulting in new keyframes, and the insertion of anticipation and overshoot phases. For each parameter, membership functions should be defined, correlated to discretization cuts acquired in the process of rough set rule generation (e.g. for animation parameters), or triangle membership functions covering ranges from 0 to 100 (e.g. for animation features). Next for each membership functions appropriate linguistic description should be subjectively selected, and used as its label. Additional tests will be needed to assign ranges between e.g. 'short', 'medium' and 'long' anticipation.

It is planned to generate sequences with the same emotional features as the ones prepared by the animator in the first stage, and verify their emotional quality and naturalness. The results obtained may be utilized for derivation of better rules, and this may increase the effectiveness of the system.

The outlined methodology can also be extended to animate other parts of the human body, and this is planned as the future aim. The practical utility of this research is to enhance computer-based animation features in order to create animation more realistic and human-like.

References

1. Bazan, J.G., Szczuka, M.S., Wróblewski, J.: A new version of rough set exploration system. In: Alpigini, J.J, Peters, J.F., Skowron, A., Zhong, N. (eds.): Third International Conference on Rough Sets and Current Trends in Computing RSCTC. Lecture Notes in Artificial Intelligence **2475**. Springer-Verlag, Malvern (2002) 397–404

2. Bazan, J.G., Szczuka, M.S.: The Rough Set Exploration System. Transactions on Rough Sets III. In: James, F., Peters, J.F., Skowron, A. (eds.): Lecture Notes in Computer Science **3400**. Springer Verlag (2005) 37–56

3. Blair, P.: Cartoon Animation. Walter Foster Publishing, Laguna Hills (1995)

4. Bruderlin, A., Williams, L.: Motion signal processing. Computer Graphics **29** (1995) 97–104

5. Bryant, D.: The Uncanny Valley. Why are monster-movie zombies so horrifying and talking animals so fascinating?
http://www.arclight.net/~pdb/nonfiction/uncanny-valley.html

6. Cassell, J.: A framework for gesture generation and interpretation. In: Cipolla, R., Pentland, A. (eds.): Computer vision in human-machine interaction. Cambridge University Press, New York (2000) 191–215

7. Fang, A.C., Pollard, N.S.: Efficient Synthesis of Physically Valid Human Motion. ACM Transactions on Graphics, Vol. 22. ACM (2003) 417–426

8. Hatice, G., Piccardi, M., Tony, J.: Face and Body Gesture Recognition for a Vision-Based Multimodal Analyzer. In: Piccardi, M., Hintz, T., He, X., Huang, M.L., Feng, D.D., Jin, J. (eds.): Pan-Sydney Area Workshop on Visual Information Processing (VIP2003), Sydney. Conferences in Research and Practice in Information Technology, vol. 36 (2003)

9. Kendon, A.: How gestures can become like words. In: Poyatos, F.(ed.): Cross-cultural perspectives in nonverbal communication. New York (1988) 131–141

10. Kostek, B.: Soft Computing in Acoustics, Applications of Neural Networks, Fuzzy Logic and Rough Sets to Musical Acoustics. Studies in Fuzziness and Soft Computing. Physica-Verlag, Heidelberg, New York (1999)

11. Kostek, B., Szczuko, P.: Analysis and Generation of Emotionally-Charged Animated Gesticulation. In: Slezak, D., Yao, J., Peters, J.F., Ziarko, W., Hu, X. (eds.): Proceedings of the 10th International Conference Rough Sets, Fuzzy Sets, Data Mining, and Granular Computing (RSFDGrC). Regina, Canada. Lecture Notes in Computer Science **3642**, Springer (2005) 333–341

12. Mehrabian, A., Communication without words. Psychol. Today **4**(2) (1968) 53–56
13. Mori, M.: The uncanny valley. Energy **7** (1970) 33–35
14. Park, J., Kang, Y., Kim, S., Cho, H.: Expressive Character Animation with Energy Constraints. Proc. Edu+Compugraphics '97. Vilamoura, Portugal (1997) 260–268
15. Pawlak, Z.: Rough Sets: Theoretical aspects of reasoning about data. Kluwer Academic, Dordrecht (1991)
16. Polkowski, L., Skowron, A. (eds.): Rough Sets in Knowledge Discovery, vol. 1, and 2. Physica verlag, Heidelberg (1998)
17. Popović, J., Seitz, S.M., Erdmann, M.: Motion Sketching for Control of Rigid-Body Simulations. ACM Transactions on Graphics, vol. 22. ACM (2003) 1034–1054
18. Terra, S.C.L., Metoyer, R.A.: Performance Timing for Keyframe Animation. Proc. Eurographics/SIGGRAPH Symposium on Computer Animation (2004) 253–258
19. The RSES Homepage, http://logic.mimuw.edu.pl/~rses
20. Thomas F., Johnston O.: Disney Animation - The Illusion of Life. Abbeville Press, New York (1981)
21. Whitepaper: Dynamic Motion Synthesis. NaturalMotion Ltd, Oxford (2004)
22. Williams, R.: The Animator's Survival Kit: A Manual of Methods, Principles, and Formulas for Classical, Computer, Games, Stop Motion, and Internet Animators. Faber & Faber (2002)
23. Zemeckis, R.: Polar Express. Warner Bros. (2004)
24. Zadeh, L.: Fuzzy logic = Computing with words. IEEE Transactions on Fuzzy Systems, vol. 4 (1996) 103–111

Introducing a Rule Importance Measure

Jiye Li[1] and Nick Cercone[2]

[1] School of Computer Science, University of Waterloo
Waterloo, Ontario N2L 3G1, Canada
j27li@uwaterloo.ca
[2] Faculty of Computer Science, Dalhousie University
Halifax, Nova Scotia B3H 1W5, Canada
nick@cs.dal.ca

Abstract. Association rule algorithms often generate an excessive number of rules, many of which are not significant. It is difficult to determine which rules are more useful, interesting and important. We introduce a rough set based Rule Importance Measure to select the most important rules. We use ROSETTA software to generate multiple reducts. Apriori association rule algorithm is then applied to generate rule sets for each data set based on each reduct. Some rules are generated more frequently than the others among the total rule sets. We consider such rules as more important. We define rule importance as the frequency of an association rule generated across all the rule sets. Rule importance is different from either rule interestingness measures or rule quality measures because of their application tasks, the processes where the measures are applied and the contents they measure. The experimental results from an artificial data set, UCI machine learning datasets and an actual geriatric care medical data set show that our method reduces the computational cost for rule generation and provides an effective measure of how important is a rule.

Keywords: Rough sets, rule importance measure, association rules.

1 Introduction

Rough sets theory was first presented by Pawlak in the 1980's [1]. He introduced an early application of rough sets theory to knowledge discovery systems, and suggested that rough sets approach can be used to increase the likelihood of correct predictions by identifying and removing redundant variables. Efforts into applying rough sets theory to knowledge discovery in databases has focused on decision making, data analysis, discovering and characterizing the inter-data relationships, and discovering interesting patterns [2].

Although the rough sets approach is frequently used on attribute selection, little research effort has been explored to apply this approach to association rules generation. The main problem of association rule algorithm is that there are usually too many rules generated, and it is difficult to process the large amount of rules by hand. In the data preprocessing stage, redundant attributes

J.F. Peters and A. Skowron (Eds.): Transactions on Rough Sets V, LNCS 4100, pp. 167–189, 2006.

can be found by a rough sets approach. By removing the redundant attributes, association rules generation will be more efficient and more effective.

Klemettinen introduced the concept of rule templates [3]. Properly defined rule templates can be helpful on generating desired association rules to be used in decision making and collaborative recommender systems [4,5].

We discuss how the rough sets theory can help generating important association rules. We propose a new rule importance measure based on rough sets to evaluate the utilities of the association rules. Rules generated from reducts are representative rules extracted from the data set; since a reduct is not unique, rule sets generated from different reducts contain different sets of rules. However, more important rules will appear in most of the rule sets; less important rules will appear less frequently than those more important ones. The frequencies of the rules can therefore represent the importance of the rules.

To test our hypothesis, we first use ROSETTA [6] rough sets toolkit to generate multiple reducts. We then use apriori association rules generation to generate rule sets for each reduct set. We are interested in applying these rules for making decisions. Therefore, the type of rules we are looking for are rules which have, on the consequent part, the decision attributes, or items that can be of interest for making decisions. Some rules are generated more frequently than the others among the total rule sets. We consider such rules more important. We define the rule importance measure according to the frequency of an association rule among the rule sets. We will show by the experimental results that our method provides diverse measures of how important are the rules, and at the same time reduces the number of rules generated. This method can be applied in both decision making and recommender system applications.

Our method is among the few attempts on applying rough sets theory to association rules generation to improve the utility of an association rule. Rule importance measure is different from either the rule interestingness measures or the rule quality measures, which are the two well-known approaches on evaluating rules. The rule importance measure is different from the rule interestingness measures. Most of the rule interestingness measures are used to evaluate classification rules, and different people have different definition for "interestingness". Rule importance measure is applied to evaluate association rules. It is a straightforward and objective measure. Rule importance measure is different from rule quality measure as well. Rule quality measure is used to evaluate the quality of an classification rule. Whereas rule importance measure is applied from the process of reduct generation to rule generation, and the rules evaluated are association rules.

In our earlier work [7], we evaluated the rule importance measures on an artificial data set and a geriatric care data set. This paper extends the experiments to include 13 data sets from UCI machine learning repository [8]. Detailed analysis on the differences between rule importance measures and rule interestingness measures, rule importance measures and rule quality measures are provided. We also conduct comparison experiments on the different effects of ranking rules between the rule importance measures and the rule interestingness measures.

A number of interesting results from the experiments on the data sets are also discussed.

We discuss related work on association rules algorithm and rough sets theory on rule discovery in section 2. In section 3 we introduce the rule importance measure. In section 4 we experiment the rule importance measure on an artificial car data set, UCI data sets and a geriatric care data set. We summarize our contributions and discuss the continuing work in section 5.

2 Related Work

2.1 Association Rules Algorithm

An association rule algorithm helps to find patterns which relate items from transactions. For example, in market basket analysis, by analyzing transaction records from the market, we could use association rule algorithms to discover different shopping behaviors such as, when customers buy bread, they will probably buy milk. Association rules can then be used to express these kinds of behaviors, thus helping to increase the number of items sold in the market by arranging related items properly. An association rule [9] is a rule of the form $\alpha \rightarrow \beta$, where α and β represent itemsets which do not share common items. The association rule $\alpha \rightarrow \beta$ holds in the transaction set L with confidence c, $c = |\alpha \cup \beta|/|\alpha|$, if $c\%$ of transactions in L that contain α also contain β. The rule $\alpha \rightarrow \beta$ has support s, $s = |\alpha \cup \beta|/|L|$, if $s\%$ of transactions in L contain $\alpha \cup \beta$. Here, we call α antecedent, and β consequent. Confidence gives a ratio of the number of transactions that the antecedent and the consequent appear together to the number of transactions the antecedent appears. Support measures how often the antecedent and the consequent appear together in the transaction set.

A well known problem for association rules generation is that too many rules are generated, and it is difficult to determine manually which rules are more useful, interesting and important. In our study of using rough sets theory to improve the utility of association rules, we propose a new rule importance measure to select the most appropriate rules. In addition to the experimentations on artificial data sets and UCI data sets, we also perform the experiments on a larger data set, a geriatric care data set, to explore the application of the proposed method.

2.2 Rough Sets Theory and Rule Discovery

Rough Sets was proposed to classify imprecise and incomplete information. Reduct and core are two important concepts in rough sets theory. A reduct is a subset of attributes that are sufficient to describe the decision attributes. Finding all the reduct sets for a data set is a NP-hard problem [10]. Approximation algorithms are used to obtain the reduct set [11]. All reducts contain the core. Core represents the most important information of the original data set. The intersection of all the possible reducts is the core.

We use ROSETTA GUI version 1.4.41 rough sets toolkit [6] for multiple reducts generation. The reducts are obtained by the Genetic Algorithm and Johnson's Algorithm with the default option of full discernibility[1] [12]. Hu et al. [13] proposed a new core generation algorithm based on rough sets theory and database operations. We use Hu's algorithm to generate core attributes and to examine the effect of core attributes on the generated rules. The algorithm is shown in Algorithm 1.

Algorithm 1. Core Generating Algorithm[13]

---

input   : Decision table $T(C, D)$, C is the condition attributes set; D is the
               decision attribute set.
output: $Core$, Core attributes set.

$Core \leftarrow \phi$;
for *each condition attribute $A \in C$* **do**
   if $Card(\Pi(C - A + D)) \neq Card(\Pi(C - A))$ **then**
   | $Core = Core \cup A$;
   end
end
return $Core$;

---

where C is the set of condition attributes, and D is the set of decision attributes. $Card$ denotes the count operation, and Π denotes the projection operation.

This algorithm is developed to consider the effect of each condition attribute on the decision attribute. If the core attribute is removed from the decision table, the rest of the attributes will bring different information to the decision making. The algorithm takes advantage of efficient database operations such as count and projection. Since the attributes of the core are contained in any reduct sets for a data set, this algorithm also provides an evaluation to justify the correctness of the reduct sets.

There have been contributions on applying rough sets theory to rule discovery. Rules and decisions generated from the reducts are representative of the data set's knowledge. In [14], two modules were used in the association rules mining procedure for supporting organizational knowledge management and decision making. Self-Organizing Map was applied to cluster sale actions based on the similarities in the characteristics of a given set of customer records. Rough sets theory was used on each cluster to determine rules for association explanations. Hassanien [15] used rough sets to find all the reducts of data that contain the minimal subset of attributes associated with a class label for classification, and

---

[1] For reduct generation, there are two options on discernibility provided by ROSETTA software, which are full discernibility and object related discernibility. With the option of full discernibility, the software will produce a set of minimal attribute subsets that can discern all the objects from each other. With object related discernibility, the software produces reducts that can discern a certain object from all the other objects.

classified the data with reduced attributes. In sections 3.5 and 3.6 we discuss other related research specific to the content of those sections.

Rough sets theory can help to determine whether there is redundant information in the data and whether we can find the essential data needed for our applications. We expect fewer rules will be generated due to fewer attributes.

3   Rule Importance Measures

3.1   Motivation

In medical diagnosis, a doctor requires a list of symptoms in order to make a diagnosis. For different diseases, there are different patient symptoms to examine. However, there are some routine exams that the doctor must perform for all the patients, such as the age of the patient, the blood pressure, the body temperature and so on. There are other symptoms that doctors may take into consideration, such as whether the patients have difficulty walking, whether the patients have bladder problems and so on. We would like to find the most important symptoms for diagnoses. We know that the symptoms that are checked more frequently are more important and essential for making diagnoses than those which are considered less frequently. However, both the symptoms that require frequent checking and the symptoms that are checked less frequently are included in the list of checkup symptoms. In this way, the doctor will make a precise diagnose based on all possible patient information.

3.2   Defining the Rule Importance Measure

The medical diagnosis process can be considered as a decision making process. The symptoms can be considered as the condition attributes. The diagnosed diseases can be considered as the decision attributes. Since not all symptoms need to be known to make a diagnosis, the essential symptoms are considered as representative. These symptoms can be selected by a reduct generation algorithm.

All the patient information can also be represented in a transaction data set, with each patient's record considered to be an item set. Association rule algorithm can be applied on this transaction data set to generate rules, which have condition attributes on the antecedent part and decision attributes on the consequent part of the rules. Rules generated from different reduct sets can contain different representative information. If only one reduct set is being considered to generate rules, other important information might be omitted. Using multiple reducts, some rules will be generated more frequently than other rules. We consider the rules that are generated more frequently more important.

We propose a new measure, *Rule Importance*, to evaluate the importance of association rules. A rule is defined to be important by the following definition.

Definition 1. *If a rule is generated more frequently across different rule sets, we say this rule is* more important *than rules generated less frequently across those same rule sets.*

Rule importance measure is defined as follows,

Definition 2

$$Rule\ Importance\ Measure = \frac{\begin{array}{c} Number\ of\ times\ a\ rule\ appears\ in\ all \\ the\ generated\ rules\ from\ the\ reduct\ sets \end{array}}{Number\ of\ reduct\ sets}.$$

The definition of the rule importance measure can be elaborated by Eq. 1. Let n be the number of reducts generated from the decision table $T(C, D)$. Let $RuleSets$ be the n rule sets generated based on the n reducts. $ruleset_j \in RuleSets$ $(1 \leq j \leq n)$ denotes individual rule sets containing rules generated based on reducts. $rule_i$ $(1 \leq i \leq m)$ denotes the individual rule from $RuleSets$. RIM_i represents the rule importance measure for the individual rule. Thus the rule importance measures can be computed by the following

$$RIM_i = \frac{|\{ruleset_j \in RuleSets | rule_i \in ruleset_j\}|}{n}. \tag{1}$$

The following example shows how to compute the rule importance measure. We use the Iris [8] data set as an example. There are $n = 4$ reducts available for rule generations. For each of the reducts, the rule sets generated based on the reduct are shown in the following.

Reducts	Rule Sets
$\{sepalLength, sepalWidth, petalLength\}$	$\{sepalLength4.4 \to setosa, sepalWidth2.9 \to versicolor, petalLength1.9 \to setosa, \ldots\}$
$\{sepalWidth, petalLength, petalWidth\}$	$\{sepalWidth2.9 \to versicolor, petalLength1.9 \to setosa, petalWidth1.1 \to versicolor, \ldots\}$
$\{sepalLength, petalLength, petalWidth\}$	$\{sepalLength4.4 \to setosa, petalLength1.9 \to setosa, petalWidth1.1 \to versicolor, \ldots\}$
$\{sepalLength, sepalWidth, petalWidth\}$	$\{sepalLength4.4 \to setosa, sepalWidth2.9 \to versicolor, petalWidth1.1 \to versicolor, \ldots\}$

Rule $sepalLength4.4 \to setosa$ is generated across 3 rule sets, therefore the rule importance is $RIM = \frac{3}{4} = 75\%$. For rules $sepalWidth2.9 \to versicolor$, $petalLength1.9 \to setosa$, $petalWidth1.1 \to versicolor$, they are all generated from 3 of the 4 rule sets, therefore their rule importance are 75%. The rule importance for the rest rules can be found in Table 5.

3.3   Modeling the Rule Importance Measure

The general model on which we compute the rule importance measure is shown in Fig. 1.

Firstly during the data preprocessing step, the inconsistent data instances and the data instances containing missing attribute values are processed. Inconsistency exists in a decision table when two or more data instances contain the same condition attribute values but different decision attribute values. These data instances must be removed. We first sort the whole data set according to the condition attributes, excluding the decision attributes. Then we select data instances that contain the same condition attributes values, but different decision attributes values. These data instances are inconsistent and they are removed during this stage. Discretizations, such as equal frequency binning or entropy algorithm [6], are also applied during this stage if necessary. Core attributes are

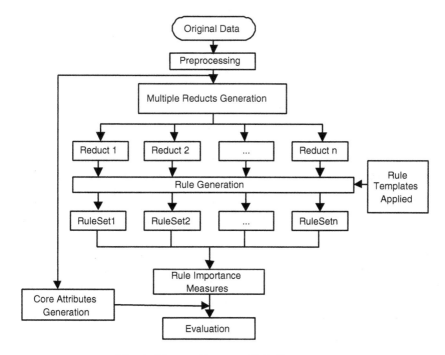

Fig. 1. How to Compute Rule Importance

generated at the end of the data preprocessing stage. It is worthwhile to mention that core generation requires no inconsistencies in the data set.

After data is preprocessed, multiple reducts are generated. Various algorithms and rough set software provide multiple reducts generation. For example, ROSETTA's genetic algorithm generates multiple reducts; RSES [16] provides a genetic algorithm for limited reducts generation (maximum 10 reducts), which is appropriate in cases of larger data sets for only generating representative reducts.

After multiple reducts are generated, the condition attributes contained in the reduct together with the decision attributes are used as the input data for rule generation. Rule templates such as

$$\langle Attribute_1, Attribute_2, \ldots, Attribute_n \rangle \rightarrow \langle DecisionAttribute \rangle,$$

are applied in the rule generation step. Depending on different applications and the expected results, rule templates for desired types of rules and for subsumed rules are defined prior to the rule generation and are applied during the rule generation process. Multiple rule sets are therefore generated after the rule generations for multiple reducts. Rule importance measures are further calculated for each generated rule by counting the rule frequencies appearing across all the rule sets. Rules with their individual importance measures are ranked according to Eq. 1 and returned from the model.

In the evaluation stage of the model, core attributes play an important role for evaluating these ranked rules. Rules with 100% importance contain all the core attributes. Rules that contain more core attributes are more important than rules that contain fewer or none core attributes. Since core attributes are the most representative among all the condition attributes, more important rules contain these more representative attributes, which are the core attributes. Therefore by checking for the presence of the core attributes in the rules, we can evaluate the ranked rules with their rule importance.

3.4   Complexity Analysis

We analyze the time complexity for the proposed approach of generating important rules. Suppose there are N data instances in the data set, and M attributes for each data instance, N' is the number of distinct values in the discernibility matrix, t is the number of multiple reducts for the data set, the time complexity in the worst case is analyzed as follows. The time complexity for multiple reducts generation is $O(N'^2)$ [17]. The core generation takes $O(NM)$ [13]. The apriori association rules generation takes $O(NM!)$ [9], therefore it takes $O(tNM!)$ to generate multiple rule sets for multiple reducts. The calculation of the rule importance for the total rules k generated by the multiple rule sets takes $O(k \log k)$. In general, t is much smaller than N, therefore the time complexity of our approach is bounded by $O(N'^2 + NM + NM! + k \log k) \approx O(NM!)$ in the worst case.

3.5   How Is Rule Importance Different from Rule Interestingness

Rule generation often brings a large amount of rules to analyze. However, only part of these rules are distinct, useful and interesting. How to select only useful, interesting rules among all the available rules has drawn the attention of many researchers. One of the approaches to help selecting rules is to rank the rules by "rule interestingness measures". Rules with higher measures are considered more interesting. The rule interestingness measures, originated from a variety of sources, have been widely used to extract interesting rules.

The rule importance measure is a new measure to rank the rules. It is different from the rule interestingness measure in the following ways.

– Rule importance measure is used to evaluate association rules. Rule interestingness measure applies to classification rules except that *support* and *confidence* are two necessary parameters used in association rules generation, and they are considered as rule interestingness measures. Rule interestingness measures usually cannot be used to measure association rules. The input data for association rule generation is transaction data, and there is usually no class label with the transaction data. However the class label is required for calculating the rule interestingness measure.
– Rule importance measure is an objective measure. Rule interestingness measure can be either objective or subjective. In order to determine whether a rule is "interesting" or not, different people may have different opinions. Therefore "domain experts" are required to help make evaluations. However the rule importance measure does not require human supervision. The rule

importance measure uses the notion of a "reduct" from rough sets theory. Recall that a reduct selects the maximally independent set of attributes from a data set, that is, the reduct contains a set of attributes that are sufficient to define all the concepts in the data. These attributes contained in the reduct are considered to be more important. Rule importance measure is thus computed across all the rule sets generated from all the possible reducts of a data set. Since the reducts contain important attributes, rule sets generated from the reducts can fully characterize the concepts of the data, rule importance thus provides an evaluation on how important these rules are. There is no subjective matters involved in this measure. Although the generation of all the reducts are NP-hard problem, ROSETTA provides the genetic algorithm to generate multiple reducts objectively. However the rule interestingness measure usually requires domain experts' evaluation. Rules that are considered interesting may not be important.

– Rule importance measure provides more direct and obvious measures. Rule interestingness measures often involve selections according to the specific applications. In [18] Hilderman showed that there is no rule interestingness measure that can always perform better than the others in all applications. Each individual rule interestingness measure is based on its selection bias on the data. In order to determine what is the best interestingness measures to use for a certain application data, all the possible measures have to be compared to determine the best measure. But the rule importance measure does not consider the type or applications of the data. It can be used directly on the data from any application field.

– Rule importance measure reduces the amount of data required for rule generation by selecting only important attributes from the original data. The number of rules generated is thus greatly reduced. Rule interestingness measure is applied after the rules are generated. Therefore it requires more computational resources.

In summary, rule importance measure is straightforward and easy to compute; it provides a direct and objective view of how important is a rule.

3.6   How Is Rule Importance Different from Rule Quality

The concept of rule quality measures was first proposed by Bruha [19]. The motivation for exploring this measure is that decision rules are different with different predicting abilities, different degrees to which people trust the rules and so on. Measures evaluating these different characteristics should be used to help people understand and use the rules more effectively. These measures have been known as rule quality measures.

The rule quality measures are often applied in the post-pruning step during the rule extraction procedure [20]. The measure is used to evaluate whether the rules overfit the data. When removing an attribute-value pair, the quality measure does not decrease in value, this pair is considered to be redundant and will be pruned. In general, rule generation system uses rule quality measures to determine the stopping criteria for the rule generations and extract high quality rules. In [21] twelve different rule quality measures were studied and compared through the ELEM2 [20] system on their classification accuracies. The measures include empirical measures, statistical measures and measures from information theory.

The rule importance measure is different from the rule quality measure because of the following.

- The rule importance measure is used to evaluate how important is an association rule. Rule quality measures explore classification tasks of data mining, and are targeted towards improving the quality of decision rules. We cannot use the rule quality measures to evaluate the association rules.
- The rule importance measure takes transaction data as input. There is no class label from the transaction data. The measure evaluates how important is an association rule without considering other information from the data. Sometimes the transaction data can be processed by organizing the data into the form of a decision table. In this situation, the rule importance measure evaluates the relations between the condition attributes and the class. However, the rule quality measures are used to evaluate the relations between the rules and the class.
- The rule importance measure takes input of multiple reducts and multiple rule sets, then calculates the frequencies of each rule across multiple rule sets. The measure is used throughout the rule generation process. The rule quality measure is often used in the post-pruning process of the rule classification system.
- The rule importance measure considers the representative attributes contained in the reducts, and rule generations are based on the reducts. Therefore, redundant attributes are removed before the rule generation, and the number of rules generated are much fewer than rules generated from the original data set. Thus the computation cost is lower. When the rule quality measures are used to remove the low quality rules from the generated rules, rule generation computation cost is greater than that of the rule importance measure.

In summary, rule importance measure is different from rule quality measure because of the differences between their application tasks, the processes where the measures are applied and the contents they measure.

4   Experiments

In this section, we explain the experiments we conducted to generate rule importance measures on an artificial car data set, UCI data sets and a geriatric care data set.

The reduct is generated from ROSETTA GUI version 1.4.41. ROSETTA provides the following approximation algorithms for reducts generation: Johnson's algorithm, Genetic algorithm, Exhaustive calculation and so on. Johnson's algorithm returns a single reduct. Genetic algorithm returns multiple reducts. Exhaustive calculation returns all possible reduct, although given a larger data set, this algorithm takes longer time to generate reduct sets [12]. In our experiment, we use the genetic algorithm to generate multiple reduct sets with the option of full discernibility. The apriori algorithm [22] for large item sets generation and rule generation is performed on Sun Fire V880, four 900Mhz UltraSPARC III processors, with 8GB of main memory.

4.1   Specifying Rule Templates

Apriori association rules algorithm is used to generate rules. Because our interest is to make decisions or recommendations based on the condition attributes, we are looking for rules with only decision attributes on the consequent part. Therefore, we specify the following 2 rule templates to extract rules we want as shown by Template 2, and to subsume rules as shown by Template 3.

$$\langle Attribute_1, Attribute_2, \ldots, Attribute_n \rangle \rightarrow \langle DecisionAttribute \rangle \qquad (2)$$

Template 2 specifies only decision attributes can be on the consequent part of a rule, and $Attribute_1$, $Attribute_2$, ..., $Attribute_n$ lead to a decision of $DecisionAttribute$, as shown by Template 2.

We specify the rules to be removed or subsumed using Template 3. For example, given rule

$$\langle Attribute_1, Attribute_2 \rangle \rightarrow \langle DecisionAttribute \rangle \qquad (3)$$

the following rules

$$\langle Attribute_1, Attribute_2, Attribute_3 \rangle \rightarrow \langle DecisionAttribute \rangle \qquad (4)$$

$$\langle Attribute_1, Attribute_2, Attribute_6 \rangle \rightarrow \langle DecisionAttribute \rangle \qquad (5)$$

can be removed because they are subsumed by Template 3.

We use the artificial car data set that is used in section 4.2 as an example to further explain how to define proper templates. Since we are interested in predicting the mileage of a car based on the model of a car, the number of doors, the compression, the weight as well as other factors related to a car, we would like to extract rules which have the decision attribute "mileage" on the consequent part of the rules. Therefore we specify the template for desired rules as shown in Eq. 6

$$\langle model, cyl, \ldots, weight \rangle \rightarrow \langle mileage \rangle. \qquad (6)$$

And if a rule

$$\langle JapanCar, weight\_medium \rangle \rightarrow \langle mileage\_High \rangle \qquad (7)$$

is generated, rules such as Eq. 8

$$\langle JapanCar, trans\_manual, weight\_medium \rangle \rightarrow \langle mileage\_High \rangle \qquad (8)$$

is subsumed, because this rule can be deduced by the previous rule.

4.2   Experiments on Artificial Car Data Set

The first data set on which we experiment is an artificial data set about cars [23], as shown in Table 1. It is used to decide the mileage of different cars. The

Table 1. Artificial Car Data Set

make_model	cyl	door	displace	compress	power	trans	weight	Mileage
USA	6	2	Medium	High	High	Auto	Medium	Medium
USA	6	4	Medium	Medium	Medium	Manual	Medium	Medium
USA	4	2	Small	High	Medium	Auto	Medium	Medium
USA	4	2	Medium	Medium	Medium	Manual	Medium	Medium
USA	4	2	Medium	Medium	High	Manual	Medium	Medium
USA	6	4	Medium	Medium	High	Auto	Medium	Medium
USA	4	2	Medium	Medium	High	Auto	Medium	Medium
USA	4	2	Medium	High	High	Manual	Light	High
Japan	4	2	Small	High	Low	Manual	Light	High
Japan	4	2	Medium	Medium	Medium	Manual	Medium	High
Japan	4	2	Small	High	High	Manual	Medium	High
Japan	4	2	Small	Medium	Low	Manual	Medium	High
Japan	4	2	Small	High	Medium	Manual	Medium	High
USA	4	2	Small	High	Medium	Manual	Medium	High

condition attributes are *make_mode, cyl, door, displace, compress, power, trans, weight*. *Mileage* is the decision attribute. There are 14 instances. The data set does not contain missing attribute values.

For the car data set, the core attributes are, *make_model* and *trans*. ROSETTA generates 4 reducts as shown in Table 2. Rule sets are generated based on these

Table 2. Reducts Generated by Genetic Algorithm for Artificial Car Data Set

No.	Reduct Sets
1	{make_model, compress, power, trans}
2	{make_model, cyl, compress, trans}
3	{make_model, displace, compress, trans}
4	{make_model, cyl, door, displace, trans, weight}

4 reduct sets with *support* = 1%, *confidence* = 100%, and we also rank their rule importance, as shown in Table 3.

Discussion. From Table 3, the first 2 rules have an importance of 100%. This observation matches our experiences on cars. The auto transmission cars usually have a lower mileage than the manual cars. Japanese cars are well known for using less gas and providing higher mileage. The rule "Door_4 → Mileage_Medium" has a lower importance because the number of doors belonging to a car does not affect car mileage. We noticed that two rules with importance of 100% contain core attributes and only core attributes to make a decision of mileage. For

Table 3. The Rule Importance for the Artificial Car Data Set

No.	Selected Rules	Rule Importance
1	Trans_Auto → Mileage_Medium	100%
2	JapanCar → Mileage_High	100%
3	USACar, Compress_Medium → Mileage_Medium	75%
4	Compress_High, Trans_Manual → Mileage_High	75%
5	Displace_Small, Trans_Manual → Mileage_High	50%
6	Cyl_6 → Mileage_Medium	50%
7	USACar, Displace_Medium, Weight_Medium → Mileage_Medium	25%
8	Power_Low → Mileage_High	25%
9	USACar, Power_High → Mileage_Medium	25%
10	Compress_Medium, Power_High → Mileage_Medium	25%
11	Displace_Small, Compress_Medium → Mileage_High	25%
12	Door_4 → Mileage_Medium	25%
13	Weight_Light → Mileage_High	25%

the rest of the rules with importance less than 100%, the attributes on the left hand side of a rule contains non-core attributes. This observation implies that core attributes are important when evaluating the importance of the rules. Our method of generating rules with reduct sets is efficient. There are 6, 327 rules generated from the original data without using reducts or rule templates. 13 rules are generated using reducts and rule templates.

4.3 Experiments on UCI Data Sets

We experiment on selected UCI data sets [8] A through M described below. In Table 4, we list for each data set, the name of the data set, the number of condition attributes, the number of instances it contains; the number of reducts returned by ROSETTA genetic algorithm, sample reducts; and core attributes returned by Algorithm 1. In Table 5, we list the number of rules generated using the original data set with certain support and confidence values without applying the rule templates or the rule importance measure; the number of rules generated from the reducts with the same support and confidence values with the rule templates following the rule importance measure procedure shown in Figure 1; and sample rules ranked by the rule importance measure.

A. Abalone Data. This data set is used to predict the age of abalone from physical measurements. There are 4, 177 instances and 8 condition attributes in this data set. There are no missing attribute values or inconsistent data instances in the data set.

B. Breast Cancer Data. This data set contains 9 condition attributes and 286 instances. The date is used to diagnose the breast cancer disease. There are missing attributes existing in the data set. We ignore all the missing attribute values, and remove 9 records, we have 277 instances in the data. There are 12 inconsistent data records removed from the data as well.

Table 4. UCI Data Sets

Data Set	Condition Attributes	No. of Instances	No. of Reducts	Sample Reducts	Core Attributes
Abalone	8	4,177	16	{WholeWeight, ShuckedWeight, ShellWeight} {Height, WholeWeight, ShuckedWeight, VisceraWeight} {Sex, Length, Height, WholeWeight, ShellWeight}	Empty
Breast Cancer	9	286	1	{age, menopause, tumor-size, deg-malig, breast, breast-quad, irradiat}	age, menopause, tumor-size, deg-malig, breast, breast-quad, irradiat
Car	6	1,728	1	{buying, maint, doors, persona, lug_boot, safety}	buying, maint, doors, persona, lug_boot, safety
Glass	9	214	21	{RI, Al} {Na, Si} {RI, Na, Mg} {Na, Mg, K, Fe}	Empty
Heart	13	303	57	{age, chol, exang} {age, trestbps, chol} {chol, thalach, slope, ca} {sex, chol, oldpeak, ca, thal}	Empty
Iris	4	150	4	{sepalLength, sepalWidth, petalLength} {sepalLength, petal Length, petalWidth} {sepalWidth, petalLength, petalWidth} {sepalLength, sepalWidth, petalWidth}	Empty
Lympho-graphy	18	148	147	{blockofaffere, hangesinnode, changesinstru, specialforms, dislocationof, noofnodesin}	Empty
Pendigits	16	7,494	246	{C3, C6, C12, C13} {C3, C7, C10, C13, C14}	Empty
Pima Diabetes	8	768	28	{blp, pedigree, age} {times, glucose, pedigree} {glucose, blp, insulin, age}	Empty
Spambase	57	4,601	110	{will, report, you, credit, hp, george, meeting re, edu, (, !, average, total} {make, all, our, mail, report, free, you, credit, your george, technology, meeting, re, edu, !, average, total}	re, meeting, george, you, !, total, edu
Wine	13	178	66	{Flavanoids, Color} {Proanthocyanins, Color} {MalicAcid, Alcalinity, Phenols}	Empty
Yeast	8	1,484	4	{mcg, alm, mit, vac}, {mcg, gvh, mit, vac} {mcg, gvh, alm, vac, nuc} {gvh, alm, mit, vac, nuc}	vac
Zoo	16	101	27	{eggs, aquatic, toothed, breathes, legs} {milk, aquatic, backbone, venomous, legs, catsize} {hair, eggs, aquatic, predator, breathes, fins, legs}	aquatic, legs

C. Car Data. The car data set contains 6 condition attributes, and 1,728 instances. We apply association rules algorithm with rule templates, and there are 9 rules generated. We first use core algorithm to generate core attributes, and all the condition attributes are the core attributes. There is only one reduct generated for this data set, and the reduct contains all the core attributes.

D. Glass Data. This data set is used for the study of classification of types of glass by criminological investigation. At the scene of the crime, the glass left can

Table 5. UCI Data Sets on Rule Importance Measures

Data set	No. Rules with Original Data	No. Rules by RIM	Sample Rules by Rule Importance Measure (% indicates the Rule Importance)
Abalone	$(s = 0.1\%,$ $c = 100\%)$ 218	17	Viscera weight=0.1730 → Rings=9 [62.50%] Infant, Height=0.12, Length=0.5 → Rings=8 [18.75%] Female, Height=0.165, Diameter=0.48 → Rings=10 [12.50%]
Breast Cancer	$(s = 1\%,$ $c = 100\%)$ $49,574$	225	age30-39, tumor-size20-24, NoIrradiat → no-recurrence-events [100%] age50-59, menopause_premeno, degmalig_3 rightbreast → recurrence-events [100%] tumor-size30-34, degmalig_3, breast-quad_rightup → recurrence-events [100%]
Car	$(s = 1\%,$ $c = 100\%)$ 341	9	BuyingPrice_v-high, Maintainance_v-high → Decision_unacceptable [100%] BuyingPrice_v-high, SizeLuggageBoot_small, Safety_med → Decision_unacceptable [100%]
Glass	$(s = 0.5\%,$ $c = 100\%)$ $9,129$	129	Si=72.19 → Type_2 [44.44%] Na=14.38 → Type_7 [33.33%] Na=13.48, Mg=3.74 → Type_1 [11.11%]
Heart	$(s = 1\%,$ $c = 100\%)$ $71,534$	237	maximum_heart_rate_179 → $class_0$ [61.40%] oldpeak_3.4 → $class_2$ [47.37%] chol_234, restecg_2 → $class_0$ [12.28%] age65, female, thal_normal → $class_0$ [3.51%] male, restingBloodPressure_130, no_exercise_induced_angina, no_major_vessels_colored_by_flourosopy → $class_0$ [1.75%]
Iris	$(s = 1\%,$ $c = 100\%)$ 352	50	petalWidth1.1 → Iris-versicolor [75%] sepalWidth2.9 → Iris-versicolor [75%] petalLength1.9 → Iris-setosa [75%] sepalLength5.4, sepalWidth3.4 → Iris-setosa [50%]
Lympho-graphy	$(s = 10\%,$ $c = 100\%)$ $75,731$	43	changesinnode=lac.margin, bloflymphc=yes → metastases [51.02%] specialforms=vesicles, lymnodesenlar=4 → malign lymph [30.61%] blockofaffere=yes, bypass=no, earlyuptakein=no → metastases [7.48%]
Pendigits	$(s = 0.5\%,$ $c = 100\%)$ 389	52	C3_0, C13_100 → Class 8 [31.30%] C3_0, C9_100, C12_100 → Class 0 [6.10%] C1_0, C12_50, C14_25 → Class 1 [0.41%]
Pima Diabetes	$(s = 0.5\%,$ $c = 100\%)$ 429	126	Diabetes pedigree function_0.237 → Tested negative [60.71%] Plasma glucose concentration_187 → Tested positive [53.57%] Pregnant_twice, insulin_0, age_25 → Tested negative [3.57%]
Spambase	$(s = 1\%,$ $c = 100\%)$ $37,374,343$	2,190	you=0, re=0, !=0, average=1 → NotSpam [100%] !=0, captialCharacterLongest=2 →NotSpam [67.27%] george=0, re=0, edu=0, !=0, longest=3 → NotSpam[67.27]
Wine	$(s = 1\%,$ $c = 100\%)$ 548	247	Nonflavanoid0.14 → $class_2$ [21.21%] Malic acid 1.64 → $class_1$ [18.18%] Nonflavanoid phenols0.53, Alcalinity of ash 21.00 → $class_3$ [10.61%] color intensity5.40, Hue 1.25 → $class_1$ [1.52%]
Yeast	$(s = 0.2\%,$ $c = 100\%)$ $20,864$	195	alm0.39, vac0.51 → ME3 [75%] alm0.51, vac0.51, gvh0.48 → CTY [50%] mcg0.43, nuc0.33 → NUC [25%] mcg0.46, vac0.51, nuc0.22 → CYT [25%]
Zoo	$(s = 10\%,$ $c = 100\%)$ $680,996$	31	aquatic, 6 legs → Type 6 [100%] no eggs, 2 legs → Type 1 [66.67%] eggs, non breathes, non fin → Type 7 [7.41%]

be used as evidence. There are 214 instances and 9 condition attributes. There are no missing attribute values or inconsistent data instances.

E. Heart Data. This data set is related to heart disease diagnosis. There are 4 databases in this data set, we use cleveland clinic foundation data in our experiment because this is the only one well processed and used by most researchers. This cleveland data contains 303 instances, and 13 condition attributes. We remove 6 missing attribute values. There is no inconsistent data existing.

F. Iris Data. For Iris data set, there are 4 condition attributes, 150 instances. There is no inconsistent data existing in the data. We first use core algorithm to generate core attributes, but the result is empty. This means none of the attributes is indispensable. There are 4 reducts generated. We apply association rules algorithm with rule templates, and there are 50 rules generated.

G. Lymphography Data. The data set contains 148 instances and 18 condition attributes. There are no missing attribute values in this data. We check that there is no inconsistent data. The core is empty for this data set. 147 reducts are generated from this data set.

H. Pendigits Data. This is a pen-based recognition of handwritten digits data set. There are 10 classes with 16 condition attributes in the data, and $7,494$ training instances and $3,498$ testing instances are in the data. We use training data to conduct our experiments. Each instance represents a hand-written digit with 16 attributes, which are coordinates information. There is no reference on the 16 condition attributes. We use Ci ($1 \leq i \leq 16$) to represent these attributes in our experiments. There are no missing attribute values, or inconsistent data in this data.

I. Pima Indians Diabetes Data. The data comes from all female patients who are at least 21 years old of the pima Indian heritage. The data is used to diagnose whether patients show signs of diabetes according to a list of criteria. There are 768 instances and 8 condition attributes in this data set. There are no missing attribute values, and no inconsistent data.

J. Spambase Data. This data set originally contains $4,601$ instances and 57 condition attributes. It is used to classify spam and non-spam emails. Most of the attributes indicate whether a certain word (such as, order, report) or character (such as !, #) appears frequently in the emails. There are no missing attribute values. There are 6 inconsistent data instances that are removed. After removing redundant data instances as well, there are $4,204$ left in this data set. There are 110 reducts and 7 core attributes generated from this data set. It is interesting to notice that, the core attributes, which are essential to determine whether an email is not a spam email, are, the word frequency of "george", "meeting", 're", "you", "edu", "!", and the total number of capital letters in the email. In addition, it is interesting to pay attention to the reducts as well. They are important information on identifying the possible spam emails.[2]

K. Wine Recognition Data. This data is about using chemical analysis to determine the origin of wines. There are 13 attributes, 178 instances, and 3 classes in the data. There are no missing attribute values or inconsistent data. The core is empty.

---

[2] The maximum number of items per itemset for apriori association rule generation is 6. Without this limitation, the rule generation gives an error of "out of memory".

Table 6. Geriatric Care Data Set

edulevel	eyesight	...	trouble	livealone	cough	hbp	heart	...	studyage	sex	livedead
0.6364	0.25	...	0.00	0.00	0.00	0.00	0.00	...	73.00	1.00	0
0.7273	0.50	...	0.50	0.00	0.00	0.00	0.00	...	70.00	2.00	0
0.9091	0.25	...	0.00	0.00	0.00	1.00	1.00	...	76.00	1.00	0
0.5455	0.25	...	0.00	1.00	1.00	0.00	0.00	...	81.00	2.00	0
0.4545	0.25	...	0.00	1.00	0.00	1.00	0.00	...	86.00	2.00	0
0.2727	0.00	...	0.50	1.00	0.00	1.00	0.00	...	76.00	2.00	0
0.0000	0.25	...	0.00	0.00	0.00	0.00	1.00	...	76.00	1.00	0
0.8182	0.00	...	0.00	0.00	0.00	1.00	0.00	...	76.00	2.00	0
...	...	...	...	...	...	...	...	...	...	...	...

L. Yeast Data. This data set is used to predict the cellular localization sites of proteins. There are 1,484 instances with 8 condition attributes in the data, and no missing attribute values. We remove 31 redundant instances.

M. Zoo Data. This artificial data set contains 7 classes of animals, 17 condition attributes, 101 data instances, and there are no missing attribute values in this data set. Since the first condition attribute "animal name" is unique for each instance, and we consider each instance a unique itemset, we do not consider this attribute in our experiment. There are no inconsistent data in this data set.

4.4   Experiments on Geriatric Care Data Set

In this experiment, a sanitized geriatric care data set is used as our test data set. Table 6 gives selected data records of this data set.

This data set is an actual data set from Dalhousie University Faculty of Medicine to determine the survival status of a patient giving all the symptoms he or she shows. The data set contains 8,547 patient records with 44 symptoms and their survival status. We use *survival status* as the decision attribute, and the 44 symptoms of a patient as condition attributes, which includes *education level, the eyesight, the age of the patient at investigation* and so on.[3] There is no missing value in this data set.

There are 12 inconsistent data entries in the medical data set. After removing these instances, the data contains 8,535 records.[4]

There are 14 core attributes generated for this data set. They are *eartroub, livealone, heart, hbp, eyetroub, hearing, sex, health, edulevel, chest, housewk, diabetes, dental, studyage*. Table 7 shows selected reduct sets among the 86 reducts generated by ROSETTA. All of these reducts contain the core attributes. For each reduct set, association rules are generated with *support* = 30%, *confidence* = 80%.[5]

---

[3] Refer to [24] for details about this data set.

[4] Notice from our previous experiments that core generation algorithm cannot return correct core attributes when the data set contains inconsistent data entries.

[5] Note that the value of support and confidence can be adjusted to generate as many or as few rules as required.

Table 7. Reduct Sets for the Geriatric Care Data Set after Preprocessing

No.	Reduct Sets
1	{edulevel,eyesight,hearing,shopping,housewk,health,trouble,livealone, cough,sneeze,hbp,heart,arthriti,eyetroub,eartroub,dental, chest,kidney,diabetes,feet,nerves,skin,studyage,sex}
2	{edulevel,eyesight,hearing,phoneuse,meal,housewk,health,trouble,livealon, cough,sneeze,hbp,heart,arthriti,evetroub,eartroub,dental, chest,bladder,diabetes,feet,nerves,skin,studyage,sex}
...	...
86	{edulevel,eyesight,hearing,shopping,meal,housewk,takemed,health, trouble,livealone,cough,tired,sneeze,hbp,heart,stroke,arthriti, eyetroub,eartroub,dental,chest,stomach,kidney,bladder,diabetes, feet,fracture,studyage,sex}

Table 8. The Rule Importance for the Geriatric Care Data Set

No.	Selected Rules	Rule Importance
1	SeriousChestProblem → Dead	100%
2	SeriousHearingProblem, HavingDiabetes → Dead	100%
3	SeriousEarTrouble → Dead	100%
4	SeriousHeartProblem → Dead	100%
5	Livealone, HavingDiabetes, HighBloodPressure → Dead	100%
...	...	...
11	Livealone, HavingDiabetes, NerveProblem → Dead	95.35%
...	...	...
14	Livealone, OftenCough, HavingDiabetes → Dead	93.02%
...	...	...
217	SeriousHearingProblem, ProblemUsePhone → Dead	1.16%
218	TakeMedicineProblem, NerveProblem → Dead	1.16%

Discussion. There are 218 rules generated and ranked according to their rule importance as shown in Table 8. We noticed there are 8 rules having importance of 100%. All attributes contained in these 8 rules are core attributes. These 8 rules are more important when compared to other rules. For example, consider rule No.5 and No.11. Rule No.11 has an importance measure of 95.35%. The difference between these two rules is that rule No.5 contains attribute *Livealone, HavingDiabetes, HighBloodPressure*, and rule No.11 contains the first 2 attributes, and instead of *HighBloodPressure*, *NerveProblem* is considered to decide whether the patient will survive. Generally high blood pressure does affect people's health condition more than nerve problem in combination with the other 2 symptoms. Rule No.11 are more important than rule No.218 because in addition to the *NerveProblem*, whether a patient is able to take medicine by himself or herself is not as fatal as whether he or she has diabetes, or lives alone without care. With the same support and confidence, 2, 626, 392 rules are generated from the original medical data set without considering reduct sets or

Table 9. Rules generated by Johnson's algorithm for the Geriatric Care Data Set

No.	Rules	Rule Importance Corresponding to Table 8
1	SeriousChestProblem → Dead	100%
2	SeriousHearingProblem, HavingDiabetes → Dead	100%
3	SeriousEarTrouble → Dead	100%
4	SeriousEyeTrouble → Dead	100%
5	SeriousHeartProblem → Dead	100%
6	Livealone, HavingDiabetes, HighBloodPressure → Dead	100%
7	VerySeriousHouseWorkProblem → Dead	100%
8	Sex_2 → Dead	100%
9	FeetProblem → Dead	96.51%
10	SeriousEyeSight → Dead	95.35%
11	Livealone, HavingDiabetes, NerveProblem → Dead	95.35%
12	TroublewithLife → Dead	81.40%
13	LostControlofBladder, HavingDiabetes → Dead	75.58%
14	Livealone, HighBloodPressure, LostControlofBladder → Dead	75.58%
15	HighBloodPressure, LostControlofBladder, NerveProblem→ Dead	72.09%
16	Livealone, LostControlofBladder, NerveProblem → Dead	72.09%

rule templates. Our method efficiently extracts important rules, and at the same time provides a ranking for important rules.

We also performed experiments using Johnson's reduct generation algorithm [6] for rule generation based on one reduct with the minimum attributes. 16 rules are generated using this reduct [24] as shown in Table 9. The 8 rules with 100% importance in Table 8 are also generated. Although the reduct generated by Johnson's algorithm can provide all the 100% importance rules, the result does not cover other important rules. For example, rule No.14 in Table 8 implies that it is important for the doctors to pay attention to some patient who lives alone, coughs often and also has diabetes. This information is not included in Table 9 by just considering the rules generated by only one reduct.

The experimental results show that considering multiple reducts gives us more diverse view of the data set, the rule importance measure provides a ranking of how important is a rule.

4.5   Comparison Experiments

Confidence is one of the interestingness measures. Given the antecedent of rule existing in the data set, the confidence measures the probabilities of both the antecedent and the consequent of the rule appear together in the data set. The higher the probability, the more interesting the rule is considered to be. Confidence is usually used to measure how frequently the items appear together in the data set, and how much associated is one item to the other item(s). Thus, if people are interested in how significant is a rule instead of how often the items contained in the rule appear together, confidence measure cannot provide such knowledge. Rule importance measure takes the semantic meaning of the data into consideration, and evaluates the significance of a rule through how significant the attributes are.

In order to show that rule importance measure is different from other existing measures on ranking the rules, e.g., confidence, we compare effects on ranking the rules from both the rule importance and confidence measures.

We take the geriatric care data set as an example. The rules ranked with their importance are shown in Table 8. These rules are generated with the minimum confidence of 80%. We list the rules ranked by their confidence in Table 10. From Table 10 we can see that what the confidence measure considers to be

Table 10. Rules Ranked with Confidence for the Geriatric Care Data Set

No.	Selected Rules	Confidence	Rule Importance
1	TroublewithLife → Dead	85.87%	81.40%
2	VerySeriousHouseWorkProblem → Dead	84.77%	100%
3	TroublewithShopping → Dead	83.03%	41.86%
4	TroublewithGetPlacesoutofWalkingDistance → Dead	81.86%	16.28%
5	SeriousHeartProblem → Dead	81.66%	100%
6	TroublePrepareMeal → Dead	81.51%	69.77%
7	EyeTrouble → Dead	80.91%	95.35%
8	Sex_2 → Dead	80.87%	100%
9	SeriousEarTrouble → Dead	80.48%	100%
10	SeriousFeetProblem → Dead	80.83%	96.51%
11	TakeMedicineProblem, KidneyProblem → Dead	80.64%	13.95%
...	...	...	...
21	SeriousEyeTrouble → Dead	80.48%	100%
...	...	...	...
36	Livealone, OftenCough, HavingDiabetes → Dead	80.40%	93.02%
37	TakeMedicineProblem, LostControlBladder → Dead	80.39%	16.28%
38	SeriousHearingProblem, HavingDiabetes → Dead	80.39%	100%
...	...	...	...
125	SeriousHearingProblem, ProblemUsePhone → Dead	80.13%	1.16%
...	...	...	...
154	SeriousChestProblem → Dead	80.07%	100%
...	...	...	...
169	Livealone, HavingDiabetes, HighBloodPressure → Dead	80.05%	100%
...	...	...	...
177	Livealone, HavingDiabetes, NerveProblem → Dead	80.04%	95.35%
...	...	...	...
218	TakeMedicineProblem, NerveProblem → Dead	80.00%	1.16%

interesting are not always important. For example, rule No. 4 and No. 5 have similar confidence, but as we all know, whether a patient has serious heart problem is more important than whether he or she can walk for a certain distance. When a patient has a heart problem, he or she normally would have trouble walking for long distances. The rules that are considered not very interesting by the confidence measure may be important. As an example, rule No. 177 has a lower confidence, and therefore is not considered to be interesting. However, whether the patient has diabetes takes an important part in diagnosing diseases, this knowledge cannot be ignored. In comparison, rule importance ranks rules containing important attribute(s) to be more significant. In certain applications, such as medical diagnoses, when the focus of knowledge discovery is on the important symptoms, rule importance measure can indeed help facilitate evaluating important knowledge.

5   Concluding Remarks

We introduce a rule importance measure which is an automatic and objective approach to extract and rank important rules. This measure is applied throughout the rule generation process. The core attributes should be taken into consideration while choosing important and useful rules. By considering as many reduct sets as possible, we try to cover all representative subsets of the original data set. This measure can also be used jointly with other measures to facilitate the evaluation of the association rules.

Rough sets theory can help with selecting representative attributes from a given data set. By removing redundant attributes, only reserving representative attributes, we achieve representative rules, at the same time the computation cost is lower comparing to the rule generation with all the attributes.

During our experiments on actual data sets, we observed some interesting results. For the UCI breast cancer data set, we extract a rule with 100% importance that if the patient is in the age of 50 to 59, premeno menopause, with degmalig of 3 and the tumor is in the right breast, then the breast cancer belongs to a recurrence event. For pima diabetes data set, it is not necessary to consider the following rule as important that if a patient has been pregnanted twice, the 2-hour serum insulin is 0, and she's 25 years old, her chance of getting diabetes is negative. For the spambase data set, one of the most important rules is when the word frequencies for "you", "re" and "!" are 0 in an email, and the average length of uninterrupted sequences of capital letters is 1, then this email is not considered possible to be a spam email. For the geriatric care data set, we found that given the same condition of a patient living alone and having lost control of bladder, high blood pressure brings more a severe effect to the patient than nerve problems.

Rule importance measures differentiate rules by indicating which rules are more important than other rules. Rule importance measures can be used in a variety of applications such as medical diagnosis, construction of spam filters, wine or glass recognitions and so on.

We observed a limitation that when there is only one reduct for a data set, such as the UCI Car data set or the Breast Cancer data set, rule importance measure returns all the rules with the importance of 100%. The result is the same as rule generation for the data set itself. So, for a given data set, if there is only one reduct, the rule importance measure does not differentiate the generated rules. More objective measures have to be explored to be combined together to further differentiate the rules.

Acknowledgements

We gratefully acknowledge the financial support of the Natural Science and Engineering Research Council (NSERC) of Canada. We thank the anonymous reviewers for helpful comments.

References

1. Pawlak, Z.: Rough Sets. In Theoretical Aspects of Reasoning about Data. Kluwer, Netherlands (1991)
2. Pawlak, Z., Grzymala-Busse, J., Slowinshi, R., Ziarko, W.: Rough Sets. Communications of the ACM, **38** No.11 November (1995)
3. Klemettinen, M., Mannila, H., Ronkainen, R., Toivonen, H., Verkamo, A.I.: Finding interesting rules from large sets of discovered association rules. In: Proceedings of the Third International Conference on Information and Knowledge Management (CIKM) (1994) 401–407
4. Li, J., Tang, B., Cercone, N.: Applying Association Rules for Interesting Recommendations Using Rule Templates. In: Proceedings of the 8th Pacific-Asia Conference in Knowledge Discovery and Data Mining (PAKDD) (2004) 166–170
5. Lin, W., Alvarez, S., Ruiz, C.: Efficient adaptive-support association rule mining for recommender systems. Data Mining and Knowledge Discovery **6** (2002) 83–105
6. Øhrn, Aleksander: Discernibility and Rough Sets in Medicine: Tools and Applications. PhD Thesis, Department of Computer and Information Science, Norwegian University of Science and Technology, Trondheim, Norway, NTNU report (1999)
7. Li, J. and Cercone, N.: A Rough Set Based Model to Rank the Importance of Association Rules. In Proceedings of 10th International Conference on Rough Sets, Fuzzy Sets, Data Mining, and Granular Computing, Regina, Canada. (2005)
8. Newman, D.J., Hettich, S., Blake, C.L. and Merz, C.J.: UCI Repository of machine learning databases. University of California, Irvine, Department of Information and Computer Seiences (1998)
   http://www.ics.uci.edu/~mlearn/MLRepository.html
9. Agrawal, R., Srikant, R.: Fast Algorithms for Mining Association Rules. In: Proceedings of 20th International Conference Very Large Data Bases, Santiago de Chile, Chile, Morgan Kaufmann (1994) 487–499
10. Kryszkiewicz, M., Rybinski, H.: Finding Reducts in Composed Information Systems, Rough Sets, Fuzzy Sets Knowldege Discovery. In W.P. Ziarko (Ed.), Proceedings of the International Workshop on Rough Sets, Knowledge Discovery, Heidelberg/Berlin: Springer-Verlag (1994) 261–273
11. Bazan, J., Nguyen, H.S., Nguyen,S.H., Synak,P.,and Wróblewski,J.: Rough set algorithms in classification problems. In Polkowski, L., Lin, T.Y., and Tsumoto, S., editors, Rough Set Methods and Applications: New Developments in Knowledge Discovery in Information Systems, Vol. 56 of Studies in Fuzziness and Soft Computing, Physica-Verlag, Heidelberg, Germany (2000) 49–88
12. Øhrn, Aleksander: ROSETTA Technical Reference Manual. Department of Computer and Information Science, Norwegian University of Science and Technology, Trondheim, Norway. May 25 (2001)

13. Hu, X., Lin, T., Han, J.: A New Rough Sets Model Based on Database Systems. Fundamenta Informaticae **59** no.2-3 (2004) 135–152
14. Huang, Z, Hu, Y.Q.: Applying AI Technology and Rough Set Theory to Mine Association Rules for Supporting Knowledge Management. In: Proceedings of the 2nd International Conference on Machine Learning and Cybernetics, Xi'an, China. November (2003)
15. Hassanien, A.E.: Rough Set Approach for Attribute Reduction and Rule Generation: A Case of Patients with Suspected Breast Cancer. Journal of The American Society for Information Science and Technology **55** 11 (2004) 954–962
16. Bazan, J.G., Szczuka, M, and Wróblewski, J.: A New Version of Rough Set Exploration System. In: Proceedings of the Third International Conference on Rough Sets and Current Trends in Computing (2002) 397–404
17. Vinterbo, S. and Øhrn, A.: Minimal Approximate Hitting Sets and Rule Templates. International Journal of Approximate Reasoning **25** 2 (2000) 123–143
18. Hilderman, R. and Hamilton, H.: Knowledge discovery and interestingness measures: A survey. Technical Report 99-04, Department of Computer Science, University of Regina, October (1999)
19. Bruha, Ivan: Quality of Decision Rules: Definitions and Classification Schemes for Multiple Rules. In Machine Learning and Statistics, The Interface, Edited by G. Nakh aeizadeh and C. C. Taylor. John Wiley & Sons, Inc. (1997) 107–131
20. An, A. and Cercone, N.: ELEM2: A Learning System for More Accurate Classifications. In: Proceedings of Canadian Conference on AI (1998) 426–441
21. An, A. and Cercone, N.: Rule Quality Measures for Rule Induction Systems: Description and Evaluation. Computational Intelligence. **17-3** (2001) 409–424.
22. Borgelt, C.: Efficient Implementations of Apriori and Eclat. Proceedings of the FIMI'03 Workshop on Frequent Itemset Mining Implementations. In: CEUR Workshop Proceedings (2003) 1613-0073 `http://CEUR-WS.org/Vol-90/borgelt.pdf`
23. Hu, X.: Knowledge Discovery in Databases: an Attribute-Oriented Rough Set Approach. PhD Thesis, University of Regina (1995)
24. Li, J. and Cercone, N.: Empirical Analysis on the Geriatric Care Data Set Using Rough Sets Theory. Technical Report, CS-2005-05, School of Computer Science, University of Waterloo (2005)

Variable Precision Bayesian Rough Set Model and Its Application to *Kansei* Engineering

Tatsuo Nishino, Mitsuo Nagamachi, and Hideo Tanaka

Department of *Kansei* Information
Faculty of Human and Social Environments
Hiroshima International University
555-36 Kurose, Higashihiroshima, Hiroshima 724-0695, Japan
{t-nishi, m-nagama, h-tanaka}@he.hirokoku-u.ac.jp

Abstract. This paper proposes a rough set method to extract decision rules from human evaluation data with much ambiguity such as sense and feeling. To handle totally ambiguous and probabilistic human evaluation data, we propose an extended decision table and a probabilistic set approximation based on a new definition of information gain. Furthermore, for our application, we propose a two-stage method to extract probabilistic *if-then* rules simply using decision functions of approximate regions. Finally, we implemented the computer program of our proposed rough set method and applied it to *Kansei* Engineering of coffee taste design and examined the effectiveness of the proposed method. The result shows that our proposed rough set method is definitely applicable to human evaluation data.

Keywords: Variable precision Bayesian rough set model, ambiguity, extraction of decision rules, human evaluation data, *Kansei* Engineering, rough sets.

1 Introduction

The original rough sets approach is restricted to the case where there exist fully correct and certain classifications derived from a decision table. Unfortunately, many cases exist where there is no lower approximation of a classification. Furthermore, if there are only very few elements in a lower approximation of some decision set, the *if-then* rules extracted from these few elements might be unreliable. Thus, it is necessary to handle a huge decision table. Consequently, the combination of rough sets approaches and probability theory can be found in many research papers [15,17,18,19,20,21,23,24,25].

On the other hand, we have applied rough set methods to *Kansei* Engineering (KE) problems [8,9,10]. Kansei Engineering is defined as the 'technology to translate human needs to product elements'[5]. Its aim is to develop customer-oriented products by using relational rules embodying design attributes of products and human evaluation data such as sensory perceptions and feelings. However, these human-product relational rules are seldom used by design experts, since their structure is very complicated. Recently, it has been shown that

J.F. Peters and A. Skowron (Eds.): Transactions on Rough Sets V, LNCS 4100, pp. 190–206, 2006.
© Springer-Verlag Berlin Heidelberg 2006

rough set approaches are very effective to extract human decision rules in KE [4,5]. However, since human evaluation data involve considerable ambiguity of the decision classes, it has been very difficult to derive effective decision rules from such human evaluation data. If we consider the properties of the human evaluation data such as ambiguity and non-linearity, we have to construct a rough set method that can treat the case where there is no lower approximation of a classification, and the case where the decision classes embody with considerable ambiguity. In such situations, we directed our attention to the variable precision rough set model (VPRS)[23], Bayesian rough set model (BRS)[18,19,20] and variable precision Bayesian rough set model (VPBRS)[17] because these models are much suitable for dealing with practical human evaluation data involving ambiguity or inconsistency.

Accordingly, in this paper, we propose a modified VPBRS suitable for analyzing human evaluation data with much ambiguity[11,12]. We defined a new information gain relative to equivalent classes suitable for handling totally ambiguous and probabilistic properties of human evaluation data. Moreover, for our application, we propose a two-stage method to extract probabilistic decision rules simply from probabilistic decision tables using decision functions of approximated classes. Next, we have designed and implemented a computer program for our rough set method, and applied the proposed rough set method to real life coffee taste design in a coffee company. Its aim is to discover effective decision rules and to develop coffee manufacturing conditions to produce a new coffee taste fitted to customers based on the extracted decision rules[13]. The results show that our proposed rough set method is more applicable to human evaluation data in KE, and it extracts 'interesting' decision rules to develop new products fitted to human sense or feeling.

The rest of the paper is organized as follows. In Section 2, Kansei Engineering and its relation with rough set is described in viewpoints of practical applications of rough sets. Preliminaries and notations to describe an extended decision table for human evaluation data are introduced in Section 3. In Section 4, concepts of information gain and probabilistic approximations to properly handle human evaluation data are introduced. In Section 5, for our applications, we present a two-stage method to simply extract probabilistic decision rules using decision functions from an approximated decision table. We show an application of our rough set method to practices of Kansei Engineering in Section 6. Finally, Section 7 presents conclusions and our future work.

2   Rough Sets and *Kansei* Engineering

The trend of a new product development emphasizes consumer orientation, namely, consumer needs and feeling are recognized as very important and invaluable in a product development for manufacturers. Kansei engineering was founded by Nagamachi, M., one of the authors, at Hiroshima University around 30 years ago [5,6,7] and it aims to implement customers' feelings and needs in new product function and design. The term 'kansei' is a Japanese adjective, which implies the customer's psychological feeling when he/she considers (evaluates) a newly coming product or situation. You are much interested in car design. When you take look at the new car picture, you will think of it as 'very fast dash' or 'good outside design' and others. These feelings are the *kansei*. Though the kansei has vague feature, it is quantified using psychological and physiological measurement. And then the qualified data are able to be transferred to the design domain and we are able to create a new product design using the design specifications. That is, kansei engineering is defined as "translation technology of kansei characteristics into design specifications". If we have the design specifications, we can design a new product fit to a customer's feeling. Most of the kansei products have sold very well in the market, because it was based on the customers' needs.

We have many examples of the kansei products developed by the kansei engineering so far in industries of automobile, construction machine, home electric appliance, housing, costume, cosmetics and others. The sports car 'Miata (MX 5)' made by Mazda is very well known in the world as one developed using the kansei engineering. Its core technology is to identify the relationships between the elements of product and human *kansei*, and then to translate human kansei into design elements using those relationships.

The main procedure of KE consists of the following three steps[5].

Step 1: human evaluation experiment to product sets.

Step 2: identification of the relationships between product attributes and its human evaluation such as sense or feeling.

Step 3: development and design of new product based on the extracted decision rules between human *kansei* and product attributes.

At Step 1, some dozens of persons evaluate product sets by using psychological measurement scales of sense or feeling words such as 'attractive', 'good design', 'functional' and so on, which are supposed to be significant to develop a new product.

At Step 2, rough sets approaches are especially able to contribute to extracting useful decision rules. Human evaluation processes include the cognition of interactions between elements of a object and the ambiguity of decisions arising from individual differences. For example, when you recognize a human face, at the beginning, you percept the interactions between eyes, mouth, face shape and so on, and then you judge if it is man or women according to your own recognition and experiences. Similarly, when you percept a product, the same process arises.

These examples indicate that there exist the interactions between the elements of a object to recognize, and the ambiguity of decision in human evaluation processes. Unfortunately, the relations between product attributes and its human evaluation are few known since the structure is very complicated.

Many different statistical analyses are used to find the solution of design specification, but there is such big problem that the kansei has not primarily linear characteristics, and that applications of statistics to find the relationship between the kansei and design specifications are problematic. Methods of computer science are utilized in the kansei engineering as well. Therefore, Fuzzy Logic and Rough Sets are very useful to find the solution related the data with non-linear characteristics.

Since rough sets are excellent in the analysis of interactions from uncertain data and its results are shown in the form of if-then decision rule which is understandable easily for human, rough sets may become a powerful tool to identify relations between human evaluation and design attributes. In the situation, the development of effective rough set method is becoming one of challenging tasks in KE[6]. At Step 3, the extracted decision rules would be useful to translate human sense or feeling to design elements of new product.

However, the original rough sets approach is restricted to the case where there exist the fully correct and certain classifications derived from the decision table. Unfortunately, in KE applications, we have many cases where there is no lower approximation of a classification. Therefore, we have developed a rough set method based on Bayesian rough set (BRS) and variable precision Bayesian rough set (VPBRS) proposed by Ślęzak, D. and Ziarko, W.[17,18,19,20], and implemented its computer program so as to suitably handle human evaluation data, and we have applied it to the extraction of effective decision rules for new product design more fitted to human sense or feeling. The extracted decision rules will be very useful because these human-product relational rules are hardly known even for design experts since the structure is very complicated.

3   Preliminaries and Notations

Let us start with a simple example of human evaluation data with respect to products shown in Table 1 where a set of products, a set of design attributes (conditional attributes) of products and human evaluation (decision attribute) to product are denoted as $E = \{E_1, E_2, E_3, E_4\}$, $A = \{a_1, a_2, a_3\}$ and d, respectively. An evaluation event of $j - th$ evaluator to $i - th$ product is denoted as x_{ji}. There are four products and five human evaluators. E_i are equivalent classes because the same product has the same attribute values.

Any attribute of A has a domain of its design attribute values, $V_{a1} = \{0, 1\}$, $V_{a2} = \{0, 1\}$ and $V_{a3} = \{0, 1\}$, which may be color, shape and size of products. Human evaluation decision d has also a domain of its evaluation values $V_d = \{0, 1, 2\}$, which may be 'very good', 'good' or 'no good'.

A set of decision classes is $D = \{D_0, D_1, D_2\}$ where $D_j = \{x \mid d(x) = j\}, j = 0, 1, 2$. The extended decision table is similar to one in [16]. It should be noted that the same product has the same attributes, but the decisions are different because the decisions are dependent on the cognition of each evaluator. Therefore, there is no lower approximation to any decision class. It is should be noticed that the ambiguity will arise in the table even when we use more conditional attributes.

Moreover, decision classes of human evaluation are assumed to occur with different prior probabilities. Thus, we have to define an approximate lower approximation of decision classes by introducing the information gain to positive region. Table 1 will be used to illustrate our approach with a numerical example.

Table 1. An example of human evaluation data

Product (E)	Event (U)	a_1	a_2	a_3	Evaluation (d)
E_1	x_{11}	0	1	1	0
	x_{21}	0	1	1	0
	x_{31}	0	1	1	0
	x_{41}	0	1	1	1
	x_{51}	0	1	1	1
E_2	x_{12}	1	0	1	1
	x_{22}	1	0	1	1
	x_{32}	1	0	1	1
	x_{42}	1	0	1	0
	x_{52}	1	0	1	2
E_3	x_{13}	0	1	0	1
	x_{23}	0	1	0	2
	x_{33}	0	1	0	2
	x_{43}	0	1	0	2
	x_{53}	0	1	0	2
E_4	x_{14}	1	1	1	0
	x_{24}	1	1	1	0
	x_{34}	1	1	1	0
	x_{44}	1	1	1	0
	x_{54}	1	1	1	1

Formally, we have $U = \{x_{11}, \ldots, x_{ji}, \ldots, x_{mn}\}$ for the universe denoted as a set of events of n evaluators to m products, $A = \{a_1, \ldots, a_k, \ldots, a_p\}$ for p conditional attributes, $U/A = \{E_1, \ldots, E_i, \ldots, E_m\}$ for m products, and $D = \{D_1, \ldots, D_j, \ldots, D_r\}$ for r decision classes where $D_j = \{x \mid d(x) = j\}$. Any conditional attribute a_k is a mapping function $a_k(x) = v_k$ and has a set of its values V_{ak}. A decision attribute d is a mapping function $d(x) = v_d$ and has V_d.

These evaluation data include at least two important probabilistic aspects. One is the probability of decision dependent on the conditional attributes of products and the other is the prior probability of decision class. Such probabilities are experientially acceptable in human evaluation data. These probabilities are well known as the conditional and prior probability, respectively. According to many literatures such as [15,17,18,19,21], the following probabilities can be defined:

$$P(D_j|E_i) = \frac{card(D_j \cap E_i)}{card(E_i)} .\qquad \text{(the conditional probability)}$$

$$P(D_j) = \frac{card(D_j)}{card(U)} .\qquad \text{(the prior probability)}$$

In the example of Table 1, we have Table 2.

Table 2. The prior and conditional probabilities

$P(D_0)=0.40$	$P(D_0|E_1)=0.6$	$P(D_0|E_2)=0.2$	$P(D_0|E_3)=0.0$	$P(D_0|E_4)=0.8$
$P(D_1)=0.35$	$P(D_1|E_1)=0.4$	$P(D_1|E_2)=0.6$	$P(D_1|E_3)=0.2$	$P(D_1|E_4)=0.2$
$P(D_2)=0.25$	$P(D_2|E_1)=0.0$	$P(D_2|E_2)=0.2$	$P(D_2|E_3)=0.8$	$P(D_2|E_4)=0.0$

4 Rough Sets Approach Based on Information Gain

According to the parameterized version of Bayesian Rough Set (BRS) model [17,18,19,20], let us consider the difference between probabilities $P(D_j)$ and $P(D_j|E_i)$ as a kind of information gain in the case that $P(D_j|E_i) > P(D_j)$. We define the information gain denoted as

$$g(i,j) = \begin{cases} 1 - \frac{P(D_j)}{P(D_j|E_i)} & \text{if } P(D_j|E_i) \neq 0 \\ 0 & \text{if } P(D_j|E_i) = 0 \end{cases} . \tag{1}$$

which means that the larger the conditional probability compared with the prior probability is, the larger the information gain is. Since the information gain enables to evaluate the influence of the set of conditional attributes on decision class relative to its prior probability, our approach based on the information gain is applicable to the human evaluation data with different prior probabilities.

The similar concept to (1) is used in market basket analysis [3] and the meaning of (1) would be clear. This information gain would be acceptable with the following numerical cases:

1) $P(D_j) = 0.6$ and $P(D_j|E_i) = 0.8 : g(i,j) = 0.25$,
2) $P(D_j) = 0.2$ and $P(D_j|E_i) = 0.4 : g(i,j) = 0.50$.

It follows from the above that the case 2) is more informative than the case 1), although the differences between $P(D_j)$ and $P(D_j|E_i)$ are the same. This fact can be acceptable for everyone. The definition of information gain by (1) corresponds with our intuition that the large increment of $P(D_j|E_i)$ being more than $P(D_j)$ should take larger information gain when $P(D_j)$ is low, while the same increment of $P(D_j|E_i)$ should take smaller information gain when $P(D_j)$ is high. The similar index is considered in [17], which can be written as

$$g_*(i,j) = \frac{P(D_j|E_i) - P(D_j)}{1 - P(D_j)} . \tag{2}$$

Thus, using (2), we have $g_*(i,j) = 0.5$ in the case 1) and $g_*(i,j) = 0.25$ in the case 2). This result is contrary to one obtained by our information gain.

Let us define the positive region by using the information gain with parameter β as

$$POS^\beta(D_j) = \bigcup \{E_i \mid g(i,j) > \beta\}$$
$$= \bigcup \left\{E_i \mid P(D_j|E_i) > \frac{P(D_j)}{1-\beta}\right\} . \tag{3}$$

which means the region that E_i would belong possibly to D_j with β. It should be noted that $0 \le \beta \le 1 - P(D_j)$. In other words, β should be less than the residual of the prior probability $P(D_j)$.

The coefficient $\frac{1}{1-\beta}$ is related to the parameter with regard to strength of the evidence $\{E_i\}$ given decision D_j. For example, if $\beta=0$, we have evidences E_i such that $P(D_j|E_i) > P(D_j)$. If $\beta=0.5$, we have evidences $\{E_i\}$ such that $P(D_j|E_i) > 2P(D_j)$. Thus, the value of β can be regarded as strength of the evidences E_i given decision D_j.

Using the duality of rough sets $NEG^\beta(D_j) = POS^\beta(\neg D_j)$, the negative region can be automatically defined as

$$NEG^\beta(D_j) = \bigcup \left\{E_i \mid P(D_j|E_i) < \frac{P(D_j) - \beta}{1-\beta}\right\} . \tag{4}$$

which means the region that E_i would not belong possibly to$D_j$withβ.

It should be noticed that $NEG^\beta(D_j)$ is defined under the condition $P(D_j|E_i)$ $< P(D_j)$ and that $0 \le \beta \le P(D_j)$. From $P(D_j|E_i) < \frac{P(D_j)-\beta}{1-\beta}$, we can derive $P(\neg D_j|E_i) > \frac{P(\neg D_j)}{1-\beta}$. Then, since $0 \le \frac{P(D_j)-\beta}{1-\beta} \le P(D_j) \le \frac{P(D_j)}{1-\beta}$, we have the following boundary region:

$$BND^\beta(D_j) = \bigcup \left\{E_i \mid P(D_j|E_i) \in \left[\frac{P(D_j) - \beta}{1-\beta}, \frac{P(D_j)}{1-\beta}\right]\right\} . \tag{5}$$

which means the region that E_i would not belong possibly to neither of D_j or $\neg D_j$with β.

The positive region means that the evidences$\{E_i\}$ show $\{E_i\} \to D_j$ being highly possible. The negative region means that the evidences$\{E_i\}$ show $\{E_i\} \to \neg D_j$. Then, the boundary region means that we cannot decide the above two regions.

As the value of β increases up to $min(1 - P(D_j), P(D_j))$, the positive and negative regions decrease, and boundary region increases. Furthermore, as the value of β increases, the information associated with E_i is strongly relevant to D_j.

It should be noted that β is similar to $1-\epsilon$ in [17]. If we take $\beta = 0$, $POS^\beta(D_j)$, $NEG^\beta(D_j)$ and $BND^\beta(D_j)$ are characterized by $P(D_j|E_i) > P(D_j)$, $P(D_j|E_i)$ $< P(D_j)$, and $P(D_j|E_i) = P(D_j)$, respectively. If we take $\beta = 1 - P(D_j)$, $POS^\beta(D_j)$ and $NEG^\beta(D_j)$ are characterized by $P(D_j|E_i) > 1$ and $P(D_j|E_i) <$ $\frac{P(D_j)-P(\neg D_j)}{P(D_j)}$, respectively. Thus, $BND^\beta(D_j)$ is characterized by $\frac{P(D_j)-P(\neg D_j)}{P(D_j)}$ $\ge P(D_j|E_i) \ge 1$. And also, if we take $\beta = P(D_j)$, $NEG^\beta(D_j)$ and $POS^\beta(D_j)$ are characterized by $P(D_j|E_i) < 0$ and $P(D_j|E_i) > \frac{P(D_j)}{P(\neg D_j)}$, respectively. Thus, $BND^\beta(D_j)$ is characterized by $0 \le P(D_j|E_i) \le \frac{P(D_j)}{P(\neg D_j)}$. It should be noticed that the properties of $POS^\beta(D_j)$ and $NEG^\beta(D_j)$ satisfy the minimal conditions $P(D_j|E_i) > P(D_j)$ and $P(\neg D_j|E_i) < P(\neg D_j)$ of the Bayesian confirmation, respectively[1,2,19,20].

Moreover, we can define:

$$UPP^\beta(D_j) = POS^\beta(D_j) \cup BND^\beta(D_j)$$
$$= \bigcup \left\{ E_i \mid P(D_j|E_i) \geq \frac{P(D_j) - \beta}{1 - \beta} \right\} . \tag{6}$$

which means the possible region except for the negative region.

This index(6) might be important in real applications in the sense that we need more samples without $\{E_i\} \to \neg D_j$ to find effective rules from much ambiguous data because of $UPP^\beta(D_j) \supseteq POS^\beta(D_j)$. Lastly it follows that

$$U = POS^\beta(D_j) \cup NEG^\beta(D_j) \cup BND^\beta(D_j) . \tag{7}$$

We can have decision rules with different certainties by changing the value of β. It should be noticed that there are orthogonal partitions with respect to decision classes $D = \{D_1, \ldots, D_r\}$.

In the example of Table 1, assuming $\beta = 0.2$, we have:

$$POS^{0.2}(D_0) = \bigcup \left\{ E_i \mid P(D_0|E_i) \geq \frac{P(D_0)}{0.8} = 0.5 \right\} = E_1 \cup E_4 ,$$
$$NEG^{0.2}(D_0) = \bigcup \{ E_i \mid P(D_0|E_i) \leq 0.25 \} = E_2 \cup E_3 ,$$
$$BND^{0.2}(D_0) = \emptyset ,$$
$$UPP^{0.2}(D_0) = E_1 \cup E_4 .$$

5   Extraction Method of Decision Rules from Approximate Regions

We presents here a two-stage method to simply extract uncertain probabilistic decision rule. The first stage extracts certain decision rules by using relative decision functions of approximation region classes. Then the second stage gives rule evaluation factors to the extracted rules. It is similar to the one proposed in the framework of the variable precision rough set model[25].

First Stage. Since approximate regions are exclusive each other from (7). we have a consistent decision table with respect to each approximate region. Thus, we can construct a decision matrix relative to each approximate class. A decision matrix with respect to $POS^\beta(D_j)$ can be described as Table 3.

Any element of the decision matrix is defined:

$$M_{ij}^\beta(D_j) = \left\{ \bigvee a_k = v_{ik} \mid a_k(E_i) \neq a_k(E_j), \forall a_k \in A \right\} , \tag{8}$$

where $\vee \, a_k = v_{ik}$ is a disjunction of attribute elements to discern E_i and E_j.

Table 3. A decision matrix with respect to approximate regions

		$NEG^\beta(D_j)$		$BND^\beta(D_j)$	
		E_{N1} ...	E_j	E_{B1} ...	E_{Bn}
$POS^\beta(D_j)$	E_{P1}				
	$\vdots$				
	E_i	...	...	$M_{ij}^\beta(D_j)$	
	$\vdots$				
	E_{Pm}				

From $POS^\beta(D_j)$, we can derive minimal decision rules in the form of if *condition* then *decision* using the following decision function.

$$POS^{\beta-rule}(D_j) = \bigvee_{E_i \in POS^\beta(D_j)} \bigwedge_{E_j \notin POS^\beta(D_j)} M_{ij}^\beta(D_j) . \tag{9}$$

Similarly, we can derive rules from $NEG^\beta(D_j)$ or $BND^\beta(D_j)$.

From $UPP^\beta(D_j)$, we can also derive minimal possible decision rules using the following decision function.

$$UPP^{\beta-rule}(D_j) = \bigvee_{E_i \in UPP^\beta(D_j)} \bigwedge_{E_j \notin UPP^\beta(D_j)} M_{ij}^\beta(D_j) . \tag{10}$$

In the example of Table 1, we have the decision matrix with respect to $POS^{0.2}(D_0)$ shown in Table 4.

Table 4. The decision matrix with respect to $POS^{0.2}(D_0)$

		$NEG^{0.2}(D_0)$	
		E_2	E_3
$POS^{0.2}(D_0)$	E_1	$a_1 = 0 \vee a_2 = 1$	$a_3 = 1$
	E_4	$a_2 = 1$	$a_1 = 1 \vee a_3 = 1$

From Table 4, we can obtain the following rules.

$$\begin{aligned}
r_1 &: if\ a_1 = 0\ and\ a_3 = 1, then\ d = 0 \quad \{E_1\} \\
r_2 &: if\ a_1 = 1\ and\ a_2 = 1, then\ d = 0 \quad \{E_4\} \\
r_3 &: if\ a_2 = 1\ and\ a_3 = 1, then\ d = 0 \quad \{E_1, E_4\}
\end{aligned} \tag{11}$$

The symbols at the end of each decision indicate the equivalent classes matching with the condition part of the rule. Notice that the condition part of the rule r_3 is matching with E_1 and E_4.

Second Stage. The second stage gives rule evaluation factors to the extracted rules. We can convert the above rule represented as certain deterministic one into uncertain probabilistic rule by giving rule evaluation factors. We extended the original rule evaluation factors proposed by Pawlak[14] to the case where is given equivalent class and its conditional probability $P(D_j|E_i)$. In the context of our applications, we propose three evaluation factors of decision rules by using the number of evaluation to products $|E_i|$ and the effects of products on decision $P(D_j|E_i)$. The extracted rule $rule_k$ can be represented in the form of if $cond_k$ then D_j $(k = 1, \ldots, m)$. Let $Cond_k$ be a set of the equivalent classes E_i matched with the condition part $cond_k$ of the extracted rule, and $|\bullet|$ denote cardinality.

The following *certainty factor* denoted as $cer(Cond_k; D_j)$ means the ratio of the number of events satisfied with $if - then$ rule to the number of events satisfied with the condition part $cond_k$ of the rule.

$$cer(Cond_k; D_j) = \frac{|Cond_k \cap D_j|}{|Cond_k|}$$
$$= \frac{\sum_{E_i \in Cond_k} |E_i| \, P(D_j|E_i)}{\sum_{E_i \in Cond_k} |E_i|}, \quad (12)$$

where $|Cond_k \cap D_j|$ referred as *support* is the number of events matched with both $cond_k$ and $d = j$ which equals $\sum_{E_i \in Cond_k} |E_i| \, P(D_j|E_i)$, and $|Cond_k|$ is the number of events matched with $cond_k$ which equals $\sum_{E_i \in Cond_k} |E_i|$.

This certainty factor shows the degree to which $cond_k \to D_j$ holds. In our applications, we can use this factor as confidence degree of decision to predict the human evaluation from any product design elements. Inversely, when we have to estimate the attribute values of the product candidates from targeted human evaluation, the following *coverage factor* denoted as $cov(Cond_k; D_j)$ will be useful.

$$cov(Cond_k; D_j) = \frac{\sum_{E_i \in Cond_k} |E_i| \, P(D_j|E_i)}{|D_j|}, \quad (13)$$

which means the ratio of the number of events satisfied with constructed rule to the number of the events satisfied with D_j. This factor shows the degree to which $D_j \to cond_k$, i.e., the inverse of rule holds.

The following *strength factor* denoted as $\sigma(Cond_k; D_j)$ can be used to evaluate the set of decision rules.

$$\sigma(Cond_k; D_j) = \frac{\sum_{E_i \in Cond_k} |E_i| \, P(D_j|E_i)}{|U|}, \quad (14)$$

which means the ratio of the number of events satisfied with $if - then$ rule to all the events.

In similar way, we can associate $if - then$ rules from $NEG^\beta(D_j)$ by using $P(\neg D_j|E_i)$ in stead of $P(D_j|E_i)$ in (12), (13) and (14). with three factors mentioned above.

For example, the rule r_1 in (11) has the following values of three factors.
Since $Cond_1 = \{E_1\}$, we have:
$cer(E_1; D_0) = \frac{|E_1|P(D_0|E_1)}{|E_1|} = 0.6$,
$cov(E_1; D_0) = 0.375$,
$\sigma(E_1; D_0) = 0.15$.

In similar way, as for r_3 we have:
$Cond_3 = \{E_1, E_4\}$,
$cer(E_1, E_4; D_0) = \frac{|E_1|P(D_0|E_1) + |E_4|P(D_0|E_4)}{|E_1| + |E_4|} = 0.7$,
$cov(E_1, E_4; D_0) = 0.875$,
$\sigma(E_1, E_4; D_0) = 0.35$.

6   Applications to Kansei Engineering

In this section, we will show the application of proposed approach to coffee taste design in a coffee company[13]. The aim is to discover effective decision rules and to develop coffee manufacturing conditions to produce new coffee taste fitted to customer based on the extracted decision rules. In this application, we extracted decision rules using decision classes by $POS^\beta(D_j)$, $NEG^\beta(D_j)$ and $BND^\beta(D_j)$ defined in Section 4.

Experiment. The purpose of the experiment is to obtain the data from which effective decision rules to design coffee taste fitted to customer feeling will be derived using our rough set method. We carried out an experiment to identify the hidden relations between significant coffee manufacturing conditions and human sensory evaluations in a coffee manufacturing company.

The manufacturing conditions were combinations of two conditional attributes of raw beans (a_1) and its roast time (a_2). $V_{a1} = \{$ Colombia Excelsio, Brazil No2 s 17/18, Mocha Lekempti $\}$, which are typical kinds coffee beans, and $V_{a2} = \{$ Light, Medium, French $\}$, which roast levels were controlled by roast time of roast machine. A coffee manufacturing expert made 9 sorts of coffees by combining V_{a1} and V_{a2}, and 10 evaluators (4 male and 6 female) evaluated them on 5 points semantic differential scale of 6 sensory words such as 'aroma', 'fruity', 'bitter', 'harsh', 'sweet', 'soft', and 6 attitudinal words such as 'want to buy' and so on, which were selected as relevant words to coffee taste.

The evaluation scores were classified into two decision classes $D = \{D_0, D_1\}$, for example, $D = \{Good\ aroma,\ No\ good\ aroma\}$ according to the value of the measurement scale. We obtained a decision table as shown in Table 5 for every sensory and attitudinal word. Although we can show the results of all words, for simple explanation, we will show the results of only two sensory words in this paper: ' aroma' and ' sweet'.

Approximation of Decision Class. The estimations of the prior probabilities of six sensory words from data set were $P(aroma) = 0.718$, $P(sour) = 0.659$, $P(sweet) = 0.430$, $P(harsh) = 0.631$, $P(fruity) = 0.362$, $P(bitter) = 0.659$. You can easily see that good aroma, bitter, harsh and sour tastes are higher

Table 5. An example of human evaluation decision table

Product (E)	Event (U)	Beans (a_1)	Roast (a_2)	Evaluation (d)
	x_{11}	Colombia	French	0
	x_{12}	Colombia	French	0
E_1	$\vdots$	$\vdots$	$\vdots$	$\vdots$
	x_{19}	Colombia	French	0
	$x_{1,10}$	Colombia	French	1
	x_{21}	Colombia	Medium	0
E_2	$\vdots$	$\vdots$	$\vdots$	$\vdots$
	$x_{2,10}$	Colombia	Medium	1
$\vdots$	$\vdots$	$\vdots$	$\vdots$	$\vdots$
	x_{51}	Brazil	Medium	1
E_5	$\vdots$	$\vdots$	$\vdots$	$\vdots$
	$x_{5,10}$	Brazil	Medium	0
$\vdots$	$\vdots$	$\vdots$	$\vdots$	$\vdots$
	x_{91}	Mocha	Light	1
	x_{92}	Mocha	Light	1
E_9	$\vdots$	$\vdots$	$\vdots$	$\vdots$
	x_{99}	Mocha	Light	0
	$x_{9,10}$	Mocha	Light	1

prior probabilities than others. We obtained the gain chart of each sensory word which shows the relation between accumulated percent of events and its gain. Decision rules were derived so that the coverage of approximation is more than 80% because the number of product samples is relatively smaller. Thus, the following values of β were set for each word: $\beta_{aroma} = 0.07$, $\beta_{fruity} = 0.14$, $\beta_{bitter} = 0.14$, $\beta_{harsh} = 0.07$, $\beta_{sweet} = 0.04$, $\beta_{sour} = 0.05$. Using these β, decision class of product was approximated for every words.

Extraction and Evaluation of Decision Rules. Although we can show every rules for each sensory and attitudinal word, for simplicity, we show only decision rules with respect to $D_0 = \{Good\ aroma,\ No\ good\ aroma\}$ and $D_0 = \{Sweet,\ No\ sweet\}$.

We had the following decision rules of 'aroma'. The evaluation factors of these rules are shown in Table 6.

$ar_1 : if\ Roast = medium\ and$ $\qquad\qquad\qquad$ then $Aroma = good$
$ar_2 : if\ Roast = french\ and\ Beans = Colombia,$ then $Aroma = good$
$ar_3 : if\ Roast = french\ and\ Beans = Brazil,$ $\quad$ then $Aroma = good$
$ar_4 : if\ Roast = light,$ $\qquad\qquad\qquad\qquad\quad$ then $Aroma = no\ good$
$ar_5 : if\ Roast = french\ and\ Beans = Mocha,$ $\quad$ then $Aroma = good\ or\ no\ good$

Table 6. The rule evaluation factors of 'aroma' rules: $\beta = 0.07$

	Certainty (cer)	Coverage (cov)	Strength (σ)
ar_1	0.81	0.38	0.27
ar_2	0.90	0.16	0.12
ar_3	1.00	0.16	0.12
ar_4	0.57	0.59	0.17
ar_5	0.70	0.13	0.09

The rules ar_1, ar_2 and ar_3 are positive rules of good aroma. Among them, the rule ar_1 with highest coverage value means that 81% of medium coffees has good aroma for person, and 38% of good aroma coffee is medium. The rule ar_3 with highest certainty means that 100% of french and Brazil coffee has good aroma for person, and only 16% of good aroma coffee is french and Brazil. The rule ar_4 is a negative rule which means that 57% of light coffee has no good aroma, and 59% of no good aroma coffee is light coffee. It is noticed that the positive and negative rules have higher certainties than their prior probabilities $P(Good\ aroma) = 0.644$ and $P(No\ good\ aroma) = 0.356$, respectively. This means that the extracted positive and negative rules satisfy the minimal condition of Bayesian confirmation of rule. The certainty of the boundary rule ar_5 is not enough higher to guarantee the information gain by $\beta = 0.07$. Total value σ_T of strength of positive and negative decision rules is 68%.

When $\beta = 0.21$ which is near maximum value we can set for this word, we obtained the following rules of 'aroma' with Table 7:

$ar_6 : if\ Roast = french,\ and\ Beans = Brazil \quad then\ Aroma = good$
$ar_7 : if\ Roast = light, \qquad\qquad\qquad\qquad then\ Aroma = no\ good$
$ar_8 : if\ Roast = medium, \qquad\qquad\qquad\ then\ Aroma = good\ or\ no\ good$
$ar_9 : if\ Roast = french,\ and\ Beans = Colombia\ then\ Aroma = good\ or\ no\ good$
$ar_{10} : if\ Roast = french,\ and\ Beans = Mochal,\ then\ Aroma = good\ or\ no\ good$

We can easily see that the number of positive and negative rules are less while the number of boundary rules are more. It is should be noticed that the positive

Table 7. The rule evaluation factors of 'aroma' rules: $\beta = 0.21$

	Certainty (cer)	Coverage (cov)	Strength (σ)
ar_6	1.00	0.16	0.12
ar_7	0.57	0.59	0.17
ar_8	0.81	0.38	0.27
ar_9	0.9	0.16	0.12
ar_{10}	0.70	0.13	0.09

and negative rules with the highest certainties are extracted among the rule set extracted by $\beta = 0.0$. Although the average of certainties is higher than when $\beta = 0.0$, σ_T decreases to 32%.

We had the following decision rules of 'sweet'. Its prior probability was 0.430. The evaluation factors of these rules are shown in Table 8.

$sr_1 : if\ Roast = medium,\ and\ Beans = Brazil$ $then\ Sweet = yes$
$sr_2 : if\ Roast = french\ \ \ and\ Beans = Brazil,$ $then\ Sweet = yes$
$sr_3 : if\ Roast = light\ \ \ \ \ and\ Beans = Mochal$ $then\ Sweet = yes$
$sr_4 : if\ Roast = french,\ and\ Beans = Colombia,\ then\ Sweet = no$
$sr_5 : if\ Roast = medium\ and\ Beans = Mocha,$ $then\ Sweet = no$
$sr_6 : if\ Roast = french\ \ \ and\ Beans = Mocha,$ $then\ Sweet = no$
$sr_7 : if\ Roast = light,\ \ \ \ and\ Beans = Colombia,\ then\ Sweet = yes\ or\ no$
$sr_8 : if\ Roast = medium\ and\ Beans = Colombia,\ then\ Sweet = yes\ or\ no$
$sr_9 : if\ Roast = light\ \ \ \ \ and\ Beans = Brazil,$ $then\ Sweet = yes\ or\ no$

Table 8. The rule evaluation factors of 'sweet' rules: $\beta = 0.07$

	Certainty (cer)	Coverage (cov)	Strength (σ)
sr_1	0.7	0.21	0.09
sr_2	0.56	0.15	0.06
sr_3	0.50	0.12	0.05
sr_4	0.63	0.11	0.06
sr_5	0.88	0.16	0.09
sr_6	0.78	0.16	0.09
sr_7	0.44	0.12	0.05
sr_8	0.44	0.12	0.05
sr_9	0.44	0.12	0.05

We obtained only the combinational rules between roast level and kinds of coffee beans. This indicates that sweet taste has more complicated structure than aroma. It is should be noticed that each positive and negative rule satisfy the minimal condition of the Bayesian confirmations, since the prior probabilities are $P(D_j) = 0.430$ and $P(\neg D_j) = 0.570$, respectively. $\sigma_T = 35\%$

When $\beta = 0.32$ which is near maximum value we can set for this word, we obtained the following rules of 'sweet' with Table 9:

$sr_{10} : if\ Roast = medium,\ \ \ and\ Beans = Brazil,\ then\ Sweet = yes$
$sr_{11} : if\ Roast = medium,\ \ \ and\ Beans = Mocha,\ then\ Sweet = no$
$sr_{12} : if\ Beans = Colombia,$ $then\ Sweet = yes\ or\ no$
$sr_{13} : if\ Roast = light,$ $then\ Sweet = yes\ or\ no$
$sr_{14} : if\ Roast = french,$ $then\ Sweet = yes\ or\ no$

Two positive and negative rules are the same as ones with highest certainty extracted when $\beta = 0.07$, sr_1 and sr_5, respectively. These rules fully cleared the confirmation condition. There are two boundary rules. The length of conditions

Table 9. The rule evaluation factors of 'sweet' rules: $\beta = 0.32$

	Certainty (cer)	Coverage (cov)	Strength (σ)
sr_{10}	0.70	0.21	0.09
sr_{11}	0.88	0.16	0.09
sr_{12}	0.42	0.32	0.14
sr_{13}	0.46	0.35	0.15
sr_{14}	0.38	0.29	0.13

of all boundary rule was one and general rule. We can easily see that as β is larger, the positive and negative rules become specific one, while boundary rule become more general one. $\sigma_T = 18\%$.

In similar way, we extracted if - then decision rules for the other words. As the relations between human feeling or sense and coffee manufacturing conditions are few known even for coffee experts, most of the extracted were new knowledge for coffee manufacturing experts. These rules would be very useful for coffee experts to produce new coffee products fitted to human taste.

7   Conclusions

This paper proposes a rough set method inspired by the BRS model or VPBRS model, and shows that it is very effective to extract effective design decision rules from human evaluation data such as sensory perception or feeling involving much ambiguity in the corresponding decision classes. We have introduced an extended decision table to represent human evaluation data together with a new information gain that better reflects the gain feature of human sensory perception and feeling evaluation. First, probabilistic set approximations method are introduced based on the new definition of information gain. Moreover, we present a two-stage method to derive probabilistic decision rules using decision functions of approximated decision classes. We have implemented our rough set model as a computer program and applied our rough set method to the extraction of decision rules in a practical coffee design problem.

The results show that the proposed rough set method to extract decision rules from human evaluation data such as sensory perception or feeling involving ambiguity or inconsistency is definitely applicable and powerful to practical problems in *Kansei* Engineering. In the near future, further, we need to refine our rough set model and apply it to more practical applications in *kansei* design and to examine its effectiveness and limitations.

References

1. Greco, S., Matarazzo, B., Pappalardo, N. Slowinski, R.: Measuring Expected Effects of Interventions based on Decision Rules, *Journal of Experimental & Theoretical Artificial Intelligence*, **17** No. 1-2, Taylor & Francis (2005), 103-118.

2. Greco, S., Matarazzo, B., Pappalardo, N. Slowinski, R.: Rough Membership and Bayesian Confirmation Measures for Parameterized Rough Sets, RSFDGrC 2005, LNAI 3641, Springer Verlag (2005), 314-324.
3. Hastie, T., Tibshirani, R. and Friedman, J.: *The Elements of Statistical Learning*, Springer Verlag (2001), 440-447.
4. Mori, N., Tanaka, H., Inoue, K. (Eds.): *Rough Sets and Kansei*, Kaibundo Publishing (2004).
5. Nagamachi, M.: *Kansei Engineering*, Kaibundo Publishing, Tokyo (1989).
6. Nagamachi, M.: Kansei engineering in consumer product design, *Ergonomics in Design*, (2002).
7. Nagamachi, M.(Eds.): *Product Development and Kansei*, Kaibundo Publishing, Tokyo (2005).
8. Nishino, T., Nagamachi, M.and Ishihara, S.: Rough Set Analysis on Kansei Evaluation of Color and Kansei Structure, *Proceedings of Quality Management and Organization Development Conf.*, (2001), 543-550.
9. Nishino, T., Nagamachi, M.and Ishihara, S.: Rough Set Analysis on Kansei Evaluation of Color, *Proceedings of The Int. Conf. on Affective Human Factors Design*, (2001), 109-115.
10. Nishino, T.and Nagamachi, M.: Extraction of Design Rules for Basic Product Designing Using Rough Set Analysis, *Proceedings of 14-th Triennial Congress of the International Ergonomics Association*, (2003), 515-518.
11. Nishino, T., Nagamachi, M.and Tanaka, H.: Variable Precision Bayesian Rough Set Model and Its Application to Human Evaluation Data, RSFDGrC 2005, LNAI 3641, Springer Verlag (2005), 294-303.
12. Nishino, T., Sakawa, M., Kato, K., Nagamachi, M.and Tanak, H.: Probabilistic Rough Set Model and Its Application to *Kansei* Engineering, *Asia Pacific Management Review*, (2005), in press.
13. Nishino, T., Nagamachi, M., Sakawa,M. : Acquisition of *Kansei* Decision Rules of Coffee Flavor Using Rough Set Method, *International Journal of Kansei Engineering*, (2005), in press.
14. Pawlak, Z.: Rough Sets Elements, In: L. Polkowski, A. Skowron(Eds.), *Rough Sets in Knowledge Discovery 1*, Physica- Verlag (1998), 10-30.
15. Pawlak, Z.: Decision rules, Bayes' rule and Rough Sets, RSFDGrC 1999, LNAI 1711, Springer Verlag (1999), 1-9.
16. Pawlak, Z.: Rough Sets and Decision Algorithms, RSCT 2000, LNAI 2005, Springer Verlag (2000), 30-45.
17. Ślęzak, D. and Ziarko, W.: Variable Precision Bayesian Rough Set Model, RSFDGrC 2003, LNAI 2639, Springer Verlag (2003), 312-315.
18. Ślęzak, D.: The Rough Bayesian Model for Distributed Decision Systems, RSCT 2004, LNAI 3066, Springer Verlag (2004), 384-393.
19. Ślęzak, D.: Rough Sets and Bayes factors, *Transactions on Rough Set III*, LNCS 3400, (2005), 202-229.
20. Ślęzak, D. and Ziarko, W.: The Investigation of the Bayesian Rough Set Model, *International Journal of Approximate Reasoning* Elsevier(2005), vol.40,2005,81-91.
21. Tsumoto, S.: Discovery of Rules about Complication, RSFDGrC 1999, LNAI 1711, Springer Verlag (1999), 29-37.
22. Stepaniuk, J.: Knowledge Discovery by Application of Rough Set Models, In: L. Polkowski, S. Tsumoto and T.Y. Lin (Eds.), *Rough Set Methods and Applications*, Physica-Verlag (2000), 137-233.
23. Ziarko, W.: Variable Precision Rough Set Model, *Journal of Computer and System Sciences*, **46** (1993) 39-59.

24. Ziarko, W.: Evaluation of Probabilistic Decision Table, RSFDGrC 2003, LNAI 2639, Springer Verlag (2003), 189-196.
25. Ziarko,W. and Xiao,X.: Computing minimal probabilistic rules from probabilistic rules from probabilistic decision tables: decision matrix approach *Proc. of the Atlantic Web Intelligence Conference*, Cancun, Mexico, LNAI 3034, Springer Verlag(2004), 84-94.
26. Ziarko, W.: Probabilistic Rough Sets, RSFDGrC 2005, LNAI 3641, Springer Verlag (2005), 283-293.

P300 Wave Detection Based on Rough Sets

Sheela Ramanna[1] and Reza Fazel-Rezai[2]

[1] Department of Applied Computer Science, University of Winnipeg,
Winnipeg, Manitoba R3B 2E9 Canada
s.ramanna@uwinnipeg.ca
[2] Department of Electrical and Computer Engineering, University of Manitoba,
Winnipeg, Manitoba R3T 5V6 Canada
fazel@ee.umanitoba.ca

Abstract. The goal of P300 wave detection is to extract relevant features from the huge number of electrical signals and to detect the P300 component accurately. This paper introduces a modified approach to P300 wave detection combined with an application of rough set methods and non-rough set based methods to classify P300 signals. The modifications include an averaging method using Mexican hat wavelet coefficients to extract features of signals. The data set has been expanded to include signals from six words and a total of 3960 objects. Experiments with a variety of classifiers were performed. The signal data analysis includes comparisons of error rates, true positives and false negatives performed using a paired t-test. It has been found that the false negatives are better indicators of efficacy of the feature extraction method rather than error rate due to the nature of the signal data. The contribution of this paper is an in-depth study P300 wave detection using a modified averaging method for feature extraction together with rough set-based classification on an expanded data set.

Keywords: Brain computer interface, EEG signal classification, Mexican hat wavelet, P300 wave detection, feature extraction, rough sets.

1 Introduction

Brain Computer Interface (BCI) involves monitoring conscious brain electrical activity, via electroencephalogram, (EEG) signals, and detecting characteristics of brain signal patterns, via digital signal processing algorithms, that the user generates in order to communicate with the outside world. BCI technology provides a direct interface between the brain and a computer for people with severe movement impairments. The goal of BCI is to liberate these individuals and to enable them to perform many activities of daily living thus improving their quality of life and allowing them more independence to play a more productive role in society and to reduce social costs. Considerable research has been done on BCI (see, e.g., [4,7,8,12,20]). One of the benefits of the P300-based BCI is that it does not require intensive user training, as P300 is one of the brain's "built-in" functions. A particular challenge in BCI is to extract the relevant signal from the huge number of electrical signals that the human brain produces each second. In

J.F. Peters and A. Skowron (Eds.): Transactions on Rough Sets V, LNCS 4100, pp. 207–223, 2006.
© Springer-Verlag Berlin Heidelberg 2006

addition, it is also critical for any BCI system to be of practical use, the number of channels used to extract the signals be kept to a minimum. Event related potentials (ERPs) are psychophysiological correlates of neurocognitive functioning that reflect the responses of the brain to changes (events) in the external or internal environment of the organism. ERPs have wide usage for clinical-diagnostic and research purposes. In addition, they have also been used in brain computer interfaces [5]. P300 is the most important and the most studied component of the ERP [4]. Previous work in classifying signal data with rough set methods has shown considerable promise [9,19]. Also, more recently a gradient boosting method for P300 detection has been used by Hoffmann *et al.* [16]. The differences between our two approaches are significant. First, Hoffmann *et al.* have designed their own experiment to collect signal data from 10 channels and to detect P300 on BCI data. Second, gradient boosting was used to stepwise maximize the Bernoulli log-likelihood of a logistic regression model. Ordinary least squares regression was used as weak learner with a classification accuracy between 90-100% for the BCI data.

In this paper, a slightly modified method for P300 wave detection is introduced. The efficacy of P300 detection consists of two components: feature extraction and classification. For feature extraction, Mexican hat wavelet coefficients provide good features when averaged over different scales [8]. We have used a slightly different averaging method based on Mexican hat wavelet coefficients. The channels used to extract the signals are also changed. The range of methods used for the classification of features values now include three rough-set based methods and three non-rough set based methods. In addition, we have used a more extensive data set as compared to our previous work [9]. The experiments were conducted on the data set provided by the BCI group at the Wadsworth Center, Albany, NY. This data set represents a complete record of P300 evoked potentials recorded with BCI2000 using a paradigm described by Donchin et al., 2000. Since the expected response to a particular character (and subsequently the word) is already known, supervised learning methods are ideal for predicting the correct character sequence. In our previous work [9] both standard supervised and a form of sequential character-by-character classification was used. More recently, layered or hierarchical learning for complex concepts has been successfully applied to data from road-traffic simulator [22] and classification of sunspot data [23]. In both these cases, complex concepts are decomposed into simpler but related sub-concepts. Learning at a higher-level is affected by learning at a lower level where sub-concepts are learned independently. Incremental learning is another form of learning [33] where the structure of the decision table is changed (updated) incrementally as new data are added to the table rather than regenerating the whole table. In other words, learning occurs on a hierarchy of decision tables rather than on a single table. Our work differs from both of these methods since (i) the concept (word recognition) is simple, and (ii) our table does not change over time. Upon further experimentation, we have found that there is no gain in using the character-by-character classification approach. It should also be noted the number of features and the number of channels used in this study are small. Our objective was to experiment with very simple features at this stage to evaluate our extraction method in the design of a practical P300 based BCI system. A more substantial discussion regarding practical issues in BCI design can be found in Sect. 4. In addition, since the classification methods used to classify signal data are well-known, details of these methods are not included in the paper.

The contribution of this paper is an in-depth study of P300 wave detection using a modified averaging method with an expanded data set. The data set includes signals from six words with a total of 3960 objects. The signal data analysis shows that for further improvements to the feature extraction method, false negatives are better indicators of efficacy rather than error rate.

This paper is organized as follows. A brief introduction to ERP is given in Section 2. ERP recording using a standard EEG technique is briefly discussed in Sect. 3. An overview of BCI, P300 and signal information is given in Sections 4, 6 respectively. Mexican hat wavelets coefficient extraction and averaging method is given in Sect. 5. Signal data classification methods and results are reported in Sect. 7. Analysis of the results are discussed in Sect. 8.

2   Event Related Potential

Event Related Potential (ERP) is a voltage fluctuation in the EEG induced within the brain that is time-locked to a sensory, motor, or cognitive event. This event is usually the response to a stimulus. The goal of ERP research is to evaluate some of the high level characteristics of information processing in the central nervous system. During the performance of a given task there is a change in the content of thought and the attentional resources that have to be used. For any ERP investigation, we assume that psychological processes that lead to completion of a given task are reflected in a measurable change of electric potentials generated by the appropriate neuronal system. In the prototypical ERP trace, there are several components as shown in Fig. 1.

The most prominent and most studied is the P3 or P300, the third positive wave with a 300 ms latency) described in the 1960s by Sutton [29]. It is a significant positive peak that occurs 300ms after an infrequent or significant stimulus. The actual origin of the P300 is still unclear. It is suggested that it is related to the end of cognitive processing, to memory updating after information evaluation or to information transfer to consciousness [4,12]. The P300 wave is evoked by a task known as the odd-ball paradigm. During this task, a series of one type of frequent stimuli is presented to the experimental subject. A different type of non-frequent (target) stimulus is sometimes presented. The task of the experimental subject is to react to the presence of target stimulus by a given motor response to the target stimuli, typically by pressing a button, or just by mental counting.

There are also alternative tasks which can be used to elicit the P300 wave. One of them is the *single stimulus paradigm*. In this task, a target tone occurs randomly in time, but the tone is sometimes replaced by silence and the subject

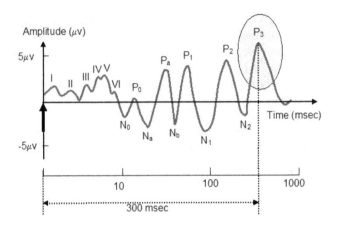

Fig. 1. ERP Components

is required to respond to every stimulus. Another possibility is to use a passive odd-ball task, where the subject does not need to respond to target stimuli [21]. Both these alternative tasks elicit the P300 wave with properties very similar to the wave recorded with the original odd-ball paradigm. The last paradigm is called the *stimulus sequence paradigm*. In this procedure, a sequence of ten tones is presented. The first six are always the standard tones and one of the next four is the target, which is random [3]. Virtually, any sensory modality can be used to elicit the response. In descending order of clinical use these are: auditory, visual, somato-sensory, olfactory or even taste stimulation [25]. The shape and latency of the P300 wave differs with each modality. For example, in auditory stimulation, the latency is shorter than in visual stimulation [18]. This indicates that the sources generating the P300 wave differ and depend on the stimulus modality [17]. Amplitude, latency and age dependency of the P300 wave also varies with electrode site. Analysis of the topographic distribution of P300 latencies has demonstrated that P300 latency is dependent on electrode location. A significant increase of P300 latency from frontal to parietal electrode sites was reported. However, maximum amplitude of the P300 wave is at the Pz electrode site and midline electrodes. Also, the amplitude and delay of P300 wave for different subjects varies significantly. For example, it was shown that for older people the amplitude of P300 is smaller and delay is more than those for younger people. This has been proposed as an index to determine age and cognitive function of brain. However, in this paper our focus is on the detection of the existence of the P300 wave. The BCI signal data used in our experiments were recorded with the odd-ball paradigm.

3   EEG Signal Recording

Traditionally, the ERP recording uses a standard EEG technique followed by averaging of traces aligned to the repeated stimulus. Target and non-target stimuli

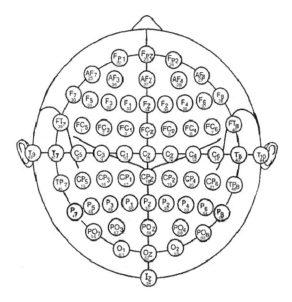

Fig. 2. Electrode Placement using the standard EEG 10-20 layout

are recorded in separate channels. Among the EEG frequency bands, for the P300 with average instantaneous frequency, the band from 0.1 to 30 Hz is used. Electrodes are placed using the standard EEG 10-20 layout as shown in Fig. 2. The layout illustrates electrode designations and channel assignment numbers used to extract signal values. The maximum amplitude of the P300 wave is seen at the parieto-occipital and fronto-central leads [25]. The succession of waves elicited by the above-mentioned oddball paradigm is P1, N1, P2, N2 and P3 (N are negative and P are positive waves shown in Fig. 1). With the exception of the last N2 and P3, the preceding waves correspond to the activation of primary sensory areas. The latency of the P3 is from 300 to 500 ms, depending on the modality. In 20 to 60 per cent of recordings, the P3 wave is composed of two separate peaks, the P3a and P3b. In these cases, the P3b is assumed to be the proper P3wave. The physical energy of the stimulus does not influence the shape, amplitude or latency of the P3 wave. The P3 is basically bilaterally symmetrical [28].

4   Brain Computer Interface and P300

A particular challenge in BCI is to extract the relevant signal from the huge number of electrical signals that the human brain produces each second. The inputs to a BCI system are EEG signals (typically 64 channels) recorded from the scalp or brain surface using a specific system of electrode placement called the International 10-20 system shown Fig. 2[2]. EEG signals are voltage changes of tens of microvolts at frequencies ranging from below 1 Hz to about 50 Hz.

Different types of brain activity reflected in EEG signals (such as visual-evoked potentials, slow cortical potentials, mu and beta rhythms, and the P300 component of ERPs) have been used in BCI [7,20]. The most widely used P300-based BCI is a spelling device designed by Donchin and Farwell in 1988 [10]. However, P300 detection for real-time applications is not easy and there are several practical issues that need to be considered when designing BCI systems:

- Transfer rate: Typically, many channels and several features should be used. In addition, the ensemble averaging of a large number of trials is required to improve the signal-to-noise ratio, because the P300 is buried in the ongoing EEG. However, the throughput of the speller may be as low as one character/min.
- Accuracy: For real-time applications, the P300-based BCI is less than 70% accurate, which is insufficient to be reliable.
- EEG pattern variability: EEG signal patterns change due to factors such as motivation, frustration, level of attention, fatigue, mental state, learning, and other nonstationarities that exist in the brain. In addition, different users might provide different EEG patterns.

In general, the more the averaged trials, the higher the accuracy and reliability will be. However, it is at the expense of longer transfer time. The communication rate determined by both accuracy and transfer time is a primary index in BCI assessment. Therefore, a good algorithm for BCI should ensure the high accuracy and reduce the transfer time.

The P300 wave has a relatively small amplitude (5-10 microvolts), and cannot be readily seen in an EEG signal (10-100 microvolts). One of the fundamental methods of detecting the P300 wave has been the EEG signal averaging. By averaging, the background EEG activity cancels, as it behaves like random noise, while the P300 wave averages to a certain distinct visible pattern. There are limitations to the averaging technique and applications for which it is not suitable. Although P300 wave is defined as a peak at 300 ms after a stimulus, it really occurs within 300 to 400 ms [1]. This latency and also amplitude of the P300 wave changes from trial to trial. Therefore, the averaging is not an accurate method for the P300 wave detection and there is a need for developing a technique based on advanced signal processing methods for this purpose. Farwell and Donchin introduced some P300 detection methods for BCI such as stepwise discriminant analysis (SWDA), peak picking, area, and covariance in 1988 [10], and discrete wavelet transform (DWT) to the SWDA [7]. A method for P300 wave detection based on averaging of Mexican hat wavelet coefficients was introduced in [8] and further elaborated in [9].

5   The Mexican Hat Wavelet and Feature Extraction

To reduce the number of trials required to extract the P300 wave and to increase the accuracy of P300 detection, we have used Mexican hat wavelets to extract features. Wavelet coefficients of a signal $x(t)$ at time point p are defined in Eq. 1.

$$c(s,p) = \int_{-\infty}^{\infty} x(t)\psi(s,p,t)dt, \tag{1}$$

where s is a scale number, t is a time point, and $\varphi(s,p,t)$ is an analyzing wavelet. The analyzing wavelet used is the Mexican hat defined in Eq. 2.

$$\psi(s,p,t) = \frac{2}{\sqrt{3}}\pi^{-\frac{1}{4}}(1-\alpha^2)e^{-\frac{1}{2}\alpha^2}. \tag{2}$$

where $\alpha = \frac{t-p}{s}$. Instead of averaging EEG signals from different trails, wavelet coefficients were extracted using different scales (s) and averaged over different scales for a trial. For each trial, the EEG signal is stored in one matrix (total number of samples by 64 channels). One of the challenges of feature extraction is to determine which of the 64 channels (scalp positions) contain useful information. In BCI literature, it has been shown that the five listed channels are typical EEG channels that have information about P300 wave [5]. In this study, signal data from channel numbers 10, 11, 12, 34, 51 have been used (a change in two channel positions from our previous study) after experimenting with different channel combinations. The average signal value is computed using Eq. 3. Subtraction of the two channels (10 and 12) represents spatial differentiation. Spatial differentiation was included to amplify the existence of P300 wave.

$$\begin{aligned}&Average\ Signal\ Value = \\ &Out(CH11) + Out(CH10) - Out(CH12) + Out(CH34) + Out(CH51).\end{aligned} \tag{3}$$

Consider the averaged wavelet coefficient over different scales shown in Fig. 3.
    Assume that the maximum of this curve has the amplitude of A_0 and occurs at time t_0. Now if we find the two local minimums, one just before A_0 and another just after A_0 with amplitudes B_1 and B_2 respectively, then we can define two heuristic features averaged over scales from 30 to 100 (shown in Fig. 3). Amplitude of the peak (f_1) and Time difference (f_2) were calculated in our earlier work using Eqs. 4 and 5.

$$\begin{aligned}&Amplitude\ of\ the\ peak = R_1 + R_2, \\ &where\ R_1 = A_0 - B_1\ and\ R_2 = A_0 - B_2\end{aligned} \tag{4}$$

$$Time\ difference = \frac{|t_0 - 300|}{300}. \tag{5}$$

Note, that to detect P300 wave, amplitude of the peak feature should have a "large" value and time difference feature should be as "small" as possible (zero is considered ideal). The two features were recalculated using Alg. 1:

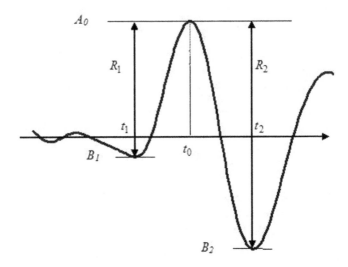

Fig. 3. Averaged Wavelet Coefficient over different scales

Algorithm 1. Algorithm for calculating f_1 and f_2

Input : Amplitudes A_0, B_1, B_2 and time t_0
Output: Features f_1 and f_2
if *($t_0 < 300$ or $t_0 > 400$)* **then**
  | $f_1 = 0$
else
end
f_1 is calculated using equation 3
if *($t_0 \leq 340$ or $t_0 \geq 360$)* **then**
  | $f_2 = 0$
else
  | $f_2 = \frac{min[|t_0 - 340|, |t_0 - 360|]}{40}$
end

---

6   BCI Signal Information

We have used the data set provided by the BCI group at the Wadsworth Center, Albany, NY. EEG was recorded from 64 scalp positions (for details, see [5]). This data set represents a complete record of P300 evoked potentials recorded with BCI2000 using a paradigm described by Donchin et al. [7], and originally by Farwell and Donchin [10]. The signals were digitized at 240Hz and collected from one subject in three sessions. Each session consisted of a number of runs. In each run, the subject focused attention on a series of characters, e.g., the word "SEND" as shown in Fig. 4(a).

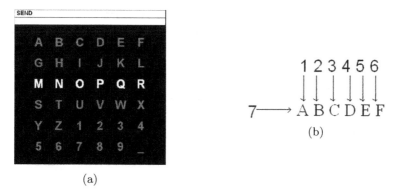

Fig. 4. (a) User Display for Spelling Paradigm, (b) Focus on a Character

In this speller, a matrix of six by six symbols, comprising all 26 letters of the alphabet and 10 digits (0-9), is presented to the user on a computer screen. The rows and columns of the matrix are intensified successively for 100 ms in a random order. After intensification of a row/column, the matrix is blank for 75 ms. At any given moment, the user focuses on the symbol he/she wishes to communicate, and counts the number of times the selected symbol flashes. In response to the counting of this oddball stimulus, the row and column of the selected symbol elicit a P300 wave, while the other 10 rows and columns do not. When a subject focuses on one character, the matrix is displayed for a 2.5 s period, and during this time each character has the same intensity (i.e., the matrix was blank). Subsequently, each row and column in the matrix was randomly intensified for 100 ms. Row/column intensifications were block randomized in blocks of 12. Sets of 12 intensifications were repeated 15 times for each character. Each sequence of 15 sets of intensifications was followed by a 2.5 s period, and during this time the matrix was blank. This period informs the user to focus on the next character in the word that was displayed on the top of the screen. In other words, for each character, 180 entries of feature values are stored: 90 for row intensification and 90 for column intensification. For example, A is recognized only when row 7 and column 1 features indicate a P300 component (see Fig. 4(b)). The steps involved in extracting signal information can be found in [9].

7   Signal Data Classification: Methods and Results

The objective of BCI signal data analysis and classification is two fold: to determine a good feature extraction method and to identify appropriate channels for signal extraction and also to find a good classifier to predict the correct characters recognized by a subject. These two objectives are intertwined, since the classifier is in essence an indicator of the effectiveness of the feature extraction method. In this work we have continued to use five features (attributes): Stimulus code (character to be recognized), amplitude of the peak, time difference, ratio of amplitude to time difference and Stimulus Type (decision). Signal

Table 1. Word "BOWL" Classification

	Classification Results			
Method	Error Rate	False Negative	True Positive	Coverage
RSRuleBased	25	75	17	94
RSTreeBased	35	50	17	81
LEM2	19	67	33	100
RuleBased(RIP)	18	100	0	100
TreeBased(C4.5)	17	0	0	100
Metalearner(bagging)	18	100	0	100

data from 6 sets of words in uppercase: BOWL, CAT, DOG HAT, FISH and WATER were used for our experiments. Each character was stored in a table with 90 entries for row intensification and 90 entries for column intensification. For example, the word BOWL has a table with 720 entries. Discretization was used since three of the feature values (i.e. amplitude of the peak, time difference, ratio of amplitude to time difference) are continuous. We have used both local and global method in RSES for discretization [2,26]. However, only the results local method were reported in this paper. The experiments were conducted using 10-fold cross-validation technique. Four different measures were used for this experiment: *error rate* (overall misclassification), *false negative* (a presence of P300 wrongly classified as zero), *false positive* (absence of P300 wrongly classified as one) and *coverage*. These measures are considerably different from the ones used in our previous work. In addition to making it possible to compare different classification methods, these measures also provide a better insight into our extraction and classification methods (discussed in Sect. 8). Experiments were performed with RSES [2,26] using rule-based (RSRuleBased) and tree-based (RSTreebased) methods. The rule-based method uses genetic algorithms in rule derivation [30]. LERS [13,14] system was also used for classification. Experiments with non-rough set based methods were performed with WEKA [31] using a rule-based(RIP), tree-based (J48) and bagging (Metalearner) methods. RIP uses a RIPPER algorithm for efficient rule induction [6]. J48 is the well-known C4.5 revision 8 algorithm [32]. We have also included a bagging technique to reduce variance and improve classification (minimize the error rate). The error rate measure is calculated in the traditional way using Eq. 6. The classification results for the six words (testing set) are shown in Tables 1,2, 3, 4, 5 and 6 respectively.

$$\text{error rate} = \frac{TP+TN}{TP+FP+TN+FN}. \tag{6}$$

8   Analysis of Classification Results

The number of entries (data) for Stimulus Type (i.e., decision = 1) were very few compared to the number of entries for Stimulus Type (decision=0). In the test

Table 2. Word "CAT" Classification

Method	Classification Results			
	Error Rate	False Negative	True Positive	Coverage
RSRuleBased	13	56	44	98
RSTreeBased	17	33	44	93
LEM2	17	60	40	100
RuleBased(RIP)	15	43	57	100
TreeBased(C4.5)	15	43	57	100
Metalearner(bagging)	13	33	67	100

Table 3. Word "DOG" Classification

Method	Classification Results			
	Error Rate	False Negative	True Positive	Coverage
RSRuleBased	19	78	22	83
RSTreeBased	30	44	22	100
LEM2	19	29	71	100
RuleBased(RIP)	19	67	33	100
TreeBased(C4.5)	17	50	50	100
Metalearner(bagging)	19	67	33	100

Table 4. Word "FISH" Classification

Method	Classification Results			
	Error Rate	False Negative	True Positive	Coverage
RSRuleBased	18	67	25	97
RSTreeBased	29	50	25	86
LEM2	19	57	43	100
RuleBased(RIP)	15	40	60	100
TreeBased(C4.5)	17	50	50	100
Metalearner(bagging)	17	50	50	100

set, the ratio is about one to four. The signal data for almost all the words were inconsistent. The inconsistency levels for some words were as high as 95%. In fact, there was no appreciable gain in prior discretization. This was particularly important for the use of LEM2 which does not work directly on numerical values. For P300 detection, the size of the rule set or the speed of classification is not important. A comparison of error rates for the six classification methods on the entire data set is shown in Fig. 5. The percentage of true positives for the six classification methods is given in Fig. 6. The percentage of false negatives for the six classification methods is compared for the six words in Fig. 7. C4.5 algorithm has a 16% error rate and LEM2 algorithm has 18% error rate when averaged

Table 5. Word "HAT" Classification

	Classification Results			
Method	Error Rate	False Negative	True Positive	Coverage
RSRuleBased	22	33	22	94
RSTreeBased	26	33	22	87
LEM2	19	32	68	100
RuleBased(RIP)	17	50	50	100
TreeBased(C4.5)	15	40	60	100
Metalearner(bagging)	15	40	60	100

Table 6. Word "WATER" Classification

	Classification Results			
Method	Error Rate	False Negative	True Positive	Coverage
RSRuleBased	20	60	27	93
RSTreeBased	33	47	27	81
LEM2	17	78	22	100
RuleBased(RIP)	18	60	40	100
TreeBased(C4.5)	17	50	50	100
Metalearner(bagging)	17	50	50	100

over six words. On an average, LEM2 algorithm has 46% and C4.5 algorithm has 45% classification accuracy for true positives. In the case of false negative rates, RSTreeBased algorithm has 43% and C4.5 algorithm has 39% false negative percentage. However, it should also be noted that the averages for non-rough set based method are somewhat misleading as C4.5 algorithm gives 0 true positives and 0 false negatives for one word (BOWL). When this word is removed from calculation, the average is 47%.

8.1   Paired Difference t-Test

A paired t-test was performed to analyze i) classification error rates ii) true positives and iii) false negatives for the RS method and non-RS method with the best results. We now give a detailed explanation of the analysis with respect to the smallest error rate for the two classes of algorithms. We want to test the hypothesis that mean difference between the two classification learning methods is zero. Let μ_d, σ_d^2 denote the mean difference and variance in the difference in error rates of a random sample of size n from a normal distribution $N(\mu_d, \sigma_d^2)$, where μ_d and σ_d^2 are both unknown. Let H_0 denote the hypothesis to be tested (i.e., $H_0 : \mu_d = 0$). This is our null hypothesis. The paired difference t-test is used to test this hypothesis and its alternative hypothesis ($H_A : \mu_d \neq 0$). We start with pairs $(\epsilon_{11}, \epsilon_{21}), , (\epsilon_{1n}, \epsilon_{2n})$, where ϵ_{1i}, ϵ_{2i} are the ith error rates resulting from the

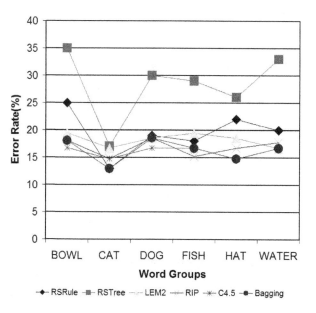

Fig. 5. Comparison of Error Rates

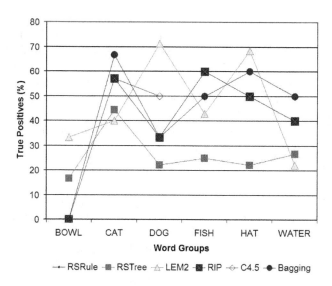

Fig. 6. Comparison of True Positives

application of the RS and non-RS classification learning algorithms, respectively, and i = 1, ..., n. Let $d_i = \epsilon_{1i} - \epsilon_{2i}$. Underlying the null hypothesis H_0 is the assumption that the d_i values are normally and independently distributed with

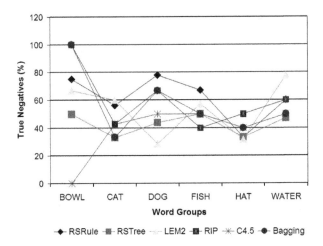

Fig. 7. Comparison of False Negatives

mean μ_d and variance σ_d^2. The t-statistic was used to test the null hypothesis is as follows:

$$t = \frac{\bar{d} - \mu_d}{S_d/\sqrt{n}} = \frac{\bar{d} - 0}{S_d/\sqrt{n}} = \frac{\bar{d}\sqrt{n}}{S_d} \tag{7}$$

where t has a student's t-distribution with $n-1$ degrees of freedom [15]. The shape of the t distribution depends on the sample size $n-1$ (number of degrees of freedom). In our case, $n-1 = 5$ relative to error rates for different six word groups. The significance level α of the test of the null hypothesis H_0 is the probability of rejecting H_0 when H_0 is true (called a Type I error). Let t_{n-1}, $\alpha/2$ denote a t-value to right of which lies $\alpha/2$ of the area under the curve of the t-distribution that has $n-1$ degrees of freedom. Next, we formulate the following decision rule:

Decision Rule: Reject H_0 ($\mu_d = 0$) at significance level α if, and only if $|t| > t_{n-1,\alpha/2}$

Probability distribution (Pr) values for $t_{n-1,\alpha/2}$ can obtained from a t-distribution table (Beyer 1968). In what follows, $\alpha = 0.10$, and $n-1 = 5$. The SAS statistical toolset called JMPIN has been used to compute t-values [27]. The paired t-test was applied for classification *error rates* (with best RS and non-RS methods) for the six words. With 5 degrees of freedom, we find that $Pr(|t| < 2.015) = 0.95$ where $t_{n-1,\alpha/2}$ for $t_{9,0.05} = 2.015$. The null hypothesis H_0 is rejected, since $|t| = |-7| > 2.015$ at the 10% significance level. The t-value for *True Positive* rates was found to be -0.17 when the paired t-test was applied (with best RS and non-RS methods) for the six words. The null hypothesis H_0 is accepted in this case, since $|t| = |-0.17| < 2.015$ at the 10% significance level. We get a similar result in the case of *False Negative* rates, where the t-value was found to be -0.42 when the paired t-test was performed. This means that the null hypothesis H_0 is accepted, since $|t| = |-0.42| < 2.015$ at the 10% significance level.

The paired t-test results show that performance of the classification algorithms differ if we consider only the error rate measure. There is no significant difference in performance if only true positive or false negative rates are considered. Also, the average error rate is about 16% which is considered reasonable for a P300 detection. However, the concern is that the percentage of *false negatives* (a presence of P300 wrongly classified as zero) is quite high (43%). This is important because of the very nature of the signal data (with too few values for d=1). Hence, the error rate measure alone is not really helpful in drawing conclusions about the efficacy of the P300 feature extraction method. This means that the feature extraction method has to be refined futher in order to obtain a sharper separation between the presence and absence of P300.

9 Conclusion

This paper introduces a modified approach to P300 wave detection combined with an application of rough set methods and non-rough set based methods used to classify P300 signals. The data set for our experiments was expanded significantly. The signal data analysis include comparisons of error rates, true positives and false negatives using a paired t-test. Our analysis shows that there is no significant difference in performance of the various classifiers. However, it has been found that for further improvements to the feature extraction method, false negatives are better indicators of efficacy rather than error rate. In addition to considering improvements to the feature extraction method for the BCI data set, we are also in the process of i) designing a new P300 based BCI system which is quite different from the one currently in use that was proposed by Farwell and Donchin [10] ii) data collection from subjects. The design of a new P300 based BCI is complete and the hardware is in place. Clinical trials for data collection is about to begin.

Acknowledgements

The authors gratefully acknowledge the experiments and data sets provided by Wadsworth Center and for helpful comments from reviewers. Sheela Ramanna's research is supported by a Natural Sciences and Engineering Research Council of Canada (NSERC) grant 194376 and the research of Reza Fazel-Rezai is supported by University of Manitoba Research Grant Program (URGP). The authors wish to thank Jerzy Grzymala-Busse, University of Kansas for his insightful comments and for making LERS available. We also wish to thank Marcin Szczuka, Warsaw University and Zdzisław S. Hippe at the University of Technology and Management, Rzeszow, Poland.

References

1. J.D. Bayliss: The use of the P3 evoked potential component for control in a virtual apartment, *IEEE Transactions on Rehabilitation Engineering*, **11**(2), 2003, 113-116.

2. J.G. Bazan, M.S. Szczuka: The Rough Set Exploration system. In: J.F. Peters, A. Skowron (Eds.), *Transactions on Rough Sets III*, Lecture Notes in Artificial Intelligence **3400**. Springer-Verlag, Berlin, 2005, 37-57.
3. J. Bennington, J. Polich: Comparison of P300 from passive and active tasks for auditory stimuli and visual stimuli, *International Journal of Psychophysiology* **34**, 1999, 171-177.
4. E. Bernat, H. Shevrin and M. Snodgrass: Subliminal visual oddball stimuli evoke a P300 component, *Clinical Neurophysiology*, **112**, 2001, 159-171.
5. B. Blankertz, K.-R. Müller, G. Curio, T. M. Vaughan, G. Schalk, J. R. Wolpaw, A. Schlögl, C. Neuper, G. Pfurtscheller, T. Hinterberger, M. Schröder, and N. Birbaumer: The BCI Competition 2003: Progress and perspectives in detection and discrimination of EEG single trials, *IEEE Trans. Biomed. Eng.*, **51**, 2004, 1044-1051.
6. W.W. Cohen: Fast effective rule induction. In A.Prieditis and S. Russell (Eds.), *Proceedings of the Twelfth International Conference on Machine Learning*, San Francisco, Morgan, Kauffman, 1995, 115-123.
7. E. Donchin, K.M. Spencer, R. Wijensighe: The mental prosthesis: Assessing the speed of a P300-based brain-computer interface, *IEEE Trans. Rehab. Eng.*, **8**, 2000, 174-179.
8. R. Fazel-Rezai, J.F. Peters: P300 Wave Feature Extraction: Preliminary Results, *Proceedings of the Canadian Conference of Electrical and Computer Engineering*, Saskatoon, SK, Canada, 2005, 376-379.
9. R. Fazel-Rezai and S. Ramanna: Brain Signals: Feature Extraction and Classification using Rough Set Methods, In: D. Ślęzak, J.T. Yao, J.F. Peters, W. Ziarko, X. Hu (Eds.), Rough Sets, Fuzzy Sets, Data Mining and Granular Computing, *Lecture Notes in Artificial Intelligence*, **3642**. Springer-Verlag, Berlin, 2005, 709-718.
10. L. A. Farwell and E. Donchin: Talking off the top of your head: Toward a mental prosthesis utilizing event-related brain potentials, *Electroencephalogr. Clin. Neurophysiol*, **70**, 1988, 510-523.
11. T. Frodl-Bauch, R. Bottlender, U. Hegerl: Neurochemical substrates and neuroanatomical generators of the event-related P300, *Neuropsychobiology*, **40**, 1999, 86-94.
12. C. J. Gonsalvez and J. Polich: P300 amplitude is determined by target-to-target interval, *Psychophysiology*, **39**, 2002, 388-396.
13. J.W. Grzymala-Busse: LERS - A system for learning from examples based on rough sets, R. Słowiński, (Ed.), *Intelligent Decision Support: Handbook of Applications and Advances of the Rough Sets Theory*, Kluwer Academic Publishers, Dordrecht, 1992, 3-18.
14. J.W. Grzymala-Busse and A.Y. Wang: Modified Algorithms LEM1 and LEM2 for rule induction from data with missing attribute values, *Proceedings of the Fifth International Workshop on Rough Sets and Soft Computing (RSSC'97) at the Third Joint Conference on Information Sciences (JCIS'97*, North Carolina, 1997, 69-72.
15. R.V. Hogg and E.A. Tanis, E.A. Probability and Statistical Inference. Macmillan Publishing Co., Inc., New York, 1977.
16. U.Hoffmann., G. Garcia., J. Vesin., K. Diserens and T. Ebrahimi: A Boosting Approach to P300 Detection with Application to Brain-Computer Interfaces, *Proceedings of the IEEE EMBS 2nd Internation Conference on Neural Engineering*, Arlington, 2005, 1-4.
17. R. Johnson: Developmental evidence for modality-dependent P300 generators: a normative study, *Psychophysiology*, **26**, 1989, 651-66.

18. J. Katayama, J. Polich: Auditory and visual P300 topography from a 3 stimulus paradigm, *Clinical Neurophysiolology*, **110**, 1999, 463-468.
19. L. Lazareck and S. Ramanna: Classification of Swallowing Sound Signals: A Rough Set Approach. In: S. Tsumoto, R. Słowiński, J. Komorowski, J.W. Grzymala-Busse (Eds.), Rough Sets and Current Trends in Computing, *Lecture Notes in Artificial Intelligence*, **2066**. Springer-Verlag, Berlin, 2004, 679-684.
20. Mason, S.G., Birch, G.E.: A general framework for brain-computer interface design, *IEEE Transactions on Neural Systems and Rehabilitation Engineering*, **11**,1 (2003) 71-85.
21. H. McIsaac, J. Polich: Comparison of infant and adult P300 from auditory stimuli. *Journal of Experimental Child Psychology*, **24**, 1992, 23-37.
22. S.H. Nguyen, J. Bazan, A. Skowron, H.S. Nguyen: Layered Learning for Concept Synthesis, *Transactions on Rough Sets*, I, LNCS 3100, 2004, 187-208.
23. T.T. Nguyen, C.P. Willis, D.J. Paddon and H.S. Nguyen: On learning of sunspot classification. In Mieczyslaw A. Klopotek, Slawomir T. Wierzchon and Krzysztof Trojanowski (Eds.), Intelligent Information Systems, Proceedings of IIPWM'04, *Advances in Soft Computing*, Springer, Berlin, 2004, 58-68.
24. Z. Pawlak: Rough sets. *International J. Comp. Inform. Science*, **11**(3),1982, 341–356.
25. J. Polich: P300 in clinical applications, In: E. Niedermayer and F. Lopes de la Silva (Eds.), *Electroencephalography: basic principles, clinical applications and related fields*, Urban and Schwartzenberger, Baltimore-Munich, 1999, 1073-1091.
26. The RSES Homepage at `http://logic.mimuw.edu.pl/~rses`
27. J. Sal, A. Lehman, L. Creighton. JMP Start Statistics: A Guide to Statistics and Data Analysis, Statistical Analysis Systems (SAS) Institute, Duxbury, Pacific Grove, CA, 2001.
28. E.M. Smith, E. Halgren, M. Sokolik, P. Baudena, A. Musolino, C. Liegeois-Chauvel, P. Chauvel: The intracranial topography of the P3 event-related potential elicited during auditory oddball, *Electroencephalogram Clinical Neurophysiology*, **76**, 1990, 235-248.
29. S. Sutton, M. Braren, J. Zubin, E.R. John: Evoked potentials correlates of stimulus uncertainty, *Science*, **150**,1965, 1187-1188.
30. J. Wróblewski: Genetic algorithms in decomposition and classification problem. Polkowski, L. and Skowron, A.(Eds.), *Rough Sets in Knowledge Discovery*, **1**. Physica-Verlag, Berlin, Germany, 1998, 471-487.
31. The WEKA Homepage at `http://www.cs.waikato.ac.nz/ml/weka`
32. Quinlan, J.R: Induction of decision trees. *Machine Learning* 1(1), 1986, 81-106.
33. W. Ziarko: Incremental Learning with Hierarchies of Rough Decision Tables, *Proc. North American Fuzzy Information Processing Society Conf.*(NAFIPS04), Banff, Alberta, 2004, 802-808.

Multimodal Classification: Case Studies

Andrzej Skowron[1], Hui Wang[2], Arkadiusz Wojna[3], and Jan Bazan[4]

[1] Institute of Mathematics
Warsaw University
Banacha 2, 02-097 Warsaw, Poland
skowron@mimuw.edu.pl
[2] School of Computing and Mathematics
University of Ulster at Jordanstown
Northern Ireland, BT37 0QB, United Kingdom
h.wang@ulst.ac.uk
[3] Institute of Informatics
Warsaw University
Banacha 2, 02-097 Warsaw, Poland
wojna@mimuw.edu.pl
[4] Institute of Mathematics
University of Rzeszów
Rejtana 16A, 35-310 Rzeszów, Poland
bazan@univ.rzeszow.pl

Abstract. Data models that are induced in classifier construction often consist of multiple parts, each of which explains part of the data. Classification methods for such multi-part models are called multimodal classification methods. The model parts may overlap or have insufficient coverage. How to deal best with the problems of overlapping and insufficient coverage? In this paper we propose a hierarchical or layered approach to this problem. Rather than seeking a single model, we consider a series of models under gradually relaxing conditions, which form a hierarchical structure. To demonstrate the effectiveness of this approach we consider two classifiers that construct multi-part models – one based on the so-called lattice machine and the other one based on rough set rule induction, and we design hierarchical versions of the two classifiers. The two hierarchical classifiers are compared through experiments with their non-hierarchical counterparts, and also with a method that combines k-nearest neighbors classifier with rough set rule induction as a benchmark. The results of the experiments show that this hierarchical approach leads to improved multimodal classifiers.

Keywords: hierarchical classification, multimodal classifier, lattice machine, rough sets, rule induction, k-nearest neighbors.

1   Introduction

Many machine learning methods are based on generation of models with separate model parts, each of which explains part of a given dataset. Examples include decision tree induction [20], rule induction [7] and the lattice machine [33].

J.F. Peters and A. Skowron (Eds.): Transactions on Rough Sets V, LNCS 4100, pp. 224–239, 2006.
© Springer-Verlag Berlin Heidelberg 2006

A decision tree consists of many branches, and each branch explains certain number of data examples. A rule induction algorithm generates a set of rules as a model of data, and each rule explains some data examples. The lattice machine generates a set of hypertuples as a model of data, and each hypertuple covers a region in the data space. We call this type of learning *multimodal learning* or *multimodal classification.*

In contrast some machine learning paradigms do not construct models with separate parts. Examples include neural networks, support vector machines and Bayesian networks.

In the multimodal learning paradigm the model parts may overlap or may have insufficient coverage of a data space, i.e., the model does not cover the whole data space. In a decision tree the branches do not overlap and cover the whole data space. In the case of rule induction, the rules may overlap and may not cover the whole data space. In the case of lattice machine the hypertuples overlap and the covering of the whole data space is not guaranteed too.

Overlapping makes it possible to label a data example by more than one class whereas insufficient coverage makes it possible that a data example is not labeled at all. How to deal best with the overlapping and insufficient coverage issues?

In this paper we consider a hierarchical strategy to answer this question. Most machine learning algorithms generate different models from data under different conditions or parameters, and they advocate some conditions for optimal models or let a user specify the condition for optimal models. Instead of trying to find the 'optimal' model we can consider a series of models constructed under different conditions. These models form a hierarchy, or a layered structure, where the bottom layer corresponds to a model with the strictest condition and the top layer corresponds to the one with the most relaxed condition. The models in different hierarchy layers correspond to different levels of pattern generalization.

To demonstrate the effectiveness of this strategy we consider two multimodal classifiers: one is the lattice machine (LM), and the other one is a rough set based rule induction algorithm RSES-O. We apply the hierarchical strategy in these two classifiers, leading to two new classification methods: HLM and RSES-H.

HLM is a hierarchical version of the lattice machine [33]. As mentioned earlier, the lattice machine generates hypertuples as model of data, but the hypertuples overlap (some objects are multiply covered) and usually only a part of the whole object space is covered by the hypertuples (some objects are not covered). Hence, for recognition of uncovered objects, we consider some more general hypertuples in the hierarchy that covers these objects. For recognition of multiply covered objects, we also consider more general hypertuples that cover (not exclusively) the objects. These covering hypertuples locate at various levels of the hierarchy. They are taken as neighborhoods of the object. A special voting strategy has been proposed to resolve conflicts between the object neighborhoods covering the classified object.

The second method, called RSES-H, is a hierarchical version of the rule-based classifier (hereafter referred to by RSES-O) in RSES [22]. RSES-O is based on rough set methods with optimization of rule shortening. RSES-H constructs a hierarchy of rule-based classifiers. The levels of the hierarchy are defined by different levels of minimal rule shortening [6,22]. A given object is classified by the classifier from the hierarchy that recognizes the object and corresponds to the minimal generalization (rule shortening) in the hierarchy of classifiers.

We compare HLM and RSES-H through a series of experiments with their non-hierarchical counterparts, LM [30,32] and RSES-O. We also compare the two algorithms with a state of the art classifier, RIONA, which is a combination of rough sets with the k-nearest neighbors (kNN) classifier [15,22]. The evaluation of described methods was done through experiments with benchmark datasets from UCI Machine Learning Repository [9] and also with some artificially generated data. The results of our experiments show that in many cases the hierarchical strategy leads to improved classification accuracy.

This paper extends the paper [24]. In this paper we provide more details on how the layers of HLM and RSES-H are constructed and a brief description of the reference algorithm RIONA. We add experimental results for artificially generated data containing noise and analyze how the hierarchical methods deal with noise. We also analyze the statistical significance of the classification accuracy improvement provided by the hierarchical approach.

It is necessary to note that our hierarchical strategy to multimodal classification is different from the classical hierarchical classification framework (see, e.g., [11,27,19,8,3,2,17]), which aims at developing methods to learn complex, usually hierarchical, concepts. In our study we do not consider the hierarchical structure of the concepts in question; therefore our study is in fact a *hierarchical approach to flat classification*.

The paper is organized as follows. Section 2.1 introduces the lattice machine classifier LM used as the basis for the hierarchical HLM. Section 2.2 describes the rough set method RSES-O used as the basis for the hierarchical RSES-H. Section 2.3 presents the algorithm RIONA used in experiments as the reference classifier. In Section 3 we introduce the hierarchical classifier HLM and in Section 4 the hierarchical RSES-H is presented. Section 5 provides experimental results obtained for the described classifiers and Section 6 concludes the paper with a brief summary.

2   Multimodal Classifiers

In this section we present in some detail three multimodal classifiers. In later sections we will present their hierarchical counterparts.

2.1   The Lattice Machine

The lattice machine [30,32,33] is a machine learning paradigm that constructs a generalized version space from data, which serves as a model (or hypothesis) of data. A model is a hyperrelation, or a set of hypertuples (patterns), such that each hypertuple in the hyperrelation is equilabeled, supported, and maximal. Being equilabeled means the model is consistent with data (i.e., matches objects with the same decision only); being maximal means the model has generalization capability; and being supported means the model does not generalize beyond the information given in the data. When data come from Euclidean space, the model is a set of hyperrectangles consistently, tightly and maximally approximating the data. Observe that, this approach is different from decision tree induction, which aims at partition of the data space. Lattice machines have two basic operations: a construction operation to build a model of data, and a classification operation that applies the model to classify data. The model is in the form of a set of hypertuples [31]. To make this paper self-contained we review the concepts of hypertuple.

Let $R = \{a_1, a_2, \cdots, a_n\}$ be a set of attributes, and y be the class (or decision) attribute; $dom(a)$ be the domain of attribute $a \in R \cup \{y\}$. In particular we let $C = dom(y)$ – the set of class labels. Let $V \stackrel{\text{def}}{=} \prod_{i=1}^n dom(a_i)$ and $L \stackrel{\text{def}}{=} \prod_{i=1}^n 2^{dom(a_i)}$. V is called the *data space* defined by R, and L an *extended data space*. A (given) *dataset* is $D \subseteq V \times C$ – a sample of V with known class labels. If we write an element $t \in V$ by $\langle v_1, v_2, \cdots, v_n \rangle$ then $v_i \in dom(a_i)$. If we write $h \in L$ by $\langle s_1, s_2, \cdots, s_n \rangle$ then $s_i \in 2^{dom(a_i)}$ or $s_i \subseteq dom(a_i)$. An element of L is called a *hypertuple*, and an element of V a *simple tuple*. The difference between the two is that a field in a simple tuple is a value (hence *value-based*) while a field in a hypertuple is a set (hence *set-based*). If we interpret $v_i \in dom(a_i)$ as a singleton set $\{v_i\}$, then a simple tuple is a special hypertuple. L is a lattice under the ordering [30]: for $s, t \in L$,

$$t \leq s \Longleftrightarrow t(x) \subseteq s(x) \tag{1}$$

with the sum and product operations given by

$$t + s = \langle t(x) \cup s(x) \rangle_{x \in R}. \tag{2}$$
$$t \times s = \langle t(x) \cap s(x) \rangle_{x \in R}. \tag{3}$$

Here $t(x)$ is the projection of t onto attribute x.

The LM algorithm [31] constructs the unique model but it is not scalable to large datasets. The efficient algorithm CASEEXTRACT, presented in [30], constructs such a model with the maximal condition relaxed. Such a model consists of a set of hypertuples which have disjoint coverage of the dataset.

Let D be a dataset, which is split into k classes: $D = \{D_1, D_2, \cdots, D_k\}$ where D_i and D_j are disjoint, $i \neq j$. The CASEEXTRACT algorithm [30] is as follows:

– For $i = 1$ to k:
   • Initialization: let $X = D_i, H_i = \emptyset$.
   • Repeat until X is empty:
     1. Let $h \in X$ and $X = X \setminus \{h\}$.
     2. For each $g \in X$, if $h+g$ is equilabeled then $h = h+g$ and $X = X \setminus \{g\}$
     3. Let $H_i = H_i \cup \{h\}$.
– $H = \bigcup_{i=1}^k H_i$ is a model of the data.

Note that $h + g$ is defined in Eq.(2). This algorithm bi-partitions X into a set of elements the sum of which is an equilabeled element, and a new X consisting of the rest of the elements. The new X is similarly bi-partitioned until X becomes empty.

When such a model is obtained, classification can be done by the C2 algorithm [32]. C2 distinguishes between two types of data: those that are covered by one and only one hypertuple (primary data), those that are covered by more than one hypertuple (secondary data) and those that are not covered (tertiary data). Classification is based on two measures. Primary data t is put in the same class as the hypertuple that covers t, and secondary and tertiary data are classified with the use of these two measures.

Let R be a set of attributes, $X \subseteq R$, V_X be the projection of V onto X, and $S \stackrel{\text{def}}{=} V_X$. V_X is the domain of X. When $X = R$, V_R is the whole data space, i.e., $V_R = V$. Consider a mass function $m : 2^S \rightarrow [0,1]$ such that $m(\emptyset) = 0$ and $\sum_{x \in 2^S} m(x) = 1$. Given $a, b \in 2^S$, where $m(b) \neq 0$, the first measure is derived by answering this question: what is the probability that b appears whenever a appears? In other words, if a appears, what is the probability that b will be regarded as appearing as well? Denoting this probability by $C_X^0(b|a)$, one solution is:

$$C_X^0(b|a) = \frac{\sum_{a \cup b \subseteq c} m(c)}{\sum_{b \subseteq c} m(c)}.$$

In the same spirit, another measure is defined as

$$C_X^1(b|a) = \frac{\sum_{c \subseteq b} m(c)}{\sum_{c \subseteq a \cup b} m(c)}.$$

$C_X^1(b|a)$ measures the degree in which merging a and b preserves the existing structure embodied by the mass function.

With the above two measures, the C2 algorithm for classification is as follows [32]. Let $t \in V$, and H be the set of hypertuples generated by the CASEEXTRACT algorithm.

– For each $s \in H$, calculate $C_R^0(s|t)$ and $C_R^1(s|t)$.
– Let A be the set of $s \in H$ which have maximal C_X^0 values. If A has only one element, namely $A = \{s\}$, then label t by the label of s. Otherwise, let B be the set of $s \in A$ which have maximal C_X^1 values. If B has only one element, namely $B = \{s\}$, then label t by the label of s. Otherwise, label t by the label of the element in B which has the highest coverage.

Some variants of C2 are discussed in [33]. C2 performed extremely well on primary data, but not desirable on secondary and tertiary data.

2.2   RSES-O

The Rough Set Exploration System (RSES) (see [6,5,22]) is a freely available software system toolset for data exploration, classification support and knowledge discovery. Many of the RSES methods have originated from rough set theory introduced by Zdzisław Pawlak during the early 1980s (see [18]). At the moment of writing this paper RSES version 2.2 is the most recent (see [5] for more details).

One of the most popular methods for classifiers construction is based on learning rules from examples. Therefore there are several methods for calculation of the decision rule sets implemented in the RSES system (see [5,22]). One of these methods generates consistent decision rules with the minimal number of descriptors. This kind of decision rules can be used for classifying new objects as a standard rough set method of classifiers construction (see e.g. [23]).

Unfortunately, the decision rules consistent with the training examples can often be inappropriate to classify unseen cases. This happens, e.g. when the number of examples supporting a decision rule is relatively small. Therefore in practice we often use approximate rules instead of consistent decision rules. In RSES we have implemented a method for computing approximate rules (see e.g. [4]). In our method we begin with algorithms for synthesis of consistent decision rules with the minimal number of descriptors from a given decision table. Next, we compute approximate rules from already calculated consistent decision rules using the consistency coefficient. For a given training table D the consistency coefficient $cons$ of a given decision rule $\alpha \to q$ (q is the decision class label) is defined by:

$$cons(\alpha \to q) = \frac{\|\{x \in D_q : x \text{ satisfies } \alpha\}\|}{\|\{x \in D : x \text{ satisfies } \alpha\}\|}$$

where D_q denotes the decision class corresponding to q. The original consistent decision rules with the minimal number of descriptors are reduced to approximate rules with consistency coefficient exceeding a fixed (optimal) threshold.

The resulting rules are shorter, more general (can be applied to more training objects) but they may lose some of their precision, i.e., may provide wrong answers (decisions) for some of the matching training objects. In exchange for this we expect to receive more general rules with higher quality of classification for new cases.

The method of classifier construction based on approximate rules is called the RSES-O method.

2.3 Rule Induction with Optimal Neighborhood Algorithm (RIONA)

RIONA [15] is a classification algorithm implemented in RSES [6,22] that combines the kNN classifier with rule induction. The method induces a distance measure and distance-based rules. For classification of a given test object the examples most similar to this object vote for decisions but first they are compared against the rules and the examples that do not match any rule are excluded from voting.

First the algorithm induces a distance measure ρ from a data sample D. The distance measure is defined by the weighted sum of the distance measures ρ_i for particular attributes a_i:

$$\rho(x, y) = \sum_{i=1}^{n} w_i \cdot \rho_i(a_i(x), a_i(y)).$$

RIONA uses the combination of the normalized city-block Manhattan metric for numerical attributes and the Simple Value Difference (SVD) metric for nominal attributes [15]. The distance between values of a numerical attribute a_i is defined by the absolute value difference between these values normalized by the range of attribute values in the data sample D:

$$\rho_i(a_i(x), a_i(y)) = \frac{|a(x) - a(y)|}{a_{max} - a_{min}}.$$

where $a_{min} = \min_{x \in D} a_i(x)$ and $a_{max} = \max_{x \in D} a_i(x)$. The SVD distance between values of a nominal attribute a_i is defined by the difference between the decision distributions for these values in the data sample D:

$$\rho_i(a_i(x), a_i(y)) = \sum_{q \in C} \left| P(z \in D_q | z \in D^{a_i(x)}) - P(z \in D_q | z \in D^{a_i(y)}) \right|.$$

where $D^{a_i(x_0)} = \{x \in D : a_i(x) = a_i(x_0)\}$. The weights w_i are optimized with the iterative attribute weighting procedure from [35].

To classify a tuple t RIONA uses the k nearest neighbors $n_1(t), \ldots, n_k(t)$ of t in the data sample D according to the previously defined distance measure ρ. Before voting the nearest neighbors are examined with consistent maximal rules derived from the data sample D [15]. If there is no consistent maximal rule that covers both a given neighbor $n_j(t)$ and the tuple t, the neighbor $n_j(t)$ is excluded from voting. The neighbors that share at least one maximal consistent rule with the tuple t are assigned with the vote weights v_j inversely proportional to square of the distance to t:

$$v_j(t) = \begin{cases} \frac{1}{\rho(n_j(t), t)^2} & \text{if there is consistent maximal rule covering } t \text{ and } n_j(t) \\ 0 & \text{otherwise} \end{cases}.$$

The tuple t is classified by q with the largest sum of nearest neighbor votes $S(t, q) = \sum_{n_j(t) \in D_q} v_j(t)$, where $1 \leq j \leq k$.

The value of k is optimized automatically in the range $1 \leq k \leq 100$ by the efficient leave-one-out procedure [15] applied to the data sample D.

3   HLM: Hierarchical Lattice Machine

In this section we present an implementation of our hierarchical approach to multimodal classification. This is a hierarchical version of the lattice machine, referred to by HLM.

We implement the hierarchical strategy in the lattice machine with the expectation that the classification accuracy of the lattice machine can be improved. Here is an outline of the solution.

We apply the CASEEXTRACT algorithm repeatedly to construct a hierarchy of hypertuples. The bottom layer is constructed by CASEEXTRACT directly from data. Then those data that are covered by the hypertuples with small coverage are marked out in the dataset, and the algorithm is applied again to construct a second layer. This process is repeated until a layer only with one hypertuple is reached. At the bottom layer all hypertuples are equilabeled, while those at higher layers may not be equilabeled.

To classify a data tuple (query) we search through the hierarchy to find a hypertuple at the lowest possible layer that covers the query. Then all data (including both marked and unmarked) covered by the hypertuple are weighted by an efficient counting-based weighting method. The weights are aggregated and used to classify the query. This is similar to the weighted kNN classifier, but it uses counting instead of distance to weigh relevant data.

3.1   Counting-Based Weighting Measure

In this section we present a counting-based weighting measure, which is suitable for use with hypertuples.

Suppose we have a neighborhood D for a query tuple (object) t and elements in D may come from any class. In order to classify the query based on the neighborhood we can take a majority voting with or without weighting. This is the essence of the well-known kNN classifier [14,12].

Weighting is usually done by the reverse of distance. Distance measures usually work for numerical data. For categorical data we need to transform the data into numerical form first. There are many ways for the transformation (see for example [26,10,34]), but most of them are task (e.g., classification) specific.

We present a general weighting method that allows us to count the number of all hypertuples, generated by the data tuples in a neighborhood of a query tuple t, that cover both t and any data tuple x in the neighborhood. Intuitively the higher the count the more relevant this x is to t, hence x should play a bigger role (higher weight). The inverse of this count can be used as a measure of distance between x and t. Therefore, by this count we can order and weight the data tuples. This counting method works for both numerical and categorical data in a conceptually uniform way. We consider next an efficient method to calculate this count.

As a measure of weighting we determine, for tuples t and x in D, the number of hypertuples that cover both t and x. We call this number the *cover* of t and x, denoted by $cov(t, x)$. The important issue here is how to calculate $cov(t, x)$ for every pair (t, x).

Consider two simple tuples $t = <t_1, t_2, \cdots, t_n>$ and $x = <x_1, x_2, \cdots, x_n>$. t is a simple tuple to be classified (query) and x is any simple tuple in D. What we want is to find all hypertuples that cover both t and x. We look at every attribute and explore the number of subsets that can be used to generate a hypertuple covering both t and x. Multiplying these numbers across all attributes gives rise to the number we require.

Consider an attribute a_i. If a_i is numerical, N_i denotes the number of intervals that can be used to generate a hypertuple covering both t_i and x_i. If a_i is categorical, N_i denotes the number of subsets for the same purpose. Assuming that all attributes have finite domains, we have [29]:

$$N_i = \begin{cases} (\max(a_i) - \max(\{x_i, t_i\}) + 1) \times (\min(\{x_i, t_i\}) - \min(a_i) + 1) \\ \qquad \text{if } a_i \text{ is numerical} \\ 2^{m_i - 1} \quad \text{if } a_i \text{ is categorical and } x_i = t_i \\ 2^{m_i - 2} \quad \text{if } a_i \text{ is categorical and } x_i \neq t_i. \end{cases} \qquad (4)$$

where $\max(a_i)$, $\min(a_i)$ are the maximal and the minimal value of a_i, respectively, if a_i is numerical, and $m_i = |dom(a_i)|$, if a_i is categorical.

The number of covering hypertuples of t and x is $cov(t, x) = \prod_i N_i$.

A simple tuple $x \in D$ is then weighted by $cov(t, x)$ in a kNN classifier. More specifically, we define

$$K(t, q) = \sum_{x \in D_q} cov(t, x).$$

where D_q is a subset of D consisting of all q class simple tuples. $K(t, q)$ is the total of the cover of all q class simple tuples. Then the weighted kNN classifier is the following rule (WKNN rule):

t is classified by q_0 that has the largest $K(t, q)$ for all q.

3.2   The Classification Procedure

We now present a classification procedure, called, *hierarchical classification based on weighting* (HLM).

Let D be a given dataset, let HH be a hierarchy of hypertuples constructed from D, and let t be a query – a simple tuple to be classified.

Step 1. Search HH in the bottom up order and stop as soon as a covering hypertuple is found at layer l. Continue searching layer l until all covering hypertuples are found. Let S be a set of all covering hypertuples from this layer;

Step 2. Let $N \leftarrow \{\underline{h} : h \in S\}$, a neighborhood of the query;

Step 3. Apply WKNN to classify t.

Note that $\underline{h}$ is the set of simple tuples covered by h.

4   RSES-H: Hierarchical Rule-Based Classifier

In this section we present another implementation of our hierarchical approach to multimodal classification. This is a hierarchical version of RSES-O, referred to by RSES-H.

In RSES-H a set of minimal decision rules [7,22] is generated. Then, different layers for classification are created by rule shortening. The algorithm works as follows:

1. At the beginning, we divide original data sets into two disjoint parts: train table and test table.
2. Next, we calculate (consistent) rules with a minimal number of descriptors for the train table (using covering method from RSES [7,22]). This set of rules is used to construct the first (the bottom) level of our classifier.

3. In the successive steps defined by the following consistency thresholds (after rule shortening): $0.95, 0.90, 0.85, 0.80, 0.75, 0.70, 0.65, 0.60$, we generate a set of rules obtained by shortening all rules generated in the previous step. The rules generated in the i-th step are used to construct the classifier with the label $i + 1$ in the classifier hierarchy.

4. Now, we can use our hierarchical classifier in the following way:
   (a) For any object from the test table, we try to classify this object using decision rules from the first level of our classifier.
   (b) If the tested object is classified by rules from the first level of classifier, we return the decision value for this object and the remaining levels of our classifier are not used.
   (c) If the tested object can not be classified by rules from the first level, we try to classify it using the second level of our hierarchical classifier, etc.
   (d) Finally, if the tested object can not be classified by rules from the level with the label 9, then our classifier can not classify the tested object. The last case happens seldom, because higher levels are usually sufficient for classifying any tested object.

The range of thresholds for the rule consistency (see Section 2.2) in the third step of the algorithm presented above have been determined on the basis of experience obtained from the previous experiments (see e.g. [1]). The step between thresholds has been determined to 0.05, because this allows us to make rule search quite precise and the number of thresholds (that have to be checked) is not too large from the computational complexity point of view.

5 Evaluation

Experiments were performed with the two hierarchical classifiers (HLM and RSES-H) described in Section 3.1 (HLM) and Section 4 (RSES-H), their non-hierarchical counterparts (LM and RSES-O based on rules) and RIONA as a benchmark classifier. The purpose of the experiment was two fold: first, we wanted to know whether the hierarchical algorithms improve their non-hierarchical counterparts. Second, we wanted to know the correspondence between the degree of improvement and distribution of data.

For this purpose we considered two types of data: real world data and artificial (or synthetic) data. The former were some popular benchmark datasets from UCI Machine Learning Repository [9], and some simple statistics are shown in Table 1. The latter were generated by Handl and Knowles [16]. The generator is based on a standard cluster model using multivariate normal distributions. We generated six datasets: three of them have two clusters labeled as two separate classes (unimodal data), and the remaining three have four clusters grouped again into two classes (multimodal data). In all cases 20% random noise were added.

In the experiment each classifier was tested 10 times on each dataset with the use of 5-fold cross-validation.

Table 1. General information on the datasets and the 5-fold cross validation success rate with standard deviation of LM, HLM, RSES-H, RSES-O and RIONA

Data	General Info			5CV success rate				
	Att	Exa	Cla	LM	HLM	RSES-O	RSES-H	RIONA
Anneal	38	798	6	95.7±0.3	96.0±0.4	94.3±0.6	96.2±0.5	92.5
Austral	14	690	2	91.9±0.3	92.0±0.4	86.4±0.5	87.0±0.5	85.7
Auto	25	205	6	73.0±1.5	76.5±1.4	69.0±3.1	73.7±1.7	76.7
Diabetes	8	768	2	70.6±0.6	72.6±0.8	73.8±0.6	73.8±1.2	75.4
Ecoli	7	336	8	79.8±1.0	85.6±0.7	72.4±2.3	76.0±1.7	84.1
German	20	1000	2	69.8±0.6	71.4±0.9	72.2±0.4	73.2±0.9	74.4
Glass	9	214	3	63.5±1.2	71.3±1.2	61.2±2.5	63.4±1.8	66.1
Heart	13	270	2	75.2±1.8	79.0±1.0	83.8±1.1	84.0±1.3	82.3
Hepatitis	19	155	2	77.2±0.7	78.7±1.2	82.6±1.3	81.9±1.6	82.0
Horse-Colic	22	368	2	78.2±0.8	76.3±0.9	85.5±0.5	86.5±0.6	84.6
Iris	4	150	3	95.0±0.4	94.1±0.4	94.9±1.5	95.5±0.8	94.4
Sonar	60	208	2	74.2±1.2	73.7±0.8	74.3±1.8	75.3±2.0	86.1
TTT	9	958	2	94.0±0.7	95.0±0.3	99.0±0.2	99.1±0.2	93.6
Vehicle	18	846	4	69.4±0.5	67.6±0.7	64.2±1.3	66.1±1.4	70.2
Vote	18	232	2	96.4±0.5	95.4±0.5	96.4±0.5	96.5±0.5	95.3
Wine	12	178	3	96.4±0.4	92.6±0.8	90.7±2.2	91.2±1.2	95.4
Yeast	8	1484	10	49.9±0.6	51.3±0.7	50.7±1.2	51.9±0.9	58.9
D20c22n0	20	522	2	85.0±0.8	89.4±0.6	88.9±0.9	88.8±1.2	91.4
D20c22n1	20	922	2	87.6±0.5	89.1±0.5	90.1±0.6	90.1±1.0	86.9
D20c22n2	20	838	2	89.2±0.5	91.2±0.3	90.3±0.4	89.9±0.9	89.4
D20c42n0	20	1370	2	81.4±0.5	85.5±0.3	83.6±1.0	84.4±1.4	90.9
D20c42n1	20	1558	2	80.1±0.3	83.8±0.3	88.5±0.4	88.7±0.6	87.1
D20c42n2	20	1524	2	77.9±0.8	79.0±0.5	79.6±0.7	79.8±1.0	83.2
Average success rate				80.53	82.05	81.41	82.3	83.77

The average results with standard deviations are shown in Table 1. HLM obtained the higher accuracy than its non-hierarchical counterpart LM on 17 data sets and it lost on 6 data sets. The best improvements were for the data sets *Glass* (7.8%) and *Ecoli* (5.8%). The difference between RSES-H and its non-hierarchical counterpart RSES-O is even more distinct: RSES-H outperformed RSES-O on 19 data sets and lost on 3 data sets only. The best improvements were for the data sets *Auto* (4.7%) and *Ecoli* (3.6%).

The supremacy of the hierarchical methods over the non-hierarchical ones can be also noticed in the total average accuracy: HLM accuracy is 1.5% higher than LM accuracy and similarly RSES-H accuracy is almost 1% higher than RSES-O accuracy. On average both hierachical methods ouperformed both non-hierarchical methods.

The benchmark classifier RIONA has the highest total average accuracy but the difference to hierarchical methods is much smaller than to non-hierarchical classifiers. The advantage of RIONA over HLM and RSES-H comes from a few specific data sets (*Sonar, Ecoli* and *Yeast*) where the nearest neighbor component helps a lot in overcoming the problem of a large number of attributes or classes.

Table 2. The levels of statistical significance of difference when comparing the hierarchical methods against the non-hierarchical methods: 5 is 99.5%, 4 is 99%, 3 is 97.5%, 2 is 95%, 1 is 90% and 0 is below 90%. Plus indicates that the average accuracy of a hierarchical method is higher than that of a non-hierarchical method and minus otherwise.

Data	General Info			Statistical significance			
				HLM vs		RSES-H vs	
	Attrib	Exampl	Classes	LM	RSES-O	LM	RSES-O
Anneal	38	798	6	+3	+5	+5	+5
Austral	14	690	2	+0	+5	−5	+3
Auto	25	205	6	+5	+5	+0	+5
Diabetes	8	768	2	+5	−5	+5	0
Ecoli	7	336	8	+5	+5	−5	+5
German	20	1000	2	+5	−4	+5	+5
Glass	9	214	3	+5	+5	−0	+3
Heart	13	270	2	+5	−5	+5	+0
Hepatitis	19	155	2	+5	−5	+5	−0
Horse-Colic	22	368	2	−5	−5	+5	+5
Iris	4	150	3	−5	−1	+2	+0
Sonar	60	208	2	−0	−0	+1	+0
TTT	9	958	2	+5	−5	+5	+1
Vehicle	18	846	4	−5	+5	−5	+4
Vote	18	232	2	−5	−5	+0	+0
Wine	12	178	3	−5	+4	−5	+0
Yeast	8	1484	10	+5	+1	+5	+3
D20c22n0	20	522	2	+5	+1	+5	−0
D20c22n1	20	922	2	+5	−5	+5	+0
D20c22n2	20	838	2	+5	+5	+3	−0
D20c42n0	20	1370	2	+5	+5	+5	+1
D20c42n1	20	1558	2	+5	−5	+5	+0
D20c42n2	20	1524	2	+5	−3	+5	+0
Wins/Losses quite probable (> 90%)				16/5	11/11	16/4	11/0
Wins/Losses certain (> 99.5%)				15/5	8/8	13/4	5/0

One could ask whether the differences in accuracy between the hierarchical and non-hierarchical methods are really significant. To answer this question in Table 2 we compared the hierarchical HLM and RSES-H against the non-hierarchical LM and RSES-O and provided the statistical significance of differences between the accuracy of classifiers on particular data sets using the one-tail unpaired Student's t-test [25].

Comparing HLM against LM (see Figure 1) one can see that for almost all the data sets the differences are significant. In other words, in 16 cases HLM provided the statistically significant improvement in accuracy over LM (for 15 data sets this improvement is practically certain) and the accuracy has significantly fallen in 5 cases. This confirms the conclusion that, in general, it is worth to apply hierarchical HLM instead of non-hierarchical LM.

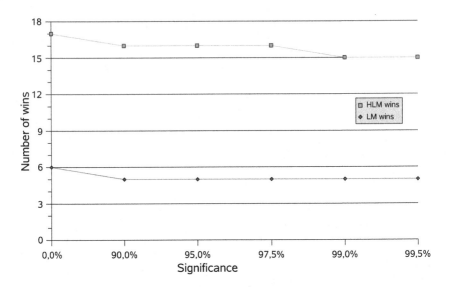

Fig. 1. The number of data sets where HLM outperforms LM and the number of datasets where HLM loses in dependence of significance level of accuracy difference

The comparison between HLM and RSES-O does not show supremacy of any method. However, for some datasets (*Australian, Auto, Ecoli* and *Glass*) HLM provided significantly better results than both RSES-H and RSES-O.

The relation between the results of RSES-H and LM is similar to the relation HLM vs LM. The differences in accuracy are significant for almost all the data sets and in most cases RSES-H outperformed LM.

The interesting relation is between the results of RSES-H and RSES-O (see Figure 2). There is no data set on which RSES-H was significantly worse than RSES-O. On other hand, in half of cases RSES-H improved significantly RSES-O. This indicates that the extension of RSES-O to RSES-H is rather stable: RSES-H keeps the level of the RSES-O accuracy. There are no risk while replacing RSES-O with RSES-H and a significant chance of improving the results.

Another interesting observation can be made when one focuses on artificial data. In comparison with LM both hierarchical methods provide significant improvement on all generated data sets. This indicates that lattice machine does not deal well with noisy data and both hierarchical methods do it better. Therefore in cases of noisy data the lattice machine is particularly recommended to be extended to hierarchical approach.

The different situation is in case of RSES-O. The comparison of both hierarchical methods with RSES-O on artificial data does not show the significant supremacy of any method. This suggests that to overcome the problem of noise the rule optimization used in RSES-O is equally effective as the hierarchical approach.

The experiments confirmed our original expectation that the performance of LM and RSES-O can be improved by our hierarchical approach. The cost is

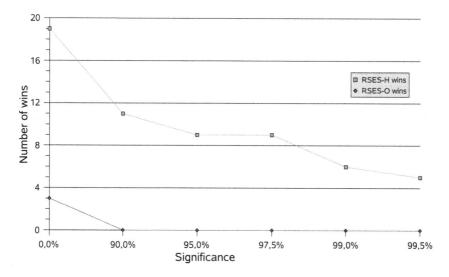

Fig. 2. The number of data sets where RSES-H outperforms RSES-O and the number of datasets where RSES-H loses in dependence of significance level of accuracy difference

some extra time to construct the hierarchy and to test some new objects using this hierarchy.

The experimental results provide also the hint which hierarchical method to use in dependence of the expected profit. RSES-H is safer: it usually provides smaller improvements than HLM but it never worsens the accuracy. Application of HLM can result in a greater improvement but there is also a higher risk of worsening the results.

6   Conclusions

In the paper we addressed the problem of balancing between accuracy and universality in multimodal classification models. The accurate model parts can be specific and cover part of data space only. On the other hand more general model parts can lose accuracy. To solve this problem we proposed the hierarchical approach to multimodal classification. The hierarchical approach introduces a number of multimodal layers instead of a single one, each with different accuracy and universality. Such a hierarchical model tries to classify data with the most accurate layer and if it fails then it moves through more and more general layers until an answer is found. Such a hierarchical approach provides a kind of adaptation to test case difficulty. The proper graduation between the successive levels allows to classify data with the optimal balance between accuracy and universality.

The experimental results confirmed that such a hierarchical model is more effective than one-layer multimodal methods: it gives higher classification accuracy. Hierarchical approach guarantees also a certain level of adaptability to data: it deals well with noise in data.

In future research we plan to develop hierarchical multimodal methods for incremental learning. Incremental learning requires reconstruction of particular layers when new examples arrive. In hierarchical model the reconstruction time of a layer can be reduced by the use of knowledge from the adjoining layers.

The other interesting research direction is the application of a hierarchical rule model in RIONA. In that classifier rules are used to validate and filter nearest neighbors (see Section 2.3). The interesting question is whether one can improve RIONA accuracy still more by replacing the single set of rules with a hierarchical rule model in the validation step.

Acknowledgments. The research has been supported by the grant 3 T11C 002 26 from Ministry of Scientific Research and Information Technology of the Republic of Poland.

References

1. Bazan, J.: A Comparison of Dynamic and non-Dynamic Rough Set Methods for Extracting Laws from Decision Table, Polkowski L., Skowron A. (eds.): *Rough Sets in Knowledge Discovery*. Heidelberg: Physica-Verlag 321-365, 1998.

2. J. Bazan. Classifiers based on two-layered learning. Lecture Notes in Artificial Intelligence 3066, Springer, Heidelberg, 356–361, 2004.

3. J. Bazan, S. Hoa Nguyen, H. Son Nguyen, A. Skowron. Rough set methods in approximation of hierarchical concepts. *Proc. of RSCTC'2004*, Lecture Notes in Artificial Intelligence 3066, Springer, Heidelberg, 346–355, 2004.

4. J. G. Bazan, H. Son Nguyen, S. Hoa Nguyen, P. Synak, J. Wróblewski. Rough Set Algorithms in Classification Problem, L. Polkowski, S. Tsumoto, T. Y. Lin, (eds.): Rough Set Methods and Applications. Heidelberg: Physica-Verlag 49-88, 2000.

5. J. G. Bazan, M. Szczuka. The Rough Set Exploration System. In *Transactions on Rough Sets* III, Lecture Notes in Computer Science 3400, 2005, 37-56.

6. J. Bazan, M. Szczuka, A. Wojna, M. Wojnarski. On the evolution of Rough Set Exploration System, Proc. of RSCTC'2004, Lecture Notes in Artificial Intelligence 3066, Springer, Heidelberg, 592–601, 2004.

7. J. G. Bazan, M. Szczuka, J. Wróblewski. A New Version of Rough Set Exploration System, Proc. of RSCTC'2002, Lecture Notes in Artificial Intelligence 2475, Springer-Verlag, Heidelberg, 397-404, 2002.

8. S. Behnke. Hierarchical Neural Networks for Image Interpretation. Lecture Notes in Artificial Intelligence 2766, Springer, Heidelberg, 2003.

9. C.L. Blake and C.J. Merz. UCI repository of machine learning databases, 1998.

10. S. Cost, S. Salzberg. A weighted nearest neighbor algorithm for learning with symbolic features. *Machine Learning*, 10:57–78, 1993.

11. T. G. Dietterich. Ensemble Learning. In M.A. Arbib (Ed.), The Handbook of Brain Theory and Neural Networks, Second edition, Cambridge, MA: The MIT Press, 405-408, 2002.

12. S. A. Dudani. The distance-weighted k-nearest-neighbor rule. *IEEE Trans. Syst. Man Cyber.*, 6:325–327, 1976.

13. Ivo Düntsch and Günther Gediga. Simple data filtering in rough set systems. *International Journal of Approximate Reasoning*, 18(1–2):93–106, 1998.
14. E. Fix and J. L. Hodges. Discriminatory analysis, nonparametric discrimination: Consistency properties. Technical Report TR4, USAF School of Aviation Medicine, Randolph Field, TX, 1951.
15. G. Góra and A. G. Wojna. RIONA: a new classification system combining rule induction and instance-based learning. *Fundamenta Informaticae*, 51(4):369–390, 2002.
16. J. Handl and J. Knowles. Cluster generators: synthetic data for the evaluation of clustering algorithms. `"http://dbkweb.ch.umist.ac.uk/handl/generators/"`.
17. S. Hoa Nguyen, J. Bazan, A. Skowron, H. Son Nguyen. Layered learning for concept synthesis. Lecture Notes in Artificial Intelligence 3100, *Transactions on Rough Sets I*:187–208, Springer, Heidelberg, 2004.
18. Z. Pawlak. Rough Sets. Theoretical Aspects of Reasoning about Data, Kluwer Academic Publishers, Dordrecht, 1991.
19. T. Poggio, S. Smale. The Mathematics of Learning: Dealing with Data. *Notices of the AMS* 50(5):537-544, 2003.
20. Ross Quinlan. Improved Use of Continuous Attributes in C4.5. Journal of Artificial Intelligence Research, 4:77-90, 1996.
21. R. Quinlan. Rulequest research data mining tools. http://www.rulequest.com/.
22. RSES: Rough set exploration system. http://logic.mimuw.edu.pl/˜rses, Institute of Mathematics, Warsaw University, Poland.
23. A. Skowron. Boolean reasoning for decision rules generation, Proc. of ISMIS'93, Lecture Notes in Artificial Intelligence 689, Springer-Verlag, Heidelberg, 295–305, 1993.
24. A. Skowron, H. Wang, A. G. Wojna, J. G. Bazan. A Hierarchical Approach to Multimodal Classification, Proc. of RSFDGrC'2005, Lecture Notes in Artificial Intelligence 3642, Springer-Verlag, Heidelberg, 2005, 119–127.
25. G. W. Snedecor, W. G. Cochran. Statisitical Methods, Iowa State University Press, Ames, IA, 2002, eighth edition.
26. C. Stanfill and D. Waltz. Toward memory-based reasoning. *Communication of ACM*, 29:1213–1229, 1986.
27. P. Stone. *Layered Learning in Multi-agent Systems: A Winning Approach to Robotic Soccer.* MIT Press, Cambridge, MA, 2000.
28. V. N. Vapnik. *Statistical learning theory.* Wiley New York, 1998.
29. H. Wang. Nearest neighbors by neighborhood counting. *IEEE Transactions on Pattern Analysis and Machine Intelligence*, 28(6), June 2006.
30. H. Wang, W. Dubitzky, I. Düntsch, and D. Bell. A lattice machine approach to automated casebase design: Marrying lazy and eager learning. In *Proc. IJCAI99*, Stockholm, Sweden, 254–259, 1999.
31. H. Wang, I. Düntsch, D. Bell. Data reduction based on hyper relations. *Proceedings of KDD98, New York*, 349–353, 1998.
32. H. Wang, I. Düntsch, G. Gediga. Classificatory filtering in decision systems. *International Journal of Approximate Reasoning*, 23:111–136, 2000.
33. H. Wang, I. Düntsch, G. Gediga, A. Skowron. Hyperrelations in version space. *International Journal of Approximate Reasoning*, 36(3):223–241, 2004.
34. D. R. Wilson and T. R. Martinez. Improved heterogeneous distance functions. *Journal of Artificial Intelligence Research*, 6:1–34, 1997.
35. A. G. Wojna. Analogy-Based Reasoning in Classifier Construction. *Transactions on Rough Sets* IV, Lecture Notes in Computer Science 3700, 2005, 277-374.

Arrow Decision Logic for Relational Information Systems*

Tuan-Fang Fan[1,2], Duen-Ren Liu[3], and Gwo-Hshiung Tzeng[4,5]

[1] Institute of Information Management
National Chiao-Tung University, Hsinchu 300, Taiwan
tffan.iim92g@nctu.edu.tw
[2] Department of Computer Science and Information Engineering
National Penghu University, Penghu 880, Taiwan
dffan@npit.edu.tw
[3] Institute of Information Management
National Chiao-Tung University, Hsinchu 300, Taiwan
dliu@iim.nctu.edu.tw
[4] Institute of Management of Technology
National Chiao-Tung University, Hsinchu 300, Taiwan
ghtzeng@cc.nctu.edu.tw
[5] Department of Business Administration
Kainan University, Taoyuan 338, Taiwan
ghtzeng@mail.knu.edu.tw

Abstract. In this paper, we propose an arrow decision logic (ADL) for relational information systems (RIS). The logic combines the main features of decision logic (DL) and arrow logic (AL). DL represents and reasons about knowledge extracted from decision tables based on rough set theory, whereas AL is the basic modal logic of arrows. The semantic models of DL are functional information systems (FIS). ADL formulas, on the other hand, are interpreted in RIS. RIS , which not only specifies the properties of objects, but also the relationships between objects. We present a complete axiomatization of ADL and discuss its application to knowledge representation in multicriteria decision analysis.

Keywords: Arrow logic, decision logic, functional information systems, multicriteria decision analysis, relational information systems, rough sets.

1 Introduction

The rough set theory proposed by Pawlak [25] provides an effective tool for extracting knowledge from data tables. To represent and reason about extracted knowledge, a decision logic (DL) is proposed in Pawlak [26]. The semantics of the logic is defined in a Tarskian style through the notions of models and satisfaction. While DL can be considered as an instance of classical logic in the context of data tables, different generalizations of DL corresponding to some non-classical logics are also desirable from the viewpoint of knowledge representation. For

---

* A preliminary version of the paper was published in [1].

J.F. Peters and A. Skowron (Eds.): Transactions on Rough Sets V, LNCS 4100, pp. 240–262, 2006.

example, to deal with uncertain or incomplete information, some generalized decision logics have been proposed [7,22,23,34,35].

In rough set theory, objects are partitioned into equivalence classes based on their attribute values, which are essentially functional information associated with the objects. Though many databases contain only functional information about objects, data about the relationships between objects has become increasingly important in decision analysis. A remarkable example is social network analysis, in which the principal types of data are attribute data and relational data.

To represent attribute data, a data table in rough set theory consists of a set of objects and a set of attributes, where each attribute is considered as a function from the set of objects to the domain of values for the attribute. Hence, such data tables are also called *functional information systems* (FIS), and rough set theory can be viewed as a theory of *functional granulation*. Recently, granulation based on relational information between objects, called *relational granulation*, has been studied by Liau and Lin [21]. To facilitate further study of relational granulation, it is necessary to represent and reason about data in *relational information systems* (RIS).

In FIS, the basic entities are objects, while DL formulas describe the properties of such objects, thus, the truth values of DL formulas are evaluated with respect to these objects. To reason about RIS, we need a language that can be interpreted in the domain of pairs of objects, since relations can be seen as properties of such pairs. Arrow logic (AL) [24,33] fulfills this need perfectly. Hence, in this paper, we propose arrow decision logic (ADL), which combines the main features of DL and AL, to represent the decision rules induced from RIS. The atomic formulas of ADL have the same descriptor form as those in DL; while the formulas of ADL are interpreted with respect to each pair of objects, just as in the pair frame of AL [24,33]. The semantic models of ADL are RIS; thus, ADL can represent knowledge induced from systems containing relational information.

The remainder of this paper is organized as follows. In Section 2, we review FIS in rough set theory and give a precise definition of RIS. We study the relationship between these two kinds of information system and present some practical examples. In Section 3, we review DL and AL to lay the foundation for ADL. In Section 4, we present the syntax and semantics of ADL. A complete axiomatization of ADL based on the combination of DL and AL axiomatic systems is presented. In Section 5, we define some quantitative measures for the rules of ADL and discuss the application of ADL to data analysis. Finally, we present our conclusions in Section 6.

2   Information Systems

Information systems are fundamental to rough set theory, in which the approximation space can be derived from attribute-value information systems [26]. In this section, we review the functional information systems used in the original rough set theory and propose a generalization of it, namely, relational information systems. We present the algebraic relationship between these two kinds of information system, and several practical examples are employed to illustrate the algebraic notions.

2.1   Functional and Relational Information Systems

In data mining problems, data is usually provided in the form of a data table, which is formally defined as an attribute-value information system and taken as the basis of the approximation space in rough set theory [26]. To emphasize the fact that each attribute in an attribute-value system is associated with a function on the set of objects, we call such systems *functional information systems*.

Definition 1. *A functional information system (FIS)*[1] *is a quadruple*

$$T_f = (U, A, \{V_i \mid i \in A\}, \{f_i \mid i \in A\}),$$

where

- *U is a nonempty set, called the universe,*
- *A is a nonempty finite set of attributes,*
- *for each $i \in A$, V_i is the domain of values for i, and*
- *for each $i \in A$, $f_i : U \to V_i$ is a total function.*

In an FIS, the information about an object is consisted of the values of its attributes. Thus, given a subset of attributes $B \subseteq A$, we can define the information function associated with B as $Inf_B : U \to \prod_{i \in B} V_i$,

$$Inf_B(x) = (f_i(x))_{i \in B}. \tag{1}$$

Example 1. One of the most popular applications in data mining is association rule mining from transaction databases [2,18]. A transaction database consists of a set of transactions, each of which includes the number of items purchased by a customer. Each transaction is identified by a transaction id (tid). Thus, a transaction database is a natural example of an FIS, where

- U: the set of transactions, $\{tid_1, tid_2, \cdots, tid_n\}$;
- A: the set of possible items to be purchased;
- V_i: $\{0, 1, 2, \cdots, max_i\}$, where max_i is the maximum quantity of item i; and
- $f_i : U \to V_i$ describes the transaction details of item i such that $f_i(tid)$ is the quantity of item i purchased in tid.

Example 2. In this paper, we take an example of multicriteria decision analysis (MCDA) as a running example [8]. Assume that Table 1 is a summary of the reviews of ten papers submitted to a journal. The papers are rated according to four criteria:

- o: originality,
- p: presentation,
- t: technical soundness, and
- d: overall evaluation (the decision attribute).

---

[1] Originally called information systems, data tables, knowledge representation systems, or attribute-value systems in rough set theory.

Table 1. An FIS of the reviews of 10 papers

$U \setminus A$	o	p	t	d
1	4	4	3	4
2	3	2	3	3
3	4	3	2	3
4	2	2	2	2
5	2	1	2	1
6	3	1	2	1
7	3	2	2	2
8	4	1	2	2
9	3	3	2	3
10	4	3	3	3

Thus, in this FIS, we have

- $U = \{1, 2, \cdots, 10\}$,
- $A = \{o, p, t, d\}$,
- $V_i = \{1, 2, 3, 4\}$ for $i \in A$, and
- f_i is specified in Table 1.

Though much information associated with individual objects is given in a functional form, it is sometimes more natural to represent information about objects in a relational form. For example, in a demographic database, it is more natural to represent the parent-child relationship as a relation between individuals, instead of an attribute of the parent or the children. In some cases, it may be necessary to use relational information simply because the exact values of some attributes may not be available. For example, we may not know the exact ages of two individuals, but we do know which one is older. These considerations motivate the following definition of an alternative kind of information system, called an RIS.

Definition 2. *A relational information system (RIS) is a quadruple*

$$T_r = (U, A, \{H_i \mid i \in A\}, \{r_i \mid i \in A\}),$$

where

- *U is a nonempty set, called the universe,*
- *A is a nonempty finite set of attributes,*
- *for each $i \in A$, H_i is the set of relational indicators for i, and*
- *for each $i \in A$, $r_i : U \times U \to H_i$ is a total function.*

A relational indicator in H_i is used to indicate the extent or degree to which two objects are related according to an attribute i. Thus, $r_i(x, y)$ denotes the extent to which x is related to y on the attribute i. When $H_i = \{0, 1\}$, then, for any $x, y \in U$, x is said to be i-related to y iff $r_i(x, y) = 1$.

Example 3. Continuing with Example 2, assume that the reviewer is asked to compare the quality of the ten papers, instead of assigning scores to them. Then, we may obtain an RIS $T_r = (U, A, \{H_i \mid i \in A\}, \{r_i \mid i \in A\})$, where U and A are defined as in Example 2, $H_i = \{0, 1\}$, and $r_i : U \times U \to \{0, 1\}$ is defined by

$$r_i(x, y) = 1 \Leftrightarrow f_i(x) \geq f_i(y)$$

for all $i \in A$.

2.2 Relationship Between Information Systems

Before exploring the relationship between FIS and RIS, we introduce the notion of information system morphism (IS-morphism).

Definition 3

1. Let $T_f = (U, A, \{V_i \mid i \in A\}, \{f_i \mid i \in A\})$, and $T_f' = (U', A', \{V_i' \mid i \in A'\}, \{f_i' \mid i \in A'\})$ be two FIS; then an IS-morphism from T_f to T_f' is a $(|A| + 2)$-tuple of functions

$$\sigma = (\sigma_u, \sigma_a, (\sigma_i)_{i \in A})$$

such that $\sigma_u : U \to U'$, $\sigma_a : A \to A'$ and $\sigma_i : V_i \to V_{\sigma_a(i)}$ $(i \in A)$ satisfy

$$f_{\sigma_a(i)}'(\sigma_u(x)) = \sigma_i(f_i(x)) \tag{2}$$

for all $x \in U$ and $i \in A$.

2. Let $T_r = (U, A, \{H_i \mid i \in A\}, \{r_i \mid i \in A\})$, and $T_r' = (U', A', \{H_i' \mid i \in A'\}, \{r_i' \mid i \in A'\})$ be two RIS; then an IS-morphism from T_r to T_r' is a $(|A| + 2)$-tuple of functions

$$\sigma = (\sigma_u, \sigma_a, (\sigma_i)_{i \in A})$$

such that $\sigma_u : U \to U'$, $\sigma_a : A \to A'$ and $\sigma_i : H_i \to H_{\sigma_a(i)}$ $(i \in A)$ satisfy

$$r_{\sigma_a(i)}'(\sigma_u(x), \sigma_u(y)) = \sigma_i(r_i(x, y)) \tag{3}$$

for all $x, y \in U$ and $i \in A$.

3. If all functions in σ are 1-1 and onto, then σ is called an IS-isomorphism.

An IS-morphism stipulates the structural similarity between two information systems of the same kind. Let T and T' be two such systems. Then we write $T \Rightarrow T'$ if there exists an IS-morphism from T to T', and $T \simeq T'$ if there exists an IS-isomorphism from T to T'. Note that $\simeq$ is an equivalence relation, whereas $\Rightarrow$ may be asymmetrical. Sometimes, we need to specify the properties of an IS-morphism. In such cases, we write $T \Rightarrow_{p_1, p_2} T'$ to indicate that there exists an IS-morphism σ from T to T' such that σ_u and σ_a satisfy properties p_1 and p_2 respectively. In particular, we need the notation $T \Rightarrow_{id, onto} T'$, which means that σ_u is the identity function of U (i.e., $\sigma_u(x) = id(x) = x$ for all $x \in U$) and σ_a is an onto function.

The relational information in an RIS may come from different sources. One of the most important sources may be the functional information. For various reasons, we may want to represent relational information between objects based on a comparison of some of the objects' attribute values. If all the relational information of an RIS is derived from an FIS, then it is said that the former is an embedment of the latter. Formally, this leads to the following definition.

Definition 4. *Let* $T_f = (U, A_1, \{V_i \mid i \in A_1\}, \{f_i \mid i \in A_1\})$ *be a FIS, and* $T_r = (U, A_2, \{H_i \mid i \in A_2\}, \{r_i \mid i \in A_2\})$ *be an RIS; then, an embedding from* T_f *to* T_r *is a* $|A_2|$*-tuple of pairs*

$$\varepsilon = ((B_i, R_i))_{i \in A_2},$$

where each $B_i \subseteq A_1$ *is nonempty and each* $R_i : \prod_{j \in B_i} V_j \times \prod_{j \in B_i} V_j \to H_i$ *satisfies*

$$r_i(x, y) = R_i(Inf_{B_i}(x), Inf_{B_i}(y)) \tag{4}$$

for all $x, y \in U$. T_r *is said to be an embedment of* T_f *if there exists an embedding from* T_f *to* T_r.

Note that the embedding relationship is only defined for two information systems with the same universe. Intuitively, T_r is an embedment of T_f if all relational informational in T_r is based on a comparison of some attribute values in T_f. Thus, for each attribute i in T_r, we can find a subset of attributes B_i in T_f such that the extent to which x is i-related to y is completely determined by comparing $Inf_{B_i}(x)$ and $Inf_{B_i}(y)$ in some particular way. We write $T_f \rhd T_r$ if T_r is an embedment of T_f.

Example 4. [**Pairwise comparison tables**] Let T_f denote the FIS in Example 2, and $T_r = (U, A, \{H_i \mid i \in A\}, \{r_i \mid i \in A\})$, where $H_i = \{-3, -2, -1, 0, 1, 2, 3\}$, and r_i is defined as $r_i(x, y) = f_i(x) - f_i(y)$ for all $x, y \in U$ and $i \in A$. Then, the embedding from T_f to T_r becomes

$$(((\{o\}, R_o), (\{p\}, R_p), (\{t\}, R_t), (\{d\}, R_d)),$$

where $R_i : V_i \times V_i \to H_i$ is defined as

$$R_i(v_1, v_2) = v_1 - v_2$$

for all $i \in A$. The resultant T_r is an instance of the pairwise comparison table (PCT) used in MCDA [10,11,12,13,14,15,16,32]. A similar embedment is used to define D-reducts (distance reducts) in [28], where a relationship between objects x and y exists iff the distance between $f_i(x)$ and $f_i(y)$ is greater than a given threshold.

Example 5. [**Dimension reduction and information compression**] If $T_f = (U, A_1, \{V_i \mid i \in A_1\}, \{f_i \mid i \in A_1\})$ is a high dimensional FIS, i.e., $|A_1|$ is very large, then we may want to reduce the dimension of the information system. Furthermore, for security reasons, we may want to compress information in the FIS. An embedment based on rough set theory that can achieve both dimension reduction and information compression is as follows. First, the set of attributes, A_1, is partitioned into k mutually disjoint subsets, $A_1 = B_1 \cup B_2 \cup \cdots \cup B_k$, where k is substantially smaller than $|A_1|$. Second, for $1 \le i \le k$, define $R_i : \prod_{j \in B_i} V_j \times \prod_{j \in B_i} V_j \rightarrow \{0, 1\}$ as $R_i(\mathbf{v}_i, \mathbf{v}_i') = 1$ iff $\mathbf{v}_i = \mathbf{v}_i'$, where $\mathbf{v}_i, \mathbf{v}_i' \in \prod_{j \in B_i} V_j$. Thus, $((B_i, R_i)_{1 \le i \le k})$ is an embedding from T_f to $T_r = (U, A_2, \{H_i \mid i \in A_2\}, \{r_i \mid i \in A_2\})$, where $A_2 = \{1, 2, \cdots, k\}$, $H_i = \{0, 1\}$, and $r_i(x, y) = 1$ iff $Inf_{B_i}(x) = Inf_{B_i}(y)$. Note that r_i is actually the characteristic function of the B_i-indiscernibility relation in rough set theory. Consequently, the dimension of the information system is reduced to k so that only the indiscernibility information with respect to some subsets of attributes is kept in the RIS.

Example 6. [**Discernibility matrices**] In [29], discernibility matrices are defined to analyze the complexity of many computational problems in rough set theory. This is especially useful in the computation of reduct in rough set theory. According to [29], given an FIS $T_f = (U, A_1, \{V_i \mid i \in A_1\}, \{f_i \mid i \in A_1\})$, its discernibility matrix is a $|U| \times |U|$ matrix D such that

$$D_{xy} = \{i \in A_1 \mid f_i(x) \ne f_i(y)\}$$

for any $x, y \in U$. In other words, the (x, y) entry of the discernibility matrix is the set of attributes that can discern between x and y. More generally, we can define a discernibility matrix $D(B)$ with respect to any subset of attributes, $B \subseteq A_1$, such that

$$D(B)_{xy} = \{i \in B \mid f_i(x) \ne f_i(y)\}$$

for any $x, y \in U$. Let $B_1, \cdots, B_k$ be a sequence of subsets of attributes. Then, the sequence of discernibility matrices, $D(B_1) \cdots, D(B_k)$, can be combined as an RIS. The RIS becomes an embedment of T_f by the embedding $((B_i, R_i)_{1 \le i \le k})$, such that $R_i : \prod_{j \in B_i} V_j \times \prod_{j \in B_i} V_j \rightarrow 2^{A_1}$ is defined by

$$R_i(\mathbf{v}_i, \mathbf{v}_i') = \{j \in B_i \mid \mathbf{v}_i(j) \ne \mathbf{v}_i'(j)\},$$

where $\mathbf{v}(j)$ denotes the j-component of the vector $\mathbf{v}$.

Next, we show that the embedding relationship is preserved by IS-morphism transformation in some conditions. In the following theorem and corollary, we assume that T_f, T_f', T_r, and T_r' have the same universe U. Thus,

$$T_f = (U, A_1, \{V_i \mid i \in A_1\}, \{f_i \mid i \in A_1\}),$$

$$T_f' = (U, A_1', \{V_i' \mid i \in A_1'\}, \{f_i' \mid i \in A_1'\}),$$

$$T_r = (U, A_2, \{H_i \mid i \in A_2\}, \{r_i \mid i \in A_2\}),$$

$$T_r' = (U, A_2', \{H_i' \mid i \in A_2'\}, \{r_i' \mid i \in A_2'\}).$$

Theorem 1

1. $T_f \rhd T_r$ and $T_r \Rightarrow_{id,onto} T'_r$ implies $T_f \rhd T'_r$.
2. $T_f \rhd T_r$ and $T'_f \Rightarrow_{id,onto} T_f$ implies $T'_f \rhd T_r$.

Proof. Let $\varepsilon = ((B_i, R_i))_{i \in A_2}$ be an embedding from T_f to T_r.

1. If $\sigma = (id, \sigma_a, (\sigma_i)_{i \in A_2})$ is an IS-morphism from T_r to T'_r such that σ_a is an onto function, then for each $j \in A'_2$, we can choose an arbitrary $i_j \in A_2$ such that $\sigma_a(i_j) = j$. Let B'_j and R'_j denote B_{i_j} and $\sigma_{i_j} \circ R_{i_j}$ respectively, then $\varepsilon' = ((B'_j, R'_j))_{j \in A'_2}$ is an embedding from T_f to T'_r. Indeed, by the definition of σ and ε, we have, for all $j \in A'_2$,

$$
\begin{aligned}
r'_j(x, y) &= r'_{\sigma_a(i_j)}(\sigma_u(x), \sigma_u(y)) && (\sigma_u = id, \sigma_a(i_j) = j) \\
&= \sigma_{i_j}(r_{i_j}(x, y)) && (Eq.\ 3) \\
&= \sigma_{i_j}(R_{i_j}(Inf_{B_{i_j}}(x), Inf_{B_{i_j}}(y))) && (Eq.\ 4) \\
&= R'_j(Inf_{B'_j}(x), Inf_{B'_j}(y))) && (R'_j = \sigma_{i_j} \circ R_{i_j})
\end{aligned}
$$

2. If $\sigma = (id, \sigma_a, (\sigma_i)_{i \in A'_1})$ is an IS-morphism from T'_f to T_f such that σ_a is an onto function, then for each $j \in A_1$, we can choose an arbitrary $k_j \in A'_1$ such that $\sigma_a(k_j) = j$. For each $i \in A_2$, let $B'_i = \{k_j \mid j \in B_i\}$ and define $R'_i : \prod_{k_j \in B'_i} V'_{k_j} \times \prod_{k_j \in B'_i} V'_{k_j} \to H_i$ by

$$
R'_i((v'_{k_j})_{j \in B_i}, (w'_{k_j})_{j \in B_i}) = R_i((\sigma_{k_j}(v'_{k_j}))_{j \in B_i}, (\sigma_{k_j}(w'_{k_j}))_{j \in B_i}). \quad (5)
$$

Then, $\varepsilon' = ((B'_i, R'_i))_{i \in A_2}$ is an embedding from T'_f to T_r. This can be verified for all $i \in A_2$ as follows:

$$
\begin{aligned}
r_i(x, y) &= R_i(Inf_{B_i}(x), Inf_{B_i}(y)) && (Eq.\ 4) \\
&= R_i((f_j(x))_{j \in B_i}, (f_j(y))_{j \in B_i}) && (Eq.\ 1) \\
&= R_i((\sigma_{k_j}(f'_{k_j}(x)))_{j \in B_i}, (\sigma_{k_j}(f'_{k_j}(y)))_{j \in B_i}) && (Eq.\ 2) \\
&= R'_i((f'_{k_j}(x))_{k_j \in B'_i}, (f'_{k_j}(y))_{k_j \in B'_i}) && (Eq.\ 5;\ B'_i,\ def.) \\
&= R'_i(Inf_{B'_i}(x), Inf_{B'_i}(y)) && (Eq.\ 1)
\end{aligned}
$$

$\square$

The theorem can be represented by the following commutative diagram notation commonly used in category theory [3].

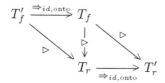

When the IS-morphism between two systems is an IS-isomorphism, we can derive the following corollary.

Corollary 1. If $T_f \rhd T_r$, $T_f \simeq T'_f$ and $T_r \simeq T'_r$, then $T'_f \rhd T'_r$

The commutative diagram of the corollary is:

$$
\begin{array}{ccc}
T_f & \xrightarrow{\;\triangleright\;} & T_r \\
\simeq \downarrow & & \downarrow \simeq \\
T_f' & \xrightarrow{\;\triangleright\;} & T_r'
\end{array}
$$

As shown in Example 5, an RIS may contain a summary of information about an FIS. Therefore, the RIS can serve as a tool for information summarization. If $T_f \triangleright T_r$, then the information in T_r is less specific than that in T_f, i.e., the information is reduced. If as much information as possible is kept during the reduction, the embedding is called a trivial embedding. Formally, a *trivial embedding* from $T_f = (U, A_1, \{V_i \mid i \in A_1\}, \{f_i \mid i \in A_1\})$ to $T_r = (U, A_2, \{H_i \mid i \in A_2\}, \{r_i \mid i \in A_2\})$ is an embedding $\varepsilon = ((B_i, R_i))_{i \in A_2}$ such that each R_i is a 1-1 function. T_r is called a trivial embedment of T_f if there exists a trivial embedding from T_f to T_r. A trivial embedment plays a similar role to the initial algebra [6] in a class of RIS with the same attributes. This is shown in the next theorem, which easily follows from the definitions.

Theorem 2. *Let* $T_f = (U, A_1, \{V_i \mid i \in A_1\}, \{f_i \mid i \in A_1\})$, $T_r = (U, A_2, \{H_i \mid i \in A_2\}, \{r_i \mid i \in A_2\})$, *and* $T_r' = (U, A_2, \{H_i' \mid i \in A_2\}, \{r_i' \mid i \in A_2\})$ *be information systems. If* $\varepsilon = ((B_i, R_i))_{i \in A_2}$ *is a trivial embedding from* T_f *to* T_r *and* $\varepsilon' = ((B_i, R_i'))_{i \in A_2}$ *is an embedding from* T_f *to* T_r', *then* $T_r \Rightarrow T_r'$.

Since embedding is an information reduction operation, many FIS may be embedded into the same RIS. Consequently, in general, it is not easy to recover an FIS that has been embedded into a given RIS. However, by applying the techniques of constraint solving, we can usually find possible candidates that have been embedded into a given RIS. More specifically, if the universe, the set of attributes, and the domain of values for each attribute of an FIS are known, then, given an embedding and the resultant embedded RIS, the problem of finding the FIS that is embedded into the given RIS is a constraint satisfaction problem (CSP). The following example illustrates this point.

Example 7. Let $U = \{1, 2, 3, 4, 5, 6\}$, $A = \{a, s\}$, $V_a = \{1, 2, \cdots, 120\}$, and $V_s = \{M, F\}$ be, respectively, the universe, the set of attributes, and the domains of values for attributes a and s, where a denotes age and s denotes sex. Assume the RIS given in Table 2 results from an embedding $(((\{a\}, R_a), (\{s\}, R_s))$, where $R_i(v_1, v_2) = 1$ iff $v_1 = v_2$.

Then, to find an FIS $T_f = (U, A, \{V_i \mid i \in A\}, \{f_i \mid i \in A\})$ such that the RIS is the embedment of T_f by the above-mentioned embedding, we have to solve the following finite domain CSP, where v_{ij} is a variable denoting the value $f_j(i)$ to be found:

$$v_{ia} \in \{1, 2, \cdots, 120\}, v_{is} \in \{M, F\}, 1 \le i \le 6,$$
$$v_{1a} = v_{2a}, v_{3a} = v_{4a}, v_{5a} = v_{6a},$$
$$v_{ia} \ne v_{ja}, (i, j) \ne (1, 2), (3, 4), \text{ or } (5, 6),$$
$$v_{1s} = v_{6s} \ne v_{2s} = v_{3s} = v_{4s} = v_{5s}.$$

Table 2. AN RIS obtained from a given embedding

a	1	2	3	4	5	6		s	1	2	3	4	5	6
1	1	1	0	0	0	0		1	1	0	0	0	0	1
2	1	1	0	0	0	0		2	0	1	1	1	1	0
3	0	0	1	1	0	0		3	0	1	1	1	1	0
4	0	0	1	1	0	0		4	0	1	1	1	1	0
5	0	0	0	0	1	1		5	0	1	1	1	1	0
6	0	0	0	0	1	1		6	1	0	0	0	0	1

3   Decision Logic and Arrow Logic

In the previous section, we demonstrated that FIS and RIS are useful formalisms for data representation. However, to represent and reason about knowledge extracted from information systems, we need a logical language. For FIS, Pawlak [26] proposed a decision logic (DL), so-called because it is particularly useful for a special kind of FIS, called *decision tables.* A decision table is an FIS whose set of attributes can be partitioned into condition and decision attributes. Decision rules that relate the condition and the decision attributes can be derived from such tables by data analysis. A rule is then represented as an implication of the formulas of the logic.

Since relations can be seen as properties of pairs of objects, to reason about RIS, we need a language that can be interpreted in the domain of pairs of objects. Arrow logic (AL) language [24,33] is designed to describe all things that may be represented in a picture by arrows. Therefore, it is an appropriate tool for reasoning about RIS.

In this section, we review the basic syntax and semantics of both DL and AL in order to lay the foundation for the development of arrow decision logic.

3.1   Decision Logic (DL)

The basic alphabet of DL consists of a finite set of attribute symbols, A; and for $i \in \mathsf{A}$, a finite set of value symbols, V_i. An atomic formula of DL is a descriptor, (i, v), where $i \in \mathsf{A}$ and $v \in \mathsf{V}_i$. The set of DL well-formed formulas (wff) is the smallest set that contains the atomic formulas and is closed under the Boolean connectives $\neg$ and $\vee$. If φ and ψ are wffs of DL, then $\varphi \longrightarrow \psi$ is a rule in DL, where φ is called the antecedent of the rule and ψ the consequent. As usual, we abbreviate $\neg\varphi \vee \psi$ as $\varphi \supset \psi$, $\neg(\neg\varphi \vee \neg\psi)$ as $\varphi \wedge \psi$, and $(\varphi \supset \psi) \wedge (\psi \supset \varphi)$ as $\varphi \equiv \psi$.

An interpretation of a given DL is an FIS $T_f = (U, A, \{V_i \mid i \in A\}, \{f_i \mid i \in A\})$ such that $A = \mathsf{A}$ and for every $i \in \mathsf{A}$, $V_i = \mathsf{V}_i$. Thus, by somewhat abusing the notation, we usually denote an atomic formula as (i, v), where $i \in A$ and $v \in V_i$ if the FIS is clear from the context. Intuitively, each element in the universe of an FIS corresponds to a data record; and an atomic formula, which is in fact an attribute-value pair, describes the value of some attribute in the data record. Thus, the atomic formulas (and therefore the wffs) can be verified or falsified in each data record. This yields a satisfaction relation between the universe and the set of wffs.

Definition 5. *Given a DL and an interpretation FIS $T_f = (U, A, \{V_i \mid i \in A\}, \{f_i \mid i \in A\})$, the satisfaction relation $\models_{T_f}$ between U and the wffs of DL is defined inductively as follows (the subscript T_f is omitted for brevity).*

1. $x \models (i, v)$ *iff* $f_i(x) = v$,
2. $x \models \neg\varphi$ *iff* $x \not\models \varphi$,
3. $x \models \varphi \vee \psi$ *iff* $x \models \varphi$ *or* $x \models \psi$.

If φ is a DL wff, the set $m_{T_f}(\varphi)$ defined by

$$m_{T_f}(\varphi) = \{x \in U \mid x \models \varphi\} \tag{6}$$

is called the meaning set of the formula φ in T_f. If T_f is understood, we simply write $m(\varphi)$.

A formula φ is said to be valid in T_f, written as $\models_{T_f} \varphi$, if and only if $m_{T_f}(\varphi) = U$. That is, φ is satisfied by all individuals in the universe. If φ is valid in T_f, then T_f is a model of φ. We write $\models \varphi$ if φ is valid in all interpretations.

3.2   Arrow Logic (AL)

AL is the basic modal logic of arrows [24,33]. An arrow can represent a state transition in a program's execution, a morphism in category theory, an edge in a directed graph, etc. In AL, an arrow is an abstract entity; however, we can usually interpret it as a concrete relationship between two objects, which results in a pair-frame model [24,33]. We now present the syntax and semantics of AL.

The basic alphabet of AL consists of a countable set of propositional symbols, the Boolean connectives $\neg$ and $\vee$, the modal constant δ, the unary modal operator $\otimes$, and the binary modal operator $\circ$. The set of AL wffs is the smallest set containing the propositional symbols and δ, closed under the Boolean connectives $\neg$ and $\vee$, and satisfying

- if φ is a wff, then $\otimes\varphi$ is a wff too;
- if φ and ψ are wffs, then $\varphi \circ \psi$ is also a wff.

In addition to the standard Boolean connectives, we also abbreviate $\neg \otimes \neg\varphi$ and $\neg(\neg\varphi \circ \neg\psi)$ as $\underline{\otimes}\varphi$ and $\varphi\underline{\circ}\psi$ respectively.

Semantically, these wffs are interpreted in arrow models.

Definition 6

1. An arrow frame is a quadruple $\mathfrak{F} = (W, C, R, I)$ such that $C \subseteq W \times W \times W$, $R \subseteq W \times W$ and $I \subseteq W$.
2. An arrow model is a pair $\mathfrak{M} = (\mathfrak{F}, \pi)$, where $\mathfrak{F} = (W, C, R, I)$ is an arrow frame and π is a valuation that maps propositional symbols to subsets of W. An element in W is called an arrow in the model $\mathfrak{M}$.

3. *The satisfaction of a wff φ on an arrow w of $\mathfrak{M}$, denoted by $w \models_\mathfrak{M} \varphi$ (as usual, the subscript $\mathfrak{M}$ can be omitted), is inductively defined as follows:*
   (a) $w \models p$ iff $w \in \pi(p)$ for any propositional symbol p,
   (b) $w \models \delta$ iff $w \in I$,
   (c) $w \models \neg\varphi$ iff $w \not\models \varphi$,
   (d) $w \models \varphi \vee \psi$ iff $w \models \varphi$ or $x \models \psi$,
   (e) $w \models \varphi \circ \psi$ iff there exist s, t such that $(w, s, t) \in C$, $s \models \varphi$, and $t \models \psi$,
   (f) $w \models \otimes\varphi$ iff there is a t with $(w, t) \in R$ and $t \models \varphi$.

Intuitively, in the arrow frame (W, C, R, I), W can be regarded as the set of edges of a directed graph; I denotes the set of identity arrows[2]; $(w, s) \in R$ if s is a reversed arrow of w; and $(w, s, t) \in C$ if w is an arrow composed of s and t. This intuition is reflected in the following definition of pair frames.

Definition 7. *An arrow frame $\mathfrak{F} = (W, C, R, I)$ is a pair frame if there exists a set U such that $W \subseteq U \times U$ and*

1. *for $x, y \in U$, if $(x, y) \in I$ then $x = y$,*
2. *for $x_1, x_2, y_1, y_2 \in U$, if $((x_1, y_1), (x_2, y_2)) \in R$, then $x_1 = y_2$ and $y_1 = x_2$,*
3. *for $x_1, x_2, x_3, y_1, y_2, y_3 \in U$, if $((x_1, y_1), (x_2, y_2), (x_3, y_3)) \in C$, then $x_1 = x_2$, $y_2 = x_3$, and $y_1 = y_3$.*

An arrow model $\mathfrak{M} = (\mathfrak{F}, \pi)$ is called a pair model if $\mathfrak{F}$ is a pair frame. A pair model is called a (full) square model if the set of arrows $W = U \times U$.

4   Arrow Decision Logic

To represent rules induced from an RIS, we propose arrow decision logic (ADL), derived by combining the main features of AL and DL. In this section, we introduce the syntax and semantics of ADL. The atomic formulas of ADL are the same as those in DL; while the formulas of ADL are interpreted with respect to each pair of objects, as in the pair frames of AL [24,33].

4.1   Syntax and Semantics of ADL

An atomic formula of ADL is a descriptor of the form (i, h), where $i \in A$, $h \in H_i$, A is a finite set of attribute symbols, and for each $i \in A$, H_i is a finite set of relational indicator symbols. In addition, the wffs of ADL are defined by the formation rules for AL, while the definition of derived connectives is the same as that in AL.

An interpretation of a given ADL is an RIS, $T_r = (U, A, \{H_i \mid i \in A\}, \{r_i \mid i \in A\})$, which can be seen as a square model of AL. Thus, the wffs of ADL are evaluated in terms of a pair of objects. More precisely, the satisfaction of a wff with respect to a pair of objects (x, y) in T_r is defined as follows (again, we omit the subscript T_r):

---

[2] An identity arrow is an arrow that has the same starting point and endpoint.

1. $(x, y) \models (i, h)$ iff $r_i(x, y) = h$,
2. $(x, y) \models \delta$ iff $x = y$,
3. $(x, y) \models \neg\varphi$ iff $(x, y) \not\models \varphi$,
4. $(x, y) \models \varphi \vee \psi$ iff $(x, y) \models \varphi$ or $(x, y) \models \psi$,
5. $(x, y) \models \otimes\varphi$ iff $(y, x) \models \varphi$,
6. $(x, y) \models \varphi \circ \psi$ iff there exists z such that $(x, z) \models \varphi$ and $(z, y) \models \psi$.

Let Σ be a set of ADL wffs; then, we write $(x, y) \models \Sigma$ if $(x, y) \models \varphi$ for all $\varphi \in \Sigma$. Also, for a set of wffs, Σ, and a wff, φ, we say that φ is an ADL consequence of Σ, written as $\Sigma \models \varphi$, if for every interpretation T_r and x, y in the universe of T_r, $(x, y) \models \Sigma$ implies $(x, y) \models \varphi$.

4.2   Axiomatization

The ADL consequence relation can be axiomatized by integrating the axiomatization of AL and specific axioms of DL. As shown in [24,33], the AL consequence relations with respect to full square models can not be finitely axiomatized by an orthodox derivation system. To develop a complete axiomatization of AL, an unorthodox inference rule based on a difference operator D is added to the AL derivation system. The use of such unorthodox rules was first proposed by Gabbay [9]. The operator is defined in shorthand as follows:

$$\mathsf{D}\varphi = \top \circ \varphi \circ \neg\delta \vee \neg\delta \circ \varphi \circ \top,$$

where $\top$ denotes any tautology. According to the semantics, $\mathsf{D}\varphi$ is true in a pair (x, y) iff there exists a pair distinct from (x, y) such that φ is true in that pair.

   The complete axiomatization of ADL consequence relations is presented in Figure 2, where $\varphi, \psi, \varphi', \psi'$, and χ are meta-variables denoting any wffs of ADL. The axiomatization consists of three parts: the propositional logic axioms; the DL and AL axioms; and the inference rules, including the classical Modus Ponens rule, the universal generalization rule for modal operators, and the unorthodox rule based on D. The operator D is also utilized in DL3 to spread the axioms DL1 and DL2 to all pairs of objects. DL3 thus plays a key role in the proof of the completeness of the axiomatization. DL1 and DL2 are exactly the specific axioms of DL in [26]. An additional axiom

$$\neg(i, h) \equiv \bigvee_{h' \in H_i, h' \neq h} (i, h')$$

is presented in [26], but it is redundant. The AL axioms and inference rules can be found in [24], where AL4 is split into two parts and an extra axiom

$$\varphi \circ (\psi \circ (\delta \wedge \chi)) \supset (\varphi \circ \psi) \circ (\delta \wedge \chi)$$

is given. The extra axiom is called the *weak associativity axiom*, since it is weaker than the associativity axiom AL5. We do not need such an axiom, as it is an instance of AL5. The only novel axiom in our system is DL3. Though it is classified as a DL axiom, it is actually a connecting axiom between DL and AL.

An ADL derivation is a finite sequence $\varphi_1, \cdots, \varphi_n$ such that every φ_i is either an instance of an axiom or obtainable from $\varphi_1, \cdots, \varphi_{i-1}$ by an inference rule. The last formula φ_n in a derivation is called an ADL theorem. A wff φ is derivable in ADL from a set of wffs Σ if there are $\varphi_1, \ldots, \varphi_n$ in Σ such that $(\varphi_1 \wedge \ldots \wedge \varphi_n) \supset \varphi$ is an ADL-theorem. We use $\vdash \varphi$ to denote that φ is an ADL theorem and $\Sigma \vdash \varphi$ to denote that φ is derivable in ADL from Σ. Also, we write $\Sigma \vdash_{AL} \varphi$ if φ is derivable in ADL from Σ without using the DL axioms.

- Axioms:
  1. P: all tautologies of propositional calculus
  2. DL axioms:
     (a) DL1: $(i, h_1) \supset \neg(i, h_2)$, for any $i \in A$, $h_1, h_2 \in H_i$ and $h_1 \neq h_2$
     (b) DL2: $\vee_{h \in H_i} (i, h)$, for any $i \in A$
     (c) DL3: $\neg D \neg \varphi_0$, where φ_0 is an instance of DL1 or DL2
  3. AL axioms:
     (a) AL0 (Distribution, DB):
         i. $(\varphi \supset \varphi') \underline{\circ} \psi \supset (\varphi \underline{\circ} \psi \supset \varphi' \underline{\circ} \psi)$
         ii. $\varphi \underline{\circ} (\psi \supset \psi') \supset (\varphi \underline{\circ} \psi \supset \varphi \underline{\circ} \psi')$
         iii. $\otimes(\varphi \supset \psi) \supset (\otimes \varphi \supset \otimes \psi)$
     (b) AL1: $\varphi \equiv \otimes \otimes \varphi$
     (c) AL2: $\otimes(\varphi \circ \psi) \supset \otimes \psi \circ \otimes \varphi$
     (d) AL3: $\otimes \varphi \circ \neg(\varphi \circ \psi) \supset \neg \psi$
     (e) AL4: $\varphi \equiv \varphi \circ \delta$
     (f) AL5: $\varphi \circ (\psi \circ \chi) \supset (\varphi \circ \psi) \circ \chi$
- Rules of Inference:
  1. R1 (Modus Ponens, MP): $\dfrac{\varphi \quad \varphi \supset \psi}{\psi}$

  2. R2 (Universal Generalization, UG): $\dfrac{\varphi}{\otimes \varphi} \qquad \dfrac{\varphi}{\varphi \underline{\circ} \psi} \qquad \dfrac{\varphi}{\psi \underline{\circ} \varphi}$

  3. R3 (Irreflexivity rule, IR_D): $\dfrac{(p \wedge \neg Dp) \supset \varphi}{\varphi}$ provided that p is an atomic formula not occurring in φ

Fig. 1. The axiomatic system for ADL

The next theorem shows that the axiomatic system is sound and complete with respect to the ADL consequence relations.

Theorem 3. *For any set of ADL wffs $\Sigma \cup \{\varphi\}$, we have $\Sigma \models \varphi$ iff $\Sigma \vdash \varphi$.*

Proof

1. Soundness: To show that $\Sigma \vdash \varphi$ implies $\Sigma \models \varphi$, it suffices to show that all ADL axioms are valid in any ADL interpretation and that the inference rules preserve validity. The fact that AL axioms are valid and the inference rules preserve validity follows from the soundness of AL, since any ADL

interpretation is an instance of a square model. Furthermore, it is clear that DL axioms are valid in any ADL interpretation.

2. Completeness: To prove completeness, we first note that the set of instances of DL axioms is finite. Let us denote χ by the conjunction of all instances of DL axioms and χ_0 by the conjunction of all instances of axioms DL1 and DL2. If $\Sigma \nvdash \varphi$, then $\Sigma \cup \{\chi\} \nvdash_{AL} \varphi$. Thus, by the completeness of AL, we have a pair model $\mathfrak{M} = (U \times U, C, R, I, \pi)$ and $x, y \in U$ such that $(x, y) \models_{\mathfrak{M}} \Sigma$, $(x, y) \models_{\mathfrak{M}} \chi$, and $(x, y) \models_{\mathfrak{M}} \neg\varphi$. Next, we show that $\mathfrak{M}$ can be transformed into an ADL interpretation. From $(x, y) \models_{\mathfrak{M}} \chi$, we can derive $(z, w) \models_{\mathfrak{M}} \chi_0$ for all $z, w \in U$ by DL3. Thus, for every $z, w \in U$ and $i \in A$, there exists exactly one $h_{i,z,w} \in H_i$ such that $(z, w) \models_{\mathfrak{M}} (i, h_{i,z,w})$. Consequently, $T_r = (U, A, \{H_i \mid i \in A\}, \{r_i \mid i \in A\})$, where $r_i(z, w) = h_{i,z,w}$ for $z, w \in U$ and $i \in A$, is an ADL interpretation such that $(x, y) \models_{T_r} \Sigma$ and $(x, y) \models_{T_r} \neg\varphi$. Thus, $\Sigma \nvdash \varphi$ implies $\Sigma \nvDash \varphi$. □

5   Discussion and Applications

5.1   Discussion

Initially, it seems that an RIS is simply an instance of FIS whose universe consists of pairs of objects. However, there is a subtle difference between RIS and FIS. In FIS, the universe is an unstructured set, whereas in RIS, an implicit structure exists in the universe. The structure is made explicit by modal operators in ADL. For example, if (x, y) is a pair in an RIS, then (x, y) and (y, x) are considered to be two independent objects from the perspective of FIS; however, from the viewpoint of RIS, they are the converse of each other.

The difference between FIS and RIS is also reflected by the definition of IS-morphisms. When σ is an IS-morphism between two RIS, then, for a pair of objects (x, y), if (x, y) is mapped to (z, w), then (y, x) must be mapped to (w, z) at the same time. However, if these two RIS's are considered simply as FIS's with pairs of objects in their universes and σ is an IS-morphism between these two FIS's, then it is possible that $\sigma_u((x, y)) = (z_1, w_1)$ and $\sigma_u((y, x)) = (z_2, w_2)$ without $z_2 = w_1$ and/or $w_2 = z_1$. In other words, the images of (x, y) and (y, x) may be totally independent if we simply view an RIS as a kind of FIS. Therefore, even though FIS and RIS are very similar in appearance, they are mathematically and conceptually different.

In fact, FIS and RIS usually represent different aspects of the information about the objects. One of the main purpose of this paper is to consider the relational structures of FIS. Sometimes, both functional and relational information about objects must be represented. Thus, to achieve full generality, we can combine these two kinds of information systems. Let us define a hybrid information system (HIS) as

$$(U, A \cup B, \{V_i \mid i \in A\}, \{H_i \mid i \in B\}, \{f_i \mid i \in A\}, \{r_i \mid i \in B\})$$

such that $(U, A, \{V_i \mid i \in A\}, \{f_i \mid i \in A\})$ is an FIS and $(U, B, \{H_i \mid i \in B\}, \{r_i \mid i \in B\})$ is an RIS. Then, a HIS can represent functional and relational information about the same set of objects simultaneously. In general, A and B are disjoint; however, this is not theoretically mandatory.

The algebraic properties of IS-morphism between FIS was previously studied in [17] under the name of O-A-D homomorphism[3]. The notion of IS-morphism between RIS is a straightforward generalization of that between FIS. In fact, if FIS and RIS are considered as many-sorted algebras [4], both IS-morphism and O-A-D homomorphism can be seen as homomorphism in universal algebra [5,6]. Indeed, we can consider information systems as a 3-sorted algebra whose sorts are the universe, the set of attributes, and the set of all attribute values (or relational indicators). Though homomorphism has been studied extensively in previous works, we define a novel notion of embedment between FIS and RIS to capture the relationship or transformation between two kinds of information related to the objects. This implies a new result, which shows that the embedding relationship can be preserved by IS-morphism under some conditions.

The investigation of RIS also facilitates a further generalization of rough set theory. In classical rough set theory, lower and upper approximations are defined in terms of indiscernibility relations based on functional information associated with the objects. However, it has been noted that many applications, such as social network analysis [27], need to represent both functional and relational information. Based on this observation, a concept of relational granulation was recently proposed in [21]. The basic idea is that even though two objects are indiscernible with respect to their attribute values, they may still be discernible because of their relationships with other objects. Consequently, the definition of lower and upper approximations must consider the finer indiscernibility relations. In a future work, we will investigate different generalized rough sets based on the relational information associated with objects.

5.2   An Application of ADL to MCDM

Relational information plays an important role in MCDA. When rough set theory is applied to MCDA, it is crucial that preference-ordered attribute domains and decision classes be dealt with [14]. The original rough set theory could not handle inconsistencies arising from violation of the dominance principle due to its use of the indiscernibility relation. In previous works on MCDA, the indiscernibility relation was replaced by a dominance relation to solve the multi-criteria sorting problem, and the FIS was replaced by a PCT to solve multi-criteria choice and ranking problems [14]. A PCT is essentially an instance of an RIS, as shown in Example 4, in which the relations are confined to preference relations. This approach is called the *dominance-based rough set approach* (DRSA). For MCDA problems, DRSA can induce a set of decision rules from sample decisions provided by decision-makers. The induced decision rules play the role of a comprehensive preference model and can make recommendations about a new decision-making environment. The process whereby the induced decision rules are used to facilitate decision-making is called *multi-criteria decision making* (MCDM).

---

[3] O, A, and D denotes objects, attributes, and the domain of values respectively.

To apply ADL to MCDM, we define a rule of ADL as $\varphi \longrightarrow \psi$, where φ and ψ are wffs of ADL, called the antecedent and the consequent of the rule respectively. As in DL, let T_r be an interpretation of an ADL. Then, the set $m_{T_r}(\varphi)$ defined by

$$m_{T_r}(\varphi) = \{(x, y) \in U \times U \mid (x, y) \models \varphi\} \qquad (7)$$

is called the *meaning set* of the formula φ in T_r. If T_r is understood, we simply write $m(\varphi)$. A wff φ is valid in T_r if $m(\varphi) = U \times U$. Some quantitative measures that are useful in data mining can be redefined for ADL rules.

Definition 8. *Let Φ be the set of all ADL rules and $T_r = (U, A, \{H_i \mid i \in A\}, \{r_i \mid i \in A\})$ be an interpretation of them. Then,*

1. *the rule $\varphi \longrightarrow \psi$ is valid in T_r iff $m_{T_r}(\varphi) \subseteq m_{T_r}(\psi)$*
2. *the absolute support function $\alpha_{T_r} : \Phi \to \mathbb{N}$ is*

$$\alpha_{T_r}(\varphi \longrightarrow \psi) = |m_{T_r}(\varphi \wedge \psi)|$$

3. *the relative support function $\rho_{T_r} : \Phi \to [0, 1]$ is*

$$\rho_{T_r}(\varphi \longrightarrow \psi) = \frac{|m_{T_r}(\varphi \wedge \psi)|}{|U|^2}$$

4. *the confidence function $\gamma_{T_r} : \Phi \to [0, 1]$ is*

$$\gamma_{T_r}(\varphi \longrightarrow \psi) = \frac{|m_{T_r}(\varphi \wedge \psi)|}{|m_{T_r}(\varphi)|}.$$

Without loss of generality, we assume that the elements of U are natural numbers from 0 to $|U| - 1$. Each wff can then be seen as a $|U| \times |U|$ Boolean matrix, called its *characteristic matrix*. Thus, we can employ matrix algebra to test the validity of a rule and calculate its support and confidence in an analogous way to that proposed in [19,20]. This is based on the intimate connection between AL and relation algebra [24,33].

In the applications, we assume that the set of relational indicators H_i is a finite set of integers for every criterion i. Under such an assumption, we can use the following shorthand:

$$(i, \geq_{h_i}) \equiv \vee_{h \geq h_i}(i, h) \text{ and } (i, \leq_{h_i}) \equiv \vee_{h \leq h_i}(i, h).$$

By using ADL, the three main types of decision rules mentioned in [14] can be represented as follows:

1. $D_{\geq}$-decision rules:

$$\bigwedge_{i \in B}(i, \geq_{h_i}) \longrightarrow (d, \geq_1),$$

2. $D_{\leq}$-decision rules:

$$\bigwedge_{i \in B}(i, \leq_{h_i}) \longrightarrow (d, \leq_{-1}),$$

3. $D_{\geq\leq}$-decision rules:

$$\bigwedge_{i \in B_1} (i, \geq_{h_i}) \wedge \bigwedge_{i \in B_2} (i, \leq_{h_i}) \longrightarrow (d, \geq_1) \vee (d, \leq_{-1}),$$

where B, B_1, and $B_2 \subseteq A$ are sets of criteria and $d \in A$ is the decision attribute. We assume that $\{-1, 1\} \subseteq H_d$ so that $r_d(x, y) = 1$ means that x outranks y, and $r_d(x, y) = -1$ means that y outranks x.

Furthermore, the modal formulas of ADL allow us to represent some properties of preference relations. For example,

1. reflexivity: $\delta \longrightarrow (i, 0)$,
2. anti-symmetry: $\otimes(i, \geq_h) \longrightarrow (i, \leq_{-h})$, and
3. transitivity: $(i, \geq_{h_1}) \circ (i, \geq_{h_2}) \longrightarrow (i, \geq_{h_1+h_2})$.

Reflexivity means that each object is similar to itself in any attribute; anti-symmetry means that if x is preferred to y by degree (at least) h, then y is inferior to x by degree (at least) h; and transitivity denotes the additivity of preference degrees. The measures α, ρ, and γ can be used to assess the degree of reflexivity, anti-symmetry, and transitivity of an induced preference relation.

Example 8. Let us continue to use the PCT in Example 4 and consider the following two ADL rules:

$$s_1 = (o, \geq_2) \longrightarrow (d, \geq_1),$$

$$s_2 = (p, \leq_{-2}) \longrightarrow (d, \leq_0).$$

Then we have

	α	ρ	γ
s_1	7	0.07	0.875
s_2	15	0.15	1

Note that rule s_2 is valid even though it only has a support value of 0.15. Also, we observe that the anti-symmetry rule $\otimes(i, \geq_h) \longrightarrow (i, \leq_{-h})$ is valid in this PCT, which means that the preference relation is anti-symmetrical.

The main advantage of the ADL representation is its deduction capability. While many data mining algorithms for rule induction have been developed for MCDA, relatively little attention has been paid to the use of induced rules. We can consider two cases of using the rules to assist decision-making in real environments. In a simple situation, it suffices to match the antecedents of the rules with the real conditions. On the other hand, if the decision-maker encounters a complex situation, the deduction capability of ADL rules may be very helpful. For example, in a dynamic environment, the decision-maker has to derive a decision plan consisting of several decision steps, each of which may be guided by a different decision model. If each decision model is represented by a set of ADL wffs or rules, then the final decision can be deduced from the union of these sets and the set of ADL wffs representing the real conditions by our axiomatic

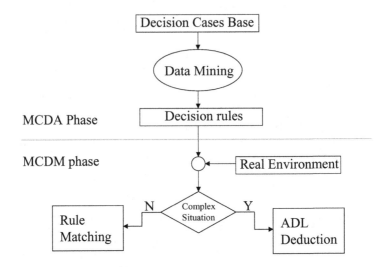

Fig. 2. The MCDA and MCDM phases

deduction system. This is illustrated in Figure 2, where MCDA and MCDM are separated into two phases. In the MCDA phase, ordinary data mining algorithms are employed to find decision rules. This is the learning phase in which previous decision experiences are summarized into rules. In the MCDM phase, the rules are applied to the real environments. As indicated above, ADL representation plays an important role in this phase when the environment is highly complex.

5.3   An Application of ADL to the Representation of Attribute Dependency

One of the most important concepts of relational databases is that of functional dependency. In a relational schema design, functional dependencies are determined by the semantics of the relation, since, in general, they cannot be determined by inspection of an instance of the relation. That is, a functional dependency is a constraint, not a property derived from a relation. However, from a viewpoint of data mining, we can indeed find functional dependencies between attributes from a data table. Such dependencies are called *attribute dependencies*, the discovery of which is essential to the computation of reduct in rough set theory.

Let $T_f = (U, A, \{V_i \mid i \in A\}, \{f_i \mid i \in A\})$ be an FIS and $B \cup \{i\} \subseteq A$ be a subset of attributes. Then, it is said that an attribute dependency between B and i, denoted by $B \longrightarrow i$, exists in T_f if for any two objects $x, y \in U$, $Inf_B(x) = Inf_B(y)$ implies $f_i(x) = f_i(y)$. Though the DL wffs can relate the values of condition attributes to those of decision attributes, the notion of attribute dependency can not be represented in DL directly. Instead, we can only simulate an attribute dependency between B and i by using a set of DL rules.

More specifically, $B \longrightarrow i$ exists in T_f if for all $j \in B$ and $v_j \in V_j$, there exists $v \in V_i$ such that $\wedge_{j \in B}(j, v_j) \longrightarrow (i, v)$ is valid in T_f.

On the other hand, we can express an attribute dependency in an FIS by an ADL rule in an embedment of T_f. This embedding is a special instance of the embedding in Example 5. That is, we embed $T_f = (U, A, \{V_i \mid i \in A\}, \{f_i \mid i \in A\})$ into the RIS $T_r = (U, A, \{H_i \mid i \in A\}, \{r_i \mid i \in A\})$, where $H_i = \{0, 1\}$ and $r_i(x, y) = 1$ iff $f_i(x) = f_i(y)$. The embedding is $((i, R_i)_{i \in A})$, where, for every $i \in A$, $R_i : V_i \times V_i \to \{0, 1\}$ is defined by

$$R_i(v_i, v_i') = 1 \text{ iff } v_i = v_i'$$

for all $v_i, v_i' \in V_i$. Note that r_i is actually the characteristic function of the $\{i\}$-indiscernibility relation in rough set theory. Then, an attribute dependency $B \longrightarrow i$ in T_f can be represented as an ADL rule

$$\wedge_{j \in B}(j, 1) \longrightarrow (i, 1),$$

i.e., the attribute dependency $B \longrightarrow i$ exists in T_f iff the rule $\wedge_{j \in B}(j, 1) \longrightarrow (i, 1)$ is valid in T_r. If we view each pair of objects in $U \times U$ as a transaction and each attribute as an item in a transaction database, then an ADL rule of the form $\wedge_{j \in B}(j, 1) \longrightarrow (i, 1)$ is in fact an association rule. Thus, by transforming an FIS into its embedded RIS, the discovery of attribute dependencies can be achieved by ordinary association rule mining algorithms [2,18].

6   Conclusions

In this paper, we present ADL by combining DL and AL. ADL is useful for representing rules induced from an RIS. An important kind of RIS is PCT, which is commonly used in MCDA. The main advantage of using ADL is its precision in syntax and semantics. As DL is a precise way to represent decision rules induced from FIS, we apply ADL to reformulate the decision rules induced by PCT in DRSA. It is shown that such reformulation makes it possible to utilize ADL deduction in the highly complex decision-making process. We also show that ADL rules can represent attribute dependencies in FIS. Consequently, ordinary association rule mining algorithms can be used to discover attribute dependencies.

While this paper is primarily concerned with the syntax and semantics of ADL, efficient algorithms for data mining based on logical representation are also important. In a future work, we will develop such algorithms.

In [30,31], it is shown that the relations between objects are induced by relational structures in the domains of attributes. The preference relations in multicriteria decision-making and time windows in time series analysis are two important examples of relational structures. The derivation of the relations between objects from such structures is simply a kind of embedment. Certainly, we could consider structures beyond binary relational structures. Then, it should be possible to generalize RIS to non-binary relational information systems. We will also explore this possibility in our future work.

Acknowledgements

We would like to thank the anonymous referees for their insightful suggestions to improve the paper.

References

1. T.F. Fan adn D.R. Liu and G.H. Tzeng. Arrow decision logic. In D. Slezak, G.Y. Wang, M. Szczuka, I. Duntsch, and Y.Y. Yao, editors, *Proceedings of the 10th International Conference on Rough Sets, Fuzzy Sets, Data Mining, and Granular Computing*, LNAI 3641, pages 651–659. Springer-Verlag, 2005.
2. R. Agrawal and R. Srikant. Fast algorithm for mining association rules. In *Proceedings of the 20th International Conference on Very Large Data Bases*, pages 487–499. Morgan Kaufmann Publishers, 1994.
3. A. Asperti and G. Longo. *Categories, Types, and Structures: An Introduction to Category Theory for the Working Computer Scientist*. MIT Press, 1991.
4. G. Birkhoff and J. Lipson. Heterogeneous algebras. *Journal of Combinatorial Theory*, 8, 1970.
5. S.N. Burris and H.P. Sankappanavar. *A Course in Universal Algebra*. Springer-Verlag, 1981.
6. P. M. Cohn. *Universal Algebra*. D. Reidel Publishing Co., 1981.
7. T.F. Fan, W.C. Hu, and C.J. Liau. Decision logics for knowledge representation in data mining. In *Proceedings of the 25th Annual International Computer Software and Applications Conference*, pages 626–631. IEEE Press, 2001.
8. T.F. Fan, D.R. Liu, and G.H. Tzeng. Rough set-based logics for multicriteria decision analysis. In *Proceedings of the 34th International Conference on Computers and Industrial Engineering*, 2004.
9. D.M. Gabbay. An irreflexivity lemma with applications to axiomatization of conditions on linear frames. In U. Mönnich, editor, *Aspects of Philosophical Logic*, pages 67–89. D. Reidel Publishing Co., 1981.
10. S. Greco, B. Matarazzo, and R. Slowinski. Rough set approach to multi-attribute choice and ranking problems. In *Proceedings of the 12th International Conference on Multiple Criteria Decision Making*, pages 318–329, 1997.
11. S. Greco, B. Matarazzo, and R. Slowinski. Rough approximation of a preference relation in a pairwise comparison table. In L. Polkowski and A. Skowron, editors, *Rough Sets in Data Mining and Knowledge Discovery*, pages 13–36. Physica-Verlag, 1998.
12. S. Greco, B. Matarazzo, and R. Slowinski. Rough approximation of a preference relation by dominance relations. *European Journal of Operational Research*, 117(1):63–83, 1999.
13. S. Greco, B. Matarazzo, and R. Slowinski. Extension of the rough set approach to multicriteria decision support. *INFOR Journal: Information Systems and Operational Research*, 38(3):161–195, 2000.
14. S. Greco, B. Matarazzo, and R. Slowinski. Rough set theory for multicriteria decision analysis. *European Journal of Operational Research*, 129(1):1–47, 2001.
15. S. Greco, B. Matarazzo, and R. Slowinski. Rough sets methodology for sorting problems in presence of multiple attributes and criteria. *European Journal of Operational Research*, 138(2):247–259, 2002.

16. S. Greco, B. Matarazzo, and R. Slowinski. Axiomatic characterization of a general utility function and its particular cases in terms of conjoint measurement and rough-set decision rules. *European Journal of Operational Research*, 158(2): 271–292, 2004.
17. J.W. Grzymala-Busse. Algebraic properties of knowledge representation systems. In *Proceedings of the ACM SIGART International Symposium on Methodologies for Intelligent Systems*, pages 432–440. ACM Press, 1986.
18. J. Han and M. Kamber. *Data Mining: Concepts and Techniques*. Morgan Kaufmann Publishers, 2001.
19. C.J. Liau. Belief reasoning, revision and fusion by matrix algebra. In J. Komorowski and S. Tsumoto, editors, *Proceedings of the 4th International Conference on Rough Sets and Current Trends in Computing*, LNAI 3066, pages 133–142. Springer-Verlag, 2004.
20. C.J. Liau. Matrix representation of belief states: An algebraic semantics for belief logics. *International Journal of Uncertainty, Fuzziness and Knowledge-based Systems*, 12(5):613–633, 2004.
21. C.J. Liau and T.Y. Lin. Reasoning about relational granulation in modal logics. In *Proc. of the First IEEE International Conference on Granular Computing*, pages 534–538. IEEE Press, 2005.
22. C.J. Liau and D.R. Liu. A logical approach to fuzzy data analysis. In J.M. Zytkow and J. Rauch, editors, *Proceedings of the 3rd European Conference on Principles of Data Mining and Knowledge Discovery*, LNAI 1704, pages 412–417. Springer-Verlag, 1999.
23. C.J. Liau and D.R. Liu. A possibilistic decision logic with applications. *Fundamenta Informaticae*, 46(3):199–217, 2001.
24. M. Marx and Y. Venema. *Multi-Dimensional Modal Logic*. Kluwer Academic Publishers, 1997.
25. Z. Pawlak. Rough sets. *International Journal of Computer and Information Sciences*, 11(15):341–356, 1982.
26. Z. Pawlak. *Rough Sets–Theoretical Aspects of Reasoning about Data*. Kluwer Academic Publishers, 1991.
27. J. Scott. *Social Network Analysis: A Handbook*. SAGE Publications, 2 edition, 2000.
28. A. Skowron. Extracting laws from decision tables. *Computational Intelligence: An International Journal*, 11:371–388, 1995.
29. A. Skowron and C. Rauszer. The discernibility matrices and functions in information systems. In R. Slowinski, editor, *Intelligent Decision Support Systems: Handbook of Applications and Advances in Rough Set Theory*, pages 331–362. Kluwer Academic Publisher, 1991.
30. A. Skowron and R. Swiniarski. Rough sets and higher order vagueness. In D. Slezak, G.Y. Wang, M. Szczuka, I. Duntsch, and Y.Y. Yao, editors, *Proceedings of the 10th International Conference on Rough Sets, Fuzzy Sets, Data Mining, and Granular Computing*, LNAI 3641, pages 33–42. Springer-Verlag, 2005.
31. A. Skowron and P. Synak. Complex patterns. *Fundamenta Informaticae*, 60(1-4):351–366, 2004.
32. R. Slowinski, S. Greco, and B. Matarazzo. Rough set analysis of preference-ordered data. In J.J. Alpigini, J.F. Peters, A. Skowron, and N. Zhong, editors, *Proceedings of the 3rd International Conference on Rough Sets and Current Trends in Computing*, LNAI 2475, pages 44–59. Springer-Verlag, 2002.
33. Y. Venema. A crash course in arrow logic. In M. Marx, L. P'olos, and M. Masuch, editors, *Arrow Logic and Multi-Modal Logic*, pages 3–34. CSLI Publications, 1996.

34. Y.Y. Yao and C.J. Liau. A generalized decision logic language for granular computing. In *Proceedings of the 11th IEEE International Conference on Fuzzy Systems*, pages 773–778. IEEE Press, 2002.
35. Y.Y. Yao and Q. Liu. A generalized decision logic in interval-set-valued information tables. In N. Zhong, A. Skowron, and S. Ohsuga, editors, *New Directions in Rough Sets, Data Mining, and Granular-Soft Computing*, LNAI 1711, pages 285–293. Springer-Verlag, 1999.

On Generalized Rough Fuzzy Approximation Operators

Wei-Zhi Wu[1], Yee Leung[2], and Wen-Xiu Zhang[3]

[1] Information College, Zhejiang Ocean University,
Zhoushan, Zhejiang, 316004, P.R. China
wuwz@zjou.net.cn
[2] Department of Geography and Resource Management
Center for Environmental Policy and Resource Management
and Institute of Space and Earth Information Science
The Chinese University of Hong Kong, Hong Kong
yeeleung@cuhk.edu.hk
[3] Institute for Information and System Sciences, Faculty of Science
Xi'an Jiaotong University, Xi'an, Shaan'xi, 710049, P.R. China
wxzhang@mail.xjtu.edu.cn

Abstract. This paper presents a general framework for the study of rough fuzzy sets in which fuzzy sets are approximated in a crisp approximation space. By the constructive approach, a pair of lower and upper generalized rough fuzzy approximation operators is first defined. The rough fuzzy approximation operators are represented by a class of generalized crisp approximation operators. Properties of rough fuzzy approximation operators are then discussed. The relationships between crisp relations and rough fuzzy approximation operators are further established. By the axiomatic approach, various classes of rough fuzzy approximation operators are characterized by different sets of axioms. The axiom sets of rough fuzzy approximation operators guarantee the existence of certain types of crisp relations producing the same operators. The relationship between a fuzzy topological space and rough fuzzy approximation operators is further established. The connections between rough fuzzy sets and Dempster-Shafer theory of evidence are also examined. Finally multi-step rough fuzzy approximations within the framework of neighborhood systems are analyzed.

Keywords: approximation operators, belief functions, binary relations, fuzzy sets, fuzzy topological spaces, neighborhood systems, rough fuzzy sets, rough sets.

1 Introduction

The theory of rough sets was originally proposed by Pawlak [26,27] as a formal tool for modelling and processing incomplete information. The basic structure of the rough set theory is an approximation space consisting of a universe of discourse and an equivalence relation imposed on it. The equivalence

J.F. Peters and A. Skowron (Eds.): Transactions on Rough Sets V, LNCS 4100, pp. 263–284, 2006.

relation is a key notion in Pawlak's rough set model. The equivalence classes in Pawlak's rough set model provide the basis of "information granules" for database analysis discussed in Zadeh's [67,68]. Rough set theory can be viewed as a crisp-set-based granular computing method that advances reaearch in this area [12,17,29,30,37,38,62].

However, the requirement of an equivalence relation in Pawlak's rough set model seems to be a very restrictive condition that may limit the applications of the rough set model. Thus one of the main directions of research in rough set theory is naturally the generalization of the Pawlak rough set approximations. There are at least two approaches for the development of rough set theory, namely the constructive and axiomatic approaches. In the constructive approach, binary relations on the universe of discourse, partitions of the universe of discourse, neighborhood systems, and Boolean algebras are all the primitive notions. The lower and upper approximation operators are constructed by means of these notions [15,23,25,26,27,28,31,39,40,46,53,56,58,59,60,61,63]. Constructive generalizations of rough set to fuzzy environment have also been discussed in a number of studies [1,2,9,10,14,19,20,21,32,50,51,52,54,57]. For example, by using an equivalence relation on U, Dubois and Prade introduced the lower and upper approximations of fuzzy sets in a Pawlak approximation space to obtain an extended notion called rough fuzzy set [9,10]. Alternatively, a fuzzy similarity relation can be used to replace an equivalence relation. The result is a deviation of rough set theory called fuzzy rough set [10,21,32,58]. Based on arbitrary fuzzy relations, fuzzy partitions on U, and Boolean subalgebras of $\mathcal{P}(U)$, extended notions called rough fuzzy sets and fuzzy rough sets have been obtained [19,20,23,50,51,52,54]. Alternatively, a rough fuzzy set is the approximation of a fuzzy set in a crisp approximation space. The rough fuzzy set model may be used to handle knowledge acquisition in information systems with fuzzy decisions [70]. And a fuzzy rough set is the approximation of a crisp set or a fuzzy set in a fuzzy approximation space. The fuzzy rough set model may be used to unravel knowledge hidden in fuzzy decision systems [55]. Employing constructive methods, extensive research has also been carried out to compare the theory of rough sets with other theories of uncertainty such as fuzzy sets and conditional events [3,19,24,46]. Thus the constructive approach is suitable for practical applications of rough sets.

On the other hand, the axiomatic approach, which is appropriate for studying the structures of rough set algebras, takes the lower and upper approximation operators as primitive notions. From this point of view, rough set theory may be interpreted as an extension theory with two additional unary operators. The lower and upper approximation operators are related respectively to the necessity (box) and possibility (diamond) operators in modal logic, and the interior and closure operators in topological space [4,5,6,13,22,44,45,47,48,60,63]. By this approach, a set of axioms is used to characterize approximation operators that are the same as the ones produced by using the constructive approach. Zakowski [69] studied a set of axioms on approximation operators. Comer [6] investigated axioms on approximation operators in relation to cylindric algebras. The investigation is made within the context of Pawlak information systems [25]. Lin and Liu [16] suggested six axioms on a pair of abstract operators on the power set of the universe of discourse within the framework of topological spaces. Under these axioms, there exists an equivalence relation such that the derived lower and upper approximations are the same as the abstract operators. Similar result was also stated earlier by Wiweger [47]. The problem of these studies is that they are restricted to the Pawlak rough set algebra defined by equivalence relations. Wybraniec-Skardowska [56] examined many axioms on various classes of approximation operators. Different constructive methods were suggested to produce such approximation operators. Thiele [41] explored axiomatic characterizations of approximation operators within modal logic for a crisp diamond and box operator represented by an arbitrary binary crisp relation. The most important axiomatic studies for crisp rough sets were made by Yao [57,59,60], Yao and Lin [63], in which various classes of crisp rough set algebras are characterized by different sets of axioms. As to the fuzzy cases, Moris and Yakout [21] studied a set of axioms on fuzzy rough sets based on a triangular norm and a residual implicator. Radzikowska and Kerre [32] defined a broad family of the so called $(\mathcal{I}, \mathcal{T})$-fuzzy rough sets which is determined by an implicator $\mathcal{I}$ and a triangular norm $\mathcal{T}$. Their studies however were restricted to fuzzy $\mathcal{T}$-rough sets defined by fuzzy $\mathcal{T}$-similarity relations which were equivalence crisp relations in the degenerated case. Thiele [42,43,44] investigated axiomatic characterizations of fuzzy rough approximation operators and rough fuzzy approximation operators within modal logic for fuzzy diamond and box operators. Wu et al. [52], Wu and Zhang [54], examined many axioms on various classes of rough fuzzy and fuzzy rough approximation operators when $\mathcal{T} = \min$. Mi and Zhang [20] discussed axiomatic characterization of a pair of dual lower and upper fuzzy approximation operators based on a residual implication. In [50], Wu et al. studied axiomatic characterization of $(\mathcal{I}, \mathcal{T})$-fuzzy rough sets corresponding to various fuzzy relations. In this paper, we mainly focus on the study of rough fuzzy approximation operators derived from crisp binary relations.

Another important direction for generalization of rough set theory is its relationship to the Dempster-Shafer theory of evidence [33] which was originated by Dempster's concept of lower and upper probabilities [7] and extended by Shafer as a theory [33]. The basic representational structure in Dempster-Shafer theory of evidence is a belief structure which consists of a family of subsets, called focal elements, with associated individual positive weights summing to one. The primitive numeric measures derived from the belief structure are a dual pair of belief and plausibility functions. Shafer's evidence theory can also be extended to the fuzzy environment [8,11,51,65]. There exist some natural connections between the rough set theory and Dempster-Shafer theory of evidence [34,35,36,49,51,64]. It is demonstrated that various belief structures are associated with various rough approximation spaces such that different dual pairs of upper and lower approximation operators induced by the rough approximation spaces may be used to interpret the corresponding dual pairs of plausibility and belief functions induced by the belief structures.

In this paper, we focus mainly on the study of mathematical structure of rough fuzzy approximation operators. We will review existing results and present some new results on generalized rough fuzzy approximation operators. In the next section, we give some basic notions of rough sets and review basic properties of generalized rough approximation operators. In Section 3, the concepts of generalized rough fuzzy approximation operators are introduced. The representation theorem of rough fuzzy approximation operators is stated and properties of the rough fuzzy approximation operators are examined. In Section 4, we present the axiomatic characterizations of rough fuzzy approximation operators. Various classes of rough fuzzy approximation operators are characterized by different sets of axioms, and the axiom sets of fuzzy approximation operators guarantee the existence of certain types of crisp relation producing the same operators. We further establish the relationship between rough fuzzy approximation operators and fuzzy topological space in Section 5. The interpretations of the rough fuzzy set theory and the Dempster-Shafer theory of evidence are discussed in Section 6. In Section 7, we build a framework for the study of k-step-neighborhood systems and rough fuzzy approximation operators in which a binary crisp relation is still used as a primitive notion. We then conclude the paper with a summary in Section 8.

2   Generalized Rough Sets

Let X be a finite and nonempty set called the universe of discourse. The class of all subsets (respectively, fuzzy subsets) of X will be denoted by $\mathcal{P}(X)$ (respectively, by $\mathcal{F}(X)$). For any $A \in \mathcal{F}(X)$, the α-level and the strong α-level set of A will be denoted by A_α and $A_{\alpha+}$, respectively, that is, $A_\alpha = \{x \in X : A(x) \geq \alpha\}$ and $A_{\alpha+} = \{x \in X : A(x) > \alpha\}$, where $\alpha \in I = [0,1]$, the unit interval. We denote by $\sim A$ the complement of A. The cardinality of A is denoted by $|A| = \sum\limits_{u \in X} A(u)$. If P is a probability measure on X, then the probability of the fuzzy set A, denoted by $P(A)$, is defined, in the sense of Zadeh [66], by

$$P(A) = \sum_{x \in X} A(x)P(x).$$

Definition 1. *Let U and W be two finite and nonempty universes of discourse. A subset $R \in \mathcal{P}(U \times W)$ is referred to as a (crisp) binary relation from U to W. The relation R is referred to as serial if for all $x \in U$ there exists $y \in W$ such that $(x,y) \in R$; If $U = W$, R is referred to as a binary relation on U. R is referred to as reflexive if for all $x \in U$, $(x,x) \in R$; R is referred to as symmetric if for all $x,y \in U$, $(x,y) \in R$ implies $(y,x) \in R$; R is referred to as transitive if for all $x,y,z \in U$, $(x,y) \in R$ and $(y,z) \in R$ imply $(x,z) \in R$; R is referred to as Euclidean if for all $x,y,z \in U$, $(x,y) \in R$ and $(x,z) \in R$ imply $(y,z) \in R$; R is referred to as an equivalence relation if R is reflexive, symmetric and transitive.*

Suppose that R is an arbitrary crisp relation from U to W. We can define a set-valued function $R_s : U \to \mathcal{P}(W)$ by:

$$R_s(x) = \{y \in W : (x,y) \in R\}, \quad x \in U.$$

$R_s(x)$ is referred to as the successor neighborhood of x with respect to R. Obviously, any set-valued function F from U to W defines a binary relation from U to W by setting $R = \{(x, y) \in U \times W : y \in F(x)\}$. From the set-valued function F, we can define a basic set assignment [51,60,64] $j : \mathcal{P}(W) \rightarrow \mathcal{P}(U)$,

$$j(A) = \{u \in U : F(u) = A\}, \quad A \in \mathcal{P}(W).$$

It is easy to verify that j satisfies the properties:

(J1) $\bigcup_{A \subseteq W} j(A) = U,$ (J2) $A \neq B \Longrightarrow j(A) \cap j(B) = \emptyset.$

Definition 2. *If R is an arbitrary crisp relation from U to W, then the triplet (U, W, R) is referred to as a generalized approximation space. For any set $A \subseteq W$, a pair of lower and upper approximations, $\underline{R}(A)$ and $\overline{R}(A)$, are defined by*

$$\begin{aligned}\underline{R}(A) &= \{x \in U : R_s(x) \subseteq A\}, \\ \overline{R}(A) &= \{x \in U : R_s(x) \cap A \neq \emptyset\}.\end{aligned} \tag{1}$$

The pair $(\underline{R}(A), \overline{R}(A))$ is referred to as a generalized crisp rough set, and $\underline{R}$ and $\overline{R} : \mathcal{F}(W) \rightarrow \mathcal{F}(U)$ are referred to as the lower and upper generalized crisp approximation operators respectively.

From the definition, the following theorem can be easily derived [59,60]:

Theorem 1. *For any relation R from U to W, its lower and upper approximation operators satisfy the following properties: for all $A, B \in \mathcal{P}(W)$,*

(L1) $\underline{R}(A) =\sim \overline{R}(\sim A),$ (U1) $\overline{R}(A) =\sim \underline{R}(\sim A);$
(L2) $\underline{R}(W) = U,$ (U2) $\overline{R}(\emptyset) = \emptyset;$
(L3) $\underline{R}(A \cap B) = \underline{R}(A) \cap \underline{R}(B),$ (U3) $\overline{R}(A \cup B) = \overline{R}(A) \cup \overline{R}(B);$
(L4) $A \subseteq B \Longrightarrow \underline{R}(A) \subseteq \underline{R}(B),$ (U4) $A \subseteq B \Longrightarrow \overline{R}(A) \subseteq \overline{R}(B);$
(L5) $\underline{R}(A \cup B) \supseteq \underline{R}(A) \cup \underline{R}(B),$ (U5) $\overline{R}(A \cap B) \subseteq \overline{R}(A) \cap \overline{R}(B).$

With respect to certain special types, say, serial, reflexive, symmetric, transitive, and Euclidean binary relations on the universe of discourse U, the approximation operators have additional properties [59,60,61].

Theorem 2. *Let R be an arbitrary crisp binary relation on U, and $\underline{R}$ and $\overline{R}$ the lower and upper generalized crisp approximation operators defined by Eq.(1). Then*

R *is serial* $\Longleftrightarrow$ (L0) $\underline{R}(\emptyset) = \emptyset,$
                         $\Longleftrightarrow$ (U0) $\overline{R}(U) = U,$
                         $\Longleftrightarrow$ (LU0) $\underline{R}(A) \subseteq \overline{R}(A), \forall A \in \mathcal{P}(U),$
R *is reflexive* $\Longleftrightarrow$ (L6) $\underline{R}(A) \subseteq A, \quad \forall A \in \mathcal{P}(U),$
                         $\Longleftrightarrow$ (U6) $A \subseteq \overline{R}(A), \quad \forall A \in \mathcal{P}(U),$

$$R \text{ is symmetric } \iff \text{(L7) } \overline{R}(\underline{R}(A)) \subseteq A, \qquad \forall A \in \mathcal{P}(U),$$
$$\iff \text{(U7) } A \subseteq \underline{R}(\overline{R}(A)), \qquad \forall A \in \mathcal{P}(U),$$
$$R \text{ is transitive } \iff \text{(L8) } \underline{R}(A) \subseteq \underline{R}(\underline{R}(A)), \forall A \in \mathcal{P}(U),$$
$$\iff \text{(U8) } \overline{R}(\overline{R}(A)) \subseteq \overline{R}(A), \forall A \in \mathcal{P}(U),$$
$$R \text{ is Euclidean } \iff \text{(L9) } \overline{R}(\underline{R}(A)) \subseteq \underline{R}(A), \forall A \in \mathcal{P}(U),$$
$$\iff \text{(U9) } \overline{R}(A) \subseteq \underline{R}(\overline{R}(A)), \forall A \in \mathcal{P}(U).$$

If R is an equivalence relation on U, then the pair (U, R) is a Pawlak approximation space and more interesting properties of lower and upper approximation operators can be derived [26,27].

3   Construction of Generalized Rough Fuzzy Approximation Operators

In this section, we review the constructive definitions of rough fuzzy approximation operators and give the basic properties of the operators.

3.1   Definitions of Rough Fuzzy Approximation Operators

A rough fuzzy set is the approximation of a fuzzy set in a crisp approximation space [10,54,58].

Definition 3. *Let U and W be two finite non-empty universes of discourse and R a crisp binary relation from U to W. For any set $A \in \mathcal{F}(W)$, the lower and upper approximations of A, $\underline{R}(A)$ and $\overline{R}(A)$, with respect to the crisp approximation space (U, W, R) are fuzzy sets of U whose membership functions, for each $x \in U$, are defined respectively by*

$$\underline{R}(A)(x) = \bigwedge_{y \in R_s(x)} A(y),$$
$$\overline{R}(A)(x) = \bigvee_{y \in R_s(x)} A(y). \tag{2}$$

The pair $(\underline{R}(A), \overline{R}(A))$ is called a generalized rough fuzzy set, and $\underline{R}$ and $\overline{R}$: $\mathcal{F}(W) \to \mathcal{F}(U)$ are referred to as the lower and upper generalized rough fuzzy approximation operators respectively.

If $A \in \mathcal{P}(W)$, then we can see that $\underline{R}(A)(x) = 1$ iff $R_s(x) \subseteq A$ and $\overline{R}(A)(x) = 1$ iff $R_s(x) \cap A \neq \emptyset$. Thus Definition 3 degenerates to Definition 2 when the fuzzy set A reduces to a crisp set.

3.2   Representations of Rough Fuzzy Approximation Operators

A fuzzy set can be represented by a family of crisp sets using its α-level sets. In [58], Yao obtained the representation theorem of rough fuzzy approximation operators derived from a Pawlak approximation space. Wu and Zhang [54] generalized Yao's representation theorem of rough fuzzy approximation operators to an arbitrary crisp approximation space. We review and summarize this idea as follows:

Definition 4. *A set-valued mapping $N : I \to \mathcal{P}(U)$ is said to be nested if for all $\alpha, \beta \in I$,*

$$\alpha \leq \beta \Longrightarrow N(\beta) \subseteq N(\alpha).$$

The class of all $\mathcal{P}(U)$-valued nested mappings on I will be denoted by $\mathcal{N}(U)$.

It is well-known that the following representation theorem holds [52,54]:

Theorem 3. *Let $N \in \mathcal{N}(U)$. Define a function $f : \mathcal{N}(U) \to \mathcal{F}(U)$ by:*

$$A(x) := f(N)(x) = \bigvee_{\alpha \in I} (\alpha \wedge N(\alpha)(x)), \quad x \in U,$$

where $N(\alpha)(x)$ is the characteristic function of $N(\alpha)$. Then f is a surjective homomorphism, and the following properties hold:

(1) $A_{\alpha+} \subseteq N(\alpha) \subseteq A_\alpha, \quad \alpha \in I,$

(2) $A_\alpha = \bigcap_{\lambda < \alpha} N(\lambda), \quad \alpha \in I,$

(3) $A_{\alpha+} = \bigcup_{\lambda > \alpha} N(\lambda), \quad \alpha \in I,$

(4) $A = \bigvee_{\alpha \in I} (\alpha \wedge A_{\alpha+}) = \bigvee_{\alpha \in I} (\alpha \wedge A_\alpha).$

Let (U, W, R) be a generalized approximation space, $\forall A \in \mathcal{F}(W)$ and $0 \leq \beta \leq 1$, the lower and upper approximations of A_β and $A_{\beta+}$ with respect to (U, W, R) are defined respectively as

$$\underline{R}(A_\beta) = \{x \in U : R_s(x) \subseteq A_\beta\}, \quad \overline{R}(A_\beta) = \{x \in U : R_s(x) \cap A_\beta \neq \emptyset\},$$
$$\underline{R}(A_{\beta+}) = \{x \in U : R_s(x) \subseteq A_{\beta+}\}, \overline{R}(A_{\beta+}) = \{x \in U : R_s(x) \cap A_{\beta+} \neq \emptyset\}.$$

It can easily be verified that the four classes $\{\underline{R}(A_\alpha) : \alpha \in I\}$, $\{\underline{R}(A_{\alpha+}) : \alpha \in I\}$, $\{\overline{R}(A_\alpha) : \alpha \in I\}$, and $\{\overline{R}(A_{\alpha+}) : \alpha \in I\}$ are $\mathcal{P}(U)$-valued nested mappings on I. By Theorem 3, each of them defines a fuzzy subset of U which equals the lower (and upper, respectively) rough fuzzy approximation operator [54].

Theorem 4. *Let (U, W, R) be a generalized approximation space and $A \in \mathcal{F}(W)$, then*

(1) $\underline{R}(A) = \bigvee_{\alpha \in I} [\alpha \wedge \underline{R}(A_\alpha)] = \bigvee_{\alpha \in I} [\alpha \wedge \underline{R}(A_{\alpha+})],$

(2) $\overline{R}(A) = \bigvee_{\alpha \in I} [\alpha \wedge \overline{R}(A_\alpha)] = \bigvee_{\alpha \in I} [\alpha \wedge \overline{R}(A_{\alpha+})].$

And

(3) $[\underline{R}(A)]_{\alpha+} \subseteq \underline{R}(A_{\alpha+}) \subseteq \underline{R}(A_\alpha) \subseteq [\underline{R}(A)]_\alpha, \quad 0 \leq \alpha \leq 1,$

(4) $[\overline{R}(A)]_{\alpha+} \subseteq \overline{R}(A_{\alpha+}) \subseteq \overline{R}(A_\alpha) \subseteq [\overline{R}(A)]_\alpha, \quad 0 \leq \alpha \leq 1.$

3.3 Properties of Rough Fuzzy Approximation Operators

By the representation theorem of rough fuzzy approximation operators we can obtain properties of rough fuzzy approximation operators [54].

Theorem 5. *The lower and upper rough fuzzy approximation operators, $\underline{R}$ and $\overline{R}$, defined by Eq.(2) satisfy the properties:* $\forall A, B \in \mathcal{F}(W), \forall \alpha \in I,$

(FL1) $\underline{R}(A) =\sim \overline{R}(\sim A),$ (FU1) $\overline{R}(A) =\sim \underline{R}(\sim A),$

(FL2) $\underline{R}(A \vee \widehat{\alpha}) = \underline{R}(A) \vee \widehat{\alpha},$ (FU2) $\overline{R}(A \wedge \widehat{\alpha}) = \overline{R}(A) \wedge \widehat{\alpha},$

(FL3) $\underline{R}(A \wedge B) = \underline{R}(A) \wedge \underline{R}(B),$ (FU3) $\overline{R}(A \vee B) = \overline{R}(A) \vee \overline{R}(B),$

(FL4) $A \subseteq B \Longrightarrow \underline{R}(A) \subseteq \underline{R}(B),$ (FU4) $A \subseteq B \Longrightarrow \overline{R}(A) \subseteq \overline{R}(B),$

(FL5) $\underline{R}(A \vee B) \supseteq \underline{R}(A) \vee \underline{R}(B),$ (FU5) $\overline{R}(A \wedge B) \subseteq \overline{R}(A) \wedge \overline{R}(B),$

where $\widehat{a}$ is the constant fuzzy set: $\widehat{a}(x) = a$, for all x.

Properties (FL1) and (FU1) show that the rough fuzzy approximation operators $\underline{R}$ and $\overline{R}$ are dual to each other. Properties with the same number may be regarded as dual properties. Properties (FL3) and (FU3) state that the lower rough fuzzy approximation operator $\underline{R}$ is multiplicative, and the upper rough fuzzy approximation operator $\overline{R}$ is additive. One may also say that $\underline{R}$ is distributive w.r.t. the intersection of fuzzy sets, and $\overline{R}$ is distributive w.r.t. the union of fuzzy sets. Properties (FL5) and (FU5) imply that $\underline{R}$ is not distributive w.r.t. set union, and $\overline{R}$ is not distributive w.r.t. set intersection. However, properties (FL2) and (FU2) show that $\underline{R}$ is distributive w.r.t. the union of a fuzzy set and a fuzzy constant set, and $\overline{R}$ is distributive w.r.t. the intersection of a fuzzy set and a constant fuzzy set. Evidently, properties (FL2) and (FU2) imply the following properties:

(FL2)$'$ $\underline{R}(W) = U,$ (FU2)$'$ $\overline{R}(\emptyset) = \emptyset.$

Analogous to Yao's study in [59], a serial rough fuzzy set model is obtained from a serial binary relation. The property of a serial relation can be characterized by the properties of its induced rough fuzzy approximation operators [54].

Theorem 6. *If R is an arbitrary crisp relation from U to W, and $\underline{R}$ and $\overline{R}$ are the rough fuzzy approximation operators defined by Eq.(2), then*

R *is serial* $\Longleftrightarrow$ (FL0) $\underline{R}(\emptyset) = \emptyset,$

$\Longleftrightarrow$ (FU0) $\overline{R}(W) = U,$

$\Longleftrightarrow$ (FL0)$'$ $\underline{R}(\widehat{\alpha}) = \widehat{\alpha},$ $\forall \alpha \in I,$

$\Longleftrightarrow$ (FU0)$'$ $\overline{R}(\widehat{\alpha}) = \widehat{\alpha},$ $\forall \alpha \in I,$

$\Longleftrightarrow$ (FLU0) $\underline{R}(A) \subseteq \overline{R}(A), \forall A \in \mathcal{F}(W).$

By (FLU0), the pair of rough fuzzy approximation operators of a serial rough set model is an interval structure. In the case of connections between other special crisp relations and rough fuzzy approximation operators, we have the following theorem which may be seen as a generalization of Theorem 2 [54]:

Theorem 7. *Let R be an arbitrary crisp relation on U, and $\underline{R}$ and $\overline{R}$ the lower and upper rough fuzzy approximation operators defined by Eq.(2). Then*

$$
\begin{aligned}
R \text{ is reflexive} &\Longleftrightarrow \text{(FL6)} \ \underline{R}(A) \subseteq A, &&\forall A \in \mathcal{F}(U), \\
&\Longleftrightarrow \text{(FU6)} \ A \subseteq \overline{R}(A), &&\forall A \in \mathcal{F}(U), \\
R \text{ is symmetric} &\Longleftrightarrow \text{(FL7)} \ \overline{R}(\underline{R}(A)) \subseteq A, &&\forall A \in \mathcal{F}(U), \\
&\Longleftrightarrow \text{(FU7)} \ A \subseteq \underline{R}(\overline{R}(A)), &&\forall A \in \mathcal{F}(U), \\
&\Longleftrightarrow \text{(FL7)}' \ \underline{R}(1_{U-\{x\}})(y) = \underline{R}(1_{U-\{y\}})(x), \ \forall (x,y) \in U \times U, \\
&\Longleftrightarrow \text{(FU7)}' \ \overline{R}(1_x)(y) = \overline{R}(1_y)(x), &&\forall (x,y) \in U \times U, \\
R \text{ is transitive} &\Longleftrightarrow \text{(FL8)} \ \underline{R}(A) \subseteq \underline{R}(\underline{R}(A)), &&\forall A \in \mathcal{F}(U), \\
&\Longleftrightarrow \text{(FU8)} \ \overline{R}(\overline{R}(A)) \subseteq \overline{R}(A), &&\forall A \in \mathcal{F}(U), \\
R \text{ is Euclidean} &\Longleftrightarrow \text{(FL9)} \ \overline{R}(\underline{R}(A)) \subseteq \underline{R}(A), &&\forall A \in \mathcal{F}(U), \\
&\Longleftrightarrow \text{(FU9)} \ \overline{R}(A) \subseteq \underline{R}(\overline{R}(A)), &&\forall A \in \mathcal{F}(U).
\end{aligned}
$$

4 Axiomatic Characterization of Rough Fuzzy Approximation Operators

In the axiomatic approach, rough sets are characterized by abstract operators. For the case of rough fuzzy sets, the primitive notion is a system $(\mathcal{F}(U), \mathcal{F}(W), \wedge, \vee, \sim, L, H)$, where L and H are unary operators from $\mathcal{F}(W)$ to $\mathcal{F}(U)$. In this section, we show that rough fuzzy approximation operators can be characterized by axioms. The results may be viewed as the generalized counterparts of Yao [57,59,60].

Definition 5. *Let $L, H : \mathcal{F}(W) \to \mathcal{F}(U)$ be two operators. They are referred to as dual operators if for all $A \in \mathcal{F}(W)$,*

$$
\begin{aligned}
\text{(fl1)} \quad & L(A) = \sim H(\sim A), \\
\text{(fu1)} \quad & H(A) = \sim L(\sim A).
\end{aligned}
$$

By the dual properties of the operators, we only need to define one operator. We state the following theorem which can be proved via the discussion on the constructive approach in [54]:

Theorem 8. *Suppose that $L, H : \mathcal{F}(W) \to \mathcal{F}(U)$ are dual operators. Then there exists a crisp binary relation R from U to W such that for all $A \in \mathcal{F}(W)$*

$$
L(A) = \underline{R}(A), \quad \text{and} \quad H(A) = \overline{R}(A)
$$

iff L satisfies axioms (flc), (fl2), (fl3), or equivalently H satisfies axioms (fuc), (fu2), (fu3):

(flc) $L(1_{W-\{y\}}) \in \mathcal{P}(U),$ $\forall y \in W,$

(fl2) $L(A \vee \widehat{\alpha}) = L(A) \vee \widehat{\alpha},$ $\forall A \in \mathcal{F}(W), \quad \forall \alpha \in I,$

(fl3) $L(A \wedge B) = L(A) \wedge L(B),$ $\forall A, B \in \mathcal{F}(W),$

(fuc) $H(1_y) \in \mathcal{P}(U),$ $\forall y \in W,$

(fu2) $H(A \wedge \widehat{\alpha}) = H(A) \wedge \widehat{\alpha},$ $\forall A \in \mathcal{F}(W), \quad \forall \alpha \in I,$

(fu3) $H(A \vee B) = H(A) \vee H(B),$ $\forall A, B \in \mathcal{F}(W),$

where 1_y denotes the fuzzy singleton with value 1 at y and 0 elsewhere.

According to Theorem 8, axioms (flc),(fl1),(fl2), (fl3), or equivalently, axioms (fuc), (fu1), (fu2), (fu3) are considered to be basic axioms of rough fuzzy approximation operators. These lead to the following definitions of rough fuzzy set algebras:

Definition 6. *Let $L, H : \mathcal{F}(W) \rightarrow \mathcal{F}(U)$ be a pair of dual operators. If L satisfies axioms (flc), (fl2), and (fl3), or equivalently H satisfies axioms (fuc), (fu2), and (fu3), then the system $(\mathcal{F}(U), \mathcal{F}(W), \wedge, \vee, \sim, L, H)$ is referred to as a rough fuzzy set algebra, and L and H are referred to as rough fuzzy approximation operators. When $U = W$, $(\mathcal{F}(U), \wedge, \vee, \sim, L, H)$ is also called a rough fuzzy set algebra, in such a case, if there exists a serial (a reflexive, a symmetric, a transitive, an Euclidean, an equivalence) relation R on U such that $L(A) = \underline{R}(A)$ and $H(A) = \overline{R}(A)$ for all $A \in \mathcal{F}(U)$, then $(\mathcal{F}(U), \wedge, \vee, \sim, L, H)$ is referred to as a serial (a reflexive, a symmetric, a transitive, an Euclidean, a Pawlak) rough fuzzy set algebra.*

Axiomatic characterization of serial rough fuzzy set algebra is summarized as the following Theorem [54]:

Theorem 9. *Suppose that $(\mathcal{F}(U), \mathcal{F}(W), \wedge, \vee, \sim, L, H)$ is a rough fuzzy set algebra, i.e., L satisfies axioms (flc), (fl1), (fl2) and (fl3), and H satisfies (fuc), (fu1), (fu2) and (fu3). Then it is a serial rough fuzzy set algebra iff one of following equivalent axioms holds:*

(fl0) $L(\widehat{\alpha}) = \widehat{\alpha},$ $\forall \alpha \in I,$

(fu0) $H(\widehat{\alpha}) = \widehat{\alpha},$ $\forall \alpha \in I,$

(fl0)$'$ $L(\emptyset) = \emptyset,$

(fu0)$'$ $H(W) = U,$

(flu0)$'$ $L(A) \subseteq H(A),$ $\forall A \in \mathcal{F}(W).$

Axiom (flu0)$'$ states that $L(A)$ is a fuzzy subset of $H(A)$. In such a case, $L, H : \mathcal{F}(W) \rightarrow \mathcal{F}(U)$ are called the lower and upper rough fuzzy approximation operators and the system $(\mathcal{F}(U), \mathcal{F}(W), \wedge, \vee, \sim, L, H)$ is an interval structure. Axiomatic characterizations of other special rough fuzzy operators are summarized in the following Theorems 10 and 11 [54]:

Theorem 10. *Suppose that $(\mathcal{F}(U), \wedge, \vee, \sim, L, H)$ is a rough fuzzy set algebra. Then*

(1) *it is a reflexive rough fuzzy set algebra iff one of following equivalent axioms holds:*

$$\text{(fl6)}\quad L(A) \subseteq A,\quad \forall A \in \mathcal{F}(U),$$
$$\text{(fu6)}\quad A \subseteq H(A),\quad \forall A \in \mathcal{F}(U).$$

(2) *it is a symmetric rough fuzzy set algebra iff one of the following equivalent axioms holds:*

$$\text{(fl7)}'\quad L(1_{U-\{x\}})(y) = L(1_{U-\{y\}})(x),\ \forall(x,y) \in U \times U,$$
$$\text{(fu7)}'\quad H(1_x)(y) = H(1_y)(x),\qquad \forall(x,y) \in U \times U,$$
$$\text{(fl7)}\quad A \subseteq L(H(A)),\qquad\qquad \forall A \in \mathcal{F}(U),$$
$$\text{(fu7)}\quad H(L(A)) \subseteq A,\qquad\qquad \forall A \in \mathcal{F}(U).$$

(3) *it is a transitive rough fuzzy set algebra iff one of following equivalent axioms holds:*

$$\text{(fl8)}\quad L(A) \subseteq L(L(A)),\quad \forall A \in \mathcal{F}(U),$$
$$\text{(fu8)}\quad H(H(A)) \subseteq H(A),\ \forall A \in \mathcal{F}(U).$$

(4) *it is an Euclidean rough fuzzy set algebra iff one of following equivalent axioms holds:*

$$\text{(fl9)}\quad H(L(A)) \subseteq L(A),\quad \forall A \in \mathcal{F}(U),$$
$$\text{(fu9)}\quad H(A) \subseteq L(H(A)),\quad \forall A \in \mathcal{F}(U).$$

Theorem 10 implies that a rough fuzzy algebra $(\mathcal{F}(U), \wedge, \vee, \sim, L, H)$ is a reflexive rough fuzzy algebra iff H is an embedding on $\mathcal{F}(U)$ [20,43] and it is a transitive rough fuzzy algebra iff H is closed on $\mathcal{F}(U)$ [20].

Theorem 11. *Suppose that* $(\mathcal{F}(U), \wedge, \vee, \sim, L, H)$ *is a rough fuzzy set algebra. Then it is a Pawlak rough fuzzy set algebra iff* L *satisfies axioms* (fl6), (fl7) *and* (fl8) *or equivalently,* H *satisfies axioms* (fu6), (fu7) *and* (fu8).

Theorem 11 implies that a rough fuzzy algebra $(\mathcal{F}(U), \wedge, \vee, \sim, L, H)$ is a Pawlak rough fuzzy algebra iff H is a symmetric closure operator on $\mathcal{F}(U)$ [14]. It can be proved that axioms (fu6), (fu7) and (fu8) in Theorem 11 can also be replaced by axioms (fu6) and (fu9).

5   Fuzzy Topological Spaces and Rough Fuzzy Approximation Operators

The relationship between topological spaces and rough approximation operators has been studied by many researchers. In [48], Wu examined the relationship between fuzzy topological spaces and fuzzy rough approximation operators. In this section we discuss the relationship between a fuzzy topological space and rough fuzzy approximation operators.

We first introduce some definitions related to fuzzy topology [18].

Definition 7. *A subset τ of $\mathcal{F}(U)$ is referred to as a fuzzy topology on U iff it satisfies*

(1) *If $\mathcal{A} \subseteq \tau$, then $\bigvee\limits_{A \in \mathcal{A}} A \in \tau$,*

(2) *If $A, B \in \tau$, then $A \wedge B \in \tau$,*

(3) *If $\widehat{\alpha} \in \mathcal{F}(U)$ is a constant fuzzy set, then $\widehat{\alpha} \in \tau$.*

Definition 8. *A map $\Psi : \mathcal{F}(U) \to \mathcal{F}(U)$ is referred to as a fuzzy interior operator iff for all $A, B \in \mathcal{F}(U)$ it satisfies:*

(1) *$\Psi(A) \subseteq A$,*

(2) *$\Psi(A \wedge B) = \Psi(A) \wedge \Psi(B)$,*

(3) *$\Psi^2(A) = \Psi(A)$,*

(4) *$\Psi(\widehat{\alpha}) = \widehat{\alpha}, \quad \forall \alpha \in I$.*

Definition 9. *A map $\Phi : \mathcal{F}(U) \to \mathcal{F}(U)$ is referred to as a fuzzy closure operator iff for all $A, B \in \mathcal{F}(U)$ it satisfies:*

(1) *$A \subseteq \Phi(A)$,*

(2) *$\Phi(A \vee B) = \Phi(A) \vee \Phi(B)$,*

(3) *$\Phi^2(A) = \Phi(A)$,*

(4) *$\Phi(\widehat{\alpha}) = \widehat{\alpha}, \quad \forall \alpha \in I$.*

The elements of a fuzzy topology τ are referred to as open fuzzy sets, and it is easy to show that a fuzzy interior operator Ψ defines a fuzzy topology $\tau_\Psi = \{A \in \mathcal{F}(U) : \Psi(A) = A\}$. So, the open fuzzy sets are the fixed points of Ψ.

By using Theorems 7 and 10, we can obtain the following theorem:

Theorem 12. *Assume that R is a binary relation on U. Then the following are equivalent:*

(1) *R is a reflexive and transitive relation;*

(2) *the upper rough fuzzy approximation operator $\Phi = \overline{R} : \mathcal{F}(U) \to \mathcal{F}(U)$ is a fuzzy closure operator;*

(3) *the lower rough fuzzy approximation operator $\Psi = \underline{R} : \mathcal{F}(U) \to \mathcal{F}(U)$ is a fuzzy interior operator.*

Theorem 12 shows that the lower and upper rough fuzzy approximation operators constructed from a reflexive and transitive crisp relation are the fuzzy interior and closure operators respectively. Thus a rough fuzzy set algebra constructed from a reflexive and transitive relation is referred to as rough fuzzy topological set algebra. Theorem 12 implies Theorem 13.

Theorem 13. *Assume that R is a reflexive and transitive crisp relation on U. Then there exists a fuzzy topology τ_R on U such that $\Psi = \underline{R} : \mathcal{F}(U) \to \mathcal{F}(U)$ and $\Phi = \overline{R} : \mathcal{F}(U) \to \mathcal{F}(U)$ are the fuzzy interior and closure operators respectively.*

By using Theorems 7, 8, 10, and 12, we can obtain following Theorems 14 and 15, which illustrate that under certain conditions a fuzzy interior (closure, resp.) operator derived from a fuzzy topological space can be associated with a reflexive and transitive fuzzy relation such that the induced fuzzy lower (upper, resp.) approximation operator is the fuzzy interior (closure, resp.) operator.

Theorem 14. *Let* $\Phi : \mathcal{F}(U) \to \mathcal{F}(U)$ *be a fuzzy closure operator. Then there exists a reflexive and transitive crisp relation on U such that $\overline{R}(A) = \Phi(A)$ for all $A \in \mathcal{F}(U)$ iff Φ satisfies the following three conditions:*

(1) $\Phi(1_y) \in \mathcal{P}(U), \quad \forall y \in U,$

(2) $\Phi(A \vee B) = \Phi(A) \vee \Phi(B), \quad \forall A, B \in \mathcal{F}(U),$

(3) $\Phi(A \wedge \widehat{\alpha}) = \Phi(A) \wedge \widehat{\alpha}, \quad \forall A \in \mathcal{F}(U), \forall \alpha \in I.$

Theorem 15. *Let* $\Psi : \mathcal{F}(U) \to \mathcal{F}(U)$ *be a fuzzy interior operator, then there exists a reflexive and transitive crisp relation on U such that $\underline{R}(A) = \Psi(A)$ for all $A \in \mathcal{F}(U)$ iff Ψ satisfies the following three conditions:*

(1) $\Psi(1_{U-\{y\}}) \in \mathcal{P}(U), \quad \forall y \in U,$

(2) $\Psi(A \wedge B) = \Psi(A) \wedge \Psi(B), \quad \forall A, B \in \mathcal{F}(U),$

(3) $\Psi(A \vee \widehat{\alpha}) = \Psi(A) \vee \widehat{\alpha}, \quad \forall A \in \mathcal{F}(U), \forall \alpha \in I.$

6 Fuzzy Belief Functions and Rough Fuzzy Approximation Operators

In this section, we present results relating evidence theory in the fuzzy environment and rough fuzzy approximation operators.

The basic representational structure in the Dempster-Shafer theory of evidence is a belief structure [8].

Definition 10. *Let W be a nonempty finite universe of discourse. A set function $m : \mathcal{P}(W) \to I = [0,1]$ is referred to as a basic probability assignment if it satisfies*

$$(\text{M1}) \ m(\emptyset) = 0, \qquad (\text{M2}) \ \sum_{A \in \mathcal{P}(W)} m(A) = 1.$$

Let $\mathcal{M} = \{A \in \mathcal{P}(W) : m(A) \neq 0\}$. Then the pair $(\mathcal{M}, m)$ is called a belief structure.

Associated with each belief structure, a pair of fuzzy belief and plausibility functions can be derived [8].

Definition 11. *A fuzzy set function $Bel : \mathcal{F}(W) \to I$ is called a fuzzy belief function iff*

$$Bel(X) = \sum_{A \in \mathcal{M}} m(A) \mathrm{N}_A(X), \quad \forall X \in \mathcal{F}(W),$$

and a fuzzy set function $Pl : \mathcal{F}(W) \to I$ is called a fuzzy plausibility function iff

$$Pl(X) = \sum_{A \in \mathcal{M}} m(A) \Pi_A(X), \quad \forall X \in \mathcal{F}(W),$$

where N_A and Π_A are respectively the fuzzy necessity and fuzzy possibility measures generated by the set A as follows [8,11]:

$$\mathrm{N}_A(X) = \bigwedge_{y \notin A} (1 - X(y)), \forall X \in \mathcal{F}(W),$$

$$\Pi_A(X) = \bigvee_{y \in A} X(y), \qquad \forall X \in \mathcal{F}(W).$$

Fuzzy belief and plausibility functions basing on the same belief structure are connected by the dual property

$$Pl(X) = 1 - Bel(\sim X), \quad \forall X \in \mathcal{F}(W).$$

When X is a crisp subset of W, it can be verified that

$$Bel(X) = \sum_{\{A \in \mathcal{M}: A \subseteq X\}} m(A), \quad Pl(X) = \sum_{\{A \in \mathcal{M}: A \cap X \neq \emptyset\}} m(A).$$

Thus fuzzy belief and plausibility functions are indeed generalizations of classical belief and plausibility functions.

In [49], the connection between a pair of fuzzy belief and plausibility functions derived from a fuzzy belief structure and a pair of lower and upper fuzzy rough approximation operations induced from a fuzzy approximation space was illustrated. Similar to the proof of analogous results in [49], we can prove the following two theorems which state that serial rough fuzzy set algebras can be used to interpret fuzzy belief and plausibility functions derived from a crisp belief structure (see [51]).

Theorem 16. *Let R be a serial relation from U to W, and $\overline{R}(X)$ and $\underline{R}(X)$ the dual pair of upper and lower rough fuzzy approximations of a fuzzy set $X \in \mathcal{F}(W)$ with respect to the approximation space (U, W, R). The qualities of the upper and lower approximations, $\overline{Q}(X)$ and $\underline{Q}(X)$, are defined by*

$$\overline{Q}(X) = |\overline{R}(X)|/|U|, \quad \underline{Q}(X) = |\underline{R}(X)|/|U|.$$

Then $\overline{Q}$ and $\underline{Q}$ are a dual pair of fuzzy plausibility and belief functions on W, and the corresponding basic probability assignment is defined by

$$m(A) = |j(A)|/|U|, \quad A \in \mathcal{P}(W),$$

where j is the basic set assignment induced by R, i.e., $j(A) = \{u \in U : R_s(u) = A\}$ for $A \in \mathcal{P}(W)$. Conversely, if Pl and $Bel : \mathcal{F}(W) \to I$ are a dual pair of fuzzy plausibility and belief functions on W induced by a belief structure $(\mathcal{M}, m)$, with $m(A)$ being a rational number of each $A \in \mathcal{M}$, then there exists a finite universe of discourse U and a serial crisp relation from U to W, such that its induced qualities of the upper and lower approximations satisfy

$$Pl(X) = \overline{Q}(X), \quad Bel(X) = \underline{Q}(X), \forall X \in \mathcal{F}(W).$$

Moreover, if Pl and $Bel : \mathcal{F}(U) \to I$ are a dual pair of fuzzy plausibility and belief functions on U induced by a belief structure $(\mathcal{M}, m)$, with $m(A)$ being equivalent to a rational number with $|U|$ as its denominator for each $A \in \mathcal{M}$, then there exists a serial crisp relation on U such that its induced qualities of upper and lower approximations satisfy

$$Pl(X) = \overline{Q}(X), \quad Bel(X) = \underline{Q}(X), \quad \forall X \in \mathcal{F}(U).$$

Theorem 17. *Let R be a serial relation from U to W and let P be a probability measure on U with $P(x) > 0$ for all $x \in U$. The quadruple $((U, P), W, R)$ is referred as a random approximation space. For $X \in \mathcal{F}(W)$, $\overline{R}(X)$ and $\underline{R}(X)$ are the dual pair of upper and lower rough fuzzy approximations of $X \in \mathcal{F}(W)$ with respect to $((U, P), W, R)$. The random qualities of the upper and lower approximations, $\overline{M}(X)$ and $\underline{M}(X)$, are defined by*

$$\overline{M}(X) = P(\overline{R}(X)), \quad \underline{M}(X) = P(\underline{R}(X)).$$

Then $\overline{M}$ and $\underline{M}$ are a dual pair of fuzzy plausibility and belief functions on W, and the corresponding basic probability assignment is defined by

$$m(A) = P(j(A)), \quad A \in \mathcal{P}(W),$$

where j is the basic set assignment induced by R. Conversely, if Pl and Bel : $\mathcal{F}(W) \to I$ are the dual pair of fuzzy plausibility and belief functions on W induced by a belief structure $(\mathcal{M}, m)$, then there exists a finite universe of discourse U, a probability measure P on U and a serial crisp relation R from U to W, such that its induced random qualities of the upper and lower approximations satisfy

$$Pl(X) = \overline{M}(X), \quad Bel(X) = \underline{M}(X), \quad \forall X \in \mathcal{F}(W).$$

7 Rough Fuzzy Approximation Operators Based on Neighborhood Systems

Studies on the relationships between rough approximation operators and neighborhood systems have been made over the years. In [61], Yao explored the relational interpretations of 1-step neighborhood operators and rough set approximation operators. Wu and Zhang [53] characterized generalized rough approximation operators under k-step neighborhood systems. In this section, we examine the relationships between rough fuzzy approximation operators and k-step neighborhood systems.

Definition 12. *For an arbitrary binary relation R on U and a positive integer k, we define a notion of binary relation R^k, called the k-step-relation of R, as follows:*

$R^1 = R$,
$R^k = \{(x, y) \in U \times U : \text{ there exists } y_1, y_2, \ldots, y_i \in U, \ 1 \leq i \leq k-1, \text{ such that } xRy_1, y_1Ry_2, \ldots, y_iRy\} \cup R^1, \quad k \geq 2$.

It is easy to see that

$$R^{k+1} = R^k \cup \{(x, y) \in U \times U : \text{there exists } y_1, y_2, \ldots, y_k \in U,$$
$$\text{such that } xRy_1, y_1Ry_2, \ldots, y_kRy\}.$$

Obviously, $R^k \subseteq R^{k+1}$, and moreover, $R^k = R^n$ for all $k \geq n$. In fact, R^n is the transitive closure of R. Of course, R^n is transitive.

Definition 13. *Let R be a binary relation. For two elements $x, y \in U$ and $k \geq 1$, if $xR^k y$, then we say that y is R^k-related to x, x is an R^k-predecessor of y, and y is an R^k-successor of x. The set of all R^k-successors of x is denoted by $r_k(x)$, i.e., $r_k(x) = \{y \in U : xR^k y\}$; $r_k(x)$ is also referred to as the k-step neighborhood of x.*

We see that $\{r_k(x) : k \geq 1\}$ is a neighborhood system of x, and $\{r_k(x) : x \in U\}$ is a k-step neighborhood system in the universe of discourse.

The k-step neighborhood system is monotone increasing with respect to k. For two relations R and R', it can be checked that

$$R \subseteq R' \iff r_1(x) \subseteq r_1'(x), \quad \text{for all } x \in U.$$

In particular,

$$r_k(x) \subseteq r_{k+1}(x), \quad \text{for all } k \geq 1 \text{ and all } x \in U.$$

Thus $\{r_k(x) : k \geq 1\}$ is a nested sequence of neighborhood system. It offers a multi-layered granulation of the object x. We can observe that

$$r_k(x) = \{y \in U : \text{there exists } y_1, y_2, \ldots, y_i \in U \text{ such that } xRy_1, y_1Ry_2, \ldots,$$
$$y_iRy, 1 \leq i \leq k-1, \text{or } xRy\}.$$

Evidently,
$$A \subseteq B \implies r_k(A) \subseteq r_k(B), \quad A, B \in \mathcal{P}(U),$$

where $r_k(A) = \cup\{r_k(x) : x \in A\}$.

It can be checked that

$$r_l(r_k(x)) \subseteq r_{k+l}(x), \text{ for all } k, l \geq 1.$$

And if R is Euclidean, then we have [53]

$$r_l(r_k(x)) = r_{k+l}(x), \text{ for all } k, l \geq 1.$$

The relationship between a special type of binary relation R and its induced k-step-relation R^k is summarized as follows [53]:

Theorem 18. *Assume that R is an arbitrary binary relation on U. Then*

$$
\begin{aligned}
R \text{ is serial} \quad &\Longleftrightarrow \quad R^k \text{ is serial for all } k \geq 1; \\
R \text{ is reflexive} \quad &\Longleftrightarrow \quad R^k \text{ is reflexive for all } k \geq 1; \\
R \text{ is symmetric} \quad &\Longleftrightarrow \quad R^k \text{ is symmetric for all } k \geq 1; \\
R \text{ is transitive} \quad &\Longleftrightarrow \quad R^k \text{ is transitive for all } k \geq 1, \text{ and } R^k = R; \\
R \text{ is Euclidean} \quad &\Longleftrightarrow \quad R^k \text{ is Euclidean for all } k \geq 1.
\end{aligned}
$$

Definition 14. *Given an arbitrary binary relation R on the universe of discourse U, for any set $X \subseteq U$ and $k \geq 1$, we may define a pair of lower and* *upper approximations of X with respect to the k-step neighborhood system as follows:*

$$\underline{R^k}(X) = \{x \in U : r_k(x) \subseteq X\},$$
$$\overline{R^k}(X) = \{x \in U : r_k(x) \cap X \neq \emptyset\}.$$

$\underline{R^k}$ and $\overline{R^k}$ *are referred to as the k-step lower and upper approximation operators respectively.*

By [53], it is not difficult to prove that the properties of a binary relation can be equivalently characterized by the properties of multi-step approximation operators.

Theorem 19. *Let R be an arbitrary binary relation on U. Then*

R *is serial*	$\Longleftrightarrow$	(KNL0)	$\underline{R^k}(\emptyset) = \emptyset,$	$k \geq 1,$
	$\Longleftrightarrow$	(KNU0)	$\overline{R^k}(U) = U,$	$k \geq 1,$
	$\Longleftrightarrow$	(KNLU0)	$\underline{R^k}(A) \subseteq \overline{R^k}(A),$	$A \in \mathcal{P}(U), k \geq 1,$
R *is reflexive*	$\Longleftrightarrow$	(KNL6)	$\underline{R^k}(A) \subseteq A,$	$A \in \mathcal{P}(U), k \geq 1,$
	$\Longleftrightarrow$	(KNU6)	$A \subseteq \overline{R^k}(A),$	$A \in \mathcal{P}(U), k \geq 1,$
R *is symmetric*	$\Longleftrightarrow$	(KNL7)	$\overline{R^l}(\underline{R^k}(A)) \subseteq A,$	$A \in \mathcal{P}(U), 1 \leq l \leq k,$
	$\Longleftrightarrow$	(KNU7)	$A \subseteq \underline{R^l}(\overline{R^k}(A)),$	$A \in \mathcal{P}(U), 1 \leq l \leq k,$
R *is transitive*	$\Longleftrightarrow$	(KNL8)	$\underline{R^k}(A) \subseteq \underline{R^l}(\underline{R^k}(A)),$	$A \in \mathcal{P}(U), 1 \leq l \leq k,$
	$\Longleftrightarrow$	(KNU8)	$\overline{R^k}(\overline{R^l}(A)) \subseteq \overline{R^k}(A),$	$A \in \mathcal{P}(U), 1 \leq l \leq k,$
R *is Euclidean*	$\Longleftrightarrow$	(KNL9)	$\overline{R^l}(\underline{R^m}(A)) \subseteq \underline{R^k}(A),$	$A \in \mathcal{P}(U), 1 \leq l \leq k \leq m,$
	$\Longleftrightarrow$	(KNU9)	$\overline{R^k}(A) \subseteq \underline{R^l}(\overline{R^m}(A)),$	$A \in \mathcal{P}(U), 1 \leq l \leq k \leq m.$

Definition 15. *Given an arbitrary binary relation R on U. For any set $X \in \mathcal{F}(U)$ and $k \geq 1$, we may define a pair of lower and upper rough fuzzy approximations of X with respect to the k-step neighborhood system as follows:*

$$\underline{R^k}(X)(x) = \bigwedge_{y \in r_k(x)} X(y), \quad x \in U,$$
$$\overline{R^k}(X)(x) = \bigvee_{y \in r_k(x)} X(y), \quad x \in U.$$

$\underline{R^k}$ and $\overline{R^k}$ *are referred to as the k-step lower and upper rough fuzzy approximation operators respectively.*

If $1 \leq l \leq k$, it is easy to verify that

(1) $\underline{R^k}(A) \subseteq \underline{R^l}(A), \forall A \in \mathcal{F}(U),$
(2) $\overline{R^l}(A) \subseteq \overline{R^k}(A), \forall A \in \mathcal{F}(U).$

In terms of Theorems 7, 18, and 19, we can verified the following theorem which shows that the structure of information granulation generated via R can be equivalently characterized by the properties of multi-step rough fuzzy approximation operators.

Theorem 20. *Let R be an arbitrary binary crisp relation on U. Then*

$$
\begin{aligned}
R \text{ is serial} \quad &\Longleftrightarrow \text{(KFL0)} \quad \underline{R}^k(\emptyset) = \emptyset, & k \geq 1, \\
&\Longleftrightarrow \text{(KFU0)} \quad \overline{R}^k(U) = U, & k \geq 1, \\
&\Longleftrightarrow \text{(KFLU0)} \underline{R}^k(A) \subseteq \overline{R}^k(A), & k \geq 1, \\
&\Longleftrightarrow \text{(KFL0)}' \quad \underline{R}^k(\widehat{\alpha}) = \widehat{\alpha}, & \alpha \in I, k \geq 1, \\
&\Longleftrightarrow \text{(KFU0)}' \quad \overline{R}^k(\widehat{\alpha}) = \widehat{\alpha}, & \alpha \in I, k \geq 1, \\
R \text{ is reflexive} \quad &\Longleftrightarrow \text{(KFL6)} \quad \underline{R}^k(A) \subseteq A, & A \in \mathcal{F}(U), k \geq 1, \\
&\Longleftrightarrow \text{(KFU6)} \quad A \subseteq \overline{R}^k(A), & A \in \mathcal{F}(U), k \geq 1, \\
R \text{ is symmetric} &\Longleftrightarrow \text{(KFL7)} \quad \overline{R}^l(\underline{R}^k(A)) \subseteq A, & A \in \mathcal{F}(U), 1 \leq l \leq k, \\
&\Longleftrightarrow \text{(KFU7)} \quad A \subseteq \underline{R}^l(\overline{R}^k(A)), & A \in \mathcal{F}(U), 1 \leq l \leq k, \\
R \text{ is transitive} &\Longleftrightarrow \text{(KFL8)} \quad \underline{R}^k(A) \subseteq \underline{R}^l(\underline{R}^k(A)), & A \in \mathcal{F}(U), 1 \leq l \leq k, \\
&\Longleftrightarrow \text{(KFU8)} \quad \overline{R}^k(\overline{R}^l(A)) \subseteq \overline{R}^k(A), & A \in \mathcal{F}(U), 1 \leq l \leq k, \\
R \text{ is Euclidean} &\Longleftrightarrow \text{(KFL9)} \quad \overline{R}^l(\underline{R}^m(A)) \subseteq \underline{R}^k(A), & A \in \mathcal{F}(U), 1 \leq l \leq k \leq m, \\
&\Longleftrightarrow \text{(KFU9)} \quad \overline{R}^k(A) \subseteq \underline{R}^l(\overline{R}^m(A)), & A \in \mathcal{F}(U), 1 \leq l \leq k \leq m.
\end{aligned}
$$

8   Conclusion

In this paper, we have reviewed and studied generalized rough fuzzy approximation operators. In our constructive method, generalized rough fuzzy sets are derived from a crisp approximation space. By the representation theorem, rough fuzzy approximation operators can be composed by a family of crisp approximation operators. By the axiomatic approach, rough fuzzy approximation operators can be characterized by axioms. Axiom sets of fuzzy approximation operators guarantee the existence of certain types of crisp relations producing the same operators. We have also established the relationship between rough fuzzy approximation operators and fuzzy topological spaces. Moreover, relationships between rough fuzzy approximation operators and fuzzy belief and fuzzy plausibility functions have been established. Multi-step rough fuzzy approximation operators can also be obtained by a neighborhood system derived from a binary relation. The relationships between binary relations and multi-step rough fuzzy approximation operators have been examined. This work may be viewed as the extension of Yao [59,60,61,64], and it may also be treated as a completion of Thiele [42,43,44]. It appears that our constructive approaches will turn out to be more useful for practical applications of the rough set theory while the axiomatic approaches will help us to gain much more insights into the mathematical structures of fuzzy approximation operators. Proving the independence of axiom sets is still an open problem. That is to say, finding the minimal axiom sets to characterize various rough fuzzy set algebras is still an outstanding problem.

Acknowledgement

This work was supported by a grant from the National Natural Science Foundation of China (No.60373078) and a grant from the Major State Basic Research Development Program of China (973 Program No.2002CB312200).

References

1. Bodjanova, S.: Approximation of a fuzzy concepts in decision making. *Fuzzy Sets and Systems.* **85** (1997) 23–29
2. Boixader, D., Jacas, J., Recasens, J.: Upper and lower approximations of fuzzy sets. *International Journal of General Systems.* **29**(4) (2000) 555–568
3. Chakrabarty, K., Biswas, R., Nanda, S.: Fuzziness in rough sets. *Fuzzy Sets and Systems.* **110** (2000) 247–251
4. Chuchro, M.: A certain conception of rough sets in topological Boolean algebras. *Bulletin of the Section of Logic.* **22**(1) (1993) 9–12
5. Chuchro, M.: On rough sets in topological Boolean algebras. In: Ziarko, W., Ed., *Rough Sets, Fuzzy Sets and Knowledge Discovery.* Springer-Verlag, Berlin (1994) 157–160
6. Comer, S.: An algebraic approach to the approximation of information. *Fundamenta Informaticae.* **14** (1991) 492–502
7. Dempster, A. P.: Upper and lower probabilities induced by a multivalued mapping. *Annals of Mathematical Statistics.* **38** (1967) 325–339
8. Denoeux, T.: Modeling vague beliefs using fuzzy-valued belief structures. *Fuzzy Sets and Systems.* **116** (2000) 167–199
9. Dubois, D., Prade, H.: Twofold fuzzy sets and rough sets—some issues in knowledge representation. *Fuzzy Sets and Systems.* **23** (1987) 3–18
10. Dubois, D., Prade, H.: Rough fuzzy sets and fuzzy rough sets. *International Journal of General Systems.* **17** (1990) 191–208
11. Dubois, D., Prade, H.: *Fundamentals of Fuzzy Sets.* Kluwer Academic Publishers, Boston (2000)
12. Inuiguchi, M., Hirano, S., Tsumoto, S., Eds.: *Rough Set Theory and Granular Computing.* Springer, Berlin (2003)
13. Kortelainen, J.: On relationship between modified sets, topological space and rough sets. *Fuzzy Sets and Systems.* **61** (1994) 91–95
14. Kuncheva, L. I.: Fuzzy rough sets: application to feature selection. *Fuzzy Sets and Systems.* **51** (1992) 147–153
15. Lin, T. Y.: Neighborhood systems—application to qualitative fuzzy and rough sets. In: Wang, P. P., Ed., *Advances in Machine Intelligence and Soft-Computing.* Department of Electrical Engineering, Duke University, Durham, NC (1997) 132–155
16. Lin, T. Y., Liu, Q.: Rough approximate operators: axiomatic rough set theory. In: Ziarko, W., Ed., *Rough Sets, Fuzzy Sets and Knowledge Discovery.* Springer, Berlin (1994) 256–260
17. Lin, T. Y., Yao, Y. Y., Zadeh, L. A., Eds.: *Data Mining, Rough Sets and Granular Computing.* Physica-Verlag, Heidelberg (2002)
18. Lowen, R.: Fuzzy topological spaces and fuzzy compactness. *Journal of Mathematical Analysis and Applications.* **56** (1976) 621–633
19. Mi, J.-S., Leung, Y., Wu, W.-Z.: An uncertainty measure in partition-based fuzzy rough sets. *International Journal of General Systems.* **34**(1) (2005) 77–90
20. Mi, J.-S., Zhang, W.-X.: An axiomatic characterization of a fuzzy generalization of rough sets. *Information Sciences.* **160** (2004) 235–249
21. Morsi, N. N., Yakout, M. M.: Axiomatics for fuzzy rough sets. *Fuzzy Sets and Systems.* **100** (1998) 327–342
22. Nakamura, A., Gao, J. M.: On a KTB-modal fuzzy logic. *Fuzzy Sets and Systems.* **45** (1992) 327–334

23. Nanda, S., Majumda, S.: Fuzzy rough sets. *Fuzzy Sets and Systems.* **45** (1992) 157–160
24. Pal, S. K.: Roughness of a fuzzy set. *Information Sciences.* **93** (1996) 235–246
25. Pawlak, Z.: Information systems, theoretical foundations. *Information Systems.* **6** (1981) 205–218
26. Pawlak, Z.: Rough sets. *International Journal of Computer and Information Science.* **11** (1982) 341–356
27. Pawlak, Z.: *Rough Sets: Theoretical Aspects of Reasoning about Data.* Kluwer Academic Publishers, Boston (1991)
28. Pei, D. W., Xu, Z.-B.: Rough set models on two universes. *International Journal of General Systems.* **33** (2004) 569–581
29. Peters, J. F., Pawlak, Z., Skowron, A.: A rough set approach to measuring information granules. In: Proceedings of COMPSAC 2002, 1135–1139
30. Peters, J. F., Skowron, A., Synak, P., Ramanna, S.: Rough sets and information granulation. Lecture Notes in Artificial Intelligence 2715. Springer, Berlin (2003) 370–377
31. Pomykala, J. A.: Approximation operations in approximation space. *Bulletin of the Polish Academy of Sciences: Mathematics.* **35** (1987) 653–662
32. Radzikowska, A. M., Kerre, E. E.: A comparative study of fuzzy rough sets. *Fuzzy Sets and Systems.* **126** (2002) 137–155
33. Shafer, G.: *A Mathematical Theory of Evidence.* Princeton University Press, Princeton (1976)
34. Skowron, A.: The relationship between the rough set theory and evidence theory. *Bulletin of Polish Academy of Science: Mathematics.* **37** (1989) 87–90
35. Skowron, A.: The rough sets theory and evidence theory. *Fundamenta Informatica.* **13** (1990) 245–162
36. Skowron, A., Grzymala-Busse, J.: From rough set theory to evidence theory. In: Yager, R., et al., Eds., *Advances in the Dempster-Shafer Theory of Evidence.* Wiley, New York (1994) 193–236
37. Skowron, A., Stepaniuk, J.: Information granules: towards foundations of granular computing. *International Journal of Intelligent Systems.* **16** (2001) 57–85
38. Skowron, A., Swiniarski, R., Synak, P.: Approximation spaces and information granulation. *Transactions on Rough Sets.* **III.** Lecture Notes in Computer Science 3400. Springer, Berlin (2005) 175–189
39. Slowinski, R., Vanderpooten, D.: Similarity relation as a basis for rough approximations. In: Wang, P. P., Ed., *Advances in Machine Intelligence and Soft-Computing.* Department of Electrical Engineering, Duke University, Durham, NC (1997) 17–33
40. Slowinski, R., Vanderpooten, D.: A Generalized definition of rough approximations based on similarity. *IEEE Transactions on Knowledge and Data Engineering.* **12**(2) (2000) 331–336
41. Thiele, H.: On axiomatic characterisations of crisp approximation operators. *Information Sciences.* **129** (2000) 221–226
42. Thiele, H.: On axiomatic characterisation of fuzzy approximation operators I, the fuzzy rough set based case. In: Proceedings of RSCTC, Banff Park Lodge, Bariff, Canada (2000) 239–247
43. Thiele, H.: On axiomatic characterisation of fuzzy approximation operators II, the rough fuzzy set based case. In: Proceedings of the 31st IEEE International Symposium on Multiple-Valued Logic, (2001) 330–335
44. Thiele, H.: On axiomatic characterization of fuzzy approximation operators III, the fuzzy diamond and fuzzy box cases. In: The 10th IEEE International Conference on Fuzzy Systems, Vol. 2 (2001) 1148–1151

45. Vakarelov, D.: A modal logic for similarity relations in Pawlak knowledge representation systems. *Fundamenta Informaticae.* **15** (1991) 61–79
46. Wasilewska, A.: Conditional knowledge representation systems—model for an implementation. *Bulletin of the Polish Academy of Sciences: Mathematics.* **37** (1990) 63–69
47. Wiweger, R.: On topological rough sets. *Bulletin of Polish Academy of Sciences: Mathematics.* **37** (1989) 89–93
48. Wu, W.-Z.: A study on relationship between fuzzy rough approximation operators and fuzzy topological spaces. In: Wang, L., Jin, Y., Eds., The 2nd International Conference on Fuzzy Systems and Knowledge Discovery. Lecture Notes in Artificial Intelligence 3613. Springer, Berlin (2005) 167–174
49. Wu, W.-Z.: Upper and lower probabilities of fuzzy events induced by a fuzzy set-valued mapping. In: Slezak, D., et al., Eds., The 10th International Conference on Rough Sets, Fuzzy Sets, Data Mining, and Granular Computing. Lecture Notes in Artificial Intelligence 3641. Springer, Berlin (2005) 345–353
50. Wu, W.-Z., Leung, Y., Mi, J.-S.: On characterizations of $(\mathcal{I}, \mathcal{T})$-fuzzy rough approximation operators. *Fuzzy Sets and Systems.* **154**(1) (2005) 76–102
51. Wu, W.-Z., Leung, Y., Zhang, W.-X.: Connections between rough set theory and Dempster-Shafer theory of evidence. *International Journal of General Systems.* **31**(4) (2002) 405–430
52. Wu, W.-Z., Mi, J.-S., Zhang, W.-X.: Generalized fuzzy rough sets. *Information Sciences.* **151** (2003) 263–282
53. Wu, W.-Z., Zhang, W.-X.: Neighborhood operator systems and approximations. *Information Sciences.* **144** (2002) 201–217
54. Wu., W.-Z., Zhang, W.-X.: Constructive and axiomatic approaches of fuzzy approximation operators. *Information Sciences.* **159** (2004) 233–254
55. Wu., W.-Z., Zhang, W.-X., Li, H.-Z.: Knowledge acquisition in incomplete fuzzy information systems via rough set approach. *Expert Systems.* **20**(5) (2003) 280–286
56. Wybraniec-Skardowska, U.: On a generalization of approximation space. *Bulletin of the Polish Academy of Sciences: Mathematics.* **37** (1989) 51–61
57. Yao, Y. Y.: Two views of the theory of rough sets in finite universes. *International Journal of Approximate Reasoning.* **15** (1996) 291–317
58. Yao, Y. Y.: Combination of rough and fuzzy sets based on α-level sets. In: Lin, T. Y., Cercone, N., Eds., *Rough Sets and Data Mining: Analysis for Imprecise Data.* Kluwer Academic Publishers, Boston (1997) 301–321
59. Yao, Y. Y.: Constructive and algebraic methods of the theory of rough sets. *Journal of Information Sciences.* **109** (1998) 21–47
60. Yao, Y. Y.: Generalized rough set model. In: Polkowski, L., Skowron, A., Eds., *Rough Sets in Knowledge Discovery 1. Methodology and Applications.* Physica-Verlag, Heidelberg (1998) 286–318
61. Yao, Y. Y.: Relational interpretations of neighborhood operators and rough set approximation operators. *Information Sciences.* **111** (1998) 239–259
62. Yao, Y.Y.: Information granulation and rough set approximation. *International Journal of Intelligent Systems.* **16** (2001) 87–104
63. Yao, Y. Y., Lin, T. Y.: Generalization of rough sets using modal logic. *Intelligent Automation and Soft Computing, an International Journal.* **2** (1996) 103–120
64. Yao, Y. Y., Lingras, P. J.: Interpretations of belief functions in the theory of rough sets. *Information Sciences.* **104** (1998) 81–106
65. Yen, J.: Generalizing the Dempster-Shafer theory to fuzzy sets. *IEEE Transaction on Systems, Man, and Cybernetics.* **20**(3) (1990) 559–570

66. Zadeh, L. A.: Probability measures of fuzzy events. *Journal of Mathematical Analysis and Applications.* **23** (1968) 421–427
67. Zadeh, L. A.: Fuzzy sets and information granularity. In: Gupta, M. M., Ragade, R. K., Yager, R. R., Eds, *Advances in Fuzzy Set Theory and Applications.* North-Holland, Amsterdam (1979) 3–18
68. Zadeh, L. A.: Towards a theory of fuzzy information granulation and its centrality in human reasoning and fuzzy logic. *Fuzzy Sets and Systems.* **19** (1997) 111–127
69. Zakowski, W.: On a concept of rough sets. *Demonstratio Mathematica.* **15** (1982) 1129–1133
70. Zhang, M., Wu, W.-Z.: Knowledge reductions in information systems with fuzzy decisions. *Journal of Engineering Mathematics.* **20**(2) (2003) 53–58

Rough Set Approximations in Formal Concept Analysis

Yiyu Yao and Yaohua Chen

Department of Computer Science, University of Regina
Regina, Saskatchewan, Canada S4S 0A2
{yyao, chen115y}@cs.uregina.ca

Abstract. A basic notion shared by rough set analysis and formal concept analysis is the definability of a set of objects based on a set of properties. The two theories can be compared, combined and applied to each other based on definability. In this paper, the notion of rough set approximations is introduced into formal concept analysis. Rough set approximations are defined by using a system of definable sets. The similar idea can be used in formal concept analysis. The families of the sets of objects and the sets of properties established in formal concept analysis are viewed as two systems of definable sets. The approximation operators are then formulated with respect to the systems. Two types of approximation operators, with respect to lattice-theoretic and set-theoretic interpretations, are studied. The results provide a better understanding of data analysis using rough set analysis and formal concept analysis.

1 Introduction

Definability deals with whether and how a set can be defined in order to be analyzed and computed [38]. A comparative examination of rough set analysis and formal concept analysis shows that each of them deals with a particular type of definability. While formal concept analysis focuses on sets of objects that can be defined by conjunctions of properties, rough set analysis focuses on disjunction of properties [33]. The common notion of definability links the two theories together. One can immediately adopt ideas from one to the other [33,34]. On the one hand, the notions of formal concepts and formal concept lattices can be introduced into rough set analysis by considering different types of formal concepts [34]. On the other hand, rough set approximation operators can be introduced into formal concept analysis by considering a different type of definability [8,35]. The combination of the two theories would produce new tools for data analysis.

An underlying notion of rough set analysis is the indiscernibility of objects [12,13]. By modelling indiscernibility as an equivalence relation, one can partition a finite universe of objects into a family of pair-wise disjoint subsets called a partition. The partition provides a granulated view of the universe. An equivalence class is considered as a whole, instead of many individuals, and is viewed as an elementary definable subset. In other words, one can only observe, measure, or characterize the equivalence classes.

J.F. Peters and A. Skowron (Eds.): Transactions on Rough Sets V, LNCS 4100, pp. 285–305, 2006.
© Springer-Verlag Berlin Heidelberg 2006

The empty set and unions of equivalence classes are also treated as definable subsets. In general, the system of such definable subsets is only a proper subset of the power set of the universe. Consequently, an arbitrary subset of universe may not necessarily be definable. It can be approximated from below and above by a pair of maximal and minimal definable subsets.

Under the rough set approximation, there is a close connection between definability and approximation. A definable set of the universe of objects must have the same approximations [2]. That is, a set of objects is definable if and only if its lower approximation equals to its upper approximation.

Formal concept analysis is developed based on a formal context given by a binary relation between a set of objects and a set of properties. From a formal context, one can construct (objects, properties) pairs known as the formal concepts [6,22]. The set of objects of a formal concept is referred to as the extension, and the set of properties as the intension. They uniquely determine each other. The family of all formal concepts is a complete lattice. The extension of a formal concept can be viewed as a definable set of objects, although in a sense different from that of rough set analysis [33,34]. In fact, the extension of a formal concept is a set of indiscernible objects with respect to the intension. Based on the properties in the intension, all objects in the extension cannot be distinguished. Furthermore, all objects in the extension share all the properties in the intension. The collection of all the extensions, sets of objects, can be considered as a different system of definable sets [35]. An arbitrary set of objects may not be an extension of a formal concept. The sets of objects that are not extensions of formal concepts are regarded as undefinable sets. Therefore, in formal concept analysis, a different type of definability is proposed.

Saquer and Deogun proposed to approximate a set of objects, a set of properties, and a pair of a set of objects and a set of properties, based on a formal concept lattice [16,17]. Hu *et al.* proposed a method to approximate a set of objects and a set of properties by using join- and meet-irreducible formal concepts with respect to set-theoretic operations [8]. However, their formulations are slightly flawed and fail to achieve such a goal. It stems from a mixed-up of the lattice-theoretic operators and set-theoretic operators. To avoid their limitation, a clear separation of two types of approximations is needed. In this paper, we propose a framework to examine the issues of rough set approximations within formal concept analysis. We concentrate on the interpretations and formulations of various notions. Two systems are examined for the definitions of approximations, the formal concept lattice and the system of extensions of all formal concepts.

The rest of the paper is organized as follows. In Section 2, we discuss three formulations of rough set approximations, subsystem based formulation, granule based formulation and element based formulation. In Section 3, formal concept analysis is reviewed. In Section 4, we apply the notion of rough set approximations into formal concept analysis. Two systems of definable sets are established. Based on each system, different definitions of approximations are examined. Section 5 discusses the existing studies and investigates their differences and connections from the viewpoint of approximations.

2   Rough Set Approximations

The rough set theory is an extension of classical set theory with two additional approximation operators [28]. It is a useful theory and tool for data analysis. Various formulations of rough set approximations have been proposed and studied [30,31,32]. In this section, we review the subsystem based formulation, granule based formulation and element based formulation, respectively. In the subsystem based formulation, a subsystem of the power set of a universe is first constructed and the approximation operators are then defined using the subsystem. In the granule based formulation, equivalence classes are considered as the elementary definable sets, and approximations can be defined directly by using equivalence classes. In the element based formulation, the individual objects in the equivalence classes are used to calculate approximations of a set of objects.

Suppose U is a finite and nonempty universe of objects. Let $E \subseteq U \times U$ be an equivalence relation on U. The equivalence relation divides the universe into a family of pair-wise disjoint subsets, called the partition of the universe and denoted by U/E. The pair $apr = (U, E)$ is referred to as an approximation space.

An approximation space induces a granulated view of the universe. For an object $x \in U$, the equivalence class containing x is given by:

$$[x]_E = \{y \mid xEy\}. \tag{1}$$

Objects in $[x]_E$ are indistinguishable from x. One is therefore forced to consider $[x]_E$ as a whole. In other words, under an equivalence relation, equivalence classes are the smallest non-empty observable, measurable, or definable subsets of U. By extending the definability of equivalence classes, we assume that the empty set and unions of some equivalence classes are definable. The family of definable subsets contains the empty set $\emptyset$ and is closed under set complement, intersection and union. It is an σ-algebra $\sigma(U/E) \subseteq 2^U$ with basis U/E, where 2^U is the power set of U.

A set of objects not in $\sigma(U/E)$ is said to be undefinable. An undefinable set must be approximated from below and above by a pair of definable sets.

Definition 1. (Subsystem based definition) *In an approximation space $apr = (U, E)$, a pair of approximation operators, $\underline{apr}, \overline{apr} : 2^U \longrightarrow 2^U$, is defined by:*

$$\underline{apr}(A) = \bigcup \{X \mid X \in \sigma(U/E), X \subseteq A\},$$
$$\overline{apr}(A) = \bigcap \{X \mid X \in \sigma(U/E), A \subseteq X\}. \tag{2}$$

The lower approximation $\underline{apr}(A) \in \sigma(U/E)$ is the greatest definable set contained in A, and the upper approximation $\overline{apr}(A) \in \sigma(U/E)$ is the least definable set containing A.

Alternatively, the approximation operators can also be defined by using equivalence classes.

Definition 2. (Granule based definition) *In an approximation space* $apr = (U, E)$, *a pair of approximation operators,* $\underline{apr}, \overline{apr} : 2^U \longrightarrow 2^U$, *is defined by:*

$$\underline{apr}(A) = \bigcup\{[x]_E \mid [x]_E \in U/E, [x]_E \subseteq A\},$$
$$\overline{apr}(A) = \bigcup\{[x]_E \mid [x]_E \in U/E, A \cap [x]_E \neq \emptyset\}. \tag{3}$$

The lower approximation is the union of equivalence classes that are subsets of A, and the upper approximation is the union of equivalence classes that have a non-empty intersection with A.

The element based definition is another way to define the lower and upper approximations of a set of objects.

Definition 3. (Element based definition) *In an approximation space* $apr = (U, E)$, *a pair of approximation operators,* $\underline{apr}, \overline{apr} : 2^U \longrightarrow 2^U$, *is defined by:*

$$\underline{apr}(A) = \{x \mid x \in U, [x]_E \subseteq A\},$$
$$\overline{apr}(A) = \{x \mid x \in U, A \cap [x]_E \neq \emptyset\}. \tag{4}$$

The lower approximation is the set of objects whose equivalence classes are subsets of A. The upper approximation is the set of objects whose equivalence classes have non-empty intersections with A.

The three formulations are equivalent, but with different forms and interpretations [32]. The lower and upper approximation operators have the following properties: for sets of objects A, A_1 and A_2,

(i). $\underline{apr}(A) = (\overline{apr}(A^c))^c$,
$\overline{apr}(A) = (\underline{apr}(A^c))^c$;

(ii). $\underline{apr}(A_1 \cap A_2) = \underline{apr}(A_1) \cap \underline{apr}(A_2)$,
$\overline{apr}(A_1 \cup A_2) = \overline{apr}(A_1) \cup \overline{apr}(A_2)$;

(iii). $\underline{apr}(A) \subseteq A \subseteq \overline{apr}(A)$;

(iv). $\underline{apr}(\underline{apr}(A)) = \underline{apr}(A)$,
$\overline{apr}(\overline{apr}(A)) = \overline{apr}(A)$;

(v). $\underline{apr}(\overline{apr}(A)) = \overline{apr}(A)$,
$\overline{apr}(\underline{apr}(A)) = \underline{apr}(A)$.

Property (i) states that the approximation operators are dual operators with respect to set complement ^{c}. Property (ii) states that the lower approximation operator is distributive over set intersection $\cap$, and the upper approximation operator is distributive over set union $\cup$. By property (iii), a set lies within its lower and upper approximations. Properties (iv) and (v) deal with the compositions of lower and upper approximation operators. The result of the composition of a sequence of lower and upper approximation operators is the same as the application of the approximation operator closest to A.

As shown by the following theorem, the approximation operators truthfully reflect the intuitive understanding of the notion of definability [12,35].

Theorem 1. *In an approximation space apr $= (U, E)$, for a set of objects A, $\underline{apr}(A) =$ $\overline{apr}(A)$ if and only if $A \in \sigma(U/E)$.*

An important implication of the theorem is that for an undefinable set $A \subseteq U$, we have $\underline{apr}(A) \neq \overline{apr}(A)$. In fact, $\underline{apr}(A)$ is a proper subset of $\overline{apr}(A)$, namely, $\underline{apr}(A) \subset \overline{apr}(A)$.

The basic ideas of subsystem based formulation can be generalized by considering different subsystems that represent different types of definability [35]. The granule based formulation and element based formulation can also be generalized by using different types of definable granules [29,32,37].

3   Formal Concept Analysis

Formal concept analysis deals with visual presentation and analysis of data [6,22]. It focuses on the definability of a set of objects based on a set of properties, and vice versa.

Let U and V be any two finite sets. Elements of U are called objects, and elements of V are called properties. The relationships between objects and properties are described by a binary relation R between U and V, which is a subset of the Cartesian product $U \times V$. For a pair of elements $x \in U$ and $y \in V$, if $(x, y) \in R$, written as xRy, we say that x has the property y, or the property y is possessed by object x. The triplet (U, V, R) is called a formal context. By the terminology of rough set analysis, a formal context is in fact a binary information table.

Based on the binary relation, we associate a set of properties to an object. An object $x \in U$ has the set of properties:

$$xR = \{y \in V \mid xRy\} \subseteq V. \tag{5}$$

Similarly, a property y is possessed by the set of objects:

$$Ry = \{x \in U \mid xRy\} \subseteq U. \tag{6}$$

By extending these notations, we can establish relationships between sets of objects and sets of properties. This leads to two operators, one from 2^U to 2^V and the other from 2^V to 2^U.

Definition 4. *Suppose (U, V, R) is a formal context. For a set of objects $A \subseteq U$, we associate it with a set of properties:*

$$\begin{aligned} A^* &= \{y \in V \mid \forall x \in U (x \in A \Longrightarrow xRy)\} \\ &= \{y \in V \mid A \subseteq Ry\} \\ &= \bigcap_{x \in A} xR. \end{aligned} \tag{7}$$

For a set of properties $B \subseteq V$, we associate it with a set of objects:

$$\begin{aligned} B^* &= \{x \in U \mid \forall y \in V (y \in B \Longrightarrow xRy)\} \\ &= \{x \in U \mid B \subseteq xR\} \\ &= \bigcap_{y \in B} Ry. \end{aligned} \tag{8}$$

For simplicity, the same symbol is used for both operators. The actual role of the operators can be easily seen from the context.

By definition, $\{x\}^* = xR$ is the set of properties possessed by x, and $\{y\}^* = Ry$ is the set of objects having property y. For a set of objects A, A^* is the *maximal* set of properties shared by *all* objects in A. For a set of properties B, B^* is the *maximal* set of objects that have *all* properties in B.

The operators * have the following properties [6,22]: for $A, A_1, A_2 \subseteq U$ and $B, B_1, B_2 \subseteq V$,

$$\text{(1).} \quad A_1 \subseteq A_2 \Longrightarrow A_1^* \supseteq A_2^*,$$
$$B_1 \subseteq B_2 \Longrightarrow B_1^* \supseteq B_2^*,$$
$$\text{(2).} \quad A \subseteq A^{**},$$
$$B \subseteq B^{**},$$
$$\text{(3).} \quad A^{***} = A^*,$$
$$B^{***} = B^*,$$
$$\text{(4).} \quad (A_1 \cup A_2)^* = A_1^* \cap A_2^*,$$
$$(B_1 \cup B_2)^* = B_1^* \cap B_2^*.$$

In formal concept analysis, one is interested in a pair of a set of objects and a set of properties that uniquely define each other. More specifically, for $(A, B) = (B^*, A^*)$, we have [33]:

$$x \in A \Longleftrightarrow x \in B^*$$
$$\Longleftrightarrow B \subseteq xR$$
$$\Longleftrightarrow \bigwedge_{y \in B} xRy;$$

$$\bigwedge_{x \in A} xRy \Longleftrightarrow A \subseteq Ry$$
$$\Longleftrightarrow y \in A^*$$
$$\Longleftrightarrow y \in B. \tag{9}$$

That is, the set of objects A is defined based on the set of properties B, and vice versa. This type of definability leads to the introduction of the notion of formal concepts [6,22].

Definition 5. *A pair* (A, B), $A \subseteq U$, $B \subseteq V$, *is called a formal concept of the context* (U, V, R), *if* $A = B^*$ *and* $B = A^*$. *Furthermore,* $extent(A, B) = A$ *is called the extension of the concept, and* $intent(A, B) = B$ *is called the intension of the concept.*

Definition 6. *For an object* x, *the pair* $(\{x\}^{**}, \{x\}^*)$ *is a formal concept and called an object concept. For a property* y, *the pair* $(\{y\}^*, \{y\}^{**})$ *is a formal concept and called a property concept.*

The set of all formal concepts forms a complete lattice called a concept lattice, denoted by $L(U, V, R)$ or simply L. The meet and join of the lattice are characterized by the following basic theorem of concept lattices [6,22].

Theorem 2. *The formal concept lattice L is a complete lattice in which the meet and join are given by:*

$$\bigwedge_{t \in T} (A_t, B_t) = (\bigcap_{t \in T} A_t, (\bigcup_{t \in T} B_t)^{**}),$$

$$\bigvee_{t \in T} (A_t, B_t) = ((\bigcup_{t \in T} A_t)^{**}, \bigcap_{t \in T} B_t). \tag{10}$$

where T is an index set and for every $t \in T$, (A_t, B_t) is a formal concept.

The order relation of the lattice can be defined based on the set inclusion relation [6,22].

Definition 7. *For two formal concepts (A_1, B_1) and (A_2, B_2), (A_1, B_1) is a sub-con-cept of (A_2, B_2), written $(A_1, B_1) \preceq (A_2, B_2)$, and (A_2, B_2) is a super-concept of (A_1, B_1), if and only if $A_1 \subseteq A_2$, or equivalently, if and only if $B_2 \subseteq B_1$.*

A more general (specific) concept is characterized by a larger (smaller) set of objects that share a smaller (larger) set of properties.

The lattice-theoretic operators of meet ($\wedge$) and join ($\vee$) of the concept lattice are defined based on the set-theoretic operators of intersection ($\cap$), union ($\cup$) and the operators $*$. However, they are not the same. An intersection of extensions (intensions) of a family of formal concepts is the extension (intension) of a formal concept. A union of extensions (intensions) of a family of formal concepts is not necessarily the extension (intension) of a formal concept.

Example 1. The ideas of formal concept analysis can be illustrated by an example taken from [35]. Table 1 gives a formal context, where the meaning of each property is given as follows: a: needs water to live; b: lives in water; c: lives on land; d: needs chlorophyll to produce food; e: two seed leaves; f: one seed leaf; g: can move around; h: has limbs; i: suckles its offspring. Figure 1 gives the corresponding concept lattice. Consider two formal concepts $(\{3, 6\}, \{a, b, c\})$ and $(\{5, 6, 7, 8\}, \{a, d\})$. Their meet is the formal concept:

$$(\{3, 6\} \cap \{5, 6, 7, 8\}, (\{a, b, c\} \cup \{a, d\})^{**}) = (\{6\}, \{a, b, c, d, f\}),$$

and their join is the formal concept:

$$((\{3, 6\} \cup \{5, 6, 7, 8\})^{**}, \{a, b, c\} \cap \{a, d\}) = (\{1, 2, 3, 4, 5, 6, 7, 8\}, \{a\}).$$

The intersection of extensions of two concepts is the extension of their meet, and the intersection of the intensions is the intension of their join. On the other hand, the union of extensions of the two concepts is $\{3, 5, 6, 7, 8\}$, which is not the extension of any formal concept. The union of the intensions is $\{a, b, c, d\}$, which is not the intension of any formal concept.

Table 1. A formal context taken from [6]

	a	b	c	d	e	f	g	h	i
1. Leech	×	×					×		
2. Bream	×	×					×	×	
3. Frog	×	×	×				×	×	
4. Dog	×		×				×	×	×
5. Spike-weed	×	×		×		×			
6. Reed	×	×	×	×		×			
7. Bean	×		×	×	×				
8. Maize	×		×	×		×			

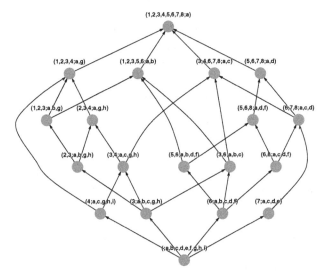

Fig. 1. Concept lattice for the context of Table 1, produced by "Formal Concept Calculator" (developed by Sören Auer, http://www.advis.de/soeren/fca/)

4   Approximations in Formal Concept Analysis

A formal concept is a pair of a definable set of objects and a definable set of properties, which uniquely determine each other. The concept lattice is the family of all concepts with respect to a formal context. Given an arbitrary subset of the universe of objects, it may not be the extension of a formal concept. The set can therefore be viewed as an undefinable set of objects. Following rough sets analysis, such a subset of the universe of objects can be approximated by definable sets of objects, namely, the extensions of formal concepts.

4.1   Approximations Based on Lattice-Theoretic Operators

One can develop the approximation operators similar to the subsystem based formulation of rough set analysis. The concept lattice is used as the system of definable concepts, and lattice-theoretic operators are used to define approximation operators.

For a set of objects $A \subseteq U$, suppose we want to approximate it by the extensions of a pair of formal concepts in the concept lattice. We can extend Definition 1 to achieve this goal. In equation (2), set-theoretic operators $\cap$ and $\cup$ are replaced by lattice-theoretic operators $\wedge$ and $\vee$, the subsystem $\sigma(U/E)$ by lattice L, and definable sets of objects by extensions of formal concepts. The extensions of the resulting two concepts are the approximations of A.

Definition 8. (Lattice-theoretic definition) *For a set of objects $A \subseteq U$, its lower and upper approximations are defined by:*

$$\underline{lapr}(A) = extent(\bigvee\{(X,Y) \mid (X,Y) \in L, X \subseteq A\})$$
$$= (\bigcup\{X \mid (X,Y) \in L, X \subseteq A\})^{**},$$
$$\overline{lapr}(A) = extent(\bigwedge\{(X,Y) \mid (X,Y) \in L, A \subseteq X\})$$
$$= \bigcap\{X \mid (X,Y) \in L, A \subseteq X\}. \tag{11}$$

The lower approximation of a set of objects A is the extension of the formal concept $(\underline{lapr}(A), (\underline{lapr}(A))^*)$, and the upper approximation is the extension of the formal concept $(\overline{lapr}(A), (\overline{lapr}(A))^*)$. The concept $(\underline{lapr}(A), (\underline{lapr}(A))^*)$ is the supremum of those concepts whose extensions are subsets of A, and $(\overline{lapr}(A), (\overline{lapr}(A))^*)$ is the infimum of those concepts whose extensions are supersets of A.

For a formal concept (X,Y), X^c may not necessarily be the extension of a formal concept. The concept lattice in general is not a complemented lattice. The approximation operators $\underline{lapr}$ and $\overline{lapr}$ are not necessarily dual operators.

Recall that the intersection of extensions is the extension of a concept, but the union of extensions may not be the extension of a concept. It follows that $(\overline{lapr}(A), (\overline{lapr}(A))^*)$ is the smallest concept whose extension is a superset of A. However, the concept $(\underline{lapr}(A), (\underline{lapr}(A))^*)$ may not be the largest concept whose extension is a subset of A. It may happen that $A \subseteq \underline{lapr}(A)$. That is, the lower approximation of A may not be a subset of A. The new approximation operators do not satisfy properties (i), (ii) and (iii). With respect to property (ii), they only satisfy a weak version known as monotonicity with respect to set inclusion:

$$(\text{vi}). \quad A_1 \subseteq A_2 \Longrightarrow \underline{lapr}(A_1) \subseteq \underline{lapr}(A_2),$$
$$A_1 \subseteq A_2 \Longrightarrow \overline{lapr}(A_1) \subseteq \overline{lapr}(A_2).$$

The following weak versions of property (iii) are satisfied:

$$(\text{vii}). \quad \underline{lapr}(A) \subseteq \overline{lapr}(A),$$
$$(\text{viii}). \quad A \subseteq \overline{lapr}(A).$$

Both $\underline{lapr}(A)$ and $\overline{lapr}(A)$ are extensions of formal concepts. It follows that the operators $\underline{lapr}$ and $\overline{lapr}$ satisfy properties (iv) and (v).

Example 2. Given the concept lattice in Figure 1, consider a set of objects $A = \{3, 5, 6\}$. The family of subsets of A that are extensions of concepts is:

$$\{\emptyset, \{3\}, \{6\}, \{3, 6\}, \{5, 6\}\}.$$

The corresponding family of concepts is:

$$\{(\emptyset, \{a, b, c, d, e, f, g, h, i\}), (\{3\}, \{a, b, c, g, h\}), (\{6\}, \{a, b, c, d, f\}),$$
$$(\{3, 6\}, \{a, b, c\}), (\{5, 6\}, \{a, b, d, f\})\}.$$

Their supremum is $(\{1, 2, 3, 5, 6\}, \{a, b\})$. The lower approximation is $\underline{lapr}(A) = \{1, 2, 3, 5, 6\}$, which is indeed a superset of A. The family of supersets of A that are extensions of concepts is:

$$\{\{1, 2, 3, 5, 6\}, \{1, 2, 3, 4, 5, 6, 7, 8\}\}.$$

The corresponding family of concepts is:

$$\{(\{1, 2, 3, 5, 6\}, \{a, b\}), (\{1, 2, 3, 4, 5, 6, 7, 8\}, \{a\})\}.$$

Their infimum is $(\{1, 2, 3, 5, 6\}, \{a, b\})$. The upper approximation is $\overline{lapr}(A) = \{1, 2, 3, 5, 6\}$, which is the smallest concept whose extension contains A. Although A is not an extension of a concept, it has the same lower and upper approximations, in contrast with Theorem 1.

With a finite set of objects and a finite set of properties, we obtain a finite lattice. The meet-irreducible and join-irreducible concepts in a concept lattice can be used as the elementary concepts. A concept in a finite concept lattice can be expressed as a join of a finite number of join-irreducible concepts and can also be expressed as a meet of a finite number of meet-irreducible concepts [1]. The extensions of meet-irreducible and join-irreducible concepts are treated as elementary definable sets of objects. Approximation operators can therefore be defined based on those elementary definable subsets.

The meet-irreducible and join-irreducible concepts can be defined as follows [1].

Definition 9. *In a concept lattice L, a concept $(A, B) \in L$ is called join-irreducible if and only if for all $(X_1, Y_1), (X_2, Y_2) \in L$, $(A, B) = (X_1, Y_1) \vee (X_2, Y_2)$ implies $(A, B) = (X_1, Y_1)$ or $(A, B) = (X_2, Y_2)$. The dual notion is called meet-irreducible for a concept $(A, B) \in L$ if and only if for all $(X_1, Y_1), (X_2, Y_2) \in L$, $(A, B) = (X_1, Y_1) \wedge (X_2, Y_2)$ implies $(A, B) = (X_1, Y_1)$ or $(A, B) = (X_2, Y_2)$.*

Let $J(L)$ be the set of all join-irreducible concepts and $M(L)$ be the set of all meet-irreducible concepts in L. A concept (A, B) can be expressed by the join of join-irreducible concepts that are the sub-concepts of (A, B) in $J(L)$. That is

$$(A, B) = \bigvee\{(X, Y) \mid (X, Y) \in J(L), (X, Y) \preceq (A, B)\}. \tag{12}$$

A concept (A, B) can also be expressed by the meet of meet-irreducible concepts that are the super-concepts of (A, B) in $M(L)$. That is

$$(A, B) = \bigwedge\{(X, Y) \mid (X, Y) \in M(L), (A, B) \preceq (X, Y)\}. \tag{13}$$

The lower and upper approximations of a set of objects can be defined based on the extensions of join-irreducible and meet-irreducible concepts [8].

Definition 10. *For a set of objects $A \subseteq U$, its lower and upper approximations are defined by:*

$$lapr(A) = extent(\bigvee\{(X,Y) \mid (X,Y) \in J(L), X \subseteq A\})$$
$$= (\bigcup\{X \mid (X,Y) \in J(L), X \subseteq A\})^{**},$$
$$\overline{lapr}(A) = extent(\bigwedge\{(X,Y) \mid (X,Y) \in M(L), A \subseteq X\})$$
$$= \bigcap\{X \mid (X,Y) \in M(L), A \subseteq X\}. \tag{14}$$

Ganter and Wille have shown that a formal concept in a concept lattice can be expressed by the join of object concepts in which the object is included in the extension of the formal concept [6]. That is, for a formal concept (A, B),

$$(A, B) = \bigvee\{(\{x\}^{**}, \{x\}^{*}) \mid x \in A\}.$$

A formal concept can also be expressed by the meet of property concepts in which the property is included in the intension of the formal concept [6]. That is, for a formal concept (A, B),

$$(A, B) = \bigwedge\{(\{y\}^{*}, \{y\}^{**}) \mid y \in B\}.$$

Therefore, the lower and upper approximations of a set of objects can be defined based on the extensions of object and property concepts.

Definition 11. *For a set of objects $A \subseteq U$, its lower and upper approximations are defined by object and property concepts:*

$$lapr(A) = extent(\bigvee\{(\{x\}^{**}, \{x\}^{*}) \mid x \in U, \{x\}^{**} \subseteq A\}),$$
$$= (\bigcup\{\{x\}^{**} \mid x \in U, x \in A\})^{**},$$
$$\overline{lapr}(A) = extent(\bigwedge\{(\{y\}^{*}, \{y\}^{**}) \mid y \in V, A \subseteq \{y\}^{*}\}),$$
$$= \bigcap\{\{y\}^{*} \mid y \in V, A \subseteq \{y\}^{*}\}. \tag{15}$$

In fact, this definition can be considered as the extension of granule based definition of rough set approximations in Definition 2.

These definitions of lower and upper approximations are the same as the ones defined in Definition 8. They are regarded as equivalent definitions with slightly different interpretations.

Example 3. In the concept lattice in Figure 1, consider the same set of objects $A = \{3, 5, 6\}$ in Example 2. The family of join-irreducible concepts is:

$$(\{1, 2, 3\}, \{a, b, g\}), \qquad (\{2, 3\}, \{a, b, g, h\}),$$
$$(\{3\}, \{a, b, c, g, h\}), \qquad (\{4\}, \{a, c, g, h, i\}),$$
$$(\{5, 6\}, \{a, b, d, f\}), \qquad (\{6\}, \{a, b, c, d, f\}),$$
$$(\{7\}, \{a, c, d, e\}), \qquad (\{6, 8\}, \{a, c, d, f\}),$$
$$(\emptyset, \{a, b, c, d, e, f, g, h, i\}).$$

The join-irreducible concepts whose extensions are subsets of A are:

$$(\{3\}, \{a, b, c, g, h\}), \quad (\{5, 6\}, \{a, b, d, f\}),$$
$$(\{6\}, \{a, b, c, d, f\}), \quad (\emptyset, \{a, b, c, d, e, f, g, h, i\}).$$

Thus, according to the Definition 10, the lower approximation is

$$\underline{lapr}(A) = extent((\{3\}, \{a, b, c, g, h\}) \bigvee (\{5, 6\}, \{a, b, d, f\})$$
$$\bigvee (\{6\}, \{a, b, c, d, f\}) \bigvee (\emptyset, \{a, b, c, d, e, f, g, h, i\}))$$
$$= \{1, 2, 3, 5, 6\}.$$

The family of meet-irreducible concepts is:

$$(\{1, 2, 3, 4, 5, 6, 7, 8\}, \{a\}), \quad (\{1, 2, 3, 5, 6\}, \{a, b\}),$$
$$(\{3, 4, 6, 7, 8\}, \{a, c\}), \quad (\{5, 6, 7, 8\}, \{a, d\}),$$
$$(\{1, 2, 3, 4\}, \{a, g\}), \quad (\{5, 6, 8\}, \{a, d, f\}),$$
$$(\{2, 3, 4\}, \{a, g, h\}), \quad (\{7\}, \{a, c, d, e\}),$$
$$(\{4\}, \{a, c, g, h, i\}).$$

The meet-irreducible concepts whose extensions are supersets of A are:

$$(\{1, 2, 3, 4, 5, 6, 7, 8\}, \{a\}), \quad (\{1, 2, 3, 5, 6\}, \{a, b\}).$$

The upper approximation is

$$\overline{lapr}(A) = extent((\{1, 2, 3, 4, 5, 6, 7, 8\}, \{a\}) \bigwedge (\{1, 2, 3, 5, 6\}, \{a, b\})),$$
$$= \{1, 2, 3, 5, 6\}.$$

For the set of objects $A = \{3, 5, 6\}$, its lower approximation equals to its upper approximation. One can see that approximations based on lattice-theoretic operators have some undesirable properties. Other possible formulations are needed.

The upper approximation operator $\overline{lapr}$ is related to the operator $*$. For any set of objects $A \subseteq U$, we can derive a set of properties A^*. For the set of properties A^*, we can derive another set of objects A^{**}. By property (3), (A^{**}, A^*) is a formal concept. By property (2), we have $A \subseteq A^{**}$. In fact, (A^{**}, A^*) is the smallest formal concept whose extension contains A. That is, for a set of objects $A \subseteq U$, its upper approximation is $\overline{lapr}(A) = A^{**}$.

Thus we can only obtain a weak version of Theorem 1.

Theorem 3. *In a concept lattice $L(U, V, R)$, if A is an extension of a concept, i.e., (A, A^*) is a concept, then $\underline{lapr}(A) = \overline{lapr}(A)$.*

As shown by the examples, the reverse implication in the theorem is not true. This is a limitation of the formulation based on lattice-theoretic operators.

The ideas of approximating a set of objects can be used to define operators that approximate a set of properties. In contract to the approximations of a set of objects, the lower approximation is defined by using meet, and the upper approximation is defined by using join.

Definition 12. (Lattice-theoretic definition) *For a set of properties $B \subseteq V$, its lower and upper approximations are defined by:*

$$lapr(B) = intent(\bigwedge\{(X,Y) \mid (X,Y) \in L, Y \subseteq B\})$$
$$= (\bigcup\{Y \mid (X,Y) \in L, Y \subseteq B\})^{**},$$
$$\overline{lapr}(B) = intent(\bigvee\{(X,Y) \mid (X,Y) \in L, B \subseteq Y\})$$
$$= \bigcap\{Y \mid (X,Y) \in L, B \subseteq Y\}. \qquad (16)$$

Definition 13. *For a set of properties $B \subseteq V$, its lower and upper approximations based on the sets of join-irreducible and meet-irreducible concepts are defined by:*

$$lapr(B) = intent(\bigwedge\{(X,Y) \mid (X,Y) \in M(L), Y \subseteq B\})$$
$$= (\bigcup\{Y \mid (X,Y) \in M(L), Y \subseteq B\})^{**},$$
$$\overline{lapr}(B) = intent(\bigvee\{(X,Y) \mid (X,Y) \in J(L), B \subseteq Y\})$$
$$= \bigcap\{Y \mid (X,Y) \in J(L), B \subseteq Y\}. \qquad (17)$$

Definition 14. *For a set of properties $B \subseteq V$, its lower and upper approximations are defined by object and property concepts:*

$$lapr(B) = intent(\bigwedge\{(\{y\}^*, \{y\}^{**}) \mid y \in V, \{y\}^{**} \subseteq B\}),$$
$$= (\bigcup\{\{y\}^{**} \mid y \in V, y \in B\})^{**},$$
$$\overline{lapr}(B) = intent(\bigvee\{(\{x\}^{**}, \{x\}^*) \mid x \in U, B \subseteq \{x\}^*\}),$$
$$= \bigcap\{\{x\}^* \mid x \in U, B \subseteq \{x\}^*\}. \qquad (18)$$

The lower approximation of a set of properties B is the intension of the formal concept $((lapr(B))^*, lapr(B))$, and the upper approximation is the intension of the formal concept $((\overline{lapr}(B))^*, \overline{lapr}(B))$.

4.2   Approximations Based on Set-Theoretic Operators

By comparing with the standard rough set approximations, one can observe two problems of the approximation operators defined by using lattice-theoretic operators. The lower approximation of a set of objects A is not necessarily a subset of A. Although a set of objects A is undefinable, i.e., A is not the extension of a formal concept, its lower and upper approximations may be the same. In order to avoid these shortcomings, we present another formulation by using set-theoretic operators.

The extension of a formal concept is a definable set of objects. A system of definable sets can be derived from a concept lattice.

Definition 15. *For a formal concept lattice L, the family of all extensions is given by:*

$$EXT(L) = \{extent(X,Y) \mid (X,Y) \in L\}. \qquad (19)$$

The system $EXT(L)$ contains the entire set U and is closed under intersection. Thus, $EXT(L)$ is a closure system [3]. Although one can define the upper approximation by extending Definition 1, one cannot define the lower approximation similarly. Nevertheless, one can still keep the intuitive interpretations of lower and upper approximations. That is, the lower approximation is a maximal set in $EXT(L)$ that are subsets of A, and the upper approximation is a minimal set in $EXT(L)$ that are supersets of A. While an upper approximation is unique (e.g., there is a smallest set in $EXT(L)$ containing A), the maximal set contained in A is generally not unique.

Definition 16. (Set-theoretic definition) *For a set of objects $A \subseteq U$, its upper approximation is defined by:*

$$\overline{sapr}(A) = \bigcap \{X \mid X \in EXT(L), A \subseteq X\}, \tag{20}$$

and its lower approximation is a family of sets:

$$\underline{sapr}(A) = \{X \mid X \in EXT(L), X \subseteq A,$$
$$\forall X' \in EXT(L)(X \subset X' \implies X' \not\subseteq A)\}. \tag{21}$$

The upper approximation $\overline{sapr}(A)$ is the same as $\overline{lapr}(A)$, namely, $\overline{sapr}(A) = \overline{lapr}(A)$. However, the lower approximation is different. An important feature is that a set can be approximated from below by several definable sets of objects. In general, for $A' \in \underline{sapr}(A)$, we have $A' \subseteq \underline{lapr}(A)$.

Example 4. In the concept lattice L of Figure 1, the family of all extensions $EXT(L)$ are:

$$EXT(L) = \{ \emptyset, \{3\}, \{4\}, \{6\}, \{7\},$$
$$\{2, 3\}, \{3, 4\}, \{3, 6\}, \{5, 6\}, \{6, 8\},$$
$$\{1, 2, 3\}, \{2, 3, 4\}, \{6, 7, 8\}, \{5, 6, 8\},$$
$$\{1, 2, 3, 4\}, \{5, 6, 7, 8\},$$
$$\{1, 2, 3, 5, 6\}, \{3, 4, 6, 7, 8\},$$
$$\{1, 2, 3, 4, 5, 6, 7, 8\} \}.$$

For a set of objects $A = \{3, 5, 6\}$, the lower approximation is given by $\underline{sapr}(A) = \{\{3, 6\}, \{5, 6\}\}$, which is a family of sets of objects. The upper approximation is given by $\overline{sapr}(A) = \{1, 2, 3, 5, 6\}$, which is a unique set of objects.

Since a concept in a finite concept lattice can be expressed as a meet of a finite number of meet-irreducible concepts, the family of extensions of meet-irreducible concepts can be used to generate the extensions of all concepts in a finite concept lattice by simply using set intersection. Hence, one can use the family of the extensions of all meet-irreducible concepts to replace the system of the extensions of all concepts in the concept lattice.

Let $EXT(M(L))$ denote the family of extensions of all the meet-irreducible concepts. $EXT(M(L))$ is a subset of $EXT(L)$. The extensions of concepts in the system

$EXT(M(L))$ are treated as elementary definable sets of objects. Therefore, the upper approximation of a set of objects is the intersection of extensions in $EXT(M(L))$ that are supersets of the set.

Definition 17. *For a set of objects $A \subseteq U$, its upper approximation is defined by:*

$$\overline{sapr}(A) = \bigcap \{X \mid X \in EXT(M(L)), A \subseteq X\}. \tag{22}$$

This definition of upper approximation is the same as Definition 16. They are equivalent but in different forms.

The lower approximation of a set of objects cannot be defined based on the system $EXT(M(L))$. The meet of some meet-irreducible concepts, whose extensions are subsets of a set of objects, is not necessarily the largest set that is contained in the set of objects.

With respect to property (iii), we have:

(ix). $A' \subseteq A \subseteq \overline{sapr}(A)$, for all $A' \in \underline{sapr}(A)$.

That is, A lies within any of its lower approximation and upper approximation. For the set-theoretic formulation, we have a theorem corresponding to Theorem 1.

Theorem 4. *In a concept lattice $L(U, V, R)$, for a subset of the universe of objects $A \subseteq U$, $\overline{sapr}(A) = A$ and $\underline{sapr}(A) = \{A\}$, if and only if A is an extension of a concept.*

In the new formulation, we resolve the difficulties with the approximation operators $\underline{lapr}$ and $\overline{lapr}$. The lower approximation offers more insights into the notion of approximations. In some situations, the union of a family of definable sets is not necessarily a definable set. It may not be reasonable to insist on a unique approximation. The approximation of a set by a family of sets may provide a better characterization of the set.

5   Related Works

In this section, we provide a review of the existing studies on the comparisons and combinations of rough set analysis and formal concept analysis and their relevance to the present study.

5.1   A Brief Review of Existing Studies

Broadly, we can classify existing studies into three groups. The first group may be labeled as the comparative studies [5,7,9,10,15,21,23,24,33]. They deal with the comparison of the two approaches with an objective to produce a more generalized data analysis framework. The second group concerns the applications of the notions and ideas of formal concept analysis into rough set analysis [5,23,34]. Reversely, the third group focuses on applying concepts and methods of rough set analysis into formal concept analysis [4,8,11,16,17,20,23,34,39]. Those studies lead to different types of abstract operators, concept lattices and approximations.

Comparative Studies

Kent examined the correspondence between similar notions used in both theories, and argued that they are in fact parallel to each other in terms of basic notions, issues and methodologies [9]. A framework of rough concept analysis was introduced as a synthesis of the two theories. Based on this framework, Ho developed a method of acquiring rough concepts [7], and Wu, Liu and Li proposed an approach for computing accuracies of rough concepts and studied the relationships between the indiscernibility relations and accuracies of rough concepts [27].

The notion of a formal context has been used in many studies under different names. Shafer used a compatibility relation to interpret the theory of evidence [18,19]. A compatibility relation is a binary relation between two universes, which is in fact a formal context. Wong, Wang and Yao investigated approximation operators over two universes with respect to a compatibility relation [25,26]. Düntsch and Gediga referred to those operators as modal-style operators and studied a class of such operators in data analysis [5]. The derivation operator in formal concept analysis is a sufficiency operator, and the rough set approximation operators are the necessity and possibility operators used in modal logics. By focusing on modal-style operators, we have a unified operator-oriented framework for the study of the two theories.

Pagliani used a Heyting algebra structure to connect concept lattices and approximation spaces together [10]. Based on the algebra structure, concept lattices and approximation spaces can be transformed into each other. Wasilewski demonstrated that formal contexts and general approximation spaces can be mutually represented [21]. Consequently, rough set analysis and formal concept analysis can be viewed as two related and complementary approaches for data analysis. It is shown that the extension of a formal concept is a definable set in the approximation space. Qi et al. argued that two theories have much in common in terms of the goals and methodologies [15]. They emphasized the basic connection and transformation between a concept lattice and a partition.

Wolski investigated Galois connections in formal concept analysis and their relations to rough set analysis [23]. A logic, called S4.t, is proposed as a good tool for approximate reasoning to reflect the formal connections between formal concept analysis and rough set analysis [24].

Yao compared the two theories based on the notions of definability, and showed that they deal with two different types of definability [33]. Rough set analysis studies concepts that are defined by disjunctions of properties. Formal concept analysis considers concepts that are definable by conjunctions of properties.

Based on those comparative studies, one can easily adopt ideas from one theory to another. The applications of rough set always lead to approximations and reductions in formal concept analysis. The approximations of formal concept analysis result in new types of concepts and concept lattices.

Approximations and Reductions in Concept Lattices

Many studies considered rough set approximations in formal concept lattice [4,8,11,16,17,20,23]. They will be discussed in Section 5.2. The present study is in fact a continuation in the same direction.

Zhang, Wei and Qi examined property (object) reduction in concept lattice using the ideas from rough set analysis [39]. The minimal sets of properties (objects) are determined based on criterions that the reduced lattice and the original lattice show certain common features or structures. For example, two lattices are isomorphic.

Concept Lattices in Rough Sets

Based on approximation operators, one can construct additional concept lattices. Those lattices, their properties, and connections to the original concept lattice are studied extensively by Düntsch and Gediga [5], and Wolski [23]. The results provide more insights into data analysis using modal-style operators.

Yao examined semantic interpretations of various concept lattices [34]. One can obtain different types of inference rules regarding objects and properties. To reflect their physical meanings, the notions of object-oriented and property-oriented concept lattices are introduced.

5.2 Approximations in Formal Concept Lattice

Saquer and Deogun suggested that all concepts in a concept lattice can be considered as definable, and a set of objects can be approximated by concepts whose extensions approximate the set of objects [16,17]. A set of properties can be similarly approximated by using intensions of formal concepts.

For a given set of objects, it may be approximated by extensions of formal concepts in two steps. The classical rough set approximations for a given set of objects are first computed. Since the lower and upper approximations of the set are not necessarily the extensions of formal concepts, they are then approximated again by using derivation operators of formal concept analysis.

At the first step, for a set of objects $A \subseteq U$, the standard lower approximation $\underline{apr}(A)$ and upper approximation $\overline{apr}(A)$ are obtained. At the second step, the lower approximation of the set of objects A is defined by the extension of the formal concept $(\underline{apr}(A)^{**}, \underline{apr}(A)^{*})$. The upper approximation of the set of objects A is defined by the extension of the formal concept $(\overline{apr}(A)^{**}, \overline{apr}(A)^{*})$. That is,

$$\underline{eapr}(A) = \underline{apr}(A)^{**},$$
$$\overline{eapr}(A) = \overline{apr}(A)^{**}.$$

If $\underline{apr}(A) = \overline{apr}(A)$, we have $\underline{apr}(A)^{**} = \overline{apr}(A)^{**}$. Namely, for a definable set A, its lower and upper formal concept approximations are the same. However, the reverse implication is not true. A set of objects that has the same lower and upper approximations may not necessarily be a definable set. This shortcoming of their definition is the same as the lattice-theoretic formulations of approximations.

Hu *et al.* suggested an alternative formulation [8]. Instead of defining an equivalence relation, they defined a partial order on the universe of objects. For an object, its principal filter, which is the set of objects "greater than or equal to" the object and is called the partial class by Hu *et al.*, is the extension of a formal concept. The family of all principal filters is the set of join-irreducible elements of the concept lattice. Similarly, a partial order relation can be defined on the set of properties. The family of meet-irreducible elements of the concept lattice can be constructed. The lower and upper approximations can be defined based on the families of meet- and join-irreducible elements in concept lattice. Their definitions are similar to our lattice-theoretic definitions. However, their definition of lower approximation has the same shortcoming of Saquer and Deogun's definition [16].

Some researchers used two different systems of concepts to approximate a set of objects or a set of properties [4,11,14,20,23]. In addition to the derivation operator, one can define the two rough set approximation operators [5,23,25,26,34].

$$A^\square = \{y \in V \mid Ry \subseteq A\},$$
$$A^\lozenge = \{y \in V \mid Ry \cap A \neq \emptyset\},$$

and

$$B^\square = \{x \in U \mid xR \subseteq B\},$$
$$B^\lozenge = \{x \in U \mid xR \cap B \neq \emptyset\}.$$

Düntsch and Gediga referred to $*$, $\square$ and $\lozenge$ as modal-style operators, called sufficiency operator, necessity operator and possibility operator, respectively [4,5].

The two operators can be used to define two different types of concepts and concept lattices [5,34]. A pair (A, B) is called an object-oriented concept if $A = B^\lozenge$ and $B = A^\square$. The family of all object-oriented concepts forms a complete lattice, denoted as $L_o(U, V, R)$. A pair (A, B) is called a property-oriented concept if $A = B^\square$ and $B = A^\lozenge$. The family of all property-oriented concepts also forms a complete lattice, denoted as $L_p(U, V, R)$. Similar to the formal concept lattice, the set of objects A is referred to as the extension of the concept, and the set of properties B is referred to as the intension of the concept. With respect to those new concept lattices, one can apply the formulation of approximations discussed previously in a similar way. For example, we may study the approximations of a set of objects by using an object-oriented concept lattice.

Another class of approximation operators can be derived by the combination of operators $\square$ and $\lozenge$. The combined operators $\square\lozenge$ and $\lozenge\square$ have following important properties [4]:

1). $\square\lozenge$ is a closure operator on U and V,

2). $\square\lozenge$ and $\lozenge\square$ are dual to each other,

3). $\lozenge\square$ is an interior operator on U and V.

Based on those properties, approximation operators can be defined [4,11,20,23]. The lower and upper approximations of a set of objects and a set of properties can be defined, respectively, based on two systems:

$$\underline{rapr}(A) = A^{\square\lozenge}, \quad \underline{rapr}(B) = B^{\square\lozenge},$$

and

$$\overline{rapr}(A) = A^{\lozenge\square}, \quad \overline{rapr}(B) = B^{\lozenge\square}.$$

The operators $\square\lozenge$ and $\lozenge\square$ and the corresponding rough set approximations have been used and studied by many authors, for example, Düntsch and Gediga [4], Pagliani [10], Pagliani and Chakraborty [11], Pei and Xu [14], Shao and Zhang [20], and Wolski [23,24].

If a set of objects A equals to its lower approximation $\underline{rapr}(A)$, we say that A is a definable set of objects in the system $L_o(U, V, R)$. If the set of objects A equals to its upper approximation $\overline{rapr}(A)$, we say that A is a definable set of objects in the system $L_p(U, V, R)$. The lower and upper approximations of a set of objects are equal if and only if the set of objects is a definable set in both systems $L_o(U, V, R)$ and $L_p(U, V, R)$. Similarly, the lower and upper approximations of a set of properties are equal if and only if the set of properties is a definable set in both systems $L_o(U, V, R)$ and $L_p(U, V, R)$.

6   Conclusion

An important issue of rough set analysis is the approximations of undefinable sets using definable sets. In the classical rough set theory, the family of definable sets is a subsystem of the power set of a universe. There are many approaches to construct subsystems of definable sets [30,36]. Formal concept analysis provides an approach for the construction of a family of definable sets. It represents a different type of definability. The notion of approximations can be introduced naturally into formal concept analysis.

Formal concepts in a formal concept lattice correspond to definable sets. Two types of approximation operators are investigated. One is based on the lattice-theoretic formulation and the other is based on the set-theoretic formulation. Their properties are studied in comparison with the properties of classical rough set approximation operators. A distinguishing feature of the lower approximation defined by set-theoretic formulation is that a subset of the universe is approximated from below by a family of definable sets, instead of a unique set in the classical rough set theory.

The theory of rough sets and formal concept analysis capture different aspects of data. They can represent different types of knowledge embedded in data sets. The introduction of the notion of approximations into formal concept analysis combines the two theories. It describes a particular characteristic of data, improves our understanding of data, and produces new tools for data analysis.

The sufficiency operators * is an example of modal-style operators [4,5,33]. One can study the notion of rough set approximations in a general framework in which various modal-style operators are defined [4,5,10,33].

Acknowledgements

The research reported in this paper was partially supported by a grant from the Natural Sciences and Engineering Research Council of Canada. Authors are grateful to the anonymous referees for their constructive comments.

References

1. Birkhoff, G. *Lattice Theory*, Third Edition, American Mathematical Society Colloquium Publications, Providence, Rhode Island, 1967.
2. Buszkowski, W. Approximation spaces and definability for incomplete information systems, *Proceedings of 1st International Conference on Rough Sets and Current Trends in Computing, RSCTC'98*, 115-122, 1998.
3. Cohn, P.M. *Universal Algebra*, Harper and Row Publishers, New York, 1965.

4. Düntsch, I. and Gediga, G. Approximation operators in qualitative data analysis, in: *Theory and Application of Relational Structures as Knowledge Instruments*, de Swart, H., Orlowska, E., Schmidt, G. and Roubens, M. (Eds.), Springer, Heidelberg, 216-233, 2003.

5. Gediga, G. and Düntsch, I. Modal-style operators in qualitative data analysis, *Proceedings of the 2002 IEEE International Conference on Data Mining*, 155-162, 2002.

6. Ganter, B. and Wille, R. *Formal Concept Analysis: Mathematical Foundations*, Springer-Verlag, New York, 1999.

7. Ho, T.B. Acquiring concept approximations in the framework of rough concept analysis, *Proceedings of 7th European-Japanese Conference on Information Modelling and Knowledge Bases*, 186-195, 1997.

8. Hu, K., Sui, Y., Lu, Y., Wang, J., and Shi, C. Concept approximation in concept lattice, *Proceedings of 5th Pacific-Asia Conference on Knowledge Discovery and Data Mining, PAKDD'01*, 167-173, 2001.

9. Kent, R.E. Rough concept analysis, *Fundamenta Informaticae*, **27**, 169-181, 1996.

10. Pagliani, P. From concept lattices to approximation spaces: algebraic structures of some spaces of partial objects, *Fundamenta Informaticae*, **18**, 1-25, 1993.

11. Pagliani, P. and Chakraborty, M.K., Information quanta and approximation spaces. I: nonclassical approximation operators, *Proceedings of IEEE International Conference on Granular Computing*, 605-610, 2005.

12. Pawlak, Z. Rough sets, *International Journal of Computer and Information Sciences*, **11**, 341-356, 1982.

13. Pawlak, Z. *Rough Sets - Theoretical Aspects of Reasoning About Data*, Kluwer Publishers, Boston, Dordrecht, 1991.

14. Pei, D.W. and Xu, Z.B. Rough set models on two universes, *International Journal of General Systems*, **33**, 569-581, 2004.

15. Qi, J.J., Wei, L. and Li, Z.Z. A partitional view of concept lattice, *Proceedings of The 10th International Conference on Rough Sets, Fuzzy Sets, Data Mining, and Granular Computing, RSFDGrC'05*, Part I, 74-83, 2005.

16. Saquer, J. and Deogun, J. Formal rough concept analysis, *Proceedings of 7th International Workshop on Rough Sets, Fuzzy Sets, Data Mining, and Granular-Soft Computing, RSFD-GrC '99*, 91-99, 1999.

17. Saquer, J. and Deogun, J. Concept approximations based on rough sets and similarity measures. *International Journal of Applied Mathematics and Computer Science*, **11**, 655-674, 2001.

18. Shafer, G. *A Mathematical Theory of Evidence*, Princeton University Press, Princeton, 1976.

19. Shafer, G. Belief functions and possibility measures, In: *Analysis of Fuzzy information, vol. 1: mathematics and logic*, Bezdek, J.C. (Ed.), CRC Press, Boca Raton, 51-84, 1987.

20. Shao, M.W. and Zhang, W.X. Approximation in formal concept analysis, *Proceedings of The 10th International Conference on Rough Sets, Fuzzy Sets, Data Mining, and Granular Computing, RSFDGrC'05*, Part I, 43-53, 2005.

21. Wasilewski, P. Concept lattices vs. approximation spaces, *Proceedings of The 10th International Conference on Rough Sets, Fuzzy Sets, Data Mining, and Granular Computing, RSFDGrC'05*, Part I, 114-123, 2005.

22. Wille, R. Restructuring lattice theory: an approach based on hierarchies of concepts, in: *Ordered sets*, Rival, I. (Ed.), Reidel, Dordecht-Boston, 445-470, 1982.

23. Wolski, M. Galois connections and data analysis, *Fundamenta Informaticae CSP*, 1-15, 2003.

24. Wolski, M. Formal concept analysis and rough set theory from the perspective of finite topological approximations, *Transaction of Rough Sets* III, LNCS 3400, 230-243, 2005.

25. Wong, S.K.M., Wang, L.S. and Yao, Y.Y., Interval structure: a framework for representing uncertain information, *Uncertainty in Artificial Intelligence: Proceedings of the 8th Conference*, 336-343, 1993.

26. Wong, S.K.M. and Wang, L.S. and Yao, Y.Y., On modeling uncertainty with interval structures, *Computational Intelligence*, **11**, 406-426, 1995.

27. Wu, Q., Liu, Z.T. and Li, Y. Rough formal concepts and their accuracies, *Proceedings of the 2004 International Conference on Services Computing, SCC'04*, 445-448, 2004.

28. Yao, Y.Y. Two views of the theory of rough sets in finite universe, *International Journal of Approximate Reasoning*, **15**, 291-317, 1996.

29. Yao, Y.Y., Generalized rough set models, in: *Rough Sets in Knowledge Discovery*, Polkowski, L. and Skowron, A. (Eds.), Physica-Verlag, Heidelberg, 286-318, 1998.

30. Yao, Y.Y. On generalizing Pawlak approximation operators, *Proceedings of 1st International Conference on Rough Sets and Current Trends in Computing, RSCTC'98*, 298-307, 1998.

31. Yao, Y.Y. Constructive and algebraic methods of the theory of rough sets, *Information Sciences*, **109**, 21-47, 1998.

32. Yao, Y.Y. On generalizing rough set theory, *Proceedings of 9th International Conference on Rough Sets, Fuzzy Sets, Data Mining, and Granular Computing, RSFDGrC'03*, 44-51, 2003.

33. Yao, Y.Y. A comparative study of formal concept analysis and rough set theory in data analysis, *Proceedings of 3rd International Conference on Rough Sets and Current Trends in Computing, RSCTC'04*, 59-68, 2004.

34. Yao, Y.Y. Concept lattices in rough set theory, *Proceedings of 23rd International Meeting of the North American Fuzzy Information Processing Society, NAFIPS'04*, 796-801, 2004.

35. Yao, Y.Y. and Chen, Y.H. Rough set approximations in formal concept analysis, *Proceedings of 23rd International Meeting of the North American Fuzzy Information Processing Society, NAFIPS'04*, 73-78, 2004.

36. Yao, Y.Y. and Chen, Y.H. Subsystem based generalizations of rough set approximations, *Proceedings of 15th International Symposium on Methodologies for Intelligent Systems (IS-MIS'05)*, 210-218, 2005.

37. Yao, Y.Y. and Lin, T.Y. Generalization of rough sets using modal logic, *Intelligent Automation and Soft Computing, An International Journal*, **2**, 103-120, 1996.

38. Zadeh, L.A. Fuzzy logic as a basis for a theory of hierarchical definability (THD), *Proceedings of the 33rd International Symposium on Multiple-Valued Logic, ISMVL'03*, 3-4, 2003.

39. Zhang, W.X., Wei, L. and Qi, J.J. Attribute reduction in concept lattice based on discernibility matrix, *Proceedings of The 10th International Conference on Rough Sets, Fuzzy Sets, Data Mining, and Granular Computing, RSFDGrC'05*, Part II, 157-165, 2005.

Motion-Information-Based Video Retrieval System Using Rough Pre-classification

Zhe Yuan, Yu Wu, Guoyin Wang, and Jianbo Li

Institute of Computer Science and Technology,
Chongqing University of Posts and Telecommunications,
Chongqing 400065, P.R. China
oxfordss4903@gmail.com, {wuyu, wanggy, icst}@cqupt.edu.cn

Abstract. Motion information is the basic element for analyzing video. It represents the change of video on the time-axis and plays an important role in describing the video content. In this paper, a robust motion-based, video retrieval system is proposed. At first, shot boundary detection is achieved by analyzing luminance information, and motion information of video is abstracted and analyzed. Then rough set theory is introduced to classify the shots into two classes, global motions and local motions. Finally, shots of these two types are respectively retrieved according to the motion types of submitted shots. Experiments show that it's effective to distinguish shots with global motions from those with local motions in various types of video, and in this situation motion-information-based video retrieval are more accurate.

Keywords: Global motion, local motion, shot boundary detection, video retrieval, rough sets.

1   Introduction

With the development of network, computer and multimedia technologies, the demands for video are becoming greater and greater. There is widespread interest in finding a quick way to obtain interesting video materials. Obviously, traditional retrieval based on text cannot meet these demands, so content-based video retrieval has been proposed as a solution. This technology uses objects such as video features, to retrieve video. This approach contrasts with identifiers that are used in traditional text-based retrieval. Video features such as colors, textures, motion types can be extracted from video for retrieval. Among these features, motion information of objects and backgrounds, as the unique feature of video, is essential in the study of video retrieval. So, motion-information-based video retrieval has had broad attention.

Video motions can be divided into two types, global and local. The global motions are caused by camera movements, and there are six motion types defined in MPEG-7 [14], including panning, titling, zooming, tracking, booming, and dollying. The local motions refers to object motions in the scene, which can be considered as parts not matching the global motion model. Referring

J.F. Peters and A. Skowron (Eds.): Transactions on Rough Sets V, LNCS 4100, pp. 306–333, 2006.
© Springer-Verlag Berlin Heidelberg 2006

to MPEG-7, a motion descriptor, called the parameter motion, can be used to describe local motions. It mainly depicts changes with the times of arbitrary object regions in video via 2-D geometry transition. In the study of global motions, several parameter estimation methods for global motions in the uncompressed domain have been proposed in recent years [4] [5] [29]. Tan and Saur proposed a quick parametric global motion estimate algorithm, in which motion vectors of macroblocks are extracted from compressed data [22]. A video retrieval system based on global motion information was founded by Tianli Yu [30]. On the other hand, in the study of local motions, there are three primary methods, which are computing motion information after video segmentation [33], segmenting video after computing motion information [1], and processing both of them at the same time [23].

Algorithms and systems mentioned above can be used to obtain good experimental results in certain video application domains. However, if video with complicated motions are applied, or motion types of shots are not the type that the systems deal with, or there are coding errors, they are unreliable. So shots need to be classified before motion analysis. Nevertheless, the classification may bring uncertainty and inconsistency. For example, there are almost similar motion features between frames of two motion types. Therefore, the theory of rough set [19] may be useful in this research. Z. Pawlak proposed rough set theory during the early 1980s as a powerful mathematical analysis tool to process incomplete data and inaccurate knowledge (see, e.g., [16,17,18]). It's a new hot spot in the artificial intelligence field at present and has been widely used in knowledge acquisition, knowledge analysis, decision analysis and so on [24]. Instead of obtaining mathematical descriptions of attributes and features of detected objects in advance, rough set methods make knowledge reduction possible and reduce the number of decision rules. Since uncertainty and inconsistency accompany classification of global and non-global motions, the rough set method is adopted to construct a better global motion model.

In this paper, a robust motion-based video retrieval system is proposed and realized to retrieve similar shots from a video database. We propose a global-motion analysis method in which feature attributions of motion information in video are computed and, then, motion types of P frames are achieved via rough-set-based video pre-classification. Furthermore, we present a method to check motion types of shots based on P frame classification.

The rest of paper is organized as follows. At first, basic theories of rough set and MPEG standard are introduced in Section 2. An extraction method of motion information from P frames is proposed and video pre-classification is performed using rough set method in Section 3. In Section 4, the method using video luminance information to detect shot boundary is given. In Section 5, shot classification based on the classification of P frames is proposed. In Section 6, we present a global-motion-information retrieval scheme by computing distances between shots. Next, in Section 7, experimental results using methods from Section 3 to Section 6 to enhance the performance of the motion based video retrieval are given. In Section 8, we give an overview of the diagram of the motion-based video retrieval and show a prototype system for video retrieval. Finally, Section 9 concludes the paper.

2   Some Basic Concepts of Rough Set and MPEG

In this section, some of the basic concepts in rough set theory and standards for dynamic image compression, are briefly considered.

2.1   Introduction of Rough Set

The expression of knowledge in rough set theory is generally expressed in terms of a so-called information system, which is a 4-tuple [18]:

$$IS = (U, A, V, f). \tag{1}$$

U is a finite set of objects (universe). A is a finite set of attributes, that is $A = \{a_1, a_2, \ldots, a_n\}$ or $A = C \bigcup D$, where C denotes the condition attributes, and D denotes the decision attributes. V_a is the domain of the attribute a. For each $a \in A$, an information function is defined as $f_a : U \rightarrow V_a$. This information system can be also described by using a two-dimensional decision information table, in which each row denotes one example of U, each column denotes one attribute of A, and each element denotes the value of the information function f_a.

   The pretreatment in rough sets includes two steps, namely, data filling, and data discretization. It is common that the information table is not fully filled. Some values are missed and there is no way to obtain them, which is the main cause of uncertainty in the information system. There are many methods to deal with the missing data. One way is to simply remove the samples with missing attribute values. In the second methods, attributes with missing data can be considered as special ones and be processed in special ways. Another way is to fill the missing data [6], which can be done by indiscernibility relation of rough set, or by analyzing distribution information of other samples in the information table according to the statistics principle.

   Because elements studied in rough set are all discrete values, data from original information need to be discretized [25]. As said above, A is made up of C and D. $r(d)$ is defined as the count of decision attributes. One broken point in V_a is defined as (a, c), where $a \in R$, c is the real number set. Any broken set $\{(a, c_1{}^a), (a, c_2{}^a), \ldots, (a, c_{k_a}{}^a)\}$ in $V_a = [l_a, r_a]$ defines one classification P_a.

$$P_a = \{[c_0{}^a, c_1{}^a), [c_1{}^a, c_1{}^a), \ldots, [c_{k_a}{}^a, c_{k_a+1}{}^a]\}. \tag{2}$$

$$l_a = c_0{}^a < c_1{}^a < c_2{}^a < \ldots < c_{k_a}{}^a < c_{k_a+1}{}^a = r_a. \tag{3}$$

$$V_a = [c_0{}^a, c_1{}^a) \bigcup [c_1{}^a, c_2{}^a) \bigcup \ldots \bigcup [c_{k_a}{}^a, c_{k_a+1}{}^a]. \tag{4}$$

   So, $P = \bigcup_{a \in R} P_a$ defines a new decision table $IS^p = (U, A, V^p, f^p), f^p(x_a) = i \Leftrightarrow f(x_a) \in [c_i{}^a, c_i + 1^a), x \in U, i \in \{0, \ldots, K_a\}$. It means that original information system is replaced by a new one after discretization.

It's known that a sample in the decision table denotes a decision rule, and all these samples that are useless for their inadaptability form a decision rule set . In order to abstract rules with high adaptability, the decision table needs to be reduced. In the rough-set-based knowledge acquisition, decision rules are achieved via reduction of the decision table, including reduction of attributes and reduction of values.

Reduction of attributes is to remove some condition attributes not important to decision attributes, so the remaining condition attributes can be used for decision making [7]. However, reduction of attributes is still not able to remove all of the redundant information from the decision table. So reduction of values is used for further reduction. We define decision rules d_x as follows:

$$d_x : des([x]_C) \Rightarrow des([x]_D), d_x(a) = a(x), a \in C \cup D, \tag{5}$$

where $des([x]_C), des([x]_D)$ are separately defined as the condition and decision of d_x.

For d_x, $[x]_C \subseteq [x]_D$, if $\forall r \in C$ and $[x]_{C \setminus \{r\}} \subseteq [x]_D$, r is not the core attribute set of d_x and it can be ignored; or else r is the core attribute set.

2.2   MPEG Standards

MPEG (Moving Picture Experts Group) [12] is an expert group established by the International Standards Organization (ISO) to develop dynamic image compression standards. There are five standards proposed by this group, including MPEG-1, MPEG-2, MPEG-4, MPEG-7, and MPEG-21, which span all aspects of compressing, authoring, identifying, and delivering multimedia. MPEG-1 and MPEG-2 standards deal with interactive video on CD-ROM and Digital Television. MPEG-4, proposed in 1999, provides the standardized technological elements enabling the integration of the production, distribution and content access paradigms of the fields of digital television, interactive graphics and interactive multimedia. MPEG-7, officially called the Multimedia Content Description Interface, is a set of rules and tools for describing content. MPEG-21 seeks to let content distributors have complete control over content at all parts of the delivery chain and on all kinds of networks and devices.

Many storage media are organized with MPEG-2 format [13]. Two coding techniques, prediction and interpolation, are adopted in the MPEG-2 video compression algorithm in order to satisfy high compression ratio and random access. There are three specific types of frames in MPEG-2, I frames (intra-coded frames), P frames (predictively coded frames) and B frames (bidirectionally predictively coded frames). I frames don't need reference frames when coded and their compression ratio is the lowest. I frames are random access points and reference frames for other types of frames. P frames are predicted by use of only one previous I or P frames. Meanwhile they are reference frames for subsequent frames. Because P frames utilize the temporal redundancy of video, their compression ratio is high. B frames are predicted referring to at most two frames(I or P frames) and can't be used as reference frames. Their compression ratio is the highest because of their bidirectional prediction with motion compensation.

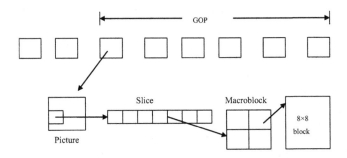

Fig. 1. Data Hierarchy of MPEG Video

In the MPEG-2 standard, there is a data structure called GOP (group of pictures), which is shown in Fig. 1. The count of frames in GOP is a constant, and the first frame in GOP must be the I frame. A frame consists of certain slices, which are composed of arbitrary numbers of macroblocks. Macroblocks appear in raster scanning order in the video stream. A macroblock, consisting luminance block or chroma block, is the unit of prediction with motion compensation. The block is the unit of DCT (discrete cosine transform). Frames are predicted with motion compensation by use of the previous decoded frames (I or P frames), then they are reconstructed via combining the prediction with the coefficient data from the IDCT (inverse discrete cosine transform) outputs.

MPEG-7, formally named "Multimedia Content Description Interface", aims to create a standard for describing the multimedia content data that will support some degree of interpretation of the informations meaning, which can be passed onto, or accessed by, a device or a computer code. MPEG-7 is not aimed at any one application in particular; rather, the elements that MPEG-7 standardizes shall support as broad a range of applications as possible. MPEG-7 is the core part of the video retrieval system. The forepart of MPEG-7 is the analysis results of multimedia data, while the rear-end of MPEG-7 is the basis of extraction of multimedia data. MPEG-7 visual description tools consist of basic structures and descriptors that cover following basic visual features: color, texture, shape, motion, localization, others. Among them, motion descriptions are relative with the work in the paper.

3   Rough-Set-Based Video Pre-classification Modeling

A rough set approach to video pre-classification is considered in this section.

3.1   Modeling of Video Pre-classification System

As said above, there are three types of frames. Motion information is contained in P frames and B frames. Since the information in adjacent B frames are nearly the same, only P frames in the GOP is analyzed in the motion-information-based video retrieval system. In this paper, video are pre-classified just by analyzing information of P frames in the compressed domain.

The video pre-classification system is shown in Fig. 2. At first, motion information of P frames is extracted from MPEG-2 video database. Then abnormal data in P frames are removed in order to make the results more exact. The remaining data are used to extract feature attributes via the algorithm proposed in the paper. After analyzing values of all feature attributes by the rough set classifier, decision rules are obtained finally.

In order to complete an information table referring to Section 2, we need to extract the data set U and determine an attribute set A.

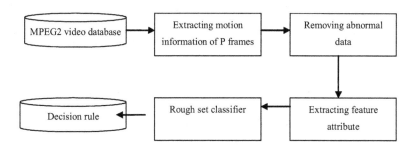

Fig. 2. Rough Set Based Video Pre-classification

For universality of samples, various test sequences are chosen, including cartoon clips, advertisement clips, and news clips. Then, after extracting many P frames, motion types of these P frames are added by hand to form information data U. There are 1367 frames in our experiments.

3.2   Determination of Attributes

Extracting Motion Information. In our information system, the analyzed data are extracted from the compressed domain of P frames. They include macroblock types and motion vectors.

The macroblock type is denoted as the field *macroblock_type* in the MPEG stream. This field is only used in video coding, but hardly in video retrieval. To exploit temporal redundancy, MPEG adopts macroblock-level motion estimation. During motion estimation, the encoder first searches for the best match of a macroblock in its neighborhood in the reference frame. If the prediction macroblock and the reference macroblock are not in the same positions of the frames, motion compensation is applied before coding.

Given that *No_MC* means no motion compensation, when a macroblock has no motion compensation, it is referred as a *No_MC* macroblock. Generally, there are two kinds of *No_MC*, which are intra-coded *No_MC* and inter-coded *No_MC*. In typical MPEG encoder architecture, there exists an inter/intra classifier. The inter/intra classifier compares the prediction error with the input picture elements. If the mean squared error of the prediction exceeds the mean squared value of the input picture, the macroblock is then intra-coded; otherwise, it is inter-coded. In fact, in a special case, when the macroblock perfectly matches its reference, it is skipped and not coded at all. So macroblocks can be classified as five types shown in the first column in Table 1. Furthermore, some types can be combined, and finally we specify three types of motions, which are motions with low change(L), motions with middle change (M) and motions with high change (H).

Table 1. Macroblock Types

Five macroblocks	Our type X
skipped macroblock	L
intra-coded macroblock	H
motion compensated and coded macroblock	M
motion compensated and not coded macroblock	M
not motion compensated and coded macroblock	L

Via the field *macroblock_type*, the ratios of macroblock types are defined as follows:

$$RateH = Hcount/Tcount, \tag{6}$$

$$RateM = Mcount/Tcount, \tag{7}$$

$$RateL = Lcount/Tcount, \tag{8}$$

where *Tcount* is the whole count of the macroblock in a P frame, and the other *Xcount* is the count of the macroblock which type is X, as given in Table 1.

Removing Abnormal Data. In order to accurately analyze motion vectors, global and local abnormal data need to be removed at first. The algorithm of removing abnormal data is shown in Fig. 3.

Assumed that all motion vectors of a P frame are based on Gaussian distributions, one motion model will be directly built up. Motion vectors here are

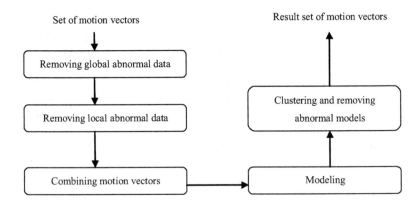

Fig. 3. Removing Abnormal Data

extracted via block matching algorithm. There are many rules to match blocks. One is to compute the minimum value of mean square deviation between the searched block and the current block. Another is to find the minimum value of mean absolute deviation between them. The search methods of the block matching algorithm include full search and logarithm search. Although the block matching algorithm tries to find the best matches, in low texture areas of frames there are random errors that are called abnormal data. For example, in Fig. 4, it denotes the zoom movement of the camera, from which one can clearly find some global abnormal data that don't conform to the whole.

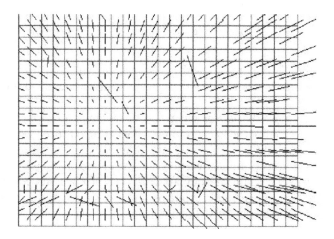

Fig. 4. Global Abnormal Data

The coordinate system for a P frame is shown in Fig. 5, in which the original point is in the center and the rectangle denotes a 16*16 macroblock with coordinate (i, j). The line emitting from the rectangle center denotes the motion vector of the macroblock. We define:

$$E_{ij} = \sqrt{v_{ijx}^2 + v_{ijy}^2}, \tag{9}$$

$$\theta_{ij} = angle, \tag{10}$$

where v_{ijx} is the horizontal motion vector of macroblock (i, j) and v_{ijy} is the vertical, E_{ij} is its movement energy and θ_{ij} is its movement direction.

The energy of the macroblock can be characterized by several levels. According to the human visual system, here it is non-uniformly quantified as five degrees: $0 \sim 8$, $8 \sim 16$, $16 \sim 32$, $32 \sim 64$, ≥ 64. Then we define:

$$ratio[i] = count[i]/sum, \tag{11}$$

where sum denotes the sum of macroblocks with motion vectors, and $count[i]$, $ratio[i]$ respectively denote the count and ratio of macroblocks in level i. After a

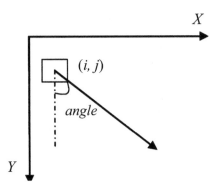

Fig. 5. Movement energy and direction of the macroblock

lot of experiments, a proper threshold 7% is set here, which means that if $ratio[i]$ is smaller than the threshold, all motion vectors in the qualified level i are set to zero.

For global motions, the movement trend of neighbor macroblocks is always changed gradually. So are the local motions sometimes, but it becomes the fastest in the regions where different objects with different motions are tangent. As a result, we use 3*3 detection matrices to trim local abnormal data. The center of matrices as a reference point is compared with other eight macroblocks by the energy and direction. Then sudden changed ones are removed during processing. After this step, the following is defined:

$$RateM1 = Mcount1/Tcount, \qquad (12)$$

where $Mcount1$ is the count of macroblocks having motion vectors after deleting abnormal data, not as the definition in Formula 7.

Abstracting Movement Models. The motion vectors inconsistent with global or local optical feeling have been removed after the above steps. The remaining motion vectors are able to accurately express the movement they belong to. In the following step we need to abstract all of the main movement models to express the motion features of video.

Sequentially, macroblocks are combined by checking whether they are neighboring. The region S_k is defined as:

$$S_k = \{(i,j)|(i,j) \text{ are neighboring and } E_{ij} \neq 0\}. \qquad (13)$$

As shown in Fig. 6, the motion vectors of P frames are divided into five regions, which are not adjacent to each other. Every region has an unique motion model, so the motion model of each region can be built up at first.

There are two ways to build up the motion mode: non-image method and image method. The non-image method is used to build up models by directly extracting operation parameters of cameras during the shoot process, but it

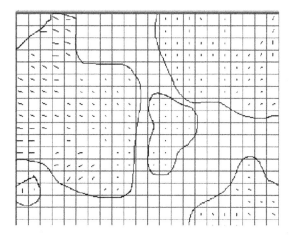

Fig. 6. Combination of motion vectors

needs extra equipments. The image method is used to build up motion-vector-based parameter models. For the parameter model, six parameters and eight parameters models are usually adopted. A six-parameter affine model can be applied for general applications. Since the motion model of each region is unique, we adopt a six-parameter affine model to describe the movement of each region. The optimal motion parameters as follows are calculated by using the least square method:

$$u = a_0 x + b_0 y + z_0, \tag{14}$$

$$v = a_1 x + b_1 y + z_1. \tag{15}$$

x is the horizontal coordinate of macroblocks. y is the vertical coordinate of macroblocks. u is the horizontal motion vector. v is the vertical motion vector. $a_0, b_0, z_0, a_1, b_1, z_1$ are six motion parameters describing models where a_0, b_0, a_1, b_1 describe the depth of motion scenes and z_0, z_1 describe 2-D global or local motions [21]. Therefore, regions with the similar motion models can be combined further by taking six parameters above into consideration. The count of final motion models is defined as *ModuleNum*.

The six-parameter affine model is used to express the movement of each region. For each region S_k, the deviation function is defined as:

$$E(k) = \sum_{(x,y) \in S_k} [(u - a_0 x - b_0 y - c_0 z)^2 + (v - a_1 x - b_1 y - c_1 z)^2]. \tag{16}$$

The optimized motion parameters of S_k make $E(k)$ in Formula 16 the minimum value.

Since the abnormal data having interference on least square method have been trimmed, the optimized motion parameters of combined regions can be obtained via the least square method.

Because the movements in video are very complex, there may be various motion models at the same time. Considering the video of a basketball match, there are scenes of players' grabbing, acclaim of the surroundings audience, and the global motion that a camera tracks the player's shooting. According to the human optical feature, we assume that at any time people usually focus their eyes on the dominant motions , which are just required in the motion-based video retrieval. For example, when the camera tracks a running player, the track of camera is the primary motion, because it indirectly reflects the running information of the player. Some information, such as the change of audiences on the background, can be ignored.

The algorithm of clustering models is to trim minor motion information and apply the remains of motion models for the content-based video retrieval and classification.

At first the rules are defined to check whether two motion models are similar. If two motion models show significant 3-D movements via analyzing a_0, b_0, a_1, b_1, they are judged to be similar. Otherwise, the ranges and directions of 2-D movements are compared via analyzing z_0, z_1; if their difference is higher than the threshold set in the system, they can't be processed as one model.

After definition of rules, each region is compared with the biggest combined region. If the two regions have inconsistent motion models and the biggest combined region is twice bigger than current region, current region should be deleted, or else its motion model will be reserved.

At last, regions with the similar motion models can be clustered further, and the count of final motion models is defined as $ModuleNum$.

Extracting Other Motion Features. A motion activity description is recommended in MPEG-7, in which spatial distributions of motion activity include motion density, motion rhythm and so on. It can also be indirectly scaled with intensity of motion vectors in the MPEG stream as follows [3]:

$$C_{mv,avg} = \frac{1}{IJ} \sum_{i=-I/2}^{I/2} \sum_{j=-J/2}^{J/2} E_{ij}, \tag{17}$$

where I is the width of a P frame in unit of a macroblock, and J is the height of it.

Referring to the literature [11], the motion centroid Com and motion radii Rog are calculated as follows:

$$m_{pq} = \sum_{i=-I/2}^{I/2} \sum_{j=-J/2}^{J/2} i^p j^q f(i,j), \text{ where } f(i,j) = \begin{cases} 0 & \text{if } E_{ij} = 0, \\ 1 & \text{if } E_{ij} \neq 0. \end{cases} \tag{18}$$

$$Com_x = \frac{m_{10}}{m_{00}}, \ Com_y = \frac{m_{01}}{m_{00}}, \ Com = \sqrt{Com_x^2 + Com_y^2}. \tag{19}$$

$$Rog_x = \sqrt{\frac{m_{20}}{m_{00}}}, \ Rog_y = \sqrt{\frac{m_{02}}{m_{00}}}, \ Rog = \sqrt{Rog_x^2 + Rog_y^2}. \tag{20}$$

Selection of Feature Attributes. Next, feature attributes of the information system need to be chosen. Due to correlations among $RateH$, $RateM$ and $RateL$ which are defined in Formula (6)(7)(8), only $RateH$ and $RateL$ are selected to describe ratios of different types of macroblocks together with $RateM1$ given in Formula (12). $ModuleNum$ given in Section 3.2 is used to denote the count of main motions in a frame. The four attributes above are computed by analyzing not only motion vectors but also the field $macroblock\_type$ in the MPEG steam. The field $macroblock\_type$ is hardly referred in literatures, but it is crucial for us to distinguish global motion frames from non-global motion frames in our experiments. So $RateH$, $RateM$, $RateL$ and $ModuleNum$ are defined for the first time. Referred to [3] and [11], Cmv, avg, Com, Rog are also selected.

Above all, seven feature attributes are defined in order to construct the video pre-classification condition attribute set C.

$$C = \{RateH, RateM1, RateL, ModuleNum, Cmv, avg, Com, Rog\}. \quad (21)$$

The seven attributes can show differences between global motion model and non-global motion model, such as ratios of different motions, the count of motions and movement characteristics. As a result, those attributes are quite enough for video classification. The decision attribute of the decision table is described that global motion type is 1 and non-global motion type is 2. Table 2 shows a example of decision table.

Table 2. Example of Decision Table in Video Pre-classification System

$RateH$	$RateM1$	$RateL$	$ModuleNum$	$C_{mv,avg}$	Com	Rog	Type
0.167	0.121	0.712	1	3.320	0.279	3.732	2
0.179	0.129	0.692	3	1.909	0.192	2.067	2
0.475	0.25	0.275	2	4.427	0.136	2.269	1
0.636	0.053	0.311	1	2.867	0.61	5.734	2
0.863	0.0542	0.083	1	0.338	0.121	0.937	1
0.106	0.053	0.819	1	0.211	0.531	1.131	2
0.159	0.25	0.591	1	0.962	0.102	0.452	1
0.004	0	0.996	0	0	0	0	2
0.285	0.135	0.579	2	0.989	0.020	1.122	2

(Condition attributes / Decision attribute)

3.3 Extraction of Decision Rules Based on Rough Set

Extraction of decision rules via the rough set theory includes several steps, which are data preprocessing, reduction of attributes, reduction of values and obtaining logic rules according to reduction of values.

Data Preprocessing. It includes deleting repeated records, filling missing data and discretization. Discretization is the process by which a set of values is grouped together into a range symbol. In this paper some discretization methods are considered.

The discretization method, Equal Interval Width, divides the range of observed values for an attribute into k equal sized intervals, where $k > 0$ is a user-supplied parameter. If an attribute a is observed to have values bounded by a_{min} and a_{max}, then this method computes the interval width $width(k) = (a_{min} - a_{max})/k$ and constructs thresholds at $a_{min} + i * width(k)$, where $i = 1, ..., k - 1$. The method is applied to each continuous attribute independently. Since this unsupervised method does not utilize decision values in setting partition boundaries, it is likely that classification information will be lost by binning as a result of combining values that are strongly associated with different classes into the same interval. In some cases this could make effective classification much more difficult.

Nguyen improved greedy algorithm is another way to discretize data. Nguyen S. H gave some detailed description about discretization in rough set in reference [15]. He gave the complexity of discretization problem and proved that it is an NP-hard problem. And at the same time he also proposed a basic heuristic algorithm based on rough set and boolean reasoning, which brought great improvement in dealing with discretization problem in rough set. For convenience of discretization, we call this algorithm basic heuristic algorithm.

In Naive Scaler algorithm [9], it implements a very straightforward and simple heuristic method that may result in very many cuts, probably far more than are desired. In the worst case, each observed value is assigned its own interval. In some cases, however, a simplistic and naive scheme may suffice. For the sake of simplifying the exposition, we will assume that all condition attributes A are numerical. For each condition attribute a we can sort its value set V_a to obtain the following ordering:

$$v_a^1 < ... < v_a^i < ... < v_a^{|v_a|}. \tag{22}$$

Let C_a denote the set of all naively generated cuts for attribute a, defined as shown below. The set C_a simply consists of all cuts midway between two observed attribute values, except for the cuts that are clearly not needed if we do not bother to discern between objects with the same decision values.

$$X_a^i = \{x \in U \mid a(x) = v_a^i\}. \tag{23}$$

$$\triangle_a^i = \{v \in V_d \mid \exists x \in X_a^i \text{ such that } d(x) = v\}. \tag{24}$$

$$C_a = \{(v_a^i + v_a^{i+1})/2 \mid |\triangle_a^i| > 1 \text{ or } |\triangle_a^{i+1}| > 1 \text{ or } \triangle_a^i \neq \triangle_a^{i+1}\}. \tag{25}$$

In essence, we place cuts midway between all v_i^a and v_i^{a+1}, except for in the situation when all objects having these values also have equal generalized decision values a that are singletons.

If no cuts are found for an attribute, the attribute is left unprocessed. Missing values are ignored in the search for cuts.

Comparison experiments show that Nguyen improved greedy algorithm achieves the best result in our case. So this algorithm is adopted.

Reduction of Attributes. The forward attributes selecting and the backward attributes deletion are two types of attributes reduction.

Algorithms of attributes reduction via forward attributes selecting usually start from the core of the condition attributes. They gradually extend the attribute set according to importance of each attribute. Different algorithms of attributes reduction have different definitions of attribute importance. Typical algorithms include that based on the condition information entropy and that based on the discernibility matrix.

The reduction algorithm based on the condition information entropy are proposed in [26]. It analyzes the information view of rough set theory and compares it with the algebra view of rough set theory. Some equivalence relations and other kinds of relations like the inclusion relation between the information view and the algebra view of rough set theory are resulted through comparing each other. Based on the above conclusion, [26] proposes a novel heuristic knowledge reduction algorithm based on the conditional information entropy.

The algorithm based on the discernibility matrix is another method to reduce attributes. In the algorithm, if one attribute is more frequent in the discernibility matrix, its classification ability is stronger. The definition of discernibility matrix is given in [7].

Unlike the forward attributes selecting algorithms, the backward attributes deletion algorithms start from the whole condition attributes. They gradually delete some attribute according to importance of each attribute until certain conditions are satisfied. Unfortunately, the backward attributes deletion is studied less.

Here the reduction algorithm based on condition information entropy [26] is adopted.

Reduction of Values. There are many algorithms to reduce values, such as the inductive reduction algorithm, the heuristic reduction algorithm, the reduction algorithm based on decision matrices and so on.

The reduction algorithm based on decision matrices [20] is adopted in the paper. It uses a variable precision rough set model. If P and Q are equivalence classes in U, the positive region of Q to P $POS_P(Q)$ is defined as:

$$POS_P(Q) = \bigcup_{X \in U/Q} P_-(X). \tag{26}$$

For an information table RED after reduction of attributes, let $X_i^+ (i=1,2,\ldots,\gamma)$, $X_j^- (j = 1,2,\ldots,\rho)$ denote the equivalence classes of the relation $R^*(RED)$, $X_i^+ \subseteq POS_{RED}^\beta(Y), X_j^- \subseteq NEG_{RED}^\beta(Y)$, the decision matrix $M = (M_{ij})_{\gamma*\beta}$ is defined as:

$$M_{ij} = \{(a, f(X_i^+, a)) : a \in RED, f(X_i^+, a) \neq f(X_j^-, a)\}. \tag{27}$$

M_{ij} contains all attribute pairs with different values in the equivalence classes X_i^+ and X_j^-. Given an equivalence class X_i^+, if each element of M_{ij} is taken as a boolean expression, the decision rule set can be expressed as follows:

$$B_i = \bigwedge_j (\bigvee M_{ij}). \qquad (28)$$

It's seen that basic implication of B_i is actually the maximal generalized rule of the equivalence class X_i^+, which belongs to the positive region $POS_{RED}^\beta(Y)$. As a result, by finding out all basic implication of the decision function $B_i (i = 1, 2, \ldots, \gamma)$, all maximal generalized rules of the positive region $POS_{RED}^\beta(Y)$ are computed.

Obtaining Logic Rules According to Reduction of Values. Combined with logic meanings of attributes, the rules derived from reduction of attributes are analyzed to form logic rules, which are validated later taking coding features of P frames into account.

4   Detection of Shot Boundary

As mentioned above, the macroblock includes luminance blocks and chroma blocks. The human visual system are more sensitive to luminance information than chroma information, so chroma blocks need not to be analyzed. The luminance information of original data is transformed to DCT coefficients via DCT and saved in a certain number of 8*8 blocks. The 64 DCT coefficients in a block are frequency coefficients. The first coefficient among them is called DC (direct coefficient), and the other 63 ones are called AC (alternating coefficients). Because the change between two close picture elements is gradual, most information is in the low frequency region, so the first coefficient DC contains the most information in the block, which is only extracted in the video retrieval system. In this way, it not only speeds up analysis of video, but also effectively expresses video information.

A shot is defined as a number of images which are continuously recorded by the same camera. Simply speaking, a shot is a set of successive frames. The abrupt transition is the primary edit type between two successive shots. It directly connects two shots with no film editing operation. In general, it represents abrupt changes between two frames, such as the change of the scene luminance, motions of object and background, the change of edges and so on. In the same scene, the abrupt change of luminance mainly results from illumination change or movements of scenery. Otherwise, it's usually caused by switch of scenes. In fact, it is difficult to know which reason mentioned above causes the luminance change. Therefore the abrupt transition detection algorithm should be able to reduce the influence of the luminance changes which are not caused by shot switches.

The boundary based method [31] is adopted, which is similar to the image segment. It makes use of discontinuity between shots and contains two important points: one is to detect each position to check whether any change happens, the other is to determine whether it is the real shot boundary according to the characteristics of abrupt transition.

The most common algorithm of boundary based method is the sliding window method [10]. The sliding window spans $2R+1$ frames. The metric used as follows is the sum of differences between successive frames:

$$D = \sum_{R}^{t=-R} |f(x,y,t) - f(x,y,t+1)|. \tag{29}$$

D is considered as a shot boundary if D is the maximum value and the second largest maximum value is D_2. $D > kD_2$, where k is a positive coefficient.

$f(x,y)$ in Formula (29) can be specified via DC values from compressed domain.

The dual-sliding-window-based detection method improves the traditional sliding window method. If traditional sliding window is applied, there are leak detection and wrong detection due to motions of big objects. The improved method can overcome these problems and performs great results. At first a big window is defined to determine some possible frames where abrupt transitions may happen. Then a small window is defined and the chosen frames in the first step are set in the middle of the window. Next, in order to avoid leak detection and wrong detection, single side detection is added in the algorithm.

The steps of this method are shown as follows:

Step 1: Define a big window whose size is W_B. Calculate the mean difference m in this window to detect the frames where abrupt transition may happen. If the differences are certain times bigger than m, the relative frames can be detected further to determine if it is the real position where abrupt transition happens.

Step 2: Define a small window whose size is $W_s = 2r - 1$. The chosen frames in the first step are set in the middle of the window.

Step 3: If the frames accord with conditions of double sides below (k_1 and k_2 are predefined thresholds), it's considered that the abrupt transition happens on frame t. Then the algorithm returns to the Step (2) and continues to detect frames starting from the frame $t + r$:

$D_t \geq k_1 m$;

$D_t \geq D_i, i = t - r + 1, t - r + 2, \ldots, t - 1, t + 1, \ldots, t + r - 2, t + r - 1$;

$D_t \geq k_2 D_2$, where D_2 is the second maximum value of difference.

Step 4: Otherwise, if they accord with the following single side condition (k_3 and k_4 are predefined thresholds), it's considered that the abrupt transition happens on frame t:

$D_t \geq k_3 m$;

$D_t \geq k_4 D_l \text{ or } D_t \geq k_4 D_r$;

where $D_t = \max(D_i), i = t - r + 1, t - r + 2, \ldots, t - 1, D_r = max(D_i), i = t - r + 1, t - r + 2, \ldots, t - 1, t + 1, \ldots, t + r - 2, t + r - 1$.

5   Shot Type Checking

It is not enough only to classify P frames based on motion information for video retrieval. A shot, as the unit of video, needs to be classified further. Generally speaking, shots can be classified as two types, global motions or local motions. If global motion is dominant in a retrieval sample, only the shots with the type of global motions in the video database need to be retrieved; otherwise only the shots with the type of local motions need to be retrieved.

In lots of experiments, it is found that motion types of shot are able to be checked by computing proportion of each type of P frames in a shot. Part of experiment data is listed in Table 3. In Table 3, CP denotes the count of P frames, CG denotes the count of P frames with global motions, CL denotes the count of P frames with local motions and CU denotes the count of P frames whose motion types can't be determined.

We define:

$$GlobalRate = \text{global motion count}/ \text{ total P frames count.} \qquad (30)$$

$$NonglobalRate = \text{non-global motion count}/ \text{ total P frames count.} \qquad (31)$$

By observing data in Table 3, it's seen that for the shots with many frames, if global motions are dominant in them, its $GlobalRate$ is bigger than a certain value, or else $NonglobalRate$ is bigger than another certain value. The two values can be equal. It's proper if the value is 0.60. Motion types of the shots with few P frames can't be determined sometimes, as exemplified by the shot with 4 P frames in Table 3. Since their analysis means no sense, their processing can be ignored.

Table 3. Count of P Frames with Different Types in Shots

Shot	CP	CG	CL	CU	$GlobalRate$	$NonglobalRate$	Shot type
1	34	12	21	1	0.35	0.62	2
2	15	13	1	1	0.87	0.07	1
3	3	3	0	0	1	0	1
4	14	4	9	1	0.29	0.64	2
5	9	1	8	0	0.07	0.89	2
6	15	13	1	1	0.87	0.07	1
7	4	1	2	1	0.25	0.50	-
8	3	0	3	0	0	1	2
9	32	28	3	1	0.88	0.09	1

6   Motion-Information-Based Shot Retrieval Scheme

In the paper, the retrieval of shots with global motions is mainly discussed. As mentioned in Sect. 3.2, a six-parameter affine model is founded to trim abnormal data. For the whole P frames with global motions, a motion model can be set up via the six-parameter affine model too. a_0, b_0, z_0, a_1, b_1, z_1 are obtained referring to Formula (14) and (15).

As the retrieval granularity is an shot, the result is a set of shots sorted by similarity with the submitted samples. So the distance between two shots needs to be defined.

If two shots are described as V_1, V_2, whose global P frames are $\{m_1(i)\}$ and $\{m_2(j)\}$, their distance is defined as:

$$D(V_1, V_2) = \min_{i,j} d(m_1(i), m_2(j)), \tag{32}$$

where $d(m_1, m_2)$ denotes the distance between two P frames.

According to the literature [11], $d(m_1, m_2)$ is be defined as follows:

$$d(m_1, m_2) = \sum_{x,y} [u_{m1}(x, y) - u_{m2}(x, y)]^2 + [v_{m1}(x, y) - v_{m2}(x, y)]^2. \tag{33}$$

Combined with formula (14) and (15), it can be defined as:

$$d(m_1, m_2) = \sum_y \sum_x [(z_{10} - z_{20}) + (a_{10} - a_{20})x + (b_{10} - b_{20})y]^2 +$$

$$[(z_{11} - z_{21}) + (a_{11} - a_{21})x + (b_{11} - b_{21})y]^2. \tag{34}$$

After reduction, the formula above becomes:

$$d(m_1, m_2) = [(a_{10} - a_{20})^2 + (a_{11} - a_{21})^2] \sum_x x^2 \sum_y 1 +$$

$$[(b_{10} - b_{20})^2 + (b_{11} - b_{21})^2] \sum_x 1 \sum_y x^2 +$$

$$[(z_{10} - z_{20})^2 + (z_{11} - z_{21})^2] \sum_x 1 \sum_y 1 +$$

$$2[(a_{10} - a_{20})(b_{10} - b_{20}) + (a_{11} - a_{21})(b_{11} - b_{21})] \sum_x x \sum_y y +$$

$$2[(b_{10} - b_{20})(z_{10} - z_{20}) + (b_{11} - b_{21})(z_{11} - z_{21})] \sum_x 1 \sum_y y +$$

$$2[(a_{10} - a_{20})(z_{10} - z_{20}) + (a_{11} - a_{21})(z_{11} - z_{21})] \sum_x x \sum_y 1. \tag{35}$$

Given the length(M) and width (N) of P frames, some values in the formula above can be computed

$$\sum_x 1 = M, \sum_y 1 = N. \tag{36}$$

$$\sum_x x = M(M + 1)/2, \sum_y y = N(N + 1)/2. \tag{37}$$

$$\sum_x x^2 = M(M + 1)(2M + 1)/6, \sum_y y^2 = N(N + 1)(2N + 1)/6. \tag{38}$$

The distance between two P frames $d(m_1, m_2)$ can be calculated by combining Formula (35)(36)(37)(38) with a_0, b_0, z_0, a_1, b_1, z_1. Finally the distance between two shots is computed according to Formula (32), then the similarity between them can be specified.

7   Experiments and Analysis of Results

7.1   Experiments of Removing Abnormal Data of P Frames

Firstly, several typical video frames are chosen to test the algorithm of deleting abnormal data of the motion vectors, compared with the method proposed in literature [11]. The chosen video frames [8], original pictures of motion vectors, result pictures of motion vectors via the algorithm proposed in literature [11] and the one adopted in the paper are shown in Fig. 7 in turn.

From the chosen video, there is the movement of a camera in the first P frame. Due to simple motion type, there are only a few global abnormal data in the original frame. From the picture, it's seen that the camera movement is obvious. Experiment results with the two algorithms are nearly the same. The second P frame includes two movements, one man going upstairs by lift in the left and another person going downstairs by lift in the top right corner. The left movement that is more obvious than the right is deleted via the method proposed in literature [11], while they are saved after the algorithm discussed in

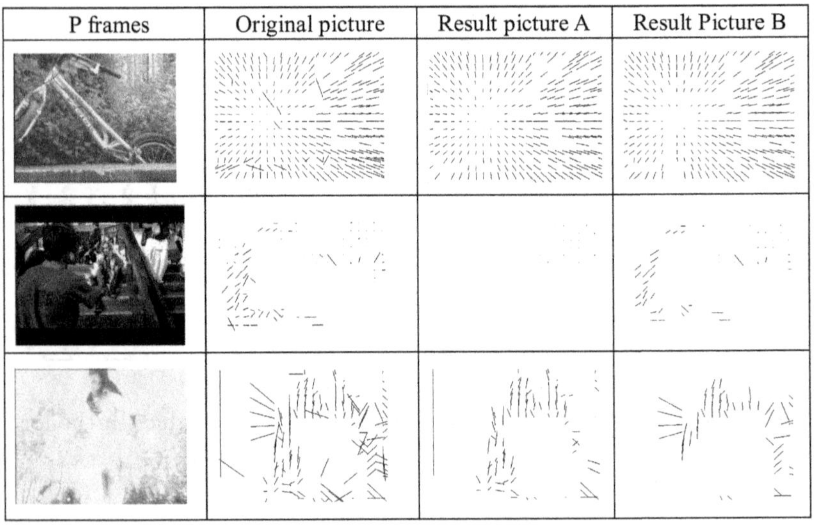

P frames	Original picture	Result picture A	Result Picture B

Fig. 7. Pictures of Motion Vectors. Original picture denotes original pictures of motion vectors; result picture A denotes result pictures of motion vectors via the algorithm proposed in literature [11]; result picture B denotes result pictures of motion vectors via the algorithm proposed in the paper.

the paper instead. The third P frame contains camera zooming and one jumping person. However, the same error occurs via the method proposed in literature [11], while the result of the algorithm proposed in the paper is right.

It's concluded that the method proposed in literature [11] is unreliable if there are global motions and local motions, especially if objects with local motions are large compared with the whole frame.

Taking universality of samples into consideration, various test sequences, including cartoon clips, advertisement clips, news clips and physical clips, are chosen. All the videos are coded with MPEG-2 and their frame sequences are organized as IBBPBBPBBIBB. All P frames in these video are analyzed, and motion information which depends on our vision is compared with remaining motion vectors and motion models via the algorithm proposed in the paper. It's considered as the wrong recognition if they are too different. The result is shown in Table 4.

From Table 4, it's seen that the accuracy of the three video is high, so the algorithm is robust if it is applied to the motion-information-based video retrieval.

Table 4. Recognition Accuracy of Motion Models

P frames	Count	Right recognition	Accuracy (%)
Physical clip	41	39	95.1
Advertisement clip	102	96	94.1
News clip	129	119	92.2

7.2 Experiments of Video Pre-classification

As said above, many kinds of video clips, including cartoon clips, advertisement clips and so on, are selected for our experiment. These video clips are available at the website of Institute of Computer Science and Technology, at Chongqing University of Posts and Telecommunication in China [8]. 1367 P frames extracted from them are taken as the data set U, 440 of which are global motion frames and the others are non-global. 400 samples are randomly selected from them as training data, in which 146 are global. Then, in order to form the video classification decision table used for rule producing, feature attributes of those frames are extracted one by one according to attribute descriptions above. The training data and test data can also be got in the website [8].

In our experiments, the RIDAS system is adopted as a data mining platform, which is developed by Institute of Computer Science and Technology, Chongqing University of Posts and Telecommunications in China [32]. The system integrates almost 30 classical algorithms regarding rough set theory.

Since values of attributes are unique and can be computed, there's nothing missing in the information table and the filling of missing data is not needed. Types of most data are float, so discretization for data is necessary. In the RIDAS system, there are 12 algorithms in all for discretization. In our experiments, we used all of the algorithms to discretize the training data. It's found that if we use any algorithm except the greedy algorithm described in Section 3.3, no matter

Table 5. Values of Intervals of Attributes after Discretization

Attribute	Count of interval	Broken set
$RateH$	2	*, 0.223, *
$RateM1$	4	*, 0.082, 0.138, 0.170, *
$RateL$	4	*, 0.330, 0.499, 0.671, *
$ModuleNum$	2	*, 1.5, *
$C_{mv,avg}$	5	*, 0.136, 0.218, 0.529, 1.243, *
Com	4	*, 0.206, 0.241, 0.441, *
Rog	6	*, 0.661, 0.889, 1.096, 1.193, 2.452, *

which method is sequentially used for reduction of attributes and values, the result is very poor. We further found that Nguyen improved greedy algorithm outperforms other greedy algorithms, so it is adopted in our experiments. Via Nguyen improved greedy algorithm, the broken set can be obtained as listed in Table 5. From each broken set, intervals can be easily got. For example, the broken set (*, 0.223, *) means that there are 2 intervals, including [*, 0.223] and [0.223, *].

All condition attributes are reserved after reduction of attributes by using all of the attributes reduction algorithms. It indicates that the attributes are all partly necessary, which means the decision attributes educed by them are accordant with that educed by all attributes.

After attribute reduction, value reduction of the training data needs to be done and rule match strategies of the test data need to be chosen. 4 algorithms of value reduction and 2 rule match strategies are used for our experiment, and the result of comparison experiments is shown in Table 6.

Table 6. Accuracy Comparison of Each Strategies

Strategy	Reduction algorithm			
	Heuristic red.	Decision matrices red.	Inductive red.	General red.
Major Acc.(%)	77.6	79.2	78.3	81.7
Minor Acc.(%)	81.1	82.4	81.1	81.7

In Table 6, "Major Acc." denotes accuracy via the majority priority strategy for test data to match rules, while "Minor Acc." denotes accuracy via the minority priority strategy [27] . It's seen from Table 6 that when value reduction based on decision matrixes and rule match the via minority priority strategy are adopted in our experiments, the result is higher than other strategies, in which the accuracy is 82.4%. So it's adopted in our system.

After value reduction, totally 97 rules are obtained from the video pre-classification information table. Some rules covering over 20 samples are listed as follows.

rule1: $C_{mv,avg}(4) \bigcup ModuleNum(1) \bigcup RateM1(1) \rightarrow D2$
rule2: $C_{mv,avg}(0) \bigcup Rog(0) \rightarrow D2$
rule3: $ModuleNum(0) \bigcup Rog(3) \bigcup RateM(3) \rightarrow D1$
rule4: $Com(0) \bigcup RateL(3) \bigcup RateM(0) \rightarrow D2$

By analyzing rules that cover many samples, we find some useful knowledge below. If the global motion is dominant in a P frame, there is much energy and few motion models, generally only one model; its motion centroid is near the center; and there are many intra-coded macroblocks in the case of few motion vectors. If the local motion is dominant in a P frame, there is little energy and more than two motion models; its motion centroid is far from the center; its motion radii is long; and the ratio of low-changed macroblocks is high.

Next 937 P frames are used as test data for our experiment. These data make rule-match via the minority priority strategy to cope with conflict and inconsistency among them. The results are listed in Table 7. Furthermore, distribution information of wrong recognition of two types of P frames is listed in Table 8.

In addition, we have designed experiments to compare rough-set-based method adopted in the paper with other solutions. One is decision tree with ID3 algorithm [28]; another is the SVMs(Support Vector Machines) classifier [2]. These classifiers are tested with the same test set. The accuracy with each classifier is shown in Table 9. By observing three groups of accuracy in Table 9, we can conclude that the rough-set-based method is better than the other two algorithms and more helpful for global-motion analysis of video.

Table 7. Distribution of Recognition Ratio

	Right recognition	Wrong recognition	unknown recognition
Recognition count	797	142	28
Percentage (%)	82.4	14.7	2.9

Table 8. Distribution of Wrong Recognition

Original type	Count	Recognition type	Count	Percentage (%)
Global motion	294	Global motion	232	79
		Non-global motion	56	19
		Unknown recognition	6	2
Non-global motion	673	Non-global motion	565	84
		Global motion	86	12.8
		Unknown recognition	22	3.2

Table 9. Accuracy with Each Classifier

	Global motion Acc.	Non-global motion Acc.	Total motion Acc.
Roughset	79%	84%	82.4%
Decision tree	74.5%	80.2%	78.4%
SVMs	71%	83.6%	79.8%

7.3 Experiments of Shot Boundary Detection

Results of the dual-sliding-window-based detection method are shown in Table 10. There are two definitions as following:

$$RateF = CR/(CR + CL), \tag{39}$$

$$RateR = CR/(CR + CW), \tag{40}$$

where $RateF$ denotes the ratio of full detection, $RateR$ denotes the ratio of right detection, CR denotes the count of right detections, CL denotes the count of leak detections and CW denotes the count of wrong detections.

In Table 10, $C1$ denotes the frame count, $C2$ denotes the count of abrupt transition, and CW denotes the count of wrong detections. Table 10 shows that the dual-sliding-window-based detection method is able to completely detect abrupt transition of these video; the ratios of full detection are 100% and the ratios of right detection are also high. Some wrong detections happening in the video of Bless and Garfield are caused by gradual transitions. However, excessive segmentation has no passive influence on shot classification and clustering. Shots produced by excessive partition are similar in their content, so shot clustering will put them together finally.

Because the shot boundary detection is the foundation of video analysis, its accuracy has great influence on the following work. The experiments show that the dual-sliding-window-based detection method is perfect, so it's adopted in our video retrieval system.

Table 10. Results of Dual-Sliding-Window-Based Detection Method

Clips	$C1$	$C2$	CR	CL	CW	$RateF$	$RateR$
Bless	1092	22	22	0	4	100%	85%
Toothpaste ad	845	26	26	0	0	100%	100%
Garfield	2332	92	92	0	5	100%	95%

8 Motion-Information-Based Video Retrieval System

An overview of a motion-information-based video retrieval system is given in this section (see the diagram in Fig. 8). The luminance information of video is extracted from MPEG-2 video database at first. Then shot boundary detection is processed via analyzing the luminance information. Motion features are extracted by analyzing P frames, and they are pre-classified on the RIDAS platform. The ratio of each type of P frames in a shot is computed to check the type of the shot; meanwhile feature attributes are restored in the motion information database. Finally, users submit the retrieval shot via the user interface, and the system returns results of shots sorted by similarity.

Shots that users submit via the user interface of the shot retrieval system are processed immediately. Users can choose the shot from video database or any

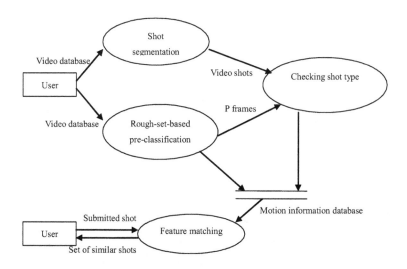

Fig. 8. Diagram of the Motion-Information-Based Video Retrieval System

shots with MPEG-2 format. If the shots are obtained from video database, motion features are directly queried from the motion information database and matched via a certain algorithm. Otherwise, types of shots still need to be checked by analyzing motion information. Shots with different motion types are processed in different ways. The process is shown in Fig. 9.

Firstly, motion information of P frames in a shot is extracted as in Section 3.2. Secondly, P frames are pre-classified via the rough set platform as in Section 3.3. Thirdly, the values of *GlobalRate* and *NonglobalRate* are computed and used to check the motion type of the shot as in Section 5. Finally, if the local motion is dominant in the shot, shots with local motions in the video database are retrieved, or else shots with global motions are retrieved as in Section 6. The retrieval of shots with local motions is relative with the objects on the foreground. Local motion vectors are the differences between original motion vectors and global motion vectors, so the global motion information should be removed at first when retrieving shots with local motions. In this system, local-motion-information-based shot retrieval has not been achieved completely, which is our future work.

A motion-information-based video retrieval system prototype has been designed. Its developing platform is Visual C++ 6.0 and its user interface is shown in Fig. 10. The topside menu bar is used to choose video files and retrieval samples and to operate the retrieval of shots. The left of the window is used to display the content of video data, the scroll bar and buttons are used to operate display modes of video. On the right, several small windows show key frames of video shots which are sorted by similarity after the retrieval of shots. If a small window is clicked, the relative shot can be shown in the left of the window.

Experiments are done in the video retrieval system. As shown in Fig. 10, a shot with the focus camera motion is taken as the retrieval sample, then six

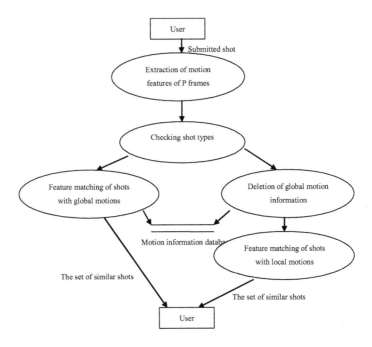

Fig. 9. Process of Motion Information-Based Shot Retrieval

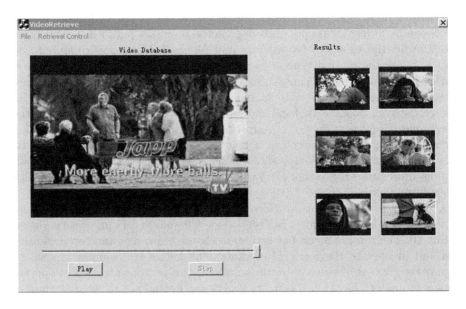

Fig. 10. Video Retrieval Prototype System

similar shots as results return to users . All result shots are separately explored and it's found that the global motions are dominant in all the shots and they are all similar with the submitted shot with the focus camera motion. So the experiment effect is well done.

9   Conclusions

Motion plays an important role as a source of unique information about video data in video retrieval. Unfortunately, video data analysis is not robust due to disturbance of coding at present, which is avoided only through transcendental knowledge. In this article, a robust video retrieval system based on motion information is proposed. In this system, video are divided into shots at first. The dual-sliding-window-based detection method is adopted to detect shot boundaries. Then video frames are pre-classified via the rough-set-based method and the motion types of shots are checked. When users input samples to retrieve video, the shots of global motions and local motions can be separately retrieved from a video database according to the motion type of retrieval samples. Experiments prove that this approach works well.

The most important algorithm in the system is the rough-set-based video pre-classification. Feature attributes of P frames are extracted by analyzing motion vectors and *macroblock_type* fields, which are crucial but usually ignored. Then P frames are classified as the global type and the non-global type via the rough set classifier, so that global motion models with more precise parameters can be built up. Furthermore, the ratio of P frames with each motion type are calculated, and used to check motion types of shots. If they are shots with global motions, similar shots are retrieved from video database and displayed as the result set by use of the shot distance algorithm. If they are shots with local motions, the analysis should be used after removing of global motion information, which is our future work. In addition, more experiments need to be done to improve the results further in the future. All data used in the paper can be obtained from the website [8].

Acknowledgements

This paper is partially supported by Program for New Century Excellent Talents in University (NCET), National Natural Science Foundation of China (No.60373111, No.60573068), Science and Technology Research Program of the Chongqing Education Commission (No.040505, No.060504), Natural Science Foundation of Chongqing of China (No.2005BA2003).

References

1. Adiv, G.: Determining Three-Dimensional Motion and Structure from Optical Flow Generated by Several Moving Objects. IEEE PAMI. (1985) 384–401
2. Chapelle, O., Haffner, P., Vapnik, V.N.: Support Vector Machines for Histogram-Based Image Classification. IEEE Trans. on Neural Networks. **10** (1999) 1055–1064

3. Divakaran, A., Sun, H.: Descriptor for Spatial Distribution of Motion Activity for Compressed Video. SPIE. **2972** (2000) 392–398
4. Dufaux, F., Konrad, J.: Efficient, Robust, and Fast Global Motion Estimation for Video Coding. IEEE Trans. on Image Process. **9** (2000) 497–501
5. Giunta, G., Mascia, U.: Estimation of Global Motion Parameters by Complex Linear Regression. IEEE Trans. on Image Process. **8** (1999) 1652–1657
6. Grzymala-Busse, J.W., Hu, M.: A Comparison of Several Approaches to Missing Attribute Values in Data Mining. Proceedings of the Second International Conference on Rough Sets and Current Trends in Computing. (2000) 378-385
7. Hu, X.H., Cercone, N.: Learning in Relational Databases: a Rough Set Approach. Computational Intelligence. **11** (1995) 323–337
8. Institute of Computer Science and Technology, Chongqing University of Posts and Telecommunications, Chongqing, China: `http://cs.cqupt.edu.cn/videoretrieval`
9. Knowledge Systems Group: Rosseta Technical Reference Manual. (1999)
10. Lu, H.B., Zhang, T.J.: An Effective Shot Cut Detection Method. Journal of Image and Graphics. (1999) 805–810
11. Ma, Y.F., Zhang, H.J.: Motion Pattern Based Video Classification and Retrieval. EURASIP JASP. **2** (2003) 199–208
12. Moving Picture Experts Group: `http://www.chiariglione.org/mpeg/`
13. MPEG Video Group: MPEG-2. ISO/IEC. (1994)
14. MPEG Video Group: Overview of the MPEG-7 standard. ISO/IEC. (2001).
15. Nguyen, S.H., Skowron, A.: Quantization of Real Value Attributes-Rough Set and Boolean Reasoning Approach. Proc of the Second Joint Conference on Information Sciences. **2** (1995) 34–37
16. Pawlak, Z.: Classification of objects by means of attributes, Research Report PAS 429, Institute of Computer Science, Polish Academy of Sciences, ISSN 138-0648, January (1981)
17. Pawlak, Z.: Rough Sets, Research Report PAS 431, Institute of Computer Science, Polish Academy of Sciences (1981)
18. Pawlak, Z.: Rough Set. International Journal of Computer and Information Sciences. **11** (1982) 341–356
19. Pawlak, Z., Grzymała-Busse, J., Slowinski, R.: Rough Sets. Communications of the ACM. **38** (1995) 89–95
20. Skowron, A., Rauszer, C.: The Discernibility Matrices and Functions in Information System. Intelligent Decision Support Handbook of Applications and Advances of the Rough Sets Theory. **2** (1992) 331–338
21. Sudhir, G., Lee, J.C.M.: Video Annotation by Motion Interpretation Using Optical Flow Streams. Journal of Visual Communication and Image Representation. **7** (1996) 354–368
22. Tan, Y.P., Saur, D.D., Kulkarni, S.R., Ramadge, P.J.: Rapid Estimation of Camera Motion from Compressed Video with Application to Video Annotation. IEEE Tans. on Circuits Syst. Video Techo. **10** (2000) 133–145
23. Tekalp, A.M.: Digital Video Processing. Prentice-Hall. (1995)
24. Wang, G.Y., Zhao, J., An, J.J., Wu, Y.: Theoretical Study on Attribute Reduction of Rough Set Theory: in Algebra View and Information View. Third International Conference on Cognitive Informatics. (2004) 148–155
25. Wang, G.Y.: Rough Set Theory and Knowledge Acquisition. Xi'an Jiaotong University. **20** (2001) 102–116
26. Wang, G.Y., Yang, D.C., Yu, H.: Condition-Information-Entropy-Based Decision Table Reduction. Chinese Journal of Computers. **2** (2002) 759–766

27. Wang, G.Y., Liu, F., Wu, Y.: Generating Rules and Reasoning under Inconsistencies. Proceedings of IEEE Intl. Conf. on Industrial Electronics, Control and Instrumentation. (2000) 646–649
28. Yin, D.S., Wang, G.Y., Wu, Y.: A Self-learning Algorithm for Decision Tree Prepruning. Proceedings of the Third International Conference on Machine Learning and Cybernetics. (2004) 2140–2145
29. Yoo, K.Y., Kim, J.K.: A New Fast Local Motion Estimation Algorithm Using Global Motion. Signal Processing. **68** (1998) 219–224
30. Yu, T.L., Zhang, S.J.: Video Retrieval Based on the Global Motion Information. Acta Electronica Sinica. **29** (2001) 1794–1798
31. Zhang, H.J., Kankanhalli, Smoliar, A.: Automatic Partitioning of Video. Multimedia System. (1993) 10–28
32. Wang, G.Y., Zheng, Z., Zhang, Y.: RIDAS– A Rough Set Based Intelligent Data Analysis System. Proceedings of the First Int. Conf. on Machine Learning and Cybernetics. (2002) 646–649
33. Zhong, D., Chang, S.F.: Video Object Model and Segmentation for Content-Based Video Indexing. Proc. ISCAS'97. (1997) 1492–1495

Approximate Boolean Reasoning:
Foundations and Applications in Data Mining

Hung Son Nguyen

Institute of Mathematics,
Warsaw University Banacha 2, 02-097
Warsaw, Poland
son@mimuw.edu.pl

Abstract. Since its introduction by George Boole during the mid-1800s, Boolean algebra has become an important part of the *lingua franca* of mathematics, science, engineering, and research in artificial intelligence, machine learning and data mining. The Boolean reasoning approach has manifestly become a powerful tool for designing effective and accurate solutions for many problems in decision-making and approximate reasoning optimization. In recent years, Boolean reasoning has become a recognized technique for developing many interesting concept approximation methods in rough set theory. The problem considered in this paper is the creation of a general framework for concept approximation. The need for such a general framework arises in machine learning and data mining. This paper presents a solution to this problem by introducing a general framework for concept approximation which combines rough set theory, Boolean reasoning methodology and data mining. This general framework for approximate reasoning is called *Rough Sets and Approximate Boolean Reasoning* (RSABR). The contribution of this paper is the presentation of the theoretical foundation of RSABR as well as its application in solving many data mining problems and knowledge discovery in databases (KDD) such as feature selection, feature extraction, data preprocessing, classification of decision rules and decision trees, association analysis.

Keywords: Rough sets, data mining, boolean reasoning, feature selection and extraction, decision rule construction, discretization, decision tree induction, association rules, large data tables.

1 Introduction

The rapidly growing volume and complexity of modern databases make the need for technologies to describe and summarize the information they contain increasingly important. Knowledge Discovery in Databases (KDD) and data mining are new research areas that try to overcome this problem. In [32], KDD was characterized as a non-trivial process of identifying valid, novel, potentially useful, and ultimately understandable patterns in data, while data mining is a process of extracting implicit, previously unknown and potentially useful patterns and relationships from data, and it is widely used in industry and business applications.

J.F. Peters and A. Skowron (Eds.): Transactions on Rough Sets V, LNCS 4100, pp. 334–506, 2006.
© Springer-Verlag Berlin Heidelberg 2006

As the main step in KDD, data mining methods are required to be not only accurate but also to deliver understandable and interpretable results for users, e.g., through visualization. The other important issue of data mining methods is their complexity and scalability. Presently, data mining is a collection of methods from various disciplines such as mathematics, statistics, logics, pattern recognition, machine learning, non-conventional models and heuristics for computing [43], [45], [155] [67].

Concept approximation is one of the most fundamental issues in machine learning and data mining. The problem considered in this paper is the creation of a general framework for concept approximation. The need for such a general framework arises in machine learning and data mining. Classification, clustering, association analysis or regression are examples of well-known problems in data mining that can be considered in the context of concept approximation. A great effort by many researchers has led to the design of newer, faster and more efficient methods for solving the concept approximation problem [100].

Rough set theory has been introduced by Zdzisław Pawlak [109] as a tool for concept approximation relative to uncertainty. Basically, the idea is to approximate a concept by three description sets, namely, *lower approximation, upper approximation* and *boundary region*. These three sets have been fundamental to the basic approach of rough set theory, since its introduction by Zdzisław Pawlak during the early 1980s (see, e.g., [107], [108], [109], [110]). The approximation process begins by partitioning a given set of objects into equivalence classes called blocks, where the objects in each block are indiscernible from each other relative to their attribute values. The approximation and boundary region sets are derived from the blocks of a partition of the available objects. The boundary region is constituted by the difference between the lower approximation and upper approximation, and provides a basis for measuring the "roughness" of an approximation. Central to the philosophy of the rough set approach to concept approximation is minimization of the boundary region. This simple but brilliant idea leads to many efficient applications of rough sets in machine learning and data mining such as feature selection, rule induction, discretization or classifier construction [57], [58], [143], [137], [142], [79].

Boolean algebra has become part of the *lingua franca* of mathematics, science, engineering, and research in artificial intelligence, machine learning and data mining ever since its introduction by George Boole during the 19th century [13]. In recent years, the combination of Boolean reasoning approach and rough set methods have provided powerful tools for designing effective as well as accurate solutions for many machine learning and data mining problems [141], [97], [91], [162], [142], [139], [164], [61], [38].

The problem considered in this paper is the creation of a general framework for concept approximation. The need for such a general framework arises in machine learning and data mining. This paper presents a solution to this problem by introducing a general framework for concept approximation which combines rough set theory, Boolean reasoning methodology and data mining. This general framework for approximate reasoning is called *Rough Sets and Approximate*

Boolean Reasonin (RSABR). The contribution of this paper is the presentation of the theoretical foundation of RSABR as well as its application in solving many data mining problems and knowledge discovery in databases (KDD) such as feature selection, feature extraction, data preprocessing, classification of decision rules and decision trees, association analysis.

1.1 Overview of Achieved Results

The discretization method based on standard Boolean reasoning approach has been described and explored in the author's Ph.D. dissertation of [79]. This section presents the assessment of advancements regarding the stated problems and summarizes the results achieved by the author from 1998.

1. **Approximate Boolean reasoning as a new approach to problem solving in rough sets and data mining:**
   As it has been mentioned in previous sections, the rough sets methods based on the straightforward application of Boolean reasoning approach were not suitable for data mining. The critical factor is the complexity and the scalability of the standard methodology. Approximate Boolean Reasoning (ABR) has been proposed as an alternative approach to overcome those problems [140], [29], [90].

   Each approximate method is characterized by two parameters: the quality of approximation and the computation time. Searching for the proper balancing between these parameters is the biggest challenge of modern heuristics. In the *Approximate Boolean Reasoning approach* not only calculation of prime implicants – which is the most time-consuming step – but every step in the original Boolean reasoning methodology can be approximately performed to achieve an approximate solution. Thus the ABR approach to problem solving consists of the following steps:

   - **Modeling:** Represent the problem or a simplified problem by a collection of boolean equations.

   - **Reduction:** Condense equations into an equivalent or approximate problem over a single boolean equation of the form $f(x_1, x_2, \ldots, x_n) = 0$ (or, dually, $\overline{f} = 1$).

   - **Development:** Generate an approximate solution of the formulated problem over f.

   - **Reasoning:** Apply a sequence of approximate reasoning steps to solve the problem.

2. **RSABR approach to discretization problem:**
   Optimal discretization problem has been investigated as an illustration of many ideas of approximate Boolean reasoning methodology. Let us survey the results achieved by application of the ABR approach to discretization.

   The greedy heuristic for the optimal discretization problem is called *MD-algorithm* (Maximal Discernibility) for discretization since it is using *discernibility measure* as a quality function. The detailed analysis of MD-algorithm has been performed in [8]. Moreover, both global and local versions of the dicretization method based on MD-heuristics were presented in [8] and implemented in the popular RSES[1] system.

Both discretization and attribute selection are data preprocessing techniques that cause a loss of some information in data. Rough set methodology to the classification problem is based on searching for some relevant sets of attributes called reducts. It has been shown that usually, fewer short reducts can make a better rough classifier than a single reduct. Unfortunately, optimal discretization leaves only one reduct in the decision table.

The relationship between discretization and short reducts has been investigated in [81] and the discretization method that preserves all short reducts (of a predefined size) has been proposed on the basis of the ABR approach.

Optimal SQL-querying method for discretization was another application of ABR approach to discretization problem [83]. The idea was based on localizing the best cut using "divide and conquer" technique. It has been shown that for data table with n objects, it is enough to use only $O(\log n)$ simple queries to find the cut that is very close to the optimal with respect to discernibility measure. This technique has been generalized for other measures [84].

3. **RSABR approach to feature selection and feature extraction problem:**
   In rough set theory, the feature selection problem is defined in term of reducts [111], i.e., irreducible subsets of most informative attributes of a given decision table or information system. The idea of *minimal description length principle* (MDLP) states that sometime we should search for the proper balance between a loss of accuracy and the more compact description of data models. Thus rough set methods are searching for short reducts to build efficient classifiers. It has been shown that every reduct exactly corresponds to one prime implicant of an encoding Boolean function called discernibility function [143].

   A set of attributes is called *approximate reduct* if it preserves a necessary information (with a satisfactory degree) to build a compatible classifier. Many experimental results are showing that approximate reducts, which are shorter than the exact ones, can construct more accurate classifiers. The problem of searching for minimal approximate reducts was investigated in [96], where the complexity of this problem has been shown and an heuristic algorithm based on the ABR approach has been proposed.

   The set of all reducts of a given decision table is an antichain on the set of attributes. Thus the maximal number of reducts is equal to

$$N(k) = \binom{k}{\lfloor k/2 \rfloor},$$

   where k is the number of attributes. The k-attribute decision table is called *malicious* if it contains exactly $N(k)$ reducts. Characterization of malicious

---

[1] RSES home page: `http://logic.mimuw.edu.pl/~rses/`

decision tables and possibility of their construction were discussed in [89] where, once again, the ABR approach plays a crucial role.

Discretization of numeric attributes can be treated not only as a data reduction process but also as a feature extraction method since it creates a new set of attributes. Some extensions of discretization methods based on the ABR schema in the context of feature extraction problem were investigated in this research. Particularly, methods of creating new features defined either by linear combinations of attributes (hyperplanes) or by sets of symbolic values were presented [93], [103].

4. **RSABR approach to decision trees:**
   The main philosophy of RSABR methodology to the classification problem is based on managing the discernible objects. Thus discernibility becomes an interesting measure for many applications of rough sets in data mining. Decision tree is one of the most popular classification methods. The decision tree construction method based on the discernibility measure has been proposed. This method, also known as MD-decision tree[2], creates binary decision tree using cuts on continuous attributes, binary partition of values for symbolic values. Properties and a detailed comparison analysis with other techniques were presented in [82], [103], [140].

5. **Soft discretization and soft decision trees:**
   Crisp partitions defined by cuts in standard discretization and decision tree methods may cause a misclassification of objects that are very close to those cuts. Soft cuts were proposed as a novel concept for mining data with numeric attributes. Unlike traditional cuts, each soft cut is defined as an interval of *possible cuts* and represents a family of possible partitions. Modified classification methods based on soft cuts and rough set theory were presented in [80], [85].

   Soft decision trees, i.e., decision trees using soft cuts, have some advantages compared to the traditional ones. Firstly, this approach can overcome the overfitting problem without pruning. Secondly, it is possible to efficiently construct soft decision trees from large data bases [84], [88]. Two techniques called *rough decision tree* and *fuzzy decision tree* were proposed in [87], [94].

6. **Relationship between rough sets and association rules:**
   Association rule discovery [3] is one of the most famous data mining techniques that can be applied to databases of transactions where each transaction consists of a set of items. In such a framework the problem is to discover all associations and correlations among data items where the presence of one set of items in a transaction implies (with a certain degree of confidence) the presence of other items. Besides market basket data, association analysis is also applicable to other application domains such as bioinformatics, medical diagnosis, web mining, and scientific data analysis.

   All existing methods for association rule generation consists of two steps (1) searching for frequent item sets, and (2) generating association rules from

---

[2] MD = maximal discernibility.

frequent item sets. The correspondence between step (2) and the problem of searching for approximate reducts has been shown in [96]. This important result implies that

- Every method for approximate reducts problem, including those methods based on the ABR approach, can be applied to association rule generation.

- All existing association rule techniques can be used to solve the reduct searching problem in rough set theory.

As an example, Apriori algorithm [3] is one of the first association rule techniques. In [86], a method based on the apriori idea for construction of lazy rough classifier has been proposed.

1.2 Organization of the Paper

This paper, as a summarization of achieved results, presents the foundation of ABR methodology and its applications in rough sets and data mining. The presentation is limited to detailed description of methods and algorithms as well as the discussion of properties of the proposed methods. Experimental results are omitted and the reader is directed to other articles in the reference list.

This paper is organized as follows. The basic theory and methods central to the application of rough sets in data mining, are presented in Sect. 2. This section includes an introduction to knowledge discovery and data mining in Sect. 2.1 and the rough set approach to data mining in Sect. 2.4. Sect. 3 provides an introduction to Boolean algebra and Boolean functions. An approach to Boolean reasoning as well as approximate Boolean reasoning is presented in Sect. 4. Sect. 5 explores the application of approximate Boolean reasoning (ABR) in feature selection and decision rule generation. The rough set and the ABR approach to the discretization is presented in Sect. 6. Application of ABR in decision tree induction is explored in Sect. 7. The ABR approach to feature extraction is given in Sect. 8. Rough sets, ABR and association analysis are presented in Sect. 9. Finally, rough set methods for mining large databases are presented in Sect. 10.

2 Basic Notions of Data Mining, Rough Set Theory and Rough Set Methodology in Data Mining

This section introduces basic jargon and definitions for a few related research disciplines including knowledge discovery from databases (KDD), data mining and rough set theory. We also characterize the basic idea of rough set methodology to data mining.

2.1 Knowledge Discovery and Data Mining

Knowledge discovery and data mining (KDD) – the rapidly growing interdisciplinary field which merges together database management, statistics, machine learning and related areas – aims at extracting useful knowledge from large collections of data.

There is a difference in understanding the terms "knowledge discovery" and "data mining" between people from different areas contributing to this new field. In this paper we adopt the following definition of these terms [32]:

> *Knowledge discovery in databases is the process of identifying valid, novel, potentially useful, and ultimately understandable patterns/models in data. Data mining is a step in the knowledge discovery process consisting of particular data mining algorithms that, under some acceptable computational efficiency limitations, finds patterns or models in data.*

Therefore, an essence of KDD projects relates to interesting patterns and/or models that exist in databases but are hidden among the volumes of data. A model can be viewed as "a global representation of a structure that summarizes the systematic component underlying the data or that describes how the data may have arisen". In contrast, "a pattern is a local structure, perhaps relating to just a handful of variables and a few cases".

Usually, a pattern is an expression ϕ in some language L describing a subset U_ϕ of the data U (or a model applicable to that subset). The term *pattern* goes beyond its traditional sense to include models or structure in data (relations between facts).

Data mining – an essential step in KDD process – is responsible for algorithmic and intelligent methods for pattern (and/or model) extraction from data. Unfortunately, not every extracted pattern becomes knowledge. To specify the notion of *knowledge* for the need of algorithms in KDD processes, we should define an *interestingness value* of patterns by combining their validity, novelty, usefulness, and simplicity. Interestingness functions should be defined to reflect the interest of users.

Given an interestingness function

$$I_D : L \to \Delta_I,$$

parameterized by a given data set D, where $\Delta_I \subseteq \mathbb{R}$ is the domain of I_D, a pattern ϕ is called knowledge if for some user defined threshold $i \in M_I$

$$I_D(\phi) > i.$$

A typical process of KDD includes an iterative sequence of the following steps:

1. data cleaning: removing noise or irrelevant data,
2. data integration: possible combining of multiple data sources,
3. data selection: retrieving of relevant data from the database,
4. data transformation,
5. data mining,

6. pattern evaluation: identifying of the truly interesting patterns representing knowledge based on some interestingness measures, and

7. knowledge presentation: presentation of the mined knowledge to the user by using some visualization and knowledge representation techniques.

The success of a KDD project strongly depends on the choice of proper data mining algorithms to extract from data those patterns or models that are really interesting for the users. Up to now, the universal recipe of assigning to each data set its proper data mining solution does not exist. Therefore, KDD must be an *iterative* and *interactive* process, where previous steps are repeated in an interaction with users or experts to identify the most suitable data mining method (or their combination) to the studied problem.

One can characterize the existing data mining methods by their goals, functionalities and computational paradigms:

- **Data mining goals:** The two primary goals of data mining in practice tend to be prediction and description. Prediction involves using some variables or fields in the database to predict unknown or future values of other variables of interest. Description focuses on finding human interpretable patterns describing the data. The relative importance of prediction and description for particular data mining applications can vary considerably.

- **Data mining functionalities:** Data mining can be treated as a collection of solutions for some predefined tasks. The major classes of knowledge discovery tasks, also called data mining functionalities, include the discovery of

  - concept/class descriptions,
  - association,
  - classification,
  - prediction,
  - segmentation (clustering),
  - trend analysis, deviation analysis, and similarity analysis.
  - dependency modeling such as graphical models or density estimation,
  - summarization such as finding the relations between fields, associations, visualization; characterization and discrimination are also forms of data summarization.

- **Data mining techniques:** the type of methods used to solve the task is called the data mining paradigm. For example, by the definition, standard statistical techniques are not data mining. However statistics can help greatly in the process of searching for patterns from data by helping to answer several important questions about data, like: "What patterns are there in my database?", "What is the chance that an event will occur?", "Which patterns are significant?" or "What is a high level summary of the data that gives me some idea of what is contained in my database?". One of the great values of statistics is in presenting a high level view of the database, e.g., pie chart or histogram, that provides some useful information without requiring every record to be understood in detail. Some typical data mining techniques are listed below:

- rule induction;
- decision tree induction;
- instancebased learning (e.g., nearest neighbors);
- clustering;
- neural networks;
- genetic algorithms/genetic programming;
- support vector machine.

Many combinations of data mining paradigm and knowledge discovery task are possible. For example the neural network approach is applicable to both predictive modeling task as well as segmentation task.

Any particular implementation of a data mining paradigm to solve a task is called a *data mining method*. Lots of data mining methods are derived by modification or improvement of existing machine learning and pattern recognition approaches to manage with large and nontypical data sets. Every method in data mining is required to be accurate, efficient and scalable. For example, in prediction tasks, the *predictive accuracy* refers to the ability of the model to correctly predict the class label of new or previously unseen data. The efficiency refers to the computation costs involved in generating and using the model. Scalability refers to the ability of the learned model to perform efficiently on large amounts of data.

For example, C5.0 is the most popular algorithm for the well-known method from machine learning and statistics called *decision tree*. C5.0 is quite fast for construction of decision tree from data sets of moderate size, but becomes inefficient for huge and distributed databases. Most decision tree algorithms have the restriction that the training samples should reside in main memory. In data mining applications, very large training sets of millions of samples are common. Hence, this restriction limits the scalability of such algorithms, where the decision tree construction can become inefficient due to swapping of the training samples in and out of main and cache memories. SLIQ and SPRINT are examples of scalable decision tree methods in data mining.

Each data mining algorithm should consist of the following components (see [32]):

1. *Model Representation* is the language L for describing discoverable patterns. Too limited a representation can never produce an accurate model what are the representational assumptions of a particular algorithm more powerful representations increase the danger of overfitting and resulting in poor predictive accuracy more complex representations increase the difficulty of search. More complex representations increase the difficulty of model interpretation

2. *Model Evaluation* estimates how well a particular pattern (a model and its parameters) meet the criteria of the KDD process. Evaluation of predictive accuracy (validity) is based on cross validation. Evaluation of descriptive quality involves predictive accuracy, novelty, utility, description length, and understandability of the fitted model. Both logical and statistical criteria can be used for model evaluation.

3. *Search Method* consists of two components:
   (a) In parameter search, the algorithm must search for the parameters which optimize the model evaluation criteria given observed data and a fixed model representation.
   (b) Model search occurs as a loop over the parameter search method: the model representation is changed so that a family of models is considered.

2.2   Approximate Reasoning Problem

Let us describe an issue of approximate reasoning that can be seen as a connection between data mining and logics. This problem occurs, e.g., during an interaction between two (human/machine) beings which are using different languages to talk about objects (cases, situations, etc.) from the same universe. The intelligence skill of those beings, called intelligent agents, is measured by the ability of understanding the other agents. This skill is performed by different ways, e.g., by learning or classification (in machine learning and pattern recognition theory), by adaptation (in evolutionary computation theory), or by recognition (in recognitive science).

Logic is a science that tries to model the way of human thinking and reasoning. Two main components of each logic are *logical language*, and *the set of inference rules*. Each logical language contains a set of formulas or well-formulated sentences in the considered logic. Usually, the meaning (semantic) of a formula is defined by a set of objects from a given universe. A subset X of a given universe $\mathfrak{U}$ is called a concept (in $\mathcal{L}$) if and only if X can be described by a formula ϕ in $\mathcal{L}$.

Therefore, it is natural to distinguish two basic problems in approximate reasoning, namely: *approximation of unknown concepts* and *approximation of reasoning scheme*.

By concept approximation problem we mean the problem of searching for description – in a predefined language $\mathcal{L}$ – of concepts definable in other language $\mathcal{L}^*$. Not every concept in $\mathcal{L}^*$ can be exactly described in $\mathcal{L}$, therefore the problem is to find an approximate description rather than exact description of unknown concepts, and the approximation is required to be as exact as possible.

In many applications, the problem is to approximate those concepts that are definable either in the natural language or by an expert or by some unknown process. For example let us consider the problem of automatic recognition of *"overweight people"* from camera pictures. This concept (in the universe of all people) is understood well in medicine and can be determined by BMI (the *Body Mass Index*)[3]. This concept can be simply defined by *weight and height* which are measurable features on each person.

$$C_{overweight} = \left\{ x : 25 \leq \frac{weight(x)}{height^2(x)} < 30 \right\}$$

---

[3] BMI is calculated as weight in kilograms divided by the square of height in meters; according to the simplest definition, people are categorized as underweight (BMI < 18.5), normal weight (BMI $\in [18.5, 25)$), overweight (BMI $\in [25, 30)$), and obese (BMI ≥ 30.0).

The more advanced definitions require more features like *sex, age and race*. In this case, the problem is to approximate the concept "overweight people" using only those features that can be calculated from their pictures.

Concept approximation problem is one of the most important issues in data mining. Classification, clustering, association analysis or regression are examples of well-known problems in data mining that can be formulated as concept approximation problems. A great effort of many researchers has been done to design newer, faster and more efficient methods for solving the concept approximation problem.

The task of concept approximation is possible only if some knowledge about the concept is available. Most methods in data mining realize the inductive learning approach, which assumes that a partial information about the concept is given by a finite sample, called *the training sample or training set*, consisting of positive and negative cases (i.e., objects belonging or not belonging to the concept). The information from training tables makes the search for patterns describing the given concept possible. In practice, we assume that all objects from the universe $\mathfrak{U}$ are perceived by means of information vectors being vectors of attribute values (information signature). In this case, the language $\mathcal{L}$ consists of boolean formulas defined over accessible (effectively measurable) attributes.

Any concept C in a universe $\mathfrak{U}$ can be represented by its characteristic function $d_C : \mathfrak{U} \to \{0, 1\}$ such that

$$d_X(u) = 1 \Leftrightarrow u \in X.$$

Let $h = \mathbb{L}(S) : \mathfrak{U} \to \{0, 1\}$ be the approximation of d_C which is inducted from a training sample S by applying an approximation algorithm $\mathbb{L}$. Formally, the approximation error is understood as

$$err_{\mathfrak{U}}^C(h) = \mu(\{x \in \mathfrak{U} : h(x) \neq d_C(x)\}),$$

where μ is the probability measure of a probability space defined on $\mathfrak{U}$. In practice, it is hard to determine the value of *exact error* $err_{\mathfrak{U}}^C(h)$ because both the function μ and its argument are unknown. We are forced to approximate this value by using an additional sample of objects called *the testing sample or the testing set*. The exact error can be estimated by using testing sample $T \subset \mathfrak{U}$ as follows:

$$err_{\mathfrak{U}}^C(h) \simeq err_T^C(h) = \frac{|\{x \in T : h(x) \neq d_C(x)\}|}{|T|}.$$

More advanced methods for evaluation of approximation algorithms are described in [119]. Let us recall some other popular measures which are very utilized by many researchers in practical applications:

- confusion matrices;
- accuracy, coverage;
- lift and gain charts;
- receiver operation characteristics (roc) curves;
- generality;
- stability of the solution.

2.3   Rough Set Preliminaries

One of the basic principles of set theory is the possibility of precise definition of any concept using only the "membership relation". The classical set theory is operating on "crisp" concepts and it is within "exact science" like mathematics. Unfortunately, in many real-life situations, we are not able to give an exact definition of the concept.

Except the imprecise or vague nature of linguistic concepts themselves (see the Sect. 2.2), the trouble may be caused by the lack of information or noise. Let us consider the photography of solar disk in Fig. 1. It is very hard to define the concept "solar disk" by giving a set of pixels. Thus nondeterminism of the membership relation "the pixel (x, y) belongs to the solar disk" is caused by uncertain information.

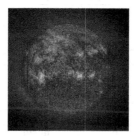

Fig. 1. Can you define the concept "solar disk" by giving a set of pixels on this picture?

Rough set theory has been introduced by Professor Z. Pawlak [109] as a tool for approximation of concepts under uncertainty. The theory is featured by operating on two subsets, a lower approximation and upper approximation. Rough set theory has two fundamental aspects:

– **Approximation of crisp concepts under uncertainty.**
   Uncertain situations are caused by incomplete information about all objects in the universe. In this case, lower and upper approximations are imprecise representations of the concept;
– **Approximation of vague concepts.**
   From the philosophical point of view the vague concepts are characterized by existing of borderline cases, which cannot be clearly classified to the concept or its complement [55]. In this sense, rough set theory realizes the idea of vagueness where borderline cases are approximated by the boundary region, i.e., the difference between upper and lower approximations.

2.3.1   Information Systems

In contrast with the classical set theory, rough set theory assumes that there are some additional information about elements. An *information system* [111] is a pair $\mathbb{S} = (U, A)$, where U is a non-empty, finite collection of *objects* and A is a non-empty, finite set, of *attributes*. Each $a \in A$ corresponds to the function

$a : U \rightarrow V_a$ called *evaluation function*, where V_a is called the *value set* of a. Elements of U could be interpreted as, e.g., cases, states, patients, observations, etc.

The above formal definition of information systems is very general and it covers many different "real information systems". Let us mention some of them.

Example 1. "Information table" is the simplest form of information systems. It can be implemented as two-dimensional array (matrix), which is standard data structure in every programming language. In information table, we usually associate its rows to objects, its columns to attributes and its cells to values of attributes on objects.

Table 1. Example of information table: a data set contains ten objects from heart-disease domain

Patient	Age	Sex	Cholesterol	Resting ECG	Heart rate	Sick
p_1	53	M	203	hyp	155	yes
p_2	60	M	185	hyp	155	yes
p_3	40	M	199	norm	178	no
p_4	46	F	243	norm	144	no
p_5	62	F	294	norm	162	no
p_6	43	M	177	hyp	120	yes
p_7	76	F	197	abnorm	116	no
p_8	62	M	267	norm	99	yes
p_9	57	M	274	norm	88	yes
p_{10}	72	M	200	abnorm	100	no

Example 2. Data base systems are also examples of information systems. The universe U is the sets of records and A is the set of attributes in data base. Usually, data bases are used to store a large amount of data and the access to data (e.g., computing the value of attribute $a \in A$ for object $x \in U$) is enabled by some data base tools like SQL queries in relational data base systems.

Given an information system $\mathbb{S} = (U, A)$, we associate with any non-empty set of attributes $B \subseteq A$ the *information signature* or *B-information vector* for any object $x \in U$ by

$$inf_B(x) = \{(a, a(x)) : a \in B\}.$$

The set $INF(\mathbb{S}) = \{inf_A(x) : x \in U\}$ is called the *A-information set*. The notions of "information vector" and "information set" have very easy interpretations. In some sense, they tabulate all "information systems $\mathbb{S}$" by information set $INF(\mathbb{S})$, where information vectors are in rows, and attributes are in columns. Hence, they are very easy to handle with arbitrary information systems.

In supervised learning problems, objects from training set are pre-classified into several *categories* or *classes*. To manipulate this type of data we use a special case of information systems called *decision systems* which are information systems of the form $\mathbb{S} = (U, A \cup \{dec\})$, where $dec \notin A$ is a distinguished attribute called *decision*. The elements of attribute set A are called *conditions*.

In practice, decision systems contain description of a finite sample U of objects from larger (possibly infinite) universe $\mathfrak{U}$, where conditions are such attributes that their values are always known for all objects from $\mathfrak{U}$, but decision is in general a hidden function except objects from the sample U. Usually decision attribute is a characteristic functions of an unknown concept, or classification of objects into several classes. As we mentioned in previous sections, the main problem of learning theory is to generalize the decision function, which is defined on the sample U, to the whole universe $\mathfrak{U}$. Below we present the example of decision system.

Example 3. "Decision table" is one of the forms of information systems which is used most often in rough set theory. It can be defined from information table by appointing some attributes to conditions and some attribute to decision. For example, from information table presented in Table 1, one can define new decision table by selecting attributes: **Age, Sex, Cholesterol, Resting ECG,** and **Heart rate** as condition attributes and **Sick** as decision.

Without loss of generality one can assume that the domain V_{dec} of the decision dec is equal to $\{1, \ldots, d\}$. The decision dec determines a partition

$$U = CLASS_1 \cup \cdots \cup CLASS_d$$

of the universe U, where

$$CLASS_k = \{x \in U : dec(x) = k\}$$

is called the k-th *decision class of* $\mathbb{S}$ for $1 \le k \le d$.

Let $X \subset U$ be an arbitrary set of objects, by "*counting table*" of X we mean the vector

$$CountTable(X) = \langle n_1, \ldots, n_d \rangle,$$

where $n_k = card(X \cap CLASS_k)$ is the number of objects from X belonging to the kth decision class.

For example, there are two decision classes in the decision table presented in Table 1:

$$CLASS_{yes} = \{p_1, p_2, p_6, p_8, p_9\}, \qquad CLASS_{no} = \{p_3, p_4, p_5, p_7, p_{10}\},$$

and the set $X = \{p_1, p_2, p_3, p_4, p_5\}$ has class distribution:

$$ClassDist(X) = \langle 2, 3 \rangle.$$

2.3.2 Standard Rough Sets

The first definition of rough approximation was introduced by Pawlak in his pioneering book on rough set theory [109], [111]. For any subset of attributes $B \subset A$, the set of objects U is divided into *equivalence classes* by the *indiscernibility relation* and the upper and lower approximations are defined as unions of corresponding equivalence classes. This definition can be called *the attribute-based rough approximation* or "standard rough sets".

Given an information system $\mathbb{S} = (U, A)$, the problem is to define a concept $X \subset U$, assuming at the moment that only some attributes from $B \subset A$ are accessible. This problem can be also described by appropriate decision table $\mathbb{S} = (U, B \cup \{dec_X\})$, where $dec_X(u) = 1$ for $u \in X$, and $dec_X(u) = 0$ for $u \notin X$.

First one can define an equivalence relation called the *B-indiscernibility relation*, denoted by $IND(B)$, as follows

$$IND(B) = \{(x, y) \in U \times U : inf_B(x) = inf_B(y)\}. \tag{1}$$

Objects x, y satisfying relation $IND(B)$ are indiscernible by attributes from B. By

$$[x]_{IND(B)} = \{u \in U : (x, u) \in IND(B)\},$$

we denote the equivalence class of $IND(B)$ defined by x.

The lower and upper approximations of X (using attributes from B) are defined by

$$\mathbf{L}_B(X) = \{x \in U : [x]_{IND(B)} \subseteq X\} \quad \text{and}$$
$$\mathbf{U}_B(X) = \{x \in U : [x]_{IND(B)} \cap X \neq \varnothing\}.$$

More generally, let $\mathbb{S} = (U, A \cup \{dec\})$ be a decision table, where $V_{dec} = \{1, \ldots, d\}$, and $B \subseteq A$. Then we can define a *generalized decision function* $\partial_B : U \rightarrow \mathcal{P}(V_{dec})$, by

$$\partial_B(x) = d\left([x]_{IND(B)}\right) = \{i : \exists_{u \in [x]_{IND(B)}} \, d(u) = i\}. \tag{2}$$

Using generalized decision function one can also define rough approximations of any decision class $CLASS_i$ (for $i \in \{1, \ldots, d\}$) by:

$$\mathbf{L}_B(CLASS_i) = \{x \in U : \partial_B(x) = \{i\}\}, \text{ and}$$
$$\mathbf{U}_B(CLASS_i) = \{x \in U : i \in \partial_B(x)\}.$$

The set

$$POS_{\mathbb{S}}(B) = \{x : |\partial_B(x)| = 1\} = \bigcup_{i=1}^{d} \mathbf{L}_B(CLASS_i)$$

is called the positive region of B, i.e., the set of objects that are uniquely defined by B.

Example 4. Let us consider again the decision table presented in Table 1, and the concept $CLASS_{no} = \{p_3, p_4, p_5, p_7, p_{10}\}$ defined by decision attribute **Sick**. Let $B = \{\mathbf{Sex}, \mathbf{Resting\ ECG}\}$, then the equivalent classes of indiscernibility relation $IND(B)$ are as following:

$$\{p_1, p_2, p_6\} \qquad inf_B(x) = [M, hyp]$$
$$\{p_3, p_8, p_9\} \qquad inf_B(x) = [M, norm]$$
$$\{p_4, p_5\} \qquad inf_B(x) = [F, norm]$$
$$\{p_7\} \qquad inf_B(x) = [F, abnorm]$$
$$\{p_{10}\} \qquad inf_B(x) = [M, abnorm].$$

Lower and upper approximations of $CLASS_{no}$ are computed as follows:

$$\mathbf{L}_B(CLASS_{no}) = \{p_4, p_5\} \cup \{p_7\} \cup \{p_{10}\} = \{p_4, p_5, p_7, p_{10}\}$$
$$\mathbf{U}_B(CLASS_{no}) = \mathbf{L}_B(CLASS_{no}) \cup \{p_3, p_8, p_9\} = \{p_3, p_4, p_5, p_7, p_8, p_9, p_{10}\}$$

Description of lower and upper approximations are extracted directly from equivalence classes, e.g.,

Certain rules:

$$[\mathbf{Sex}(x), \mathbf{Resting\ ECG}(x)] = [F, norm] \implies \mathbf{Sick}(x) = no$$
$$[\mathbf{Sex}(x), \mathbf{Resting\ ECG}(x)] = [F, abnorm] \implies \mathbf{Sick}(x) = no$$
$$[\mathbf{Sex}(x), \mathbf{Resting\ ECG}(x)] = [M, abnorm] \implies \mathbf{Sick}(x) = no$$

Possible rules:

$$[\mathbf{Sex}(x), \mathbf{Resting\ ECG}(x)] = [M, norm] \implies \mathbf{Sick}(x) = no.$$

The positive region of B can be calculated as follows:

$$POS_B = \mathbf{L}_B(CLASS_{no}) \cup \mathbf{L}_B(CLASS_{yes}) = \{p_1, p_2, p_6, p_4, p_5, p_7, p_{10}\}.$$

2.3.3 Rough Set Space

Rough set spaces are structures which allow to formalize the notions of lower and upper approximations.

Let $\mathcal{L}$ be a description language and $\mathfrak{U}$ be a given universe of objects. Assume that semantics of formulas from $\mathcal{L}$ are defined by subsets of objects from $\mathfrak{U}$. Let $[[\alpha]] \subset \mathfrak{U}$ be the semantics of the formula α in $\mathcal{L}$, i.e., $[[\alpha]]$ is the set of objects satisfying α. A set $A \subset \mathfrak{U}$ is called $\mathcal{L}$-definable if there exists a formula $\alpha \in \mathcal{L}$ such that $[[\alpha]] = A$.

For any set Φ of formulas from $\mathcal{L}$, we denote by $\mathfrak{U}_\Phi$ the collection of all subsets of $\mathfrak{U}$ which are definable by formulas from Φ:

$$\mathfrak{U}_\Phi = \{[[\alpha]] \subset \mathfrak{U} : \text{ for } \phi \in \mathcal{L}\}.$$

Definition 1 (rough set space). *A rough set space over a language $\mathcal{L}$ is a structure $\mathbb{R} = \langle \mathfrak{U}, \Phi, \mathcal{P} \rangle$, where Φ is a non-empty set of formulas from $\mathcal{L}$, and $\mathcal{P} \subset \mathfrak{U}_\Phi^2$ is a non-empty collection of pairs of definable sets from $\mathfrak{U}_\Phi$ such that*

1. $A \subseteq B$ for every $(A, B) \in \mathcal{P}$, and
2. $(A, A) \in \mathcal{P}$ for each $A \in \mathfrak{U}_\Phi$.

Let us assume that there is a *vague inclusion function* (see Skowron and Stepaniuk [144]):

$$\nu : \mathfrak{U} \times \mathfrak{U} \to [0,1]$$

measuring the degree of inclusion between subsets of $\mathfrak{U}$. Vague inclusion must be *monotone* with respect to the second argument, i.e., if $Y \subseteq Z$ then $\nu(X,Y) \leq \nu(X,Z)$ for $X, Y, Z \subseteq U$.

Definition 2 (rough approximation of concept). *Let* $\mathbb{R} = \langle \mathfrak{U}, \Phi, \mathcal{P} \rangle$ *be a rough set space. Any pair of formulas* (α, β) *from* Φ *is called the* rough *approximation of a given concept* $X \subset \mathfrak{U}$ *in* $\mathbb{R}$ *if an only if*

1. $([[\alpha]], [[\beta]]) \in \mathcal{P}$,
2. $\nu([[\alpha]], X) = 1$ *and* $\nu(X, [[\beta]]) = 1$.

The sets $[[\alpha]]$ *and* $[[\beta]]$ *are called the* lower *and the* upper *approximation of* X, *respectively. The set* $[[\beta]] - [[\alpha]]$ *is called the boundary region.*

Let us point out that many rough approximations of the same concept may exist, and conversely, the same pair of definable sets can be the rough approximations of many different concepts. The concept $X \subset \mathfrak{U}$ is called $\mathbb{R}$-crisp if there exists such a formula $\alpha \in \Phi$ that (α, α) is one of the rough approximations of X. Rough set methods tend to search for the optimal rough approximation with respect to a given evaluation criterion. For example, if we know that the target concept is crisp then we should find the rough approximation with the thinnest boundary region.

Rough set theory can be also understood as an extension of the classical set theory where the classical "membership relation" is replaced by "rough membership function".

Definition 3 (rough membership function). *A function* $f : \mathfrak{U} \to [0,1]$ *is called a* rough membership function *of a concept* X *in a rough set space* $\mathbb{AS} = \langle \mathfrak{U}, \Phi, \mathcal{P} \rangle$ *if and only if the pair* (L_f, U_f), *where*

$$L_f = \{x \in \mathfrak{U} : f(x) = 1\} \quad and \quad U_f = \{x \in \mathfrak{U} : f(x) > 0\},$$

establishes a rough approximation of X *in* $\mathbb{AS} = \langle \mathfrak{U}, \Phi, \mathcal{P} \rangle$.

In practice, the set of formulas Φ is not specified explicitly and the task of searching for rough approximations of a concept can be formulated as the problem of searching for "constructive" or "computable" function $f : \mathfrak{U} \to [0,1]$ such that $\nu(L_f, X) = 1$, $\nu(X, U_f) = 1$, and f is optimal with respect to a given optimization criterion. Usually, the quality of a rough membership function is evaluated by the sets L_f, U_f and the complexity of the function f.

In case of the classical rough set theory, any set of attributes B determines a rough membership function $\mu_X^B : U \to [0,1]$ as follows:

$$\mu_X^B(x) = \frac{|X \cap [x]_{IND(B)}|}{|[x]_{IND(B)}|}. \tag{3}$$

This function defines rough approximations of the concept X:

$$\mathbf{L}_B(X) = \mathbf{L}_{\mu_X^B} = \left\{x : \mu_X^B(x) = 1\right\} = \left\{x \in U : [x]_{IND(B)} \subseteq X\right\}, \text{ and}$$

$$\mathbf{U}_B(X) = \mathbf{U}_{\mu_X^B} = \left\{x : \mu_X^B(x) > 0\right\} = \left\{x \in U : [x]_{IND(B)} \cap X \neq \varnothing\right\},$$

which are compatible with the definition of B-*lower* and the B-*upper approximation* of X in $\mathbb{S}$ in Sect. 2.3.1.

2.4   Rough Set Approach to Data Mining

In recent years, rough set theory has attracted attention of many researchers and practitioners all over the world, who have contributed essentially to its development and applications. With many practical and interesting applications rough set approach seems to be of fundamental importance to AI and cognitive sciences, especially in the areas of machine learning, knowledge acquisition, decision analysis, knowledge discovery from databases, expert systems, inductive reasoning and pattern recognition [112], [42], [41], [139], [142].

2.4.1   Inductive Searching for Rough Approximations

As it has been described in Sect. 2.2, the concept approximation problem can be formulated in the form of a teacher-learner interactive system where the learner have to find (learn) an approximate definition of concepts (that are used by the teacher) in his own language. The complexity of this task is caused by the following reasons:

1. **Poor expressiveness of the learner's language:** Usually the learner, which is a computer system, is assumed to use a very primitive description language (e.g., the language of propositional calculus or a simplified first order language) to approximate compound linguistic concepts.
2. **Inductive assumption:** the target concept $\mathcal{X}$ is unknown on the whole universe $\mathfrak{U}$ of objects but is partially given on a finite training set $U \subsetneq \mathfrak{U}$ of positive examples $X = U \cap \mathcal{X}$ and negative examples $\overline{X} = U - X$;

In the classical rough set theory, rough approximations of a concept are determined for objects from U only. More precisely, classical rough approximations are extracted from the restricted rough set space $\mathbb{AS}|_U = \langle U, \Phi, \mathcal{P}|_U \rangle$, where Φ is the set of formulas representing the information vectors over subsets of attributes and $\mathcal{P}|_U$ consists of pairs of definable sets restricted to the objects from U. The aim of rough set methods is to search for rough approximations of a concept in $\mathbb{AS}|_U$ with minimal boundary region. Thus classical rough sets do not take care about unseen objects from $\mathfrak{U} - U$.

It is a big challenge to construct high quality rough approximations of concept from a sample of objects. The inductive learning approach to rough approximations of concepts can be understood as the problem of searching for an extension of *rough membership function*:

$$
\begin{array}{ccc}
U & \longrightarrow & f : U \to [0,1] \\
\downarrow & & \downarrow \\
\mathfrak{U} & \longrightarrow & \mathcal{F} : \mathfrak{U} \to [0,1]
\end{array}
$$

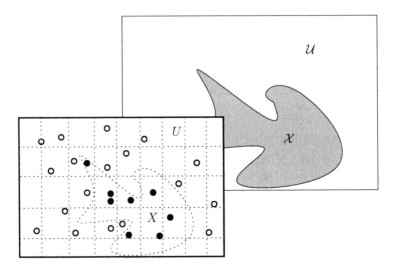

Fig. 2. Illustration of inductive concept approximation problem

Function $\mathcal{F}$ should be constructed using information from the training set U and should satisfy the following conditions:

C1: $\mathcal{F}$ should be determined for all objects from $\mathfrak{U}$;
C2: $\mathcal{F}$ should be a rough membership function for $\mathcal{X}$;
C3: $\mathcal{F}$ should be the "best" rough membership function, in the sense of approximation accuracy, satisfying previous conditions.

This is only a kind of "wish list", because it is either very hard (**C1**) or even impossible (**C2,C3**) to find a function that satisfy all of those conditions. Thus, instead of **C1, C2, C3**, the function $\mathcal{F}$ is required to satisfy some weaker conditions over the training set U, e.g.,

C2 $\Longrightarrow$ **C4:** $\mathcal{F}$ should be an inductive extension of an efficient rough membership function $f : U \to [0,1]$ for the restricted concept $X = U \cap \mathcal{X}$. In other words, instead of being rough membership function of the target concept $\mathcal{X}$, the function $\mathcal{F}$ is required to be rough membership function for X over U.
C1 $\Longrightarrow$ **C5:** $\mathcal{F}$ should be determined for as many objects from $\mathfrak{U}$ as possible.
C3 $\Longrightarrow$ **C6:** Rough approximations defined by $f = \mathcal{F}|_U$ should be an accurate approximation of X over U.

Many extensions of classical rough sets have been proposed to deal with this problem. Let us mention some of them:

– **Variable Rough Set Model (VRSM):** This method (see [163]) proposed a generalization of approximations by introducing a special non-decreasing function $f_\beta : [0,1] \to [0,1]$ (for $0 \le \beta < 0.5$) satisfying properties:

$$f_\beta(t) = 0 \iff 0 \le t \le \beta \text{ and } f_\beta(t) = 1 \iff 1 - \beta \le t \le 1.$$

The generalized membership function called f_β-*membership function* is then defined as

$$\mu_X^{f_\beta}(x) = f_\beta(\mu_R(x)),$$

where μ_R is an arbitrary membership function defined by a relation R. For example, μ_R can be classical rough membership function μ_X^B from Eqn. (3). In this case, with $\beta = 0$ and f_β equal to identity on $[0, 1]$, we have the case of classical rough set [111];

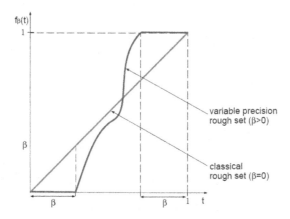

Fig. 3. Example of f_β for variable precision rough set $(\beta > 0)$ and classical rough set $(\beta = 0)$

- **Tolerance-based and similarity-based Rough Sets:** Another idea was based on tolerance or similarity relation [144], [148]. The tolerance approximation space [144] was defined by two functions:
  1. An uncertainty function

  $$I : \mathfrak{U} \to \mathcal{P}(\mathfrak{U}),$$

  determines *tolerance class* for each object from $\mathfrak{U}$. Intuitively, if we look at objects from $\mathfrak{U}$ through "lenses" of available information vectors then $I(x)$ is the set of objects that "look" *similar* to x.
  2. A vague inclusion function

  $$\nu_U : \mathcal{P}(\mathfrak{U}) \times \mathcal{P}(\mathfrak{U}) \to [0, 1],$$

  measures the degree of inclusion between two sets.
  Together with uncertainty function I, vague inclusion function ν defines the *rough membership function* for any concept $\mathcal{X} \subset \mathfrak{U}$ as follows:

  $$\mu_{\mathcal{X}}^{I,\nu}(x) = \nu(I(x), \mathcal{X}).$$

  Obviously, the right hand side of this equation is not effectively defined fpr all objects x because the target concept $\mathcal{X}$ is known only on a finite sample $U \subset \mathfrak{U}$. In practice, we have to induce an estimation of this function from its restriction to U.

2.4.2   Applications of Rough Sets

The simple, but brilliant idea of using lower and upper approximations, leads to many efficient applications of rough sets in machine learning and data mining like feature selection, rule induction, discretization or classifier construction [57]. The most illustrative example of application relates to classification problem.

Learning to classify is one of most important tasks in machine learning and data mining (see [77]). Consider an universe $\mathbb{X}$ of objects. Assume that objects from $\mathbb{X}$ are partitioned into d disjoint subsets $\mathbb{X}_1, \cdots, \mathbb{X}_d$ called *decision classes* (or briefly classes). This partition is performed by a decision function $dec : \mathbb{X} \to V_{dec} = \{1, \cdots, d\}$ which is unknown for the learner. Every object from $\mathbb{X}$ is characterized by attributes from A, but the decision dec is known for objects from some sample set $U \subset \mathbb{X}$ only. The information about function dec is given by the decision table $\mathbb{A} = (U, A \cup \{dec\})$.

The problem is to construct from $\mathbb{A}$ a function $L_{\mathbb{A}} : INF_A \to V_{dec}$ in such a way, that the prediction accuracy, i.e., the probability

$$\mathbf{P}(\{u \in \mathbb{X} : dec(u) = L_{\mathbb{A}}(inf_A(u))\})$$

is sufficiently high. The function $L_{\mathbb{A}}$ is called *decision algorithm* or *classifier* and the methods of its construction from decision tables are called *classification methods*.

It is obvious that the classification can be treated as a concept approximation problem. The above description of classification problem can be understood as a problem of multi–valued concept approximation.

The standard (attribute-based) rough approximations are fundamental for many reasoning methods under uncertainty (caused by the lack of attributes) and are applicable for the classification problem. However, it silently assumes that the information system $\mathbb{S}$ contains all objects of the universe. This is a kind of "closed world" assumption, because we are not interested in the generalization ability of the obtained approximation. Thus classifiers based on standard rough approximations often have a tendency to give an "unknown" answer to those objects $x \in \mathbb{X} - U$, for which $[x]_{IND(B)} \cap U = \varnothing$. A great effort of many researchers in rough set society has been made to modify and to improve this classical approach. One can find many interesting methods for construction of rough classifiers such as variable precision rough set model [163], approximation space [144], tolerance-based rough approximation [148], or classifier-based rough approximations [9].

Rough set based methods for classification are highly acknowledged in many practical applications, particularly in medical data analysis, as they can extract many meaningful and human–readable decision rules from data.

Classification is not the only example of the concept approximation problem. Many tasks in data mining can be formulated as concept approximation problems. For example

- Clustering: the problem of searching for approximation of the concept of "being similar" in the universe of object pairs;

– Basket data analysis: looking for approximation of customer behavior in terms of association rules from the universe of transactions.

Rough set theory has an overlap with many other theories. However we will refrain to discuss these connections here. Despite of the above mentioned connections rough set theory may be considered as the independent discipline on its own.

The main advantage of rough set theory in data analysis is that it does not need any preliminary or additional information about data – like probability in statistics, or basic probability assignment in Dempster-Shafer theory, grade of membership or the value of possibility in fuzzy set theory. The proposed approach

– provides efficient algorithms for finding hidden patterns in data,
– finds minimal sets of data (data reduction),
– evaluates significance of data,
– generates sets of decision rules from data,
– is easy to understand,
– offers straightforward interpretation of obtained results,
– enables parallel processing.

3   Boolean Functions

This section contains some basic definitions, notations and terminology that will be used in the next sections.

The main subject of this section is related to the notion of *Boolean functions*. We consider two equivalent representations of Boolean functions, namely *the truth table form*, and *the boolean expressions form*. The latter representation is derived from the George Boole's formalism (1854) that eventually became *Boolean algebra* [14]. We also discuss some special classes of boolean expressions that are useful in practical applications.

3.1   Boolean Algebra

Boolean algebra was an attempt to use algebraic techniques to deal with expressions in the propositional calculus. Today, these algebras find many applications in electronic design. They were first applied to switching by Claude Shannon in the 20th century [135], [136]. Boolean algebra is also a convenient notation for representing Boolean functions.

Boolean algebras are algebraic structures which "capture the essence" of the logical operations AND, OR and NOT as well as the corresponding set-theoretic operations intersection, union and complement. As Huntington recognized, there are various equivalent ways of characterizing Boolean algebras [49]. One of the most convenient definitions is the following.

Definition 4 (Boolean algebra). *The Boolean algebra is a tuple*

$$\mathcal{B} = (B, +, \cdot, 0, 1),$$

where B is a non-empty set, $+$ and $\cdot$ are binary operations, 0 and 1 are distinct elements of B that satisfy the following axioms:

Commutative laws: *For all elements a, b in B:*

$$a + b = b + a \quad and \ a \cdot b = b \cdot a. \tag{4}$$

Distributive laws: *$\cdot$ is distributive over $+$ and $+$ is distributive over $\cdot$, i.e., for all elements a, b, c in B:*

$$a \cdot (b + c) = (a \cdot b) + (a \cdot c), \ and \ a + (b \cdot c) = (a + b) \cdot (a + c). \tag{5}$$

Identity elements: *For all a in B:*

$$a + 0 = a \quad and \quad a \cdot 1 = a. \tag{6}$$

Complement: *To any element a in B there exists an element $\bar{a}$ in B such that*

$$a + \bar{a} = 1 \ and \ a \cdot \bar{a} = 0. \tag{7}$$

The operations "$+$" (*boolean "addition"*), "$\cdot$" (*boolean "multiplication"*) and "$\overline{(.)}$" (*boolean complementation*) are known as *boolean operations*. The set B is called *the universe* or *the carrier*. The elements 0 and 1 are called the *zero and unit elements* of B, respectively.

A Boolean algebra is called *finite* if its universe is a finite set. Although Boolean algebras are quintuples, it is customary to refer to a Boolean algebra by its carrier.

Example 5. The following structures are most popular Boolean algebras:

1. Two-value (or binary) Boolean algebra $\mathcal{B}_2 = (\{0, 1\}, +, \cdot, 0, 1)$ is the smallest, but the most important, model of general Boolean algebra. It has only two elements, 0 and 1. The binary operations $+$ and $\cdot$ and the unary operation $\neg$ are defined as follows:

x	y	$x + y$	$x \cdot y$
0	0	0	0
0	1	1	0
1	0	1	0
1	1	1	1

x	$\neg x$
0	1
1	0

2. The power set of any given set S forms a Boolean algebra with the two operations $+ := \cup$ (union) and $\cdot := \cap$ (intersection). The smallest element 0 is the empty set and the largest element 1 is the set S itself.
3. The set of all subsets of S that are either finite or cofinite is a Boolean algebra.

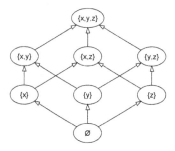

Fig. 4. The Boolean algebra of subsets

3.2 Some Properties of Boolean Algebras

Let us mention some well-known properties of Boolean algebras that are useful for further consideration [16].

Below we list some identities that are valid for any elements x, y, z of an arbitrary Boolean algebra $\mathcal{B} = (B, +, \cdot, 0, 1)$.

Associative law:

$$(x + y) + z = x + (y + z) \text{ and } (x \cdot y) \cdot z = x \cdot (y \cdot z) \tag{8}$$

Idempotence:

$$x + x = x \quad and \quad x \cdot x = x \tag{9}$$

Operations with 0 and 1:

$$x + 1 = 1 \quad and \quad x \cdot 0 = 0 \tag{10}$$

Absorption laws:

$$(y \cdot x) + x = x \quad and \quad (y + x) \cdot x = x \tag{11}$$

Involution:

$$\overline{(\overline{x})} = x \tag{12}$$

DeMorgan laws:

$$\overline{(x + y)} = \overline{x} \cdot \overline{y} \quad and \quad \overline{(x \cdot y)} = \overline{x} + \overline{y} \tag{13}$$

Consensus laws:

$$(x + y) \cdot (\overline{x} + z) \cdot (y + z) = (x + y) \cdot (\overline{x} + z) \text{ and} \tag{14}$$
$$(x \cdot y) + (\overline{x} \cdot z) + (y \cdot z) = (x \cdot y) + (\overline{x} \cdot z) \tag{15}$$

Duality principle: Any algebraic equality derived from the axioms of Boolean algebra remains true when the operations $+$ and $\cdot$ are interchanged and the identity elements 0 and 1 are interchanged. For example, $x + 1 = 1$ and $x \cdot 0 = 0$ are dual equations. Because of the duality principle, for any given theorem we immediately get also its dual.

The proofs of those properties can be derived from axioms in Definition 4. For example, the absorption law can be proved as follows:

$$
\begin{aligned}
(y \cdot x) + x &= (y \cdot x) + (x \cdot 1) & (Identity)\\
&= (y \cdot x) + (1 \cdot x) & (Commutative)\\
&= (y + 1) \cdot x & (Distributive)\\
&= 1 \cdot x & (Operations\ with\ 0\ and\ 1)\\
&= x & (Identity)
\end{aligned}
$$

It is not necessary to provide a separate proof for the dual because of the principle of duality.

Let us define for every Boolean algebra $\mathcal{B} = (B, +, \cdot, 0, 1)$ a relation "$\leq$" on B by setting

$$
x \leq y \quad \text{iff} \quad x = x \cdot y
$$

One can show that this relation is reflexive, antisymmetric and transitive, therefore "$\leq$" is a partial order. Furthermore, in this relation, $x + y$ is the least upper bound of x and y and $x \cdot y$ is the greatest lower bound of x and y. These properties indicate that every Boolean algebra is also a bounded lattice.

Another well-known results are related to the Stone representation theorem for Boolean algebras. It has been shown in [152] that every Boolean algebra is isomorphic to the algebra of clopen (i.e., simultaneously closed and open) subsets of its Stone space. Due to the properties of Stone space for finite algebras, this result means that every finite Boolean algebra is isomorphic to the Boolean algebra of subsets of some finite set S.

3.3   Boolean Expressions

Statements in Boolean algebras are represented by boolean expressions which can be defined inductively, starting with constants, variables and three elementary operations as building blocks.

Definition 5. *Given a Boolean algebra $\mathcal{B}$, the set of boolean expressions (or boolean formulas) on the set of n symbols $\{x_1, x_2, \ldots, x_n\}$ is defined by the following rules:*

(1) *The elements of $\mathcal{B}$ are boolean expressions;*
(2) *The symbols $x_1, x_2, \ldots, x_n$ are boolean expressions;*
(3) *If ϕ and ψ are boolean expressions, then $(\phi) + (\psi)$, $(\phi) \cdot (\psi)$ and $\overline{(\phi)}$ are boolean expressions;*
(4) *A string is a boolean expression if and only if it is formed by applying a finite number of rules (1), (2) and (3).*

By other words, boolean expressions involve constants, variables, boolean operations and corresponding parentheses. The notation $\phi(x_1, \ldots, x_n)$ denotes that ϕ is a boolean expression over $\{x_1, \ldots, x_n\}$.

As in ordinary algebra, we may omit the symbol "$\cdot$" in boolean expressions, except the places where emphasis is desired. Also we may reduce the number of parentheses in a boolean expression by assuming that multiplication "·" are performed before addition and by removing some unnecessary parentheses. For example, the well-formed boolean expression $((a) \cdot (x_1)) + ((b) \cdot (\overline{(x_2)}))$ can be simplified into more friendly form: $ax_1 + b\overline{x_2}$.

The discussion so far was related only to the syntax of boolean expressions, i.e., rules for the formation of string of symbols. Sometimes, instead of using the notion of sum and product, it is more convenient to call boolean expressions $(\phi) + (\psi)$ and $(\phi) \cdot (\psi)$ the *disjunction* and the *conjunction*, respectively. We will denote boolean expressions by Greek letters like ψ, ϕ, ζ, etc.

3.4   Boolean Functions

Every boolean expression $\psi(x_1, \ldots, x_n)$ can be interpreted as a definition of an n-ary boolean operation, i.e., a mapping

$$f_\psi^{\mathcal{B}} : \mathcal{B}^n \to \mathcal{B}$$

where $\mathcal{B}$ is an arbitrary Boolean algebra. The mapping $f_\psi^{\mathcal{B}}$ can be defined by the composition: for every point $(\alpha_1, \ldots, \alpha_n) \in \mathcal{B}^n$, the value of $f_\psi^{\mathcal{B}}(\alpha_1, \ldots, \alpha_n)$ is obtained by recursively applying Definition 5 to the expression ψ.

Definition 6. *An n-variable mapping $f : \mathcal{B}^n \to \mathcal{B}$ is called a Boolean function if and only if it can be expressed by a boolean expression.*

Without use of the notion of boolean expressions, n-variable Boolean functions can be also defined by the following rules:

1. For any $b \in B$, the *constant function*, defined by

$$f(x_1, \ldots, x_n) = b \quad \text{for all } x_1, \ldots, x_n \in B$$

   is an n-variable Boolean function;
2. For any $i \in \{1, \ldots, n\}$, the ith *projection function*, defined by

$$p_i(x_1, \ldots, x_n) = x_i \quad \text{for all } x_1, \ldots, x_n \in B$$

   is an n-variable Boolean function;
3. If f and g are n-variable Boolean functions, then n-variable Boolean functions are the functions $f + g$, fg and $\overline{f}$, which are defined by

$$\begin{array}{lll} (a) & (f+g)(x_1, \ldots, x_n) = & f(x_1, \ldots, x_n) + g(x_1, \ldots, x_n) \\ (b) & (fg)(x_1, \ldots, x_n) = & f(x_1, \ldots, x_n) \cdot g(x_1, \ldots, x_n) \\ (c) & (\overline{f})(x_1, \ldots, x_n) = & \overline{f(x_1, \ldots, x_n)} \end{array}$$

   for all $x_1, \ldots, x_n \in B$.
4. Only functions which can be defined by finitely many applications of rules 1., 2. and 3. are n-variable Boolean functions.

Therefore, n-variable Boolean functions establish a smallest set of mappings $f : \mathcal{B}^n \to \mathcal{B}$ containing constant functions, projection functions, and close under sum, product and complementary operations.

It is important to understand that every Boolean function can be represented by numerous boolean expressions, whereas every boolean expression represents a unique function. As a matter of fact, for a given finite Boolean algebra $\mathcal{B}$, the number of n-variables Boolean functions is bounded by $|\mathcal{B}|^{|\mathcal{B}|^n}$, whereas the number of n-variable boolean expressions is infinite. These remarks motivate the distinction that we draw between functions and expressions. We say that two boolean expressions ϕ and ψ are semantically equivalent if they represent the same Boolean function over a Boolean algebra $\mathcal{B}$. When this is the case, we write $\phi =_\mathcal{B} \psi$.

3.5   Representations of Boolean Functions

An important task in many applications of Boolean algebra is to select a "good" formula, with respect to a pre-defined criterion, to represent a Boolean function.

The simplest, most elementary method to represent a Boolean function over a finite Boolean algebra $\mathcal{B}$ is to provide its *function-table*, i.e., to give a complete list of all points in boolean hypercube $\mathcal{B}^n$ together with the value of the function in each point. If $\mathcal{B}$ has k elements, then the number of rows in the function-table of an n-variable Boolean function is k^n. We will show that every Boolean function over a finite Boolean algebra can be represented in a more compact way.

The following fact, called *Shannon's expansion theorem* [135], is a fundamental for many computations with Boolean functions.

Theorem 1 (Shannon's expansion theorem). *If $f : \mathcal{B}^n \to \mathcal{B}$ is a Boolean function, then*

$$f(x_1, \ldots, x_{n-1}, x_n) = x_n f(x_1, \ldots, x_{n-1}, 1) + \overline{x}_n f(x_1, \ldots, x_{n-1}, 0)$$

for all $(x_1, \ldots, x_{n-1}, x_n)$ in $\mathcal{B}^n$.

The proof of this fact follows from the recursive definition of Boolean functions. For example, using expansion theorem, any 3-variable Boolean function (over an arbitrary Boolean algebra) can be expanded as follows:

$$\begin{aligned}
f(x_1, x_2, x_3) =& f(0,0,0)\overline{x_1}\overline{x_2}\overline{x_3} + f(0,0,1)\overline{x_1}\overline{x_2}x_3 \\
& f(0,1,0)\overline{x_1}x_2\overline{x_3} + f(0,1,1)\overline{x_1}x_2x_3 \\
& f(1,0,0)x_1\overline{x_2}\overline{x_3} + f(1,0,1)x_1\overline{x_2}x_3 \\
& f(1,1,0)x_1x_2\overline{x_3} + f(1,1,1)x_1x_2x_3
\end{aligned}$$

For our convenience, let us introduce the notation x^a for $x \in \mathcal{B}$ and $a \in \{0,1\}$, where

$$x^0 = \overline{x}, \qquad x^1 = x.$$

For any sequence $\mathbf{b} = (b_1, b_2, \ldots, b_n) \in \{0, 1\}^n$ and any vector of boolean variables $\mathbf{X} = (x_1, x_2, \ldots, x_n)$ we define the *minterm* of $\mathbf{X}$ by

$$m_{\mathbf{b}}(\mathbf{X}) = X^{\mathbf{b}} = x_1^{b_1} x_2^{b_2} \ldots x_n^{b_n}$$

and *the maxterm* of $\mathbf{X}$ by

$$s_{\mathbf{b}}(\mathbf{X}) = \overline{m_{\mathbf{b}}(\mathbf{X})} = x_1^{\overline{b_1}} + x_2^{\overline{b_2}} + \cdots + x_n^{\overline{b_n}}.$$

This notation enables us to formulate the following characterization of Boolean functions.

Theorem 2 (minterm canonical form). *A function $f : \mathcal{B}^n \to \mathcal{B}$ is a Boolean function if and only if it can be expressed in the minterm canonical form:*

$$f(X) = \sum_{\mathbf{b} \in \{0,1\}^n} f(\mathbf{b}) X^{\mathbf{b}} \tag{16}$$

The proof of this result follows from Shannon's expansion theorem. For any $\mathbf{b} = (b_1, \ldots, b_n) \in \{0, 1\}^n$, the value $f(\mathbf{b}) \in \mathcal{B}$ is called *the discriminant* of the function f.

Theorem 2 indicates that any Boolean function is completely defined by its discriminants. The minterms, which are independent of f, are only standardized functional building blocks. Therefore, an n-variable Boolean function can be represented by 2^n rows, corresponding to all 0, 1 assignments of arguments, of its function-table. This sub-table of all 0, 1 assignments is called the *truth table*.

Example 6. Let us consider a Boolean function $f : \mathcal{B}^2 \to \mathcal{B}$ defined over $\mathcal{B} = (\{0, 1, a, \overline{a}\}, +, \cdot, 0, 1)$ by the formula $\psi(x, y) = \overline{a}x + a\overline{y}$. The function-table of this Boolean function should contain 16 rows. Table 2 shows the corresponding truth-table that contains only 4 rows. Thus, the minterm canonical form of this function is represented by

$$f(x, y) = a\overline{x}\overline{y} + x\overline{y} + \overline{a}xy.$$

Table 2. Truth-table for $\psi(x, y) = \overline{a}x + a\overline{y}$

x	y	$f(x, y)$
0	0	a
0	1	0
1	0	1
1	1	$\overline{a}$

3.6   Binary Boolean Algebra

In this paper we concentrate on applications of the binary Boolean algebra only. Let us show that the binary Boolean algebra plays a crucial role for the verification problem.

A statement involving constants and arguments $x_1, \ldots, x_n$ is called *an identity* in a Boolean algebra $\mathcal{B}$ if and only if it is valid for all substitutions of arguments on $\mathcal{B}^n$. The problem is to verify whether an identity is valid in all Boolean algebras.

One of verification methods is based on searching for a proof of the identity by repeated use of axioms (2.1)–(2.4) and other properties (2.5)–(2.14). The other method is based on Theorem 2, which states that any Boolean function is uniquely determined by its 0, 1 assignments of variables. Therefore, any identity can be verified by all 0, 1 substitutions of arguments. This result, called the *Löwenheim-Müller Verification Theorem* [128],[16], can be formulated as follows:

Theorem 3 (Löwenheim-Müller Verification Theorem).   *An identity expressed by Boolean expressions is valid in all Boolean algebras if and only if it is valid in the binary Boolean algebra (which can always be checked by a trivial brute force algorithm using truth tables).*

For example, DeMorgan laws can be verified by checking in the binary Boolean algebra as it is shown in Table 3.

Table 3. A proof of DeMorgan law by using truth-table. Two last columns are the same, therefore $\overline{(x + y)} = \overline{x} \cdot \overline{y}$.

x	y	$x + y$	$\overline{(x + y)}$
0	0	0	1
0	1	1	0
1	0	1	0
1	1	1	0

x	y	$\overline{x}$	$\overline{y}$	$\overline{x} \cdot \overline{y}$
0	0	1	1	1
0	1	1	0	0
1	0	0	1	0
1	1	0	0	0

Binary Boolean algebra has also applications in propositional calculus, interpreting 0 as false, 1 as true, $+$ as logical OR (disjunction), $\cdot$ as logical AND (conjunction), and $\overline{(.)}$ as logical NOT (complementation, negation).

Any mapping

$$f : \{0, 1\}^n \rightarrow \{0, 1\}$$

is called an n-variable *switching function*. Some properties of switching functions are listed below:

– The function-table of any switching function is the same as its truth-table.
– Since there are 2^n rows in the truth-table of any Boolean function over n variables, therefore the number of n-variable switching functions is equal to 2^{2^n}.
– Every switching function is a Boolean function.

3.7   Some Classes of Boolean Expressions, Normal Forms

Expressions in binary Boolean algebras are quite specific, because almost all of them are constant–free (excepts the two constant expressions 0 and 1). Let us recall the definitions of some common subclasses of boolean expressions like literals, terms, clauses, CNF and DNF that will be used later in this paper.

Boolean expressions are formed from *letters*, i.e., constants and variables using boolean operations like conjunction, disjunction and complementation.

- A *literal* is a letter or its complement.
- A *term* is either 1 (the unit element), a single literal, or a conjunction of literals in which no letter appears more than once. Some example terms are $x_1 x_3$ and $x_1 x_2 \overline{x_4}$. The size of a term is the number of literals it contains. The examples are of sizes 2 and 3, respectively. A *monomial* is a Boolean function that can be expressed by a term. It is easy to show that there are exactly 3^n possible terms over n variables.
- An *clause* is either 0 (the zero element), a single literal, or a conjunction of literals in which no letter appears more than once. Some example clauses are $x_3 + x_5 + x_6$ and $x_1 + \overline{x_4}$. The size of a term is the number of literals it contains. The examples are of sizes 2 and 3, respectively. There are 3^n possible clauses. If f can be represented by a term, then (by De Morgan laws) $\overline{f}$ can be represented by a clause, and vice versa. Thus, terms and clauses are dual of each other.

From psychological experiments it follows that conjunctions of literals seem easier for humans to learn than disjunctions of literals.

A boolean expression is said to be in disjunctive normal form (DNF) if it is a disjunction of terms. Some examples in DNF are:

$$\phi_1 = x_1 x_2 + x_2 x_3 x_4;$$
$$\phi_2 = x_1 x_3 + x_2 x_3 + \overline{x_1 x_2} x_3.$$

A DNF expression is called a "k–term DNF" expression if it is a disjunction of k terms; it is in the class "k-DNF" if the size of its largest term is k. The examples above are 2-term and 3-term expressions, respectively. Both expressions are in the 3-DNF class.

Disjunctive normal form has a dual conjunctive normal form (CNF). A Boolean function is said to be in CNF if it can be written as a conjunction of clauses. An example in CNF is:

$$f = (x_1 + x_2)(x_2 + x_3 + x_4).$$

A CNF expression is called a $k$clause CNF expression if it is a conjunction of k clauses; it is in the class kCNF if the size of its largest clause is k. The example is a 2clause expression in 3CNF.

Any Boolean function can be represented in both CNF and DNF. One of possible DNF representations of a Boolean function implies from Theorem 2. In case of the binary Boolean algebra, the minterm canonical form of a switching function is represented by

$$f(\mathbf{X}) = \sum_{\mathbf{b} \in f^{-1}(1)} m_{\mathbf{b}}(\mathbf{X}).$$

The dual representation of minterm canonical form is called *the maxterm canonical form* and it is written as follows:

$$f(\mathbf{X}) = \prod_{\mathbf{a} \in f^{-1}(0)} s_{\mathbf{a}}(\mathbf{x})$$

For example, let a switching function f be given in the form of a truth table represented in Table 4.

The minterm and maxterm canonical forms of this function are as follow:

$$\phi_1 = xy\overline{z} + x\overline{y}z + \overline{x}yz + xyz$$
$$\phi_2 = (x+y+z)(\overline{x}+y+z)(x+\overline{y}+z)(x+y+\overline{z})$$

Table 4. Example of switching function

x	y	z	f
0	0	0	0
1	0	0	0
0	1	0	0
1	1	0	1
0	0	1	0
1	0	1	1
0	1	1	1
1	1	1	1

3.8 Implicants and Prime Implicants

Given a function f and a term t, we define the *quotient* of f with respect to t, denoted by f/t, to be the function formed from f by imposing the constraint $t = 1$. For example, let a Boolean function f be given by

$$f(x_1, x_2, x_3, x_4) = \overline{x_1}x_2x_4 + x_2\overline{x_3}\overline{x_4} + x_1\overline{x_2}x_4.$$

The quotient of f with respect to $x_1\overline{x_3}$ is

$$f/x_1\overline{x_3} = f(1, x_2, 0, x_4) = x_2\overline{x_4} + \overline{x_2}x_4.$$

It is clear that the function f/t can be represented by a formula that does not involve any variable appearing in t. Let us define two basic notions in Boolean function theory called *implicant* and *prime implicant*.

Definition 7. *A term* t *is* an implicant *of a function* f *if* $f/t = 1$. *An implicant* t *is* a prime implicant *of* f *if the term* t' *formed by taking any literal out of* t *is no longer an implicant of* f *(the prime implicant cannot be "divided" by any term and remain an implicant).*

Let us observe that each term in a DNF expression of a function is an implicant because it "implies" the function (if the term has value 1, so does the DNF expression).

In a general Boolean algebra, for two Boolean functions h and g we write $h \ll g$ if and only if the identity $h\overline{g} = 0$ is satisfied. This property can be verified by checking whether $h(\mathbf{X})\overline{g(\mathbf{X})} = 0$ for any zero-one vector $\mathbf{X} = (\alpha_1, \ldots, \alpha_n) \in \{0, 1\}^n$. A term t is an *implicant* of a function f if and only if $t \ll f$.

Thus, both $x_2\overline{x_3}$ and $\overline{x_1}\overline{x_3}$ are prime implicants of $f = x_2\overline{x_3} + \overline{x_1}\overline{x_3} + x_2x_1\overline{x_3} + x_1x_2x_3$, but $x_2x_1\overline{x_3}$ is not.

The relationship between implicants and prime implicants can be geometrically illustrated using the cube representation for Boolean functions. To represent an n-variable Boolean function we need an n-dimensional hypercube with 2^n vertices corresponding to 2^n zero-one vectors in $\{0, 1\}^n$. In fact, cube representation of a Boolean function is a regular graph having 2^n vertices with degree $n - 1$ each.

Given a Boolean function f and a term t. Let $C = (V, E)$ be the cube representation of f. Then the subgraph $C|_t = (V_t, E_t)$ generated by

$$V_t = \{\mathbf{x} \in \{0, 1\}^n : t(\mathbf{x}) = 1\}$$

is a surface of C and it is the cube representation of the quotient f/t. It is clear that $C|_t$ is a $(n - k)$-dimensional surface where k is the number of literals in t. The term t is an implicant if and only if the surface $C|_t$ contains the vertices having value 1 only.

For example, the illustration of function

$$f = x_2\overline{x_3} + \overline{x_1}\overline{x_3} + x_2x_1\overline{x_3} + x_1x_2x_3$$

and the cube representations of two quotients f/x_2 and $f/\overline{x_1}\overline{x_3}$ are shown in Fig. 5. Vertices having value 1 are labeled by solid circles, and vertices having value 0 are labeled by hollow circles.

The function, written as the disjunction of some terms, corresponds to the union of all the vertices belonging to all of the surfaces. The function can be written as a disjunction of a set of implicants if and only if the corresponding surfaces create a set covering the set of truth values of f.

In this example, the term x_2 is not an implicant because $C|_{x_2}$ contains a vertex having value 0 (i.e., $f/x_2(011) = 0$). In this way we can "see" only 7 implicants of the function f:

$$\overline{x_1}x_2\overline{x_3}, \ \overline{x_1}x_2\overline{x_3}, \ x_1x_2\overline{x_3}, \ x_1x_2x_3, \ x_2\overline{x_3}, \ \overline{x_1}\overline{x_3}, \ x_1x_2$$

the function f can be represented in various forms, e.g.,

$$f = \overline{x_1}\overline{x_3} + x_1x_2$$
$$= \overline{x_1}x_2\overline{x_3} + x_2\overline{x_3} + x_1x_2x_3$$

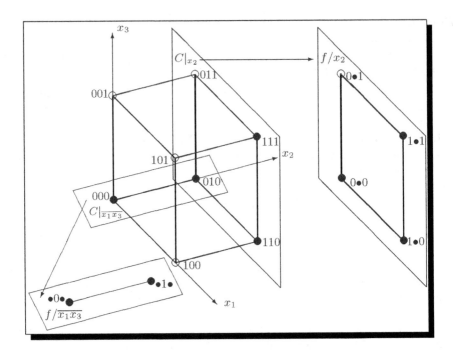

Fig. 5. An illustration of $f = x_2\overline{x_3} + \overline{x_1 x_3} + x_2 x_1 \overline{x_3} + x_1 x_2 x_3$. Vertices having value 1 are labeled by solid circles.

Geometrically, an implicant is prime if and only if its corresponding surface is the largest dimensional surface that includes all of its vertices and no other vertices having value 0. In the previous example only $x_2\overline{x_3}$, $\overline{x_1 x_3}$, $x_1 x_2$ are prime implicants.

4 Boolean Reasoning and Approximate Boolean Reasoning

As Boolean algebra plays a fundamental role in computer science, the boolean reasoning approach is a crucial methodology for problem solving in Artificial Intelligence. For more details on this topic, we refer the reader to Chang and Lee [19], Gallaire and Minker [34], Loveland [65], Kowalski [59], Hayes-Roth, Waterman and Lenat [46], Jeroslow [51], Anthony and Biggs [6].

Boolean reasoning approach is a general framework for solving many complex decision or optimization problems. We recall the standard Boolean reasoning approach called the syllogistic reasoning which was described by Brown in [16]. Next, we describe the famous application of Boolean reasoning in planning problem. This elegant method, called SAT-planning was proposed by Henry Kautz and Bart Selman [53], [54]. We also present the general framework of approximate Boolean reasoning methodology which is the main subject of this contribution.

4.1   Boolean Reasoning Methodology

The greatest idea of Boole's algebraic approach to logic was to reduce the processes of reasoning to processes of calculation. In Boolean algebras, a system of logical equations can be transformed to a single equivalent boolean equation.

Boole and other 19th-century logicians based their symbolic reasoning on an equation of 0-normal form, i.e.,

$$f(x_1, x_2, \ldots, x_n) = 0.$$

Blake [12] showed that the consequents of this equation are directly derived from the prime implicants of f. Thus the representation of f as a disjunction of all its prime implicants is called the *Blake Canonical Form* of a Boolean function f and denoted by $BCF(f)$, i.e.,

$$BCF(f) = t_1 + t_2 + \cdots + t_k,$$

where $\{t_1, \ldots, t_k\}$ is the collection of all prime implicants of the function f. This observation enables to develop an interesting Boolean reasoning method called the *syllogistic reasoning* that extract conclusions from a collection of boolean data (see Example 7). Quine [121], citequine1952, [122] also appreciated the importance of the concept of prime implicants in his research related to the problem of minimizing the complexity of boolean formulas.

The main steps of Boolean reasoning methodology to the problem solving are presented in Fig. 6.

---

1. **Modeling:** Represent the problem by a collection of boolean equations. The idea is to represent constraints and facts in the clausal form.
2. **Reduction:** Condense the equations into a problem over a single boolean equation of form

$$f(x_1, x_2, \ldots, x_n) = 0 \qquad (17)$$

   (or, dually, $\overline{f} = 1$).
3. **Development:** Generate a set of all or some prime implicants of f, depending on the formulation of the problem.
4. **Reasoning:** Apply a sequence of reasoning to solve the problem.

Fig. 6. The Boolean reasoning methodology

Analogically to symbolic approaches in other algebras, Step 1 is performed by introducing some variables and describing the problem in the language of Boolean algebra. After that, obtained description of the problem is converted into boolean equations using following laws in Boolean algebra theory:

$$a \le b \quad \Leftrightarrow \quad a\overline{b} = 0$$
$$a \le b \le c \quad \Leftrightarrow \quad a\overline{b} + b\overline{c} = 0$$
$$a = b \quad \Leftrightarrow \quad a\overline{b} + \overline{a}b = 0$$
$$a = 0 \text{ and } b = 0 \quad \Leftrightarrow \quad a + b = 0$$
$$a = 1 \text{ and } b = 1 \quad \Leftrightarrow \quad ab = 1$$

where a, b, c are elements of a Boolean algebra $\mathcal{B}$.

Steps 2 and 3 are independent of the problem to be solved and are more or less automated. In Step 2, three types of problems over boolean equation are considered:

- Search for all solutions (all prime implicants) of Eqn. (17);
- (SAT) Check whether any solution of (17) exists;
- Search for the shortest prime implicant of (17);

The complexity of Step 4 depends on the problem and the encoding method in Step 1. Let us illustrate the Boolean reasoning approach by the following examples.

4.1.1 Syllogistic Reasoning

The following example of syllogistic reasoning was considered in [16]:

Example 7. Consider the following logical puzzle:

Problem: Four friends Alice, Ben, Charlie, David are considering going to a party. The following social constraints hold:
- If Alice goes than Ben will not go and Charlie will;
- If Ben and David go, then either Alice or Charlie (but not both) will go;
- If Charlie goes and Ben does not, then David will go but Alice will not.

First, to apply the Boolean reasoning approach to this problem, we have to introduce some variables as follows:

$$A : \text{Alice will go}$$
$$B : \text{Ben will go}$$
$$C : \text{Charlie will go}$$
$$D : \text{David will go}$$

1. Problem modeling:

$$A \implies \neg B \wedge C \qquad\qquad \rightsquigarrow \quad A(B + \overline{C}) \qquad = 0$$
$$B \wedge D \implies (A \wedge \neg C) \vee (C \wedge \neg A) \quad \rightsquigarrow \quad BD(AC + \overline{AC}) \quad = 0$$
$$C \wedge \neg B \implies D \wedge \neg A \qquad\qquad \rightsquigarrow \quad \overline{B}C(A + \overline{D}) \quad = 0$$

2. **After reduction:**

$$f = A(B + \overline{C}) + BD(AC + \overline{AC}) + \overline{B}C(A + \overline{D}) = 0$$

3. **Development:** The function f has three prime implicants: $B\overline{C}D, \overline{B}C\overline{D}, A$. Therefore, the equation, after transformation to the Blake canonical form, is rewritten as follows:

$$f = B\overline{C}D + \overline{B}C\overline{D} + A = 0$$

Solutions of this equation are derived from prime implicants, i.e.,

$$f = 0 \Leftrightarrow \begin{cases} B\overline{C}D & = 0 \\ \overline{B}C\overline{D} & = 0 \\ A & = 0 \end{cases}$$

4. **Reasoning:** The Blake's reasoning method was based on clausal form. The idea is based on the fact that any equation of form

$$x_1 \ldots x_n \overline{y_1} \ldots \overline{y_m} = 0 \tag{18}$$

can be transformed to the equivalent propositional formula in the clausal form

$$x_1 \cdots \cdot x_n \implies y_1 \vee \cdots \vee y_m \tag{19}$$

Thus, given information of the problem is equivalent to the following facts:

$$\begin{array}{ll} B \wedge D \longrightarrow C & \text{“if Ben and David go then Charlie will”} \\ C \longrightarrow B \vee D & \text{“if Charlie goes then Ben or David will go”} \\ A \longrightarrow 0 & \text{“Alice will not go”} \end{array}$$

The obtained facts can be treated as an input to the automated theorem proving systems. E.g., one can show that "nobody will go alone".

4.1.2   Application of Boolean Reasoning Approach in AI

Another application of Boolean reasoning approach is related to the planning problem. Generally, planning is encoded as a synthesis problem; given an initial state, a desired final condition and some possible operators that can be used to change state, a planning algorithm will output a sequence of actions which achieves the final condition. Each action is a full instantiation of the parameters of an operator. This sequence of actions is called a plan.

Henry Kautz and Bart Selman [53], [54] proposed a planning method which is also known as "satisfiability planning" (or SAT planning) since it is based on the satisfiability problem. In this method, the specification of the studied problem is encoded by a Boolean function in such a way that the encoding function is satisfiable if and only if there exists a correct plan for the given specification. Let us illustrate this method by the famous "blocks world" problem.

Example 8 (Blocks world planning problem). One of the most famous planning domains is known as the blocks world. This domain consists of a set of cube-shaped blocks sitting on a table. The blocks can be stacked, but only one

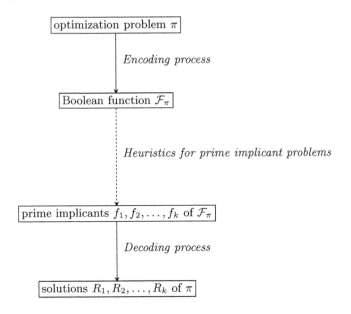

Fig. 7. The Boolean reasoning scheme for optimization problems

block can fit directly on top of another. A robot arm can pick up a block and move it to another position, either on the table or on top of another block. The arm can pick up only one block at a time, so it cannot pick up a block that has another one on it. The goal will always be to build one or more stacks of blocks, specified in terms of what blocks are on top of what other blocks. For example, Fig. 8 presents a problem, where a goal is to get block E on C and block D on B.

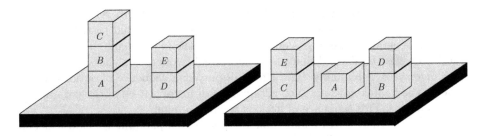

Fig. 8. An example of blocks world planning problem. The initial situation is presented in the left and the final situation is presented in the right.

The goal is to produce a set of boolean variables and a set of rules that they have to obey. In the blocks–world problem, all following statements are boolean variables:

- "$on(x, y, i)$" means that block x is on block y at time i;
- "$clear(x, i)$" means that there is room on top of block x at time i;
- "$move(x, y, z, i)$" means that block x is moved from block y to block z between i and $i + 1$.

The encoding formula is a conjunction of clauses. There are four different parts of the plan that must be converted by hand into axiom schemas:

Initial state: the state that is assumed to hold at time 1;

Goal condition: holds at time $n + 1$ where n is the expected number of actions required to achieve the plan;

For each operator: two families of axioms are defined:

- The *effect axioms* are asserting that an operator which executes at time i implies its preconditions at time i and its effects at time $i + 1$.
- The *frame axioms* state that anything that holds at time i and is not changed by the effects must also hold at time $i + 1$.

Exclusion axioms: these axioms are denoted at-least-one (ALO) and at-most-one (AMO), and are used to prevent the problem of actions that have conflicting preconditions or effects executing simultaneously.

For the aforementioned blocks-world example these axiom schemas might be:

initial state:	$on(C, B, 1) \land on(B, A, 1) \land on(A, table, 1) \land clear(C, 1) \land$ $on(E, D, 1) \land on(D, table, 1) \land clear(E, 1)$
goal condition:	$on(A, table, 6) \land on(B, table, 6) \land on(C, table, 6) \land$ $on(E, C, 6) \land on(D, B, 6) \land clear(A, 6) \land clear(D, 6) \land$ $clear(E, 6)$
effect axiom schemas:	For the move operator: **preconditions:** $\forall_{x,y,z,i} move(x, y, z, i) \implies clear(x, i) \land on(x, y, i) \land$ $clear(z, i)$ **effects:** $\forall_{x,y,z,i} move(x, y, z, i) \implies on(x, z, i+1) \land clear(y, i+1) \land \neg on(x, y, i+1) \land \neg clear(z, i+1)$
frame axiom schemas	For the move operator: 1. $\forall_{w,x,y,z,i} move(x, y, z, i) \land w \neq y \land w \neq z \land clear(w, i)$ $\implies clear(w, i+1)$ 2. $\forall_{v,w,x,y,z,i} move(x, y, z, i) \land v \neq x \land w \neq x \land w \neq y \land w \neq z \land on(v, w, i) \implies on(v, w, i+1)$
exclusion axiom schemas:	Exactly one action occurs at each time step: **AMO:** $\forall_{x,x',y,y',z,z',i}(x \neq x' \lor y \neq y' \lor z \neq z') \implies$ $\neg move(x, y, z, i) \lor \neg move(x', y', z', i)$ **ALO:** $\forall_i \exists_{x,y,z} move(x, y, z, i)$.

The number of clauses produced by this schema is tk^6, where t is the number of time steps and k is the number of blocks. For a trivial problem with 2 time steps and 2 blocks, this schema derives 128 clauses.

4.1.3 Complexity of Prime Implicant Problems

Calculating a set of prime implicants is the most time–consuming step in Boolean reasoning schema. It is known that there are n–variable Boolean functions with $\Omega(3^n/n)$ prime implicants (see, e.g., [18]) and the maximal number of prime implicants of n–variable Boolean functions does not exceed $O(3^n/\sqrt{n})$. Thus many problems related to calculation of prime implicants are hard [37].

In the complexity theory, the most famous problem connected with Boolean functions is the *satisfiability problem* (SAT). It is based on deciding whether there exists an evaluation of variables that satisfies a given Boolean formula. In other words, the problem is related to the boolean equation $f(x_1, \ldots, x_n) = 1$ and the existence of its solution.

SAT is the first decision problem which has been proved to be NP–complete (the Cook's theorem). This important result is used to prove the NP–hardness of many other problems by showing the polynomial transformation of SAT to the studied problem.

The relationship between SAT and prime implicants is obvious, a valid formula has the empty monomial 0 as its only prime implicant. An unsatisfiable formula has no prime implicants at all. In general, a formula ϕ has a prime implicant if and only if ϕ is satisfiable. Therefore, the question of whether a formula has a prime implicant is NP–complete, and it is in L (logarithmic space) for monotone formulas.

> SAT: Satisfiability problem
> *input*: A boolean formula ϕ of n variables.
> *question*: Does ϕ have a prime implicant?

Let us consider the problem of checking whether a term is a prime implicant of a Boolean function.

> ISPRIMI:
> *input*: A boolean formula ϕ and a term t.
> *question*: Is t a prime implicant of ϕ?

It has been shown that the complexity of ISPRIMI is intermediate between NP∪coNP and Σ_2^p. Another problem that is very useful in the Boolean reasoning approach relates to the size of prime implicants.

> PRIMISIZE:
> *input*: A boolean formula ϕ of n variables, an integer k.
> *question*: Does ϕ have a prime implicant consisting of at most k variables?

This problem was shown to be Σ_2^p–complete [153].

4.1.4   Monotone Boolean Functions

A boolean function $\phi : \{0,1\}^n \to \{0,1\}$ is called *"monotone"* if

$$\forall_{\mathbf{x},\mathbf{y} \in \{0,1\}^n} (\mathbf{x} \leq \mathbf{y}) \Rightarrow (\phi(\mathbf{x}) \leq \phi(\mathbf{y})).$$

It has been shown that monotone functions can be represented by a boolean expression without negations. Thus, a *monotone expression* is an expression without negation.

One can show that if ϕ is a positive boolean formula of n variables $x_1, \ldots, x_n$, then for each variable x_i

$$\phi(x_1, \ldots, x_n) = x_i \cdot \phi/x_i + \phi/\overline{x}_i, \tag{20}$$

where

$$\phi/x_i = \phi(x_1, \ldots, x_{i-1}, 1, x_{i+1}, \ldots, x_n) \quad \text{and}$$
$$\phi/\overline{x}_i = \phi(x_1, \ldots, x_{i-1}, 0, x_{i+1}, \ldots, x_n)$$

are obtained from ϕ by replacing x_i by constants 1 and 0, respectively. One can prove Equation (20) by the truth–table method. Let us consider two cases:

– if $x_i = 0$ then

$$x_i \cdot \phi/x_i + \phi/\overline{x}_i = \phi/\overline{x}_i$$
$$= \phi(x_1, \ldots, 0, \ldots, x_n);$$

– if $x_i = 1$ then

$$x_i \cdot \phi/x_i + \phi/\overline{x}_i = \phi/x_i + \phi/\overline{x}_i$$
$$= \phi/x_i \qquad \qquad (monotonicity).$$

The last identity holds because ϕ is monotone, hence $\phi/x_i \geq \phi/\overline{x}_i$. Therefore, Equation (20) is valid for each $x_1, \ldots, x_n \in \{0,1\}^n$.

A monotone formula ϕ in disjunctive normal form is *irredundant* if and only if no term of ϕ covers another term of ϕ. For a monotone formula, the disjunction of all its prime implicants yields an equivalent monotone DNF. On the other hand, every prime implicant must appear in every equivalent DNF for a monotone formula. Hence, the smallest DNF for a monotone formula is unique and equals the disjunction of all its prime implicants. This is not the case for non-monotone formulas, where the smallest DNF is a subset of the set of all prime implicants. It is NP-hard to select the right prime implicants [Mas79]. See also [Czo99] for an overview on the complexity of calculating DNFs.

Many calculations of prime implicants for monotone boolean formulas are much easier than for general formulas. For the example of IsPRIMI problem, it can be checked in logarithmic space whether an assignment corresponding to the term satisfies the formula. The PRIMISIZE problem for monotone formulas is NP-complete only. Table 5 summarizes the complexity of discussed problems.

Table 5. Computational complexity of some prime implicant problems

Problem	Arbitrary formula	Monotone formula
SAT	NP-complete	L
ISPRIMI	between $NP \cup coNP$ and $\sum_2^p$	L
PRIMISIZE	$\sum_2^p$-complete	NP-complete

This result implies that the problem of searching for prime implicant of minimal size (even for monotone formulas) is NP–hard.

MINPRIMI$_{mon}$: minimal prime implicant of monotone formulas
input: Monotone boolean formula ϕ of n variables.
output: A prime implicant of minimal size.

We have mentioned that SAT plays a fundamental role in the computation theory, as it is used to prove NP–hardness of other problems. From practical point of view, any SAT-solver (a heuristic algorithm for SAT) can be used to design heuristic solutions for other problems in the class NP. Therefore, instead of solving a couple of hard problems, the main effort may be limited to create efficient heuristics for the SAT problem.

Every boolean formula can be transformed into a monotone boolean formula such that satisfying assignments of the basic formula are similar to prime implicants of the monotone formula. The transformation is constructed as follows:

Let ϕ be a boolean formula in negation normal form [4] (NNF) with n variables $x_1, \ldots, x_n$.

- Let $r(\phi)$ denote the formula obtained by replacing all appearances of $\bar{x}_i$ in ϕ by the new variable y_i (for $i = 1, 2, \ldots, n$).
- Let $c(\phi)$ denote the conjunction $\prod_{i=1}^{n}(x_i + y_i)$.

One can show that can ϕ is satisfied if and only if monotone boolean formula $r(\phi) \cdot c(\phi)$ has a prime implicant consisting of at most n variables. The main idea is to prove that a vector $\mathbf{a} = (a_1, \ldots, a_n) \in \{0,1\}^n$ is a satisfied evaluation for ϕ (i.e., $f_\phi(\mathbf{a}) = 1$) if and only if the term

$$t_\mathbf{a} = \prod_{a_i=1} x_i \cdot \prod_{a_i=0} y_i$$

is a prime implicant of $r(\phi) \cdot c(\phi)$.

Every heuristic algorithm $\mathcal{A}$ for MINPRIMI$_{mon}$ problem can be used to solve (in approximate way) the SAT problem for an arbitrary formula ϕ as follows:

1. calculate the minimal prime implicant t of the monotone formula $r(\phi) \cdot c(\phi)$;
2. return the answer "YES" if and only if t consists of at most n variables.

---

[4] In the negation normal form, negations are attached to atomic formulas. Any boolean formula can be converted into NNF by the recursive using of evolution and DeMorgan law.

4.2   Approximate Boolean Reasoning Method

The high computational complexity of prime implicant problems means that the Boolean reasoning approach is not applicable in many real-world problems, particularly in data mining, where large amount of data is one of the major challenges.

The natural approach to managing hard problems is to search for an approximate instead of an exact or optimal solution. The first attempt might be related to calculation of prime implicants, as it is the most complex step in the Boolean reasoning schema. In the next section we describe some well-known approximate algorithms for prime implicant problems.

Each approximate method is characterized by two parameters: the quality of approximation and the computation time. Searching for the proper balance between those parameters is the biggest challenge of modern heuristics. We have proposed a novel method, called the *approximate Boolean reasoning method*, to extend this idea. In the approximate Boolean reasoning approach to problem solving, not only calculation of prime implicants, but every step in the original scheme (Fig. 9) is approximately performed to achieve an approximate solution.

- **Modeling:** Represent the problem Π (or its simplification) by a collection of Boolean equations.
- **Reduction:** Condense the equations into an equivalent problem encoded by a single boolean equation of the form $f_\Pi(x_1, \dots, x_n) = 1$ or a simplified problem encoded by an equation $f'_\Pi = 1$.
- **Development:** Generate an approximate solution of the formulated problem over f_Π (or f'_Π).
- **Reasoning:** Apply an approximate reasoning method to solve the original problem.

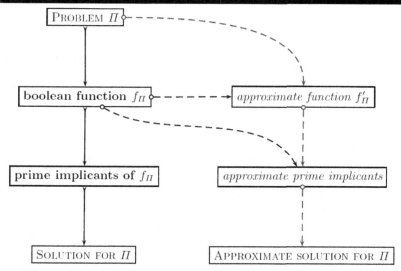

Fig. 9. General scheme of approximate Boolean reasoning approach

The general scheme of the approximate Boolean reasoning approach is presented in Fig. 9.

Since the minimal prime implicant problem is NP–hard, it cannot be solved (in general case) by exact methods only. It is necessary to create some heuristics to search for short prime implicants of large and complicated boolean functions.

Most of the problems in data mining are formulated as optimization problems. We will show in the next sections many applications of the Boolean reasoning approach to optimization problem where the minimal prime implicants play a crucial role.

4.2.1   Heuristics for Prime Implicant Problems

Minimal prime implicants of a given function can be determined from the set of all its prime implicants. One of the well-known methods was proposed by Quine-McCluskey [121]. This method is featured by possible exponential time and space complexity as it is based on using consensus laws to minimalize the canonical DNF of a function (defined by true points) into the DNF in which each term is a prime implicant.

DPLL procedures. The close relationship between SAT and MINPRIMI$_{mon}$ (as it was described in the previous section) implies the similarity between their solutions. Let us mention some most important *SAT-solvers*, i.e., the solving methods for SAT problem.

The first SAT-solvers, which still remain the most popular ones, are the Davis-Putnam (DP) [27] and Davis-Logemann-Loveland (DLL) algorithms [26]. These methods are featured by possible exponential space (in case of DP) and time (in both cases) complexity, therefore they have a limited practical applicability. But compound idea of both methods are very useful; it is known as DPLL for historical reasons. DPLL is a basic framework for many modern SAT solvers.

CDLC methods. The DPLL algorithm remained dominant among complete methods until the introduction of clause learning solvers like GRASP in 1996 [69], Chaff [66], BerkMin [36], Siege [129] and many others. This new method is a variation of DPLL with two additional techniques called the backtracking algorithm and the corresponding refutation.

1. The first technique, called "clause learning" and "clause recording", or just "learning", is that, if we actually derive the clauses labeling the search tree, we can add some of them to the formula. If later in the execution the assignment at some node falsifies one of these clauses, the search below that node is avoided with possible time savings.
2. The second technique is based on the method called "conflict directed back-jumping" (CBJ) in the constraint satisfaction literature [118]. The idea is based on effective recognition of those clauses for which no second recursive call is necessary.

Algorithm 1. procedure DPLL(ϕ, t)

---

begin
    // SAT:
    if ϕ/t *is empty* **then**
       | return SATISFIABLE;
    end
    // Conflict:
    if ϕ/t *contains an empty clause* **then**
       | return UNSATISFIABLE;
    end
    // Unit Clause:
    if ϕ/t *contains a unit clause* $\{p\}$ **then**
       | return DPLL(ϕ, tp);
    end
    // Branch:
    Let p be a literal from a minimal size clause of ϕ/t;
    if *DPLL(ϕ, tp)* **then**
       | return SATISFIABLE;
    else
       | return DPLL($\phi, t\bar{p}$);
    end
end

Therefore, this class of solvers is sometimes called "conflict driven clause learning" (CDCL) algorithm.

Max-Sat based methods. Another noticeable method for SAT was proposed by Selman [132]. The idea is to treat SAT as a version of MAX-SAT problem, where the task is to find an assignment that satisfies the most number of clauses. Any local search algorithm can be employed to the search space containing all assignments, and the cost function for a given assignment is set by a number of unsatisfied clauses.

4.2.2 Searching for Prime Implicants of Monotone Functions

In case of minimal prime implicant for monotone functions, the input boolean function is assumed to be given in the CNF form, i.e., it is presented as a conjunction of clauses, e.g.,

$$\psi = (x_1 + x_2 + x_3)(x_2 + x_4)(x_1 + x_3 + x_5)(x_1 + x_5)(x_5 + x_6)(x_1 + x_2) \quad (21)$$

Searching for the minimal prime implicant can be treated as the minimal hitting set problem, i.e., the problem of searching for minimal set of variables X such that for every clause of the given formula, at least one of its variables must occur in X.

Let us recall that every monotone Boolean function can be expanded by a variable x_i as follows (see Eqn. (20)):

$$\phi(x_1, \ldots, x_n) = x_i \cdot \phi/x_i + \phi/\overline{x}_i$$
$$= x_i\phi(x_1, \ldots, x_{i-1}, 1, x_{i+1}, \ldots, x_n) + \phi(x_1, \ldots, x_{i-1}, 0, x_{i+1}, \ldots, x_n)$$

The basic problem in many existing heuristics for minimal prime implicant is to evaluate the chance that the variable x_i belongs to the minimal prime implicant of ϕ. We can do that by defining an evaluation function $Eval(x_i; \phi)$ which takes under consideration two formulas ϕ/x_i and $\phi/\overline{x}_i$, i.e.

$$Eval(x_i; \phi) = F(\phi/x_i, \phi/\overline{x}_i) \tag{22}$$

The algorithm should decide either to select the best variable x_{best} and continue with ϕ/x_{best} or to remove the worst variable x_{worst} and continue with $\phi/\overline{x}_{worst}$.

One can apply an idea of DPLL algorithms to solve the minimal prime implicant for monotone formulas. The algorithm of searching for minimal prime implicant starts from an empty term t, and in each step, it might choose one of the following actions:

unit clause: If $\phi/t\overline{x}_i$ degenerates for some variable x_i, i.e., $\phi/t\overline{x}_i = 0$, then x_i must occur in every prime implicant of ϕ/t. Such variable is called *the core variable*. The core variable can be quickly recognized by checking whether there exists a unit clause, i.e., a clause that consists of one variable only. If x_i is core variable, then the algorithm should continue with ϕ/tx_i;

final step: If there exists variable x_i such that ϕ/tx_i degenerates, then x_i is the minimal prime implicant of ϕ/t, the algorithm should return tx_i as a result and stop here;

heuristic decision: If none of previous rules cannot be performed, the algorithm should use the evaluation function (Equation (22)) to decide how to continue the searching process. The decision is related to adding a variable x_i to t and continuing the search with formula ϕ/tx_i or rejecting a variable x_j and continuing the search with formula $\phi/t\overline{x}_j$.

Let us mention some most popular heuristics that have been proposed for minimal prime implicant problem for monotone Boolean functions:

1. **Greedy algorithm:**
   This simple method (see [35]) is using the number of unsatisfied clauses as a heuristic function. In each step, the greedy method selects the variable that most frequently occurs within clauses of the given function and removes all those clauses which contain the selected variable. For the function in Eqn. (21) x_1 is the most preferable variable by the greedy algorithm. The result of greedy algorithm for this function might be $x_1x_4x_6$, while the minimal prime implicant is x_2x_5.

2. **Linear programming:** The minimal prime implicant can also be resolved by converting the given function into a system of linear inequations and

applying the Integer Linear Programming (ILP) approach to this system, see [115], [68].

Assume that an input monotone boolean formula is given in CNF. The idea is to associate with each boolean variable x_i an integer variable t_i. Each monotone clause $x_{i_1} + \cdots + x_{i_k}$ is replaced by an equivalent inequality:

$$t_{i_1} + \cdots + t_{i_k} \geq 1$$

and the whole CNF formula is replaced by a set of inequalities $\mathbf{A} \cdot \mathbf{t} \geq \mathbf{b}$. The problem is to minimize the number of variables with the value one assigned. The resulting ILP model is as follows:

$$\min(t_1 + t_2 + \cdots + t_n)$$
$$s.t. \quad \mathbf{A} \cdot \mathbf{t} \geq \mathbf{b}$$

3. **Simulated annealing:** many optimization problems are resolved by a Monte-Carlo search method called simulated annealing. In case of minimal prime implicant problem, the search space consists of all subsets of variables and the cost function for a given subset X of boolean variables is defined by two factors: (1) the number of clauses that are uncovered by X, and (2) the size of X, see [133].

4.2.3 Ten Challenges in Boolean Reasoning

In 1997, Selman et al. [131] present an excellent summary of the state of the art in propositional (Boolean) reasoning, and sketches challenges for the next 10 years:

SAT problems. Two specific open SAT problems:

Challenge 1 Prove that a hard 700 variable random 3-SAT formula is unsatisfiable.

Challenge 2 Develop an algorithm that finds a model for the DIMACS 32-bit parity problem.

Proof systems. Are there stronger proof systems than resolution?

Challenge 3 Demonstrate that a propositional proof system more powerful than resolution can be made practical for satisfiability testing.

Challenge 4 Demonstrate that integer programming can be made practical for satisfiability testing.

Local search. Can local search be made to work for proving unsatisfiability?

Challenge 5 Design a practical stochastic local search procedure for proving unsatisfiability.

Challenge 6 Improve stochastic local search on structured problems by efficiently handling variable dependencies.

Challenge 7 Demonstrate the successful combination of stochastic search and systematic search techniques, by the creation of a new algorithm that outperforms the best previous examples of both approaches.

Encodings. different encodings of the same problem can have vastly different computational properties.

> **Challenge 8** Characterize the computational properties of different encodings of a real-world problem domain, and/or give general principles that hold over a range of domains.
>
> **Challenge 9** Find encodings of real-world domains which are robust in the sense that "near models" are actually "near solutions".
>
> **Challenge 10** Develop a generator for problem instances that have computational properties that are more similar to real-world instances

In the next sections, we present some applications of approximate Boolean reasoning approach to rough set methods and data mining. We will show that in many cases, the domain knowledge is very useful for designing effective and efficient solutions.

5   Application of ABR in Rough Sets

In this section, we recall two famous applications of Boolean reasoning methodology in rough set theory. The first is related to the problem of searching for reducts, i.e., subsets of most informative attributes of a given decision table or information system. The second application concerns the problem of searching for decision rules which are building units of many rule-based classification methods.

5.1   Rough Sets and Feature Selection Problem

Feature selection has been an active research area in pattern recognition, statistics, and data mining communities. The main idea of feature selection is to select a subset of most relevant attributes for classification task, or to eliminate features with little or no predictive information. Feature selection can significantly improve the comprehensibility of the resulting classifier models and often build a model that generalizes better to unseen objects [63]. Further, it is often the case that finding the correct subset of predictive features is an important problem in its own right.

In rough set theory, the feature selection problem is defined in terms of reducts [111]. We will generalize this notion and show an application of the ABR approach to this problem.

In general, reducts are minimal subsets (with respect to the set inclusion relation) of attributes which contain a necessary portion of *information* about the set of all attributes. The notion of information is as abstractive as the notion of energy in physics, and we will not able to define it exactly. Instead of explicit information, we have to define some *objective properties* for all subsets of attributes. Such properties can be expressed in different ways, e.g., by logical formulas or, as in this section, by a *monotone evaluation function* which is described as follows.

For a given information system $\mathbb{S} = (U, A)$, the function

$$\mu_{\mathbb{S}} : \mathcal{P}(A) \longrightarrow \Re^+$$

where $\mathcal{P}(A)$ is the power set of A, is called *the monotone evaluation function* if the following conditions hold:

1. the value of $\mu_{\mathbb{S}}(B)$ can be computed using information set $INF(B)$ for any $B \subset A$;
2. for any $B, C \subset A$, if $B \subset C$, then $\mu_{\mathbb{S}}(B) \leq \mu_{\mathbb{S}}(C)$.

Definition 8 (μ-reduct). *Any set $B \subseteq A$ is called* the reduct relative to a monotone evaluation function μ, *or briefly μ-reduct, if B is the smallest subset of attributes that $\mu(B) = \mu(A)$, i.e., $\mu(B') \not\geq \mu(B)$ for any proper subset $B' \subsetneq B$.*

This definition is general for many different definition of reducts. Let us mention some well-known types of reducts used in rough set theory.

5.1.1 Basic Types of Reducts in Rough Set Theory

In Sect. 2.3, we have introduced the *B-indiscernibility relation* (denoted by $IND_{\mathbb{S}}(B)$) for any subset of attributes $B \subset A$ of a given information system $\mathbb{S} = (U, A)$ by

$$IND_{\mathbb{S}}(B) = \{(x, y) \in U \times U : inf_B(x) = inf_B(y)\}.$$

Relation $IND_{\mathbb{S}}(B)$ is an equivalence relation. Its equivalence classes can be used to define the lower and upper approximations of concepts in rough set theory [109], [111].

The complement of indiscernibility relation is called *B-discernibility relation* and is denoted by $DISC_{\mathbb{S}}(B)$. Hence,

$$DISC_{\mathbb{S}}(B) = U \times U - IND_{\mathbb{S}}(B)$$
$$= \{(x, y) \in U \times U : inf_B(x) \neq inf_B(y)\}$$
$$= \{(x, y) \in U \times U : \exists_{a \in B} a(x) \neq a(y)\}.$$

It is easy to show that $DISC_{\mathbb{S}}(B)$ is monotone, i.e., for any $B, C \subset A$

$$B \subset C \Longrightarrow DISC_{\mathbb{S}}(B) \subset DISC_{\mathbb{S}}(C).$$

Intuitively, any reduct (in rough set theory) is a minimal subset of attributes that preserves the discernibility between information vectors of objects. The following notions of reducts are often used in rough set theory.

Definition 9 (information reducts). *Any minimal subset B of A such that $DISC_{\mathbb{S}}(A) = DISC_{\mathbb{S}}(B)$ is called the* information reduct *(or reduct, for short) of $\mathbb{S}$. The set of all reducts of a given information system $\mathbb{S}$ is denoted by $RED(\mathbb{S})$*

In the case of decision tables, we are interested in the ability of describing decision classes by using subsets of condition attributes. This ability can be expressed in terms of *generalized decision function* $\partial_B : U \to \mathcal{P}(V_{dec})$, where

$$\partial_B(x) = \{i : \exists_{x' \in U} [(x' IND(B)x) \wedge (d(x') = i)]\}$$

(see Equation (2) in Sect. 2.3).

Definition 10 (decision-relative reducts). *The set of attributes $B \subseteq A$ is called a* relative reduct *(or simply a* decision reduct*) of decision table $\mathbb{S}$ if and only if*

- $\partial_B(x) = \partial_A(x)$ *for all object $x \in U$;*
- *any proper subset of B does not satisfy the previous condition;*

i.e., B is a minimal subset (with respect to the inclusion relation $\subseteq$) of the attribute set satisfying the property $\forall_{x \in U} \partial_B(x) = \partial_A(x)$.

The set $C \subset A$ of attributes is called *super-reduct* if there exists a reduct B such that $B \subset C$. One can prove the following theorem:

Theorem 4 (the equivalency of definitions).

1. *Information reducts for a given information system $\mathbb{S} = (U, A)$ are exactly those reducts with respect to discernibility function, which is defined for arbitrary subset of attributes $B \subset A$ as a number pairs of objects discerned by attributes from B, i.e.,*

$$disc(B) = \frac{1}{2} card(DISC_{\mathbb{S}}(B)).$$

2. *Relative reducts for decision tables $\mathbb{S} = (U, A \cup \{dec\})$ are exactly those reducts with respect to* relative discernibility function*, which is defined by*

$$disc_{dec}(B) = \frac{1}{2} card(DISC_{\mathbb{S}}(B) \cap DISC_{\mathbb{S}}(\{dec\})).$$

The relative discernibility function returns the number of pairs of objects from different classes, which are discerned by attributes from B.

Many other types of reducts, e.g., frequency based reducts [145] or entropy reducts in [146], can be defined by selection of different monotone evaluation functions.

Example 9. Let us consider the "weather" problem, which is represented by decision table (see Table 6). Objects are described by four conditional attributes and are divided into 2 classes. Let us consider the first 12 observations. In this example, $U = \{1, 2, \ldots, 12\}$, $A = \{a_1, a_2, a_3, a_4\}$, $CLASS_{no} = \{1, 2, 6, 8\}$, $CLASS_{yes} = \{3, 4, 5, 7, 9, 10, 11, 12\}$.

The equivalence classes of indiscernibility relation $IND_{\mathbb{S}}(B)$ for some sets of attributes are given in Table 7.

Table 6. The exemplary "weather" decision table

date	outlook	temperature	humidity	windy	play
ID	a_1	a_2	a_3	a_4	dec
1	sunny	hot	high	FALSE	no
2	sunny	hot	high	TRUE	no
3	overcast	hot	high	FALSE	yes
4	rainy	mild	high	FALSE	yes
5	rainy	cool	normal	FALSE	yes
6	rainy	cool	normal	TRUE	no
7	overcast	cool	normal	TRUE	yes
8	sunny	mild	high	FALSE	no
9	sunny	cool	normal	FALSE	yes
10	rainy	mild	normal	FALSE	yes
11	sunny	mild	normal	TRUE	yes
12	overcast	mild	high	TRUE	yes
13	overcast	hot	normal	FALSE	yes
14	rainy	mild	high	TRUE	no

Table 7. Indiscernibility classes of $IND_\mathbb{S}(B)$ for some sets of attributes

The set of attributes B	Equivalent classes of $IND_\mathbb{S}(B)$
$B = \{a_1\}$	$\{1,2,8,9,11\},\{3,7,12\},\{4,5,6,10\}$
$B = \{a_1, a_2\}$	$\{1,2\},\{3\},\{4,10\},\{5,6\},\{7\},\{8,11\},\{9\},\{12\}$
$B = A = \{a_1, a_2, a_3, a_4\}$	$\{1\},\{2\},\{3\},\{4\},\{5\},\{6\},\{7\},\{8\},\{9\},\{10\},\{11\},\{12\}$

Table 8. The discernibility function of different subsets of attributes

Attribute sets B	$disc_{dec}(B)$
$B = \varnothing$	0
$B = \{a_1\}$	23
$B = \{a_2\}$	23
$B = \{a_3\}$	18
$B = \{a_4\}$	16
$B = \{a_1, a_2\}$	30
$B = \{a_1, a_3\}$	31
$B = \{a_1, a_4\}$	29
$B = \{a_2, a_3\}$	27
$B = \{a_2, a_4\}$	28
$B = \{a_3, a_4\}$	25
$B = \{a_1, a_2, a_3\}$	31
$B = \{a_1, a_2, a_4\}$	**32**
$B = \{a_1, a_3, a_4\}$	**32**
$B = \{a_2, a_3, a_4\}$	29
$B = \{a_1, a_2, a_3, a_4\}$	**32**

The values of relative discernibility function $disc_{dec}(B)$ for all subsets $B \subset A$ are given in Table 8. One can see that there are two relative reducts for this table: $R_1 = \{a_1, a_2, a_4\}$ and $R_2 = \{a_1, a_3, a_4\}$.

5.1.2 Boolean Reasoning Approach for Reduct Problem

There are two problems related to the notion of "reduct", which have been intensively explored in rough set theory by many researchers (see, e.g., [7], [50], [60], [145], [146], [157]. The first problem is related to searching for reducts with the minimal cardinality called *the shortest reduct problem*. The second problem is related to searching for all reducts. It has been shown that the first problem is NP-hard (see [143]) and second is at least NP-hard. Some heuristics have been proposed for those problems. Here we present the approach based on Boolean reasoning as proposed in [143] (see Fig. 10).

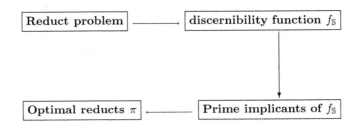

Fig. 10. The Boolean reasoning scheme for solving reduct problem

Given a decision table $\mathbb{S} = (U, A \cup \{dec\})$, where $U = \{u_1, u_2, \dots, u_n\}$ and $A = \{a_1, \dots, a_k\}$. By *discernibility matrix* of the decision table $\mathbb{S}$ we mean the $(n \times n)$ matrix

$$\mathbf{M}(\mathbb{S}) = [M_{i,j}]_{ij=1}^n$$

where $M_{i,j} \subset A$ is the set of attributes discerning u_i and u_j, i.e.,

$$M_{i,j} = \{a_m \in A : a_m(u_i) \neq a_m(u_j)\}. \tag{23}$$

Let us denote by $VAR_{\mathbb{S}} = \{x_1, \dots, x_k\}$ a set of boolean variables corresponding to attributes $a_1, \dots, a_k$. For any subset of attributes $B \subset A$, we denote by $X(B)$ the set of boolean variables corresponding to attributes from B. We will encode reduct problem as a problem of searching for the corresponding set of variables.

For any two objects $u_i, u_j \in U$, the boolean clause χ_{u_i,u_j}, called *discernibility clause*, is defined as follows:

$$\chi_{u_i,u_j}(x_1, \dots, x_k) = \begin{cases} \displaystyle\sum_{a_m \in M_{i,j}} x_m & \text{if } M_{i,j} \neq \varnothing \\ 1 & \text{if } M_{i,j} = \varnothing \end{cases}$$

The objective is to create a boolean function $f_{\mathbb{S}}$ such that a set of attributes is a reduct of $\mathbb{S}$ if and only if it corresponds to a prime implicant of $f_{\mathbb{S}}$. This function is defined as follows:

1. for information reduct problem:

$$f_{\mathbb{S}}(x_1,\ldots,x_k) = \prod_{i\neq j} \left(\chi_{u_i,u_j}(x_1,\ldots,x_k) \right) \tag{24}$$

The function $f_{\mathbb{S}}$ was defined by Skowron and Rauszer [143] and it is called the *discernibility function*.

2. for relative reduct problem:

$$f_{\mathbb{S}}^{dec}(x_1,\ldots,x_k) = \prod_{i,j:dec(u_i)\neq dec(u_j)} \left(\chi_{u_i,u_j}(x_1,\ldots,x_k) \right) \tag{25}$$

The function $f_{\mathbb{S}}^{dec}$ is called the *decision oriented discernibility function*.

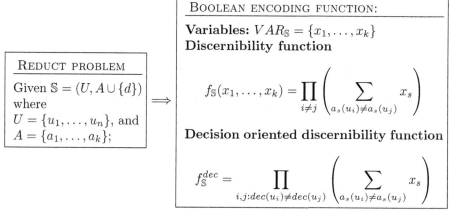

REDUCT PROBLEM

Given $\mathbb{S} = (U, A \cup \{d\})$ where $U = \{u_1,\ldots,u_n\}$, and $A = \{a_1,\ldots,a_k\}$;

$\Longrightarrow$

BOOLEAN ENCODING FUNCTION:

Variables: $VAR_{\mathbb{S}} = \{x_1,\ldots,x_k\}$

Discernibility function

$$f_{\mathbb{S}}(x_1,\ldots,x_k) = \prod_{i\neq j} \left(\sum_{a_s(u_i)\neq a_s(u_j)} x_s \right)$$

Decision oriented discernibility function

$$f_{\mathbb{S}}^{dec} = \prod_{i,j:dec(u_i)\neq dec(u_j)} \left(\sum_{a_s(u_i)\neq a_s(u_j)} x_s \right)$$

Let us associate with every subset of attributes $B \subset A$ an assignment $\mathbf{a}_B \in \{0,1\}^k$ as follows:

$$\mathbf{a}_B = (v_1,\ldots,v_k), \text{ where } v_i = 1 \Leftrightarrow a_i \in B,$$

i.e., $\mathbf{a}_B$ is the characteristic vector of B. We have the following propositions:

Proposition 1. *For any $B \subset A$ the following conditions are equivalent:*

1. $f_{\mathbb{S}}(\mathbf{a}_B) = 1$;
2. $disc(B) = disc(A)$, *i.e., B is a super-reduct of $\mathbb{S}$.*

Proof:
Assume that $\mathbf{a}_B = (v_1,\ldots,v_k)$ we have:

$$f_{\mathbb{S}}(\mathbf{a}_B) = 1 \Leftrightarrow \forall_{u_i,u_j\in U} \ \chi_{u_i,u_j}(\mathbf{a}_B) = 1$$
$$\Leftrightarrow \forall_{u_i,u_j\in U}\exists_{a_m\in M_{i,j}} \ v_m = 1$$
$$\Leftrightarrow \forall_{u_i,u_j\in U} \ M_{i,j} \neq \varnothing \implies B \cap M_{i,j} \neq \varnothing$$
$$\Leftrightarrow disc(B) = disc(A)$$

∎

Now we are ready to show the following theorem:

Theorem 5. *A subset of attributes B is a reduct of* $\mathbb{S}$ *if and only if the term*

$$T_{X(B)} = \prod_{a_i \in B} x_i$$

is a prime implicant of the discernibility function $f_{\mathbb{S}}$.

Proof: Let us assume that B is a reduct of $\mathbb{S}$. It means that B is the smallest set of attributes such that $disc(B) = disc(A)$. Using Proposition 1, we have $f_{\mathbb{S}}(\mathbf{a}_B) = 1$. We will show that $T_{X(B)}$ is a prime implicant of $f_{\mathbb{S}}$.

To do so, let us assume that $T_{X(B)}(\mathbf{a}) = 1$ for some $\mathbf{a} \in \{0, 1\}^k$. Because $\mathbf{a}_B$ is the minimal assignment satisfying $T_{X(B)}$, hence $\mathbf{a}_B \leq \mathbf{a}$. Consequently we have

$$1 = f_{\mathbb{S}}(\mathbf{a}_B) \leq f_{\mathbb{S}}(\mathbf{a}) \leq 1$$

Thus $T_{X(B)} \leq f_{\mathbb{S}}$, i.e., $T_{X(B)}$ is a implicant of $f_{\mathbb{S}}$.

To prove that $T_{X(B)}$ is prime implicant of $f_{\mathbb{S}}$, we must show that there is no smaller implicant than $T_{X(B)}$. To do so, let us assume conversely that there is an implicant T, which is really smaller than $T_{X(B)}$. It means that $T = T_{X(C)}$ for some subset of attributes $C \subsetneq B$. Since $T_{X(C)}$ is implicant of $f_{\mathbb{S}}$, then we have $f_{\mathbb{S}}(\mathbf{a}_C) = 1$. This condition is equivalent to $disc(C) = disc(A)$ (see Proposition 1). Hence, we have a contradiction to the assumption that B is a reduct of $\mathbb{S}$.

The proof of the second implication (i.e., $T_{X(B)}$ is prime implicant $\Rightarrow B$ is the reduct of $\mathbb{S}$) is very similar and we will omit it.   ∎

One can slightly modify the previous proof to show the following theorem.

Theorem 6. *A subset of attributes B is a relative reduct of decision table* $\mathbb{S}$ *if and only if* $T_{X(B)}$ *is the prime implicant of the relative discernibility function* $f_{\mathbb{S}}^{dec}$.

Example 10. Let us consider again the decision table "weather" presented in Table 6. This table consists of 4 attributes: a_1, a_2, a_3, a_4, hence the set of corresponding boolean variables consists of

$$VAR_{\mathbb{S}} = \{x_1, x_2, x_3, x_4\}$$

The discernibility matrix is presented in Table 9.

The discernibility matrix can be treated as a board containing $n \times n$ boxes. Noteworthy is the fact that discernibility matrix is symmetrical with respect to the main diagonal, because $M_{i,j} = M_{j,i}$, and that sorting all objects according to their decision classes causes a shift off all empty boxes nearby to the main diagonal. In case of decision table with two decision classes, the discernibility matrix can be rewritten in a more compact form as shown in Table 10.

The discernibility function is constructed from discernibility matrix by taking a conjunction of all discernibility clauses. After reducing of all repeated clauses we have:

$$\begin{aligned}
f(x_1, x_2, x_3, x_4) = &(x_1)(x_1 + x_4)(x_1 + x_2)(x_1 + x_2 + x_3 + x_4)(x_1 + x_2 + x_4) \\
&(x_2 + x_3 + x_4)(x_1 + x_2 + x_3)(x_4)(x_2 + x_3)(x_2 + x_4) \\
&(x_1 + x_3)(x_3 + x_4)(x_1 + x_2 + x_4).
\end{aligned}$$

Table 9. Discernibility matrix

M	1	2	3	4	5	6	7	8	9	10	11	12
1			a_1	a_1, a_2	a_1, a_2, a_3		a_1, a_2, a_3, a_4		a_2, a_3	a_1, a_2, a_3	a_2, a_3, a_4	a_1, a_2, a_4
2			a_1, a_4	a_1, a_2, a_4	a_1, a_2, a_3, a_4		a_1, a_2, a_3		a_2, a_3, a_4	a_1, a_2, a_3, a_4	a_2, a_3	a_1, a_2
3	a_1	a_1, a_4				a_1, a_2, a_3, a_4		a_1, a_2				
4	a_1, a_2	a_1, a_2, a_4				a_2, a_3, a_4		a_1				
5	a_1, a_2, a_3	a_1, a_2, a_3, a_4				a_4		a_1, a_2, a_3				
6			a_1, a_2, a_3, a_4	a_2, a_3, a_4	a_4		a_1		a_1, a_4	a_2, a_4	a_1, a_2	a_1, a_2, a_3
7	a_1, a_2, a_3, a_4	a_1, a_2, a_3				a_1		a_1, a_2, a_3, a_4				
8			a_1, a_2	a_1	a_1, a_2, a_3		a_1, a_2, a_3, a_4		a_2, a_3	a_1, a_3	a_3, a_4	a_1, a_4
9	a_2, a_3	a_2, a_3, a_4				a_1, a_4		a_2, a_3				
10	a_1, a_2, a_3	a_1, a_2, a_3, a_4				a_2, a_4		a_1, a_3				
11	a_2, a_3, a_4	a_2, a_3				a_1, a_2		a_3, a_4				
12	a_1, a_2, a_4	a_1, a_2				a_1, a_2, a_3		a_1, a_4				

One can find relative reducts of the decision table by searching for its prime implicants. The straightforward method calculates all prime implicants by translation to DNF. One can do it as follow:

- remove those clauses that are absorbed by some other clauses (using absorption rule: $p(p + q) \equiv p$):

$$f = (x_1)(x_4)(x_2 + x_3)$$

- Translate f from CNF into DNF

$$f = x_1 x_4 x_2 + x_1 x_4 x_3.$$

- Every monomial corresponds to a reduct. Thus we have 2 reducts: $R_1 = \{a_1, a_2, a_4\}$ and $R_2 = \{a_1, a_3, a_4\}$.

5.2   Approximate Algorithms for Reduct Problem

Every heuristic algorithm for the prime implicant problem can be applied to the discernibility function to solve the minimal reduct problem. One of such

Table 10. The compact form of discernibility matrix

M	1	2	6	8
3	a_1	a_1, a_4	a_1, a_2, a_3, a_4	a_1, a_2
4	a_1, a_2	a_1, a_2, a_4	a_2, a_3, a_4	a_1
5	a_1, a_2, a_3	a_1, a_2, a_3, a_4	a_4	a_1, a_2, a_3
7	a_1, a_2, a_3, a_4	a_1, a_2, a_3	a_1	a_1, a_2, a_3, a_4
9	a_2, a_3	a_2, a_3, a_4	a_1, a_4	a_2, a_3
10	a_1, a_2, a_3	a_1, a_2, a_3, a_4	a_2, a_4	a_1, a_3
11	a_2, a_3, a_4	a_2, a_3	a_1, a_2	a_3, a_4
12	a_1, a_2, a_4	a_1, a_2	a_1, a_2, a_3	a_1, a_4

heuristics was proposed in [143] and was based on the idea of greedy algorithm (see Sect. 4.2), where each attribute is evaluated by its discernibility measure, i.e., the number of pairs of objects which are discerned by the attribute, or, equivalently, the number of its occurrences in the discernibility matrix.

Let us illustrate the idea by using discernibility matrix (Table 10) from the previous section.

- First we have to calculate the number of occurrences of each attributes in the discernibility matrix:

$$eval(a_1) = disc_{dec}(a_1) = 23 \qquad eval(a_2) = disc_{dec}(a_2) = 23$$
$$eval(a_3) = disc_{dec}(a_3) = 18 \qquad eval(a_4) = disc_{dec}(a_4) = 16$$

  Thus a_1 and a_2 are the two most preferred attributes.
- Assume that we select a_1. Now we are taking under consideration only those cells of the discernibility matrix which are not containing a_1. There are 9 such cells only, and the number of occurrences are as the following:

$$eval(a_2) = disc_{dec}(a_1, a_2) - disc_{dec}(a_1) = 7$$
$$eval(a_3) = disc_{dec}(a_1, a_3) - disc_{dec}(a_1) = 7$$
$$eval(a_4) = disc_{dec}(a_1, a_4) - disc_{dec}(a_1) = 6$$

- If this time we select a_2, then the are only 2 remaining cells, and, both are containing a_4;
- Therefore, the greedy algorithm returns the set $\{a_1, a_2, a_4\}$ as a reduct of sufficiently small size.

There is another reason for choosing a_1 and a_4, because they are *core attributes*[5]. It has been shown that an attribute is a core attribute if and only if occurs in the discernibility matrix as a singleton [143]. Therefore, core attributes can be recognized by searching for all single cells of the discernibility matrix. The pseudo-code of this algorithm is presented in Algorithm 2.

---

[5] An attribute is called core attribute if and only if it occurs in every reduct.

Algorithm 2. Searching for short reduct

---

begin
  | $B := \emptyset$;
  | // Step 1. Initializing B by core attributes
  | **for** $a \in A$ **do**
  | | **if** $isCore(a)$ **then**
  | | | $B := B \cup \{a\}$;
  | | **end**
  | **end**
  | // Step 2. Including attributes to B
  | **repeat**
  | | $a_{\max} := \arg\max\limits_{a \in A - B} disc_{dec}(B \cup \{a\})$;
  | | $eval(a_{\max}) := disc_{dec}(B \cup \{a_{\max}\}) - disc_{dec}(B)$;
  | | **if** $(eval(a_{\max}) > 0)$ **then**
  | | | $B := B \cup \{a\}$;
  | | **end**
  | **until** $(eval(a_{\max}) == 0)$ *OR* $(B == A)$;
  | // Step 3. Elimination
  | **for** $a \in B$ **do**
  | | **if** $(disc_{dec}(B) = disc_{dec}(B - \{a\}))$ **then**
  | | | $B := B - \{a\}$;
  | | **end**
  | **end**
end

The reader may have a feeling that the greedy algorithm for reduct problem has quite a high complexity, because two main operations:

- $disc(B)$ – number of pairs of objects discerned by attributes from B;
- $isCore(a)$ – check whether a is a core attribute;

are defined by the discernibility matrix which is a complex data structure containing $O(n^2)$ cells, and each cell can contain up to $O(m)$ attributes, where n is the number of objects and m is the number of attributes of the given decision table. This suggests that the two main operations need at least $O(mn^2)$ computational time.

Fortunately, both operations can be performed more efficiently. It has been shown [101] that both operations can be calculated in time $O(mn \log n)$ without the necessity to store the discernibility matrix. In Sect. 10, we present an effective implementation of this heuristics that can be applied to large data sets.

5.3   Malicious Decision Tables

In this section, we consider a class of decision tables with maximal number of reducts. In some sense, such tables are the hardest decision tables for reduct problems. We are interesting in the structure of such tables and we will present a solution based on Boolean reasoning approach.

Let $\mathbb{S} = (U, A \cup \{d\})$ be an arbitrary decision table containing m attributes, i.e., $A = \{a_1, \ldots, a_m\}$, and n objects, i.e., $U = \{u_1, \ldots, u_n\}$, and let $\mathbf{M}(\mathbb{S}) = [C_{i,j}]_{ij=1}^n$ be the discernibility matrix of $\mathbb{S}$.

We denote by $RED(\mathbb{S})$ the set of all relative reducts of decision table $\mathbb{S}$. Let us recall some properties of the set $RED(\mathbb{S})$:

1. If $B_1 \in RED(\mathbb{S})$ is a reduct of the system $\mathbb{S}$, then there is no such reduct $B_2 \in RED(\mathbb{S})$ that $B_1 \subsetneq B_2$.
2. The elements of $RED(\mathbb{S})$ create an antichain with respect to the inclusion between subsets of A.
3. If $|A| = m$ is an even positive integer, i.e., $m = 2k$, then

$$\mathcal{C} = \{B \subset A : |B| = k\} \qquad (26)$$

is the only antichain containing maximal number of subsets of A.
4. If $|A| = m$ is an odd positive integer, i.e., $m = 2k + 1$, then there are two antichains containing the maximal number of subsets:

$$\mathcal{C}_1 = \{B \subset A : |B| = k\}; \quad \mathcal{C}_2 = \{B \subset A : |B| = k + 1\} \qquad (27)$$

We have

Proposition 2. *The number of reducts for any decision table $\mathbb{S}$ with m attributes is bounded by*

$$N(m) = \binom{m}{\lfloor m/2 \rfloor}.$$

A decision table $\mathbb{S}$ is called *malicious* if it contains exactly $N(m)$ reducts. The problem is to construct a malicious decision table containing m attributes for each integer m.

Let

$$f_{\mathbb{S}} = \mathbf{C}_1 \cdot \mathbf{C}_2 \cdots \mathbf{C}_M$$

be the discernibility function of decision table $\mathbb{S}$, where $\mathbf{C}_1, \ldots, \mathbf{C}_M$ are clauses defined on boolean variables from $VAR = \{x_1, \ldots, x_m\}$ corresponding to attributes $a_1, \ldots, a_m$ (see Sect. 5.1.1).

From (26) and (27) one can prove the following propositions:

Proposition 3. *A decision table $\mathbb{S}$ with m attributes is malicious if and only if the discernibility function $f_{\mathbb{S}}$ has exactly $N(m)$ prime implicants. In particular,*

- *if m is even, then $f_{\mathbb{S}}$ can be transformed to the form:*

$$f^* = \sum_{X \subset VAR: |X| = m/2} \mathbf{T}_X$$

- *if k is odd, then $f_{\mathbb{S}}$ can be transformed to one of the forms:*

$$f_1^* = \sum_{X \subset VAR: |X| = (k-1)/2} \mathbf{T}_X$$

or

$$f_2^* = \sum_{X \subset VAR: |X| = (k+1)/2} \mathbf{T}_X.$$

The next proposition describes how discernibility functions of malicious decision tables look like.

Proposition 4. *If $\mathbb{S}$ is a malicious decision table, then its discernibility function function $f_\mathbb{S}$ must consist of at least $\Omega(N(k))$ clauses.*

Proof: Let

$$f_\mathbb{S} = \mathbf{C}_1 \cdot \mathbf{C}_2 \cdot \cdots \cdot \mathbf{C}_M$$

be an irreducible CNF of the discernibility function $f_\mathbb{S}$. We will prove the following facts:

Fact 1. A term $\mathbf{T}_X$ is an implicant of $f_\mathbb{S}$ if and only if $X \cap VAR(\mathbf{C}_i) \neq \varnothing$ for any $m \in \{1, \dots, M\}$.
Ad 1. This fact has been proved in Sect. 5.1.1.
Fact 2. If m is an even integer, then $|VAR(\mathbf{C}_i)| \geq m/2 + 1$ for any $i \in \{1, \dots, M\}$.
Ad 2. Let us assume that there is an index $i \in \{1, \dots, M\}$ such that $|VAR(\mathbf{C}_i)| \leq m/2$. We will show that $f_\mathbb{S} \neq f^*$ (in contrary to Proposition 3).
 In fact, because $|VAR \setminus VAR(\mathbf{C}_i)| \geq m/2$ then there exists a set of variables $X \subset VAR \setminus VAR(\mathbf{C}_i)$ such that $|X| = m/2$, which implies that $\mathbf{T}_X$ is not an implicant of $f_\mathbb{S}$, because $X \cap VAR(\mathbf{C}_i) = \varnothing$. Therefore $f_\mathbb{S} \neq f^*$.
Fact 3. If m is an even integer, then for any subset of variables $X \subset VAR$ such that $|X| = m/2 + 1$, there exists $i \in \{1, \dots, M\}$ such that $VAR(\mathbf{C}_i) = X$.
Ad 3. Let us assume conversely that there exists such X that $|X| = m/2 + 1$ and $X \neq VAR(\mathbf{C}_i)$ for any $i \in \{1, \dots, M\}$. Let $Y = VAR \setminus X$, we have $|Y| = m/2 - 1$. Recall that $|VAR(\mathbf{C}_i)| \geq m/2 + 1$, thus

$$|Y| + |VAR(\mathbf{C}_i)| \geq m. \tag{28}$$

Moreover, for any $i \in \{1, \dots, M\}$, we have

$$Y = VAR \setminus X \neq VAR \setminus VAR(\mathbf{C}_i). \tag{29}$$

From (28) and (29) we have

$$Y \cap VAR(\mathbf{C}_i) \neq \varnothing.$$

Therefore, $\mathbf{T}_Y$ is an implicant of $f_\mathbb{S}$, which is contradictory to Proposition 3.
Fact 4. If m is an odd integer and $f_\mathbb{S}$ is transformable to f_1^*, then for any subset of variables $X \subset VAR$ such that $|X| = (m-1)/2 + 2$, there exists $i \in \{1, \dots, M\}$ such that $VAR(\mathbf{C}_i) = X$.
Fact 5. If m is an odd integer and $f_\mathbb{S}$ is transformable to f_2^*, then for any subset of variables $X \subset VAR$ such that $|X| = (m-1)/2 + 1$, there exists $i \in \{1, \dots, M\}$ such that $VAR(\mathbf{C}_i) = X$.

The proofs of Fact 4 and 5 are analogical to the proof of Fact 3.

From Fact 3, 4 and 5 we have:

$$M \geqslant \begin{cases} \binom{m}{m/2+1} = \frac{m}{m+2}N(m) & \text{if } f_{\mathbb{S}} \text{ is transformable to } f^* \\ \binom{m}{(m+1)/2+1} = \frac{m-1}{m+3}N(m) & \text{if } f_{\mathbb{S}} \text{ is transformable to } f_1^* \\ \binom{m}{(m+1)/2} = N(m) & \text{if } f_{\mathbb{S}} \text{ is transformable to } f_2^* \end{cases}$$

Therefore, $M \geqslant \Omega(N(m))$ in all cases.   ∎

Let n be the number of objects of $\mathbb{S}$, we have $n \cdot (n-1)/2 \geqslant M$. From Proposition 4 we have $M \geqslant \Omega(N(k))$, therefore $n \geqslant \Omega(\sqrt{N(k)})$. Thus we obtain the following theorem.

Theorem 7. *If a decision table $\mathbb{S}$ is malicious, then it contains at least Ω $(\sqrt{N(k)})$ objects.*

This result means that, even if malicious decision tables consist of exponential number of reducts, they are not really terrible because they must contain also an exponential number of objects.

5.4   Rough Sets and Classification Problems

Classification is one of the most important data mining problem types that occurs in a wide range of various applications. Many data mining tasks can be solved by classification methods. The objective is to build from the given decision table a classification algorithm (sometimes called *classification model* or *classifier*), which assigns the correct decision class to previously unseen objects.

A number of classification methods have been proposed to solve the classification problem. In this section, we are dealing with the rule based approach, which is preferred by many rough set based classification methods [8], [149], [159], [163].

In general, decision rules are logical formulas that indicate the relationship between condition and decision attributes. Let us begin with the description language which is a basic tool to define different kinds of description rules.

Definition 11 (The description language). *Let A be a set of attributes. The description language for A is a triple*

$$\mathcal{L}(A) = (\mathbf{D}_A, \{\vee, \wedge, \neg\}, \mathbf{F}_A),$$

where

- $\mathbf{D}_A$ *is the set of all descriptors of the form:*

$$\mathbf{D}_A = \{(a = v) : a \in A \text{ and } v \in Val_a\};$$

- $\{\vee, \wedge, \neg\}$ *is the set of standard propositional Boolean connectives;*
- $\mathbf{F}_A$ *is the set of boolean expressions defined over* $\mathbf{D}_A$*, called* formulas.

Formulas from $\mathbf{F}_A$ can be treated as a syntactic definition of the description logics. Their semantics is related to the sample of objects from the universe that is given by the information table (or decision table). Intuitively, the semantics of a given formula is defined by the set of all objects that match (satisfy) the formula. Therefore, semantics can be understood as a function $[[.]] : \mathbf{F} \to 2^U$.

Definition 12 (The semantics). *Let* $\mathbb{S} = (U, A)$ *be an information table describing a sample* $U \subset \mathbb{X}$*. The semantics of any formula* $\phi \in \mathbf{F}$*, denoted by* $[[\phi]]_{\mathbb{S}}$*, is defined inductively as follows:*

$$[[(a = v)]]_{\mathbb{S}} = \{x \in U : a(x) = v\} \tag{30}$$

$$[[\phi_1 \vee \phi_2]]_{\mathbb{S}} = [[\phi_1]]_{\mathbb{S}} \cup [[\phi_2]]_{\mathbb{S}} \tag{31}$$

$$[[\phi_1 \wedge \phi_2]]_{\mathbb{S}} = [[\phi_1]]_{\mathbb{S}} \cap [[\phi_2]]_{\mathbb{S}} \tag{32}$$

$$[[\neg\phi]]_{\mathbb{S}} = U \setminus [[\phi]]_{\mathbb{S}} \tag{33}$$

Let us emphasize that the formula can be defined by an information system $\mathbb{S}$, but one can compute its semantics in another information system $\mathbb{S}' \neq \mathbb{S}$. In such cases, some well defined descriptors which are interpretable in $\mathbb{S}$ can have an empty semantics in $\mathbb{S}'$.

The following theorem shows the correctness of the definition of semantics.

Theorem 8. *If* $\phi_1 \implies \phi_2$ *is a tautology of the propositional calculus, then* $[[\phi_1]]_{\mathbb{S}'} \subseteq [[\phi_2]]_{\mathbb{S}'}$ *for any information system* $\mathbb{S}'$*.*

In the terminology of data mining, every formula $\phi \in \mathbf{F}$ can be treated as a pattern, since it describes a set of objects, namely $[[\phi]]_{\mathbb{S}}$, with some similar features. We associate with every formula ϕ the following numeric features:

- $length(\phi)$ = the number of descriptors that occur in ϕ;
- $support(\phi) = |[[\phi]]_{\mathbb{S}}|$ = the number of objects that match the formula.

Thus, one can define the interestingness of a formula by its length and support [98]. Now we are ready to define decision rules for a given decision table.

Definition 13 (Decision rule). *Let* $\mathbb{S} = \{U, A \cup \{dec\}\}$ *be a decision table. Any implication of a form* $\phi \Rightarrow \delta$*, where* $\phi \in \mathbf{F}_A$ *and* $\delta \in \mathbf{F}_{dec}$*, is called a decision rule in* $\mathbb{S}$*. Formula* ϕ *is called the premise and* δ *is called the consequence of the decision rule* $\mathbf{r} := \phi \Rightarrow \delta$*. We denote the premise and the consequence of a given decision rule* $\mathbf{r}$ *by* $pre(\mathbf{r})$ *and* $cons(\mathbf{r})$*, respectively.*

Example 11. Let us note that every object $x \in U$ of the decision table $\mathbb{S} = \{U, A \cup \{dec\}\}$ can be interpreted as a decision rule $\mathbf{r}(x)$ defined by:

$$\mathbf{r}(x) \equiv \bigwedge_{a_i \in A} (a_i = a_i(x)) \Rightarrow (dec = dec(x))$$

Definition 14 (Generic decision rule). *The decision rule* **r** *with the premise as a boolean monomial of descriptors, i.e.,*

$$\mathbf{r} \equiv (a_{i_1} = v_1) \wedge \cdots \wedge (a_{i_m} = v_m) \Rightarrow (dec = k) \qquad (34)$$

is called the generic decision rule.

In this paper, we will consider generic decision rules only. For a simplicity, we will talk about decision rules keeping in mind the generic ones.

Every decision rule **r** of the form (34) can be characterized by the following features:

$length(\mathbf{r})$ – the number of descriptors in the premise of **r**

$[\mathbf{r}]$ – the carrier of **r**, i.e., the set of objects from U satisfying the premise of **r**

$support(\mathbf{r})$ – the number of objects satisfying the premise of **r**: $support(\mathbf{r}) = card([\mathbf{r}])$

$conf(\mathbf{r})$ – the confidence of **r**: $confidence(\mathbf{r}) = \frac{|[\mathbf{r}] \cap DEC_k|}{|[\mathbf{r}]|}$

The decision rule **r** is called *consistent* with $\mathbb{A}$ if $confidence(\mathbf{r}) = 1$.

5.4.1   Rule Based Classification Approach

In data mining, decision rules are treated as a form of patterns that are discovered from data. We are interested in *short, strong* decision rules with *high confidence*. The linguistic features like "short", "strong" or "high confidence" of decision rules can be formulated by means of their length, support and confidence. Such rules can be treated as interesting, valuable and useful patterns in data.

Any rule-based classification method consists of three phases (Fig. 11):

1. Learning phase: generates a set of decision rules $RULES(\mathbb{A})$ (satisfying some predefined conditions) from a given decision table $\mathbb{A}$.
2. Rule selection phase: selects from $RULES(\mathbb{A})$ the set of such rules that can be supported by x. We denote this set by $MatchRules(\mathbb{A}, x)$.
3. Classifying phase: makes a decision for x using some voting algorithm for decision rules from $MatchRules(\mathbb{A}, x)$ with respect to the following cases:
    (a) $MatchRules(\mathbb{A}, x)$ is empty: in this case the decision for x is $dec(x) = $ "$UNKNOWN$", i.e., we have no idea how to classify x;
    (b) $MatchRules(\mathbb{A}, x)$ consists of decision rules for the same decision class, say k^{th} decision class: in this case $dec(x) = k$;
    (c) $MatchRules(\mathbb{A}, x)$ consists of decision rules for the different decision classes: in this case the decision for x should be made using some voting algorithm for decision rules from $MatchRules(\mathbb{A}, x)$.

The main trouble when we apply the rule-based approach to classification problem is related to the fact that the number of all decision rules can be exponential with respect to the size of the given decision table [137], [8], [139]. In

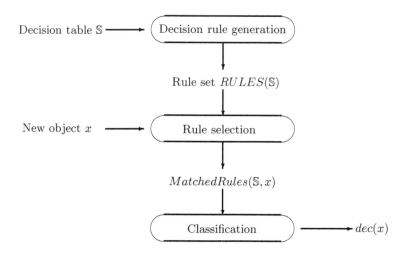

Fig. 11. Rule-based classification system

practice, we are forced to apply some heuristics to generate a subset of decision rules which are, in some sense, most interesting.

The most well-known rule induction methods are CN2 [23] [22], AQ [73], [74], [156], RIPPER [24], [25], LERS [39], [40]. In the next section we will present the method based on the Boolean reasoning approach.

5.4.2   Boolean Reasoning Approach to Decision Rule Inductions

Let us recall that every formula in the description language $\mathcal{L}(A)$ (determined by the set of attributes A) describes a set of objects and every decision rule describes the relationship between objects and decision classes. In this section, we concentrate on generic decision rules only.

Consider the collection $\mathbf{M}(A)$ of all monomials in $\mathcal{L}(A)$ together with the partial order $\ll$ where, for formulas ϕ_1 and ϕ_2 of $\mathbf{M}(A)$, $\phi_1 \ll \phi_2$ if $VAR(\phi_1) \subseteq VAR(\phi_2)$, i.e., ϕ_1 is created by removing some descriptors of ϕ_2. The relation $\phi_1 \ll \phi_2$ can be read as "ϕ_1 is a shortening of ϕ_2" or "ϕ_2 is a lengthening of ϕ_1".

For any object $u \in U$ and any subset of attributes $B \subset A$, the information vector $inf_B(u)$ of u can be interpreted as a formula

$$v_B(u) = \bigwedge_{a_i \in A} (a_i = a_i(u)).$$

We have the following proposition.

Proposition 5. *The collection $\mathbf{M}(A)$ of all monomials over the description language $\mathcal{L}(A)$ together with the relation $\ll$ is a partial order. Single descriptors are the minimal elements, and information vectors $v_A(u)$ for $u \in U$, are the maximum elements of $(\mathbf{M}(A), \ll)$.*

The relation between $\ll$ and other characteristics of decision rules is expressed by the following proposition.

Proposition 6. *Assume that ϕ_1 and $\phi_2 \in \mathbf{M}(A)$, and ϕ_2 is a lengthening of ϕ_1, i.e., $\phi_1 \ll \phi_2$, then the following facts hold:*

- *$length(\phi_1) \leq length(\phi_2)$;*
- *$[[\phi_2]]_{\mathbb{S}} \leq [[\phi_1]]_{\mathbb{S}}$ for any information table $\mathbb{S}$;*
- *$support(\phi_1) \geq support(\phi_2)$;*
- *If $\phi_1 \Rightarrow (dec = i)$ is a consistent decision rule then $\phi_2 \Rightarrow (dec = i)$ is also consistent.*

Many rule generation methods have been developed on the base of rough set theory. Let us recall the rule induction method based on the Boolean reasoning approach. This method uses the notion of *minimal consistent decision rules* which is defined as follows.

Definition 15 (Minimal consistent rules). *For a given decision table $\mathbb{S} = (U, A \cup \{dec\})$, a consistent rule:*

$$\mathbf{r} \equiv \phi \Rightarrow (dec = k)$$

is called the minimal consistent decision rule if any decision rule $\phi' \Rightarrow (dec = k)$ (where ϕ' is a shortening of ϕ) is not consistent with $\mathbb{S}$.

The boolean reasoning approach for computing minimal consistent decision rules has been presented in [137]. Similarly to the reduct problem, let $Var = \{x_1, \ldots, x_k\}$ be the set of boolean variables corresponding to attributes $a_1, \ldots, a_k$ from A. We have defined the discernibility function for $u, v \in U$ as follows:

$$disc_{u,v}(x_1, \ldots, x_k) = \sum \{x_i : a_i(u) \neq a_i(v)\}.$$

For any object $u \in U$ in a given decision table $\mathbb{S} = (U, A \cup \{dec\})$, we define a function $f_u(x_1, \ldots, x_k)$, called the *discernibility function for u* by

$$f_u(x_1, \ldots, x_k) = \prod_{v:dec(v) \neq dec(u)} disc_{u,v}(x_1, \ldots, x_k). \tag{35}$$

The set of attributes B is called the *object-oriented reduct (relative to the object u)* if the implication

$$v_B(u) \Rightarrow (dec = dec(u))$$

is a minimal consistent rule. It has been shown that every prime implicant of f_u corresponds to an "object-oriented reduct" for object u and such reducts are associated with a minimal consistent decision rules that are satisfied by u [137] [150]. This fact can be described by the following theorem.

Theorem 9. *For any set of attributes $B \subset A$ the following conditions are equivalent:*

1. *The monomial* $\mathbf{T}_{X(B)}$ *is the prime implicant of the discernibility function* f_u.
2. *The rule* $v_B(u) \Rightarrow (dec = dec(u))$ *is minimal consistent decision rule.*

The proof of this theorem is very similar to the proof of Theorem 6, therefore it has been omitted here. One can see the idea of this proof through the following example.

Example 12. Let us consider the decision table which is shown in Example 9. Consider the object number 1:

1. The discernibility function is determined as follows:

$$f_1(x_1, x_2, x_3, x_4) = x_1(x_1 + x_2)(x_1 + x_2 + x_3)(x_1 + x_2 + x_3 + x_4)$$
$$(x_2 + x_3)(x_1 + x_2 + x_3)(x_2 + x_3 + x_4)(x_1 + x_2 + x_4)$$

2. After transformation into DNF we have

$$f_1(x_1, x_2, x_3, x_4) = x_1(x_2 + x_3)$$
$$= x_1 x_2 + x_1 x_3$$

3. Hence, there are two object oriented reducts, i.e., $\{a_1, a_2\}$ and $\{a_1, a_3\}$. The corresponding decision rules are

$$(a_1 = \text{sunny}) \wedge (a_2 = \text{hot}) \Rightarrow (dec = \text{no})$$
$$(a_1 = \text{sunny}) \wedge (a_3 = \text{high}) \Rightarrow (dec = \text{no})$$

Let us notice that all rules have the same decision class, precisely the class of the considered object. If we wish to obtain minimal consistent rules for the other decision classes, we should repeat the algorithm for another object. Let us demonstrate once again the application of the Boolean reasoning approach to decision rule induction for the object number 11.

1. The discernibility function:

$$f_{11}(x_1, x_2, x_3, x_4) = (x_2 + x_3 + x_4)(x_2 + x_3)(x_1 + x_2)(x_3 + x_4)$$

2. After transformation into DNF we have

$$f_{11}(x_1, x_2, x_3, x_4) = (x_2 + x_3)(x_1 + x_2)(x_3 + x_4)$$
$$= (x_2 + x_1 x_3)(x_3 + x_4)$$
$$= x_2 x_3 + x_2 x_4 + x_1 x_3 + x_1 x_3 x_4$$
$$= x_2 x_3 + x_2 x_4 + x_1 x_3$$

3. Hence, there are three object oriented reducts, i.e., $\{a_2, a_3\}$, $\{a_2, a_4\}$ and $\{a_1, a_3\}$. The corresponding decision rules are

$$(a_2 = \text{mild}) \wedge (a_3 = \text{normal}) \Rightarrow (dec = \text{yes})$$
$$(a_2 = \text{mild}) \wedge (a_4 = \text{TRUE}) \Rightarrow (dec = \text{yes})$$
$$(a_1 = \text{sunny}) \wedge (a_3 = \text{normal}) \Rightarrow (dec = \text{yes})$$

Let us denote by $MinConsRules(\mathbb{S})$ the set of all minimal consistent decision rules for a given decision table $\mathbb{S}$ and denote by $MinRules(u|\mathbb{S})$ the set of all minimal consistent decision rules that are supported by object u. We have

$$MinConsRules(\mathbb{S}) = \bigcup_{u \in U} MinRules(u).$$

In practice, instead of $MinConsRules(\mathbb{S})$, we use the set of short, strong, and of high accuracy decision rules defined by

$$MinRules(\mathbb{S}, \lambda_{\max}, \sigma_{\min}, \alpha_{\min}) = \{\mathbf{r} : (length(\mathbf{r}) \leq \lambda_{\max}) \text{ AND}$$
$$\text{AND } (support(\mathbf{r}) \geq \sigma_{\min}) \text{ AND } (confidence(\mathbf{r}) \geq \alpha_{\min})\}. \quad (36)$$

Any heuristics for object oriented reducts can be modified to extract decision rules from $MinRules(\mathbb{A}, \lambda_{\max}, \sigma_{\min}, \alpha_{\min})$. Some of those algorithms were described in [8] and implemented in ROSETTA [106] and RSES [11] systems.

5.4.3   Rough Classifier

The rule-based rough approximation of concept (or rough-classifier) has been proposed in [9]. Let us recall how to construct rough membership functions for decision classes from a given set of decision rules.

For any object $x \in \mathfrak{U}$, let $MatchRules(\mathbb{S}, x) = \mathbf{R}_{yes} \cup \mathbf{R}_{no}$, where $\mathbf{R}_{yes}$ is the set of all decision rules for a decision class C and $\mathbf{R}_{no}$ is the set of decision rules for other classes. We assign two real values w_{yes}, w_{no} called "for" and "against" weights to the object x.

The values w_{yes}, w_{no} are defined by

$$w_{yes} = \sum_{\mathbf{r} \in \mathbf{R}_{yes}} strength(\mathbf{r}); \qquad w_{no} = \sum_{\mathbf{r} \in \mathbf{R}_{no}} strength(\mathbf{r})$$

where $strength(\mathbf{r})$ is a normalized function which depends on $length(\mathbf{r})$, $support(\mathbf{r})$, $confidence(\mathbf{r})$ and some global information about the decision table $\mathbb{S}$ such as table size, global class distribution, etc.

One can define the value of $\mu_C(x)$ by

$$\mu_C(x) = \begin{cases} \text{undetermined if } \max(w_{yes}, w_{no}) < \omega \\ 0 \qquad \text{if } w_{no} - w_{yes} \geq \theta \text{ and } w_{no} > \omega \\ 1 \qquad \text{if } w_{yes} - w_{no} \geq \theta \text{ and } w_{yes} > \omega \\ \frac{\theta + (w_{yes} - w_{no})}{2\theta} \text{ in other cases} \end{cases}$$

where ω, θ are parameters set by users. These parameters allow us for flexible control of the size of the boundary region.

The illustration of how the rough membership function is determined by the values of w_{yes} and w_{no} is presented in Fig. 12.

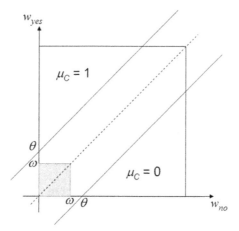

Fig. 12. Illustration of $\mu_C(x)$

6 Rough Sets and ABR Approach to Discretization

Discretization of real value attributes is an important task in data mining, particularly for the classification problem. Empirical results show that the quality of classification methods depends on the discretization algorithm used in the preprocessing step. In general, discretization is a process of searching for partition of attribute domains into intervals and unifying the values over each interval. Hence, the discretization problem can be defined as a problem of searching for a relevant set of cuts (i.e., boundary points of intervals) on attribute domains.

6.1 Discretization as Data Transformation Process

Let $S = (U, A \cup \{dec\})$ be a given decision table where $U = \{x_1, x_2, \ldots, x_n\}$. Any attribute $a \in A$ is called a *numeric attribute* if its domain is a subset of real numbers. Without loss of generality we will assume that $V_a = [l_a, r_a) \subset \mathbb{R}$ where $\mathbb{R}$ is the set of real numbers. Moreover, we will assume that the decision table S is consistent, i.e., every two objects that have distinct decision classes are discernible by at least one attribute.

Any pair $(a; c)$, where $a \in A$ and $c \in \mathbb{R}$, defines a partition of V_a into *left-hand-side* and *right-hand-side interval*. In general, if we consider an arbitrary set of cuts on an attribute $a \in A$

$$\mathbf{C}_a = \{(a; c_1^a), (a; c_2^a), \ldots, (a; c_{k_a}^a)\}$$

where $k_a \in \mathbb{N}$ and $c_0^a = l_a < c_1^a < c_2^a < \cdots < c_{k_a}^a < r_a = c_{k_a+1}^a$, one can see that $\mathbf{C}_a$ defines a partition on V_a into sub-intervals as follow:

$$V_a = [c_0; c_1^a) \cup [c_1^a; c_2^a) \cup \cdots \cup [c_{k_a}^a; c_{k_a+1}^a).$$

Therefore, we can say that the set of cuts $\mathbf{C}_a$ defines a discretization of a, i.e., creates a new discrete attribute $a|_{\mathbf{C}_a} : U \to \{0, \ldots, k_a\}$ such that

$$
a|_{\mathbf{C}_a}(x) = \begin{cases} 0 & \text{if } a(x) < c_1^a, \\ 1 & \text{if } a(x) \in [c_1^a, c_2^a), \\ \ldots & \ldots \\ k_a - 1 & \text{if } a(x) \in [c_{k_a-1}^a, c_{k_a}^a), \\ k_a & \text{if } a(x) \geq c_{k_a}^a. \end{cases} \tag{37}
$$

In other words, $a|_{\mathbf{C}_a}(x) = i \Leftrightarrow a(x) \in [c_i^a; c_{i+1}^a)$ for any $x \in U$ and $i \in \{0, \ldots, k_a\}$ (see Fig. 13).

Fig. 13. The discretization of real value attribute $a \in A$ defined by the set of cuts $\{(a; c_1^a), (a; c_2^a), \ldots, (a; c_{k_a}^a)\}$

Analogously, any collection of cuts on a family of real value attributes $\mathbf{C} = \bigcup_{a \in A} \mathbf{C}_a$ determines a *global discretization* of the whole decision table. Particularly, a collection of cuts

$$
\mathbf{C} = \bigcup_{a_i \in A} \mathbf{C}_{a_i} = \{(a_1; c_1^1), \ldots, (a_1; c_{k_1}^1)\} \cup \{(a_2; c_1^2), \ldots, (a_2; c_{k_2}^2)\} \cup \ldots
$$

transforms the original decision table $\mathbb{S} = (U, A \cup \{dec\})$ into a new decision table $\mathbb{S}|_{\mathbf{C}} = (U, A_{\mathbf{C}} \cup \{dec\})$, where $A|_{\mathbf{C}} = \{a|_{\mathbf{C}_a} : a \in A\}$ is the set of discretized attributes. The table $\mathbb{S}|_{\mathbf{C}}$ is also called *the **C**-discretized decision table of* $\mathbb{S}$.

Example 13. Let us consider again the weather data which has been discussed in the previous section (see Table 6) but this time attribute a_2 measures the temperature in Fahrenheit degrees, and a_3 measures the humidity (in %).

The collection of cuts $\mathbf{C} = \{(a_2; 70), (a_2; 80), (a_3; 82)\}$ creates two new attributes $a_2|_{\mathbf{C}}$ and $a_3|_{\mathbf{C}}$ defined by

$$
a_2|_{\mathbf{C}}(x) = \begin{cases} 0 & \text{if } a_2(x) < 70; \\ 1 & \text{if } a_2(x) \in [70, 80); \\ 2 & \text{if } a_2(x) \geq 80. \end{cases} \qquad a_3|_{\mathbf{C}}(x) = \begin{cases} 0 & \text{if } a_3(x) < 82; \\ 1 & \text{if } a_3(x) \geq 82. \end{cases}
$$

The discretized decision table is presented in Table 13. The reader can compare this discretized decision table with the decision table presented in Table 6. One can see that these tables are equivalent. In particular, if we assign to the values 0, 1, 2 of $a_2|_{\mathbf{C}}$ names "cool", "mild" and "hot" and map the values 0, 1 of $a_3|_{\mathbf{C}}$ into "normal" and "high", respectively, then we will have the same decision tables.

Table 11. An example of decision table with two symbolic and two continuous attributes (on the left) and its discretized decision table using cut set $\mathbf{C} = \{(a_2; 70), (a_2; 80), (a_3; 82)\}$ (on the right side)

outlook	temp.	hum.	windy	play		outlook	temp.	hum.	windy	play
a_1	a_2	a_3	a_4	dec		a_1	$a_2\|_{\mathbf{C}}$	$a_3\|_{\mathbf{C}}$	a_4	dec
sunny	85	85	FALSE	no		sunny	2	1	FALSE	no
sunny	80	90	TRUE	no		sunny	2	1	TRUE	no
overcast	83	86	FALSE	yes		overcast	2	1	FALSE	yes
rainy	70	96	FALSE	yes		rainy	1	1	FALSE	yes
rainy	68	80	FALSE	yes		rainy	0	0	FALSE	yes
rainy	65	70	TRUE	no	$\Longrightarrow$	rainy	0	0	TRUE	no
overcast	64	65	TRUE	yes		overcast	0	0	TRUE	yes
sunny	72	95	FALSE	no		sunny	1	1	FALSE	no
sunny	69	70	FALSE	yes		sunny	0	0	FALSE	yes
rainy	75	80	FALSE	yes		rainy	1	0	FALSE	yes
sunny	75	70	TRUE	yes		sunny	1	0	TRUE	yes
overcast	72	90	TRUE	yes		overcast	1	1	TRUE	yes
overcast	81	75	FALSE	yes		overcast	2	0	FALSE	yes
rainy	71	91	TRUE	no		rainy	1	1	TRUE	no

The previous example illustrates discretization as the data transformation process. Sometimes, it is more convenient to denote the discretization as an operator on the domain of decision tables. Hence, instead of $\mathbb{S}|_{\mathbf{C}}$ we will sometimes use the notion $Discretize(\mathbb{S}, \mathbf{C})$ to denote the discretized decision table.

6.1.1 Classification of Discretization Methods

One can distinguish among existing discretization (quantization) methods using different criteria [28]:

1. *Local versus global methods*: Local methods produce partitions that are applied to localized regions of the object space (e.g., decision tree). Global methods produce a mesh over k-dimensional real space, where each attribute value set is partitioned into intervals independent of the other attributes.

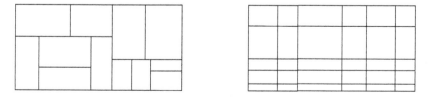

Fig. 14. Illustration of the local and global discretization

2. *Static versus dynamic methods*: One can distinguish between static and dynamic discretization methods. Static methods perform one discretization pass for each attribute and determine the maximal number of cuts for this attribute independently of the others. Dynamic methods are realized by searching through the family of all possible cuts for all attributes simultaneously.
3. *Supervised versus unsupervised methods*: Several discretization methods do not make use of decision values of objects in discretization process. Such methods are called *unsupervised discretization methods*. In contrast, methods that utilize the decision attribute are called *supervised discretization methods*.

According to this classification, the discretization method described in the next section is dynamic and supervised.

6.1.2   Optimal Discretization Problem

It is obvious, that the discretization process is associated with a loss of information. Usually, the task of discretization is to determine the set of cuts $\mathbf{C}$ of a minimal size from a given decision table $\mathbb{S}$ such that, in spite of losing information, the $\mathbf{C}$-discretized table $\mathbb{S}|_{\mathbf{C}}$ still keeps some usefull properties of $\mathbb{S}$. In [79], we have presented a discretization method based on the rough set and the Boolean reasoning approach that guarantees the discernibility between objects. This method makes use of the notion of consistent, irreducible and optimal sets of cuts. Let us recall the basic definition of discernibility between objects.

Definition 16. *Let $\mathbb{S} = (U, A \cup \{dec\})$ be a given decision table. We say that a cut $(a; c)$ on an attribute $a \in A$ discerns a pair of objects $x, y \in U$ (or objects x and y are discernible by $(a; c)$) if*

$$(a(x) - c)(a(y) - c) < 0.$$

Two objects are discernible by a set of cuts $\mathbf{C}$ if they are discernible by at least one cut from $\mathbf{C}$.

Intuitively, the cut $(a; c)$ on a discerns objects x and y if and only if $a(x)$ and $a(y)$ are lying on distinct sides of c on the real axis (see Fig. 15).

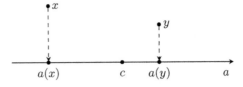

Fig. 15. Two objects x, y are discernible by a cut $(a; c)$

Let us point out that the notion of discernibility and indiscernibility was introduced in Sect. 2.3. Two objects $x, y \in U$ are said to be discernible by a set of attributes $B \subset A$ if $inf_B(x) \neq inf_B(y)$. One can see that there are some analogies between the attribute-based discernibility and the cut-based discernibility. In fact, the discernibility determined by cuts implies the discernibility determined by attributes. The inverse implication does not always hold, therefore we have the following definition.

Definition 17. *A set of cuts* **C** *is* consistent with $\mathbb{S}$ *(or $\mathbb{S}$ -consistent, for short) if and only if for any pair of objects $x, y \in U$ such that $dec(x) \neq dec(y)$, the following condition holds:*

IF *x, y are discernible by A* **THEN** *x, y are discernible by* **C**.

The discretization process made by consistent set of cuts is called the *compatible discretization*. We are interested in searching for consistent sets of cuts of the size as small as possible. Let us specify some special types of consistent sets of cuts.

Definition 18. *A consistent set* **C** *of cuts is $\mathbb{S}$-irreducible if every proper subset* **C'** *of* **C** *is not $\mathbb{S}$-consistent. A consistent set* **C** *of cuts is $\mathbb{S}$-optimal if for any $\mathbb{S}$-consistent set of cuts* **C'**:

$$card\left(\mathbf{C}\right) \leq card\left(\mathbf{C'}\right),$$

i.e., **C** *contains a smallest number of cuts among $\mathbb{S}$-consistent sets of cuts.*

The irreducibility can be understood as a type of reducts. Irreducible sets of cuts are minimal, w.r.t. the set inclusion $\subseteq$, in the family of all consistent sets of cuts. In such interpretation, optimal sets of cuts can be treated as minimal reducts. Formally, the optimal discretization problem is defined as follows:

OPTIDISC: optimal discretization problem
 input: A decision table $\mathbb{S}$.
 output: $\mathbb{S}$-optimal set of cuts.

The corresponding decision problem can be formulated as:

DISCSIZE: k-cuts discretization problem
 input: A decision table $\mathbb{S}$ and an integer k.
 question: Decide whether there exists a $\mathbb{S}$-irreducible set of cuts **P** such that $card(\mathbf{P}) < k$.

The following fact has been shown in [79].

Theorem 10. *The problem* DISCSIZE *is polynomially equivalent to the* PRIMISIZE *problem.*

As a corollary, we can prove the following Theorem.

Theorem 11 (Computational complexity of discretization problems).

1. *The problem* DISCSIZE *is NP-complete.*
2. *The problem* OPTIDISC *is NP-hard.*

This result means that we can not expect a polynomial time searching algorithm for optimal discretization, unless P = NP.

6.2 Discretization Method Based on Rough Set and Boolean Reasoning

Any cut $(a; c)$ on an attribute $a \in A$ defines a partition of V_a into *left-hand-side* and *right-hand-side intervals* and also defines a partition of U into two disjoint subsets of objects $U_{left}(a; c)$ and $U_{right}(a; c)$ as follows:

$$U_{left}(a; c) = \{x \in U : a(x) < c\}, \quad U_{right}(a; c) = \{x \in U : a(x) \geq c\}.$$

Two cuts $(a; c_1)$ and $(a; c_2)$ on the same attribute a are called *equivalent* if they define the same partition of U, i.e.,

$$(U_{left}(a; c_1), U_{right}(a; c_1)) = (U_{left}(a; c_2), U_{right}(a; c_2)).$$

We denote this equivalence relation by $c_1 \equiv_a c_2$.

For a given decision table $\mathbb{S} = \{U, A \cup \{dec\}\}$ and a given attribute $a \in A$, we denote by

$$a(U) = \{a(x) : x \in U\} = \{v_1^a, v_2^a, \ldots, v_{n_a}^a\}$$

the set of all values of attribute a occurring in the table $\mathbb{S}$. Additionally, let us assume that these values are sorted in increasing order, i.e., $v_1^a < v_2^a < \cdots < v_{n_a}^a$.

One can see that two cuts $(a; c_1)$ and $(a; c_2)$ are equivalent if and only if there exists $i \in \{1, n_a - 1\}$ such that $c_1, c_2 \in (v_i^a, v_{i+1}^a]$. In this section, we will not distinguish between equivalent cuts. Therefore, we will unify all cuts in the interval $(v_i^a, v_{i+1}^a]$ by one representative cut $\left(a; \frac{v_i^a + v_{i+1}^a}{2}\right)$ which is also called the *generic cut*.

The set of all possible generic cuts on a, with respect to the equivalence relation, is denoted by

$$\mathbf{GCuts}_a = \left\{\left(a; \frac{v_1^a + v_2^a}{2}\right), \left(a; \frac{v_2^a + v_3^a}{2}\right), \ldots, \left(a; \frac{v_{n_a-1}^a + v_{n_a}^a}{2}\right)\right\}. \tag{38}$$

The set of all candidate cuts of a given decision table is denoted by

$$\mathbf{GCuts}_\mathbb{S} = \bigcup_{a \in A} \mathbf{GCuts}_a \tag{39}$$

In an analogy to the equivalence relation between two single cuts, two sets of cuts $\mathbf{C}'$ and $\mathbf{C}$ are *equivalent with respect to decision table* $\mathbb{S}$ and denoted by $\mathbf{C}' \equiv_\mathbb{S} \mathbf{C}$, if and only if $\mathbb{S}|_\mathbf{C} = \mathbb{S}|_{\mathbf{C}'}$. The equivalence relation $\equiv_\mathbb{S}$ has finite number of equivalence classes. In the sequel, we will consider only those discretization processes which are made by a subset $\mathbf{C} \subset \mathbf{GCuts}_\mathbb{S}$ of candidate cuts.

Example 14. For the decision table from Example 13, the sets of all values of continuous attributes are as follows:

$$a_2(U) = \{64, 65, 68, 69, 70, 71, 72, 75, 80, 81, 83, 85\}$$
$$a_3(U) = \{65, 70, 75, 80, 85, 86, 90, 91, 95, 96\}$$

Therefore, we have 11 candidate cuts on a_2 and 9 candidate cuts on a_3:

$$\mathbf{GCuts}_{a_2} = \{(a_2; 64.5), (a_2; 66.5), (a_2; 68.5), (a_2; 69.5), (a_2; 70.5),$$
$$(a_2; 71.5), (a_2; 73.5), (a_2; 77.5), (a_2; 80.5), (a_2; 82), (a_2; 84)\};$$
$$\mathbf{GCuts}_{a_3} = \{(a_3; 67.5), (a_3; 72.5), (a_3; 77.5), (a_3; 82.5), (a_3; 85.5),$$
$$(a_3; 88), (a_3; 90.5), (a_3; 93), (a_3; 95.5)\}.$$

The set of cuts $\mathbf{C} = \{(a_2; 70), (a_2; 80), (a_3; 82)\}$ in the previous example is equivalent to the following set of generic cuts:

$$\mathbf{C} = \{(a_2; 70), (a_2; 80), (a_3; 82)\} \equiv_{\mathbb{S}} \{(a_2; 69.5), (a_2; 77.5), (a_3; 82.5)\}.$$

Fig. 16 presents the application of the Boolean reasoning approach to the optimal discretization problem. As usual, let us begin with the encoding method.

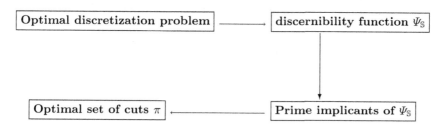

Fig. 16. The Boolean reasoning approach to the optimal discretization problem

6.2.1 Encoding of Optimal Discretization Problem by Boolean Functions

Consider a decision table $\mathbb{S} = (U, A \cup \{d\})$ where $U = \{u_1, u_2, \ldots, u_n\}$ and $A = \{a_1, \ldots, a_k\}$. We encode the optimal discretization problem for $\mathbb{S}$ as follows:

Boolean variables: Let $\mathbf{C} = \sum_{a_m \in A} \mathbf{C}_{a_m}$ be a set of candidate cuts. Candidate cuts are defined either by an expert/user or by taking all generic cuts (i.e., by setting $\mathbf{C} = \mathbf{GCuts}_{\mathbb{S}}$). Assume that

$$\mathbf{C}_{a_m} = \{\underbrace{(a_m, c_1^m)}_{p_1^{a_m}}, \ldots, \underbrace{(a_m, c_{n_m}^m)}_{p_{n_m}^{a_m}}\}$$

are candidate cuts on the attribute $a_m \in A$.

Let us associate with each cut $(a_m, c_i^m) \in \mathbf{C}_{a_m}$ a boolean variable p_i^m and let us denote by $\mathbf{BCuts}_{a_m} = \{p_1^{a_m}, \ldots, p_{n_m}^{a_m}\}$ the set of boolean variables corresponding to candidate cuts on the attribute a_m. For any set of cuts $\mathbf{X} \subset \mathbf{C}$ we denote by $\Sigma_{\mathbf{X}}$ (and $\Pi_{\mathbf{X}}$) the Boolean function being disjunction (and conjunction) of boolean variables corresponding to cuts from $\mathbf{X}$, respectively.

Encoding function: The optimal discretization problem is encoded by a Boolean function over the set of boolean variables $\mathbf{P} = \bigcup \mathbf{P}_{a_m}$ as follows:

Firstly, for any pair of objects $u_i, u_j \in U$ we denote by $\mathbf{X}_{i,j}^a$ the set of cuts from $\mathbf{C}_a$ discerning u_i and u_j, i.e.

$$\mathbf{X}_{i,j}^a = \{(a; c_k^a) \in \mathbf{C}_a : (a(u_i) - c_k^a)(a(u_j) - c_k^a) < 0\}.$$

Let $\mathbf{X}_{i,j} = \bigcup_{a \in A} \mathbf{X}_{i,j}^a$. The *discernibility function* $\psi_{i,j}$ for a pair of objects u_i, u_j is defined by disjunction of variables corresponding to cuts from $\mathbf{X}_{i,j}$, i.e.,

$$\psi_{i,j} = \begin{cases} \Sigma \mathbf{X}_{i,j} & \text{if } \mathbf{X}_{i,j} \neq \varnothing \\ 1 & \text{if } X_{i,j} = \varnothing \end{cases} \tag{40}$$

For any set of cuts $X \subset \mathbf{C}$ let $\mathcal{A}_X : \mathbf{P} \to \{0, 1\}$ be an assignment of variables corresponding to the characteristic function of X, i.e.,

$$\mathcal{A}_X(p_k^{a_m}) = 1 \Leftrightarrow (a; c_k^{a_m}) \in X.$$

We can see that a set of cuts $X \subset \mathbf{C}$ satisfies $\psi_{i,j}$, i.e., $\psi_{i,j}(\mathcal{A}_X) = 1$ if and only if u_i and u_j are discernible by at least one cut from X. The *discernibility Boolean function of* $\mathbb{S}$ is defined by:

$$\Phi_{\mathbb{S}} = \prod_{d(u_i) \neq d(u_j)} \psi_{i,j}. \tag{41}$$

One can prove the following theorem [79].

Theorem 12. *For any set of cuts X:*

1. X *is $\mathbb{S}$-consistent if and only if $\Phi_{\mathbb{S}}(\mathcal{A}_X) = 1$;*
2. X *is $\mathbb{S}$-irreducible if and only if the monomial Π_X is a prime implicant of $\Phi_{\mathbb{S}}$;*
3. X *is $\mathbb{S}$-optimal if and only if the monomial Π_X is the shortest prime implicant of the function $\Phi_{\mathbb{S}}$.*

As a corollary we can obtain that the problem of searching for an optimal set of cuts for a given decision table is polynomially reducible to the problem of searching for the minimal prime implicant of a monotone Boolean function.

Example 15. The following example illustrates main ideas of the construction.

We consider the decision table with two conditional attributes a, b and seven objects $u_1, \ldots, u_7$. The values of attributes on these objects and the values of the decision d are presented in Table 12(a). Geometrical interpretation of objects and decision classes are shown in Fig. 17.

The sets of values of a and b on objects from U are given by

$$a(U) = \{0.8, 1, 1.3, 1.4, 1.6\},$$
$$b(U) = \{0.5, 1, 2, 3\},$$

Table 12. The discretization process: (a)The original decision table $\mathbb{S}$. (b)The **C**-discretization of $\mathbb{S}$, where $\mathbf{C} = \{(a; 0.9), (a; 1.5), (b; 0.75), (b; 1.5)\}$.

$\mathbb{S}$	a	b	d
u_1	0.8	2	1
u_2	1	0.5	0
u_3	1.3	3	0
u_4	1.4	1	1
u_5	1.4	2	0
u_6	1.6	3	1
u_7	1.3	1	1

(a)

$\Longrightarrow$

$\mathbb{S}_C$	a_C	b_C	d
u_1	0	2	1
u_2	1	0	0
u_3	1	2	0
u_4	1	1	1
u_5	1	2	0
u_6	2	2	1
u_7	1	1	1

(b)

and the cardinalities of $a(U)$ and $b(U)$ are equal to $n_a = 5$ and $n_b = 4$, respectively. The set of boolean variables defined by $\mathbb{S}$ is equal to

$$\mathbf{BCuts}_\mathbb{S} = \left\{ p_1^a, p_2^a, p_3^a, p_4^a, p_1^b, p_2^b, p_3^b \right\},$$

where $p_1^a \sim [0.8; 1)$ of a (i.e., p_1^a corresponds to the interval $[0.8; 1)$ of attribute a); $p_2^a \sim [1; 1.3)$ of a; $p_3^a [1.3; 1.4)$ of a; $p_4^a \sim [1.4; 1.6)$ of a; $p_1^b \sim [0.5; 1)$ of b; $p_2^b \sim [1; 2)$ of b; $p_3^b \sim [2; 3)$ of b.

The discernibility formulas $\psi_{i,j}$ for different pairs (u_i, u_j) of objects from U are as following:

$$\begin{aligned}
\psi_{2,1} &= p_1^a + p_1^b + p_2^b; & \psi_{2,4} &= p_2^a + p_3^a + p_1^b; \\
\psi_{2,6} &= p_2^a + p_3^a + p_4^a + p_1^b + p_2^b + p_3^b; & \psi_{2,7} &= p_2^a + p_1^b; \\
\psi_{3,1} &= p_1^a + p_2^a + p_3^b; & \psi_{3,4} &= p_2^a + p_2^b + p_3^b; \\
\psi_{3,6} &= p_3^a + p_4^a; & \psi_{3,7} &= p_2^b + p_3^b; \\
\psi_{5,1} &= p_1^a + p_2^a + p_3^a; & \psi_{5,4} &= p_2^b; \\
\psi_{5,6} &= p_4^a + p_3^b; & \psi_{5,7} &= p_3^a + p_2^b.
\end{aligned}$$

The discernibility formula $\Phi_\mathbb{S}$ in CNF form is given by

$$\begin{aligned}
\Phi_\mathbb{S} = &\left(p_1^a + p_1^b + p_2^b \right) \left(p_1^a + p_2^a + p_3^b \right) \left(p_1^a + p_2^a + p_3^a \right) \left(p_2^a + p_3^a + p_1^b \right) p_2^b \\
&\left(p_2^a + p_2^b + p_3^b \right) \left(p_2^a + p_3^a + p_4^a + p_1^b + p_2^b + p_3^b \right) \left(p_3^a + p_4^a \right) \left(p_4^a + p_3^b \right) \\
&\left(p_2^a + p_1^b \right) \left(p_2^b + p_3^b \right) \left(p_3^a + p_2^b \right).
\end{aligned}$$

Transforming the formula $\Phi_\mathbb{S}$ into its DNF form we obtain four prime implicants:

$$\Phi_\mathbb{S} = p_2^a p_4^a p_2^b + p_2^a p_3^a p_2^b p_3^b + p_3^a p_1^b p_2^b p_3^b + p_1^a p_4^a p_1^b p_2^b.$$

If we decide to take, e.g., the last prime implicant $t = p_1^a p_4^a p_1^b p_2^b$, we obtain the following set of cuts

$$\mathbf{C}(t) = \{(a; 0.9), (a; 1.5), (b; 0.75), (b; 1.5)\}.$$

The new decision table $\mathbb{S}^{\mathbf{P}(S)}$ is represented in Table 12 (b).

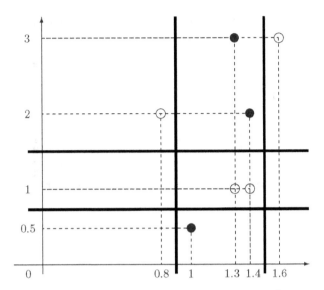

Fig. 17. Geometrical representation of data and cuts

6.2.2   Discretization by Reduct Calculation

First we show that discretization problems for a given decision table $\mathbb{S} = (U, A \cup \{d\})$ are polynomially equivalent to some problems related to reduct computation of a decision table $\mathbb{S}^*$ built from $\mathbb{S}$. The construction of decision table $\mathbb{S}^* = (U^*, A^* \cup \{d^*\})$ is as follows:

- $U^* = \{(u_i, u_j) \in U \times U : (i < j) \wedge (d(u_i) \neq d(u_j))\} \cup \{new\}$, where $new \notin U \times U$ is an artificial element which is useful in the proof of Proposition 7 presented below;
- $d^* : U^* \rightarrow \{0, 1\}$ is defined by $d^*(x) = \begin{cases} 0 \text{ if } x = new \\ 1 \text{ otherwise}; \end{cases}$
- $A^* = \{p_s^a : a \in A \text{ and } s \text{ corresponds to the } s^{th} \text{ interval } [v_s^a, v_{s+1}^a) \text{ for } a\}.$

For any $p_s^a \in A^*$ the value $p_s^a((u_i, u_j))$ is equal to 1 if

$$[v_s^a, v_{s+1}^a) \subseteq [\min\{a(u_i), a(u_j)\}, \max\{a(u_i), a(u_j)\})$$

and 0 otherwise. We also put $p_s^a(new) = 0$. The following proposition has been proved in [79].

Proposition 7. *The problem of searching for an irreducible set of cuts is polynomially equivalent to the problem of searching for a relative reduct for a decision table.*

6.2.3   Basic Maximal Discernibility Heuristic

In the previous section, the optimal discretization problem has been transformed to the minimal reduct problem (see Proposition 7). According to the proof of

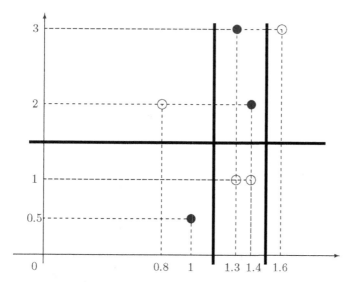

Fig. 18. The minimal set of cuts of decision table $\mathbb{S}$

this fact, every cut can be associated with a set of pairs of objects which are discernible by this cuts. Therefore, any optimal set of cuts can be treated as the minimal covering of the set of all conflict pairs of objects, i.e., objects from different decision classes. The *"MD heuristic"*, in fact, is the greedy algorithm for the minimal set covering (or minimal set hitting problem). It is based on the best-first searching strategy [95], [102]. The MD heuristic always makes the choice of the cut that discerns maximal number of conflict pairs of objects. This step is repeated until all conflict pairs are discerned by selected cuts. This idea is formulated in Algorithm 3.

Algorithm 3. MD-heuristic for the optimal discretization problem

Input: Decision table $\mathbb{S} = (U, A, dec)$.
Output: The semi-optimal set of cuts.
begin
1   Construct the table $\mathbb{S}^*$ from $\mathbb{S}$ and set **B** :=$\mathbb{S}^*$;
2   Select the column of **B** with the maximal number of occurrences of 1's;
3   Delete from **B** the selected column in Step 2 together with all rows marked in this column by 1;
4   **if B** *consists of more than one row* **then**
     | Go to Step 2;
     else
     | Return the set of selected cuts as a result;
     | Stop;
     end
end

Example 16. Let us demonstrate the MD-heuristic for the decision table from Example 15.

In Table 13 the decision table $\mathbb{S}^*$ is presented. Objects in this table are all pairs (x_i, x_j) discernible by the decision d. One more object is included, namely *new* with all values of attributes equal to 0. This allows us formally to keep the condition: "at least one occurrence of 1 (for conditional attributes) appears in any row for any subset of columns corresponding to any prime implicant".

Table 13. Table $\mathbb{S}^*$ constructed from table $\mathbb{S}$

$\mathbb{S}^*$	p_1^a	p_2^a	p_3^a	p_4^a	p_1^b	p_2^b	p_3^b	d^*
(u_1, u_2)	1	0	0	0	1	1	0	1
(u_1, u_3)	1	1	0	0	0	0	1	1
(u_1, u_5)	1	1	1	0	0	0	0	1
(u_4, u_2)	0	1	1	0	1	0	0	1
(u_4, u_3)	0	0	1	0	0	1	1	1
(u_4, u_5)	0	0	0	0	0	1	0	1
(u_6, u_2)	0	1	1	1	1	1	1	1
(u_6, u_3)	0	0	1	1	0	0	0	1
(u_6, u_5)	0	0	0	1	0	0	1	1
(u_7, u_2)	0	1	0	0	1	0	0	1
(u_7, u_3)	0	0	0	0	0	1	1	1
(u_7, u_5)	0	0	1	0	0	1	0	1
new	0	0	0	0	0	0	0	0

Relative reducts of this table correspond exactly to prime implicants of the function $\Phi^{\mathbb{S}}$ (Proposition 7). Our algorithm is choosing first p_2^b, next p_2^a, and finally p_4^a. Hence, $S = \{p_2^a, p_4^a, p_2^b\}$, and the resulting set of cuts $\mathbf{C} = \mathbf{C}(S) = \{(a; 1.15), (a; 1.5), (b; 1.5)\}$. According to Example 15 we know that this result is the optimal set of cuts.

Fig. 18 presents the geometrical interpretation of the constructed set of cuts (marked by bold lines).

6.2.4   Complexity of MD-heuristic for Discretization

MD-heuristic is a global, dynamic and supervised discretization method. Unlike local methods which can be efficiently implemented by the decision tree approach, global methods are very challenging for programmers because in each iteration the quality function strictly depends on the distribution of objects into the mesh made by the set of actual cuts.

In Algorithm 3, the size of the table $\mathbb{S}^*$ is $O(nk \cdot n^2)$ where n is the number of objects and k is the number of columns in $\mathbb{S}$. Hence, the time complexity of Step 2 and Step 3 is $O(n^3k)$. Therefore, the pessimistic time complexity of the straightforward implementation of MD heuristic is $O(n^3k \times |\mathbf{C}|)$, where $\mathbf{C}$ is the result set of cuts returned by the algorithm. Moreover, it requires $O(n^3k)$ of memory space to store the table $\mathbb{S}^*$.

Algorithm 4. Implementation of **MD-heuristic** using *DTree* structure

---

Input: Decision table $\mathbb{S} = (U, A, dec)$.
Output: The semi-optimal set of cuts.
begin
 $\quad$ *DTree* **D** := new *DTree*();
 $\quad$ **D**.Init($\mathbb{S}$);
 $\quad$ **while** (**D**.*Conflict()*> 0) **do**
 $\qquad$ *Cut c* := **D**.GetBestCut();
 $\qquad$ **if** (*c.quality*$== 0$) **then**
 $\qquad$ $\mid$ **break**;
 $\qquad$ **end**
 $\qquad$ **D**.InsertCut(c.attribute,c.cutpoint);
 $\quad$ **end**
 $\quad$ **D**.PrintCuts();
end

We have shown that the presented MD-heuristic can be implemented more efficiently. The idea is based on a special data structure for efficient storing the partition of objects made by the actual set of cuts. This data structure, called *DTree* – a shortcut of discretization tree, is a modified decision tree structure. It contains the following methods:

- *Init*($\mathbb{S}$): initializes the data structure for the given decision table;
- *Conflict*(): returns the number of pairs of undiscerned objects;
- *GetBestCut*(): returns the best cut point with respect to the discernibility measure;
- *InsertCut*(a, c): inserts the cut $(a; c)$ and updates the data structure.

It has been shown that except *Init*($\mathbb{S}$) the time complexity of all other methods is $O(nk)$, where n is the number of objects and k is the number of attribute, see [79], [105]. The method *Init*($\mathbb{S}$) requires $O(nk \log n)$ computation steps, because it prepares each attribute by sorting objects with respect to this attribute.

MD-heuristic (Algorithm 3) can be efficiently implemented using *DTree* structure as follows:

This improved algorithm has been implemented in ROSETTA [106] and RSES [11], [8] systems.

6.3   More Complexity Results

The NP-hardness of the optimal discretization problem was proved for the family of arbitrary decision tables. In this section, we will show much stronger fact that the optimal discretization problem restricted to 2-dimensional decision tables is also NP-hard.

We consider a family of decision tables consisting of exactly two real value condition attributes a, b and a binary decision attribute $d : U \rightarrow \{0, 1\}$. Any such decision table is denoted by $\mathbb{S} = (U, \{a, b\} \cup \{d\})$ and represents a set of colored points $S = \{P(u_i) = (a(u_i), b(u_i)) : u_i \in U\}$ on the plane $\mathbb{R}^2$, where black and white colors are assigned to points according to their decision. Any cut $(a; c)$ on a (or $(b; c)$ on b), where $c \in \mathbb{R}$, can be represented by a vertical (or horizontal) line. A set of cuts is $\mathbb{S}$-consistent if the set of lines representing them defines a partition of the plane into regions in such a way that all points in the same region have the same color. The discretization problem for a decision table with two condition attributes can be defined as follows:

> DISCSIZE2D: k-cuts discretization problem in $\mathbb{R}^2$
> *input:* The set S of black and white points $P_1, \ldots, P_n$ on the plane, and an integer k.
> *question:* Decide whether there exists a consistent set of at most k lines..

We also consider the corresponding optimization problem:

> OPTIDISC2D: Optimal discretization in $\mathbb{R}^2$
> *input:* The set S of black and white points $P_1, \ldots, P_n$ on the plane, and an integer k.
> *output:* $\mathbb{S}$-optimal set of cuts.

The next two theorems about the complexity of discretization problem in $\mathbb{R}^2$ were presented in [20] and [99]. We would like to recall the proofs of those theorems to demonstrate the power of the Boolean reasoning approach.

Theorem 13. *The decision problem* DISCSIZE2D *is NP-complete and the optimization version of this problem is NP-hard.*

Proof: Assume that an instance I of SETCOVER problem consists of

$$S = \{u_1, u_2, \ldots, u_n\}, \mathcal{F} = \{S_1, S_2, \ldots, S_m\}, \text{ and an integer } K$$

where $S_j \subseteq S$ and $\bigcup_{i=1}^{m} S_i = S$, and the question is if there are K sets from $\mathcal{F}$ with the sum containing all elements of S. We need to construct an instance I' of DISCSIZE2D such that I has a positive answer iff I' has a positive answer.

The construction of I' is quite similar to the construction described in the previous section. We start by building a grid-line structure consisting of vertical and horizontal strips. The regions are in rows labeled by $y_{u_1}, \ldots, y_{u_n}$ and columns labeled by $x_{S_1}, \ldots, x_{S_m}, x_{u_1}, \ldots, x_{u_n}$ (see Fig. 19). In the first step, for any element $u_i \in S$ we define a family $\mathcal{F}_i = \{S_{i_1}, S_{i_2}, \ldots, S_{i_{m_i}}\}$ of all subsets containing the element u_i.

If $\mathcal{F}_i$ consists of exactly $m_i \leq m$ subsets, then subdivide the row y_{u_i} into m_i strips, corresponding to the subsets from $\mathcal{F}_i$. For each $S_j \in \mathcal{F}_i$ place one pair of black and white points in the strip labeled by $u_i \in S_j$ inside a region (x_{u_i}, y_{u_j}) and the second pair in the column labeled by x_{S_j} (see Fig. 19). In each region (x_{u_i}, y_{u_j}) add a special point in the top left corner with a color different from the

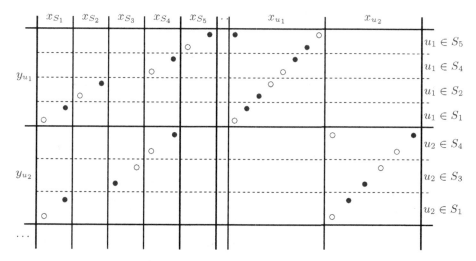

Fig. 19. Construction of configurations R_{u_1} and R_{u_2} where $\mathcal{F}_{u_1} = \{S_1, S_2, S_4, S_5\}$ and $\mathcal{F}_{u_2} = \{S_1, S_3, S_4\}$

color of the point on the top right corner. This point is introduced to force at least one vertical line across a region. Place the configuration R_{u_i} for u_i in the region labeled by (x_{u_i}, y_{u_i}). Examples of R_{u_1} and R_{u_2} where $\mathcal{F}_{u_1} = \{S_1, S_2, S_4, S_5\}$ and $\mathcal{F}_{u_2} = \{S_1, S_3, S_4\}$, are depicted in Fig. 19.

The configuration R_{u_i} requires at least m_i lines to be separated, among them at least one vertical. Thus, the whole construction for u_i requires at least $m_i + 1$ lines. Let I' be an instance of DISCSIZE2D defined by the set of all points forcing the grid and all configurations R_{u_i} with $K = k + \sum_{i=1}^{n} m_i + (2n + m + 2)$ as the number, where the last component $(2n + m + 2)$ is the number of lines defining the grid. If there is a covering of S by k subsets $S_{j_1}, S_{j_2}, \ldots, S_{j_k}$, then we can construct K lines that separate well the set of points, namely $(2n + m + 2)$ grid lines, k vertical lines in columns corresponding to $S_{j_1}, S_{j_2}, \ldots, S_{j_k}$ and m_i lines for the each element u_i $(i = 1, \ldots, n)$.

On the other hand, let us assume that there are K lines separating the points from instance I'. We show that there exists a covering of S by k subsets. There is a set of lines such that for any $i \in \{1, \ldots, n\}$ there are exactly m_i lines passing across the configuration R_{u_i} (i.e., the region labeled by (x_{u_i}, y_{u_i})), among them exactly one vertical line. Hence, there are at most k vertical lines on rows labeled by $x_{S_1}, \ldots, x_{S_m}$. These lines determine k subsets which cover the whole S.   ∎

Next we will consider the discretization problem that minimizes the *number of homogeneous regions* defined by a set of cuts. This discretization problem is called the Optimal Splitting problem. We will show that the optimal splitting problem is NP-hard, even when the number of attributes is fixed by 2. The optimal splitting problem is defined as follows:

OPTISPLIT2D: Optimal splitting in $\mathbb{R}^2$
input: The set S of black and white points $P_1, \ldots, P_n$ on the plane, and an integer k.
question: Is there a consistent set of cuts partitioning the plane into at most k regions

Theorem 14. OPTISPLIT2D *is NP-complete.*

Proof: It is clear that OPTISPLIT2D is in NP. The NP-hardness part of the proof is done by reducing 3SAT to OPTISPLIT2D (cf. [35]).

Let $\Phi = C_1 \cdots C_k$ be an instance of 3SAT. We construct an instance I_Φ of OPTISPLIT2D such that Φ is satisfiable iff there is a sufficiently small consistent set of lines for I_Φ. The description of I_Φ will specify a set of points S, which will be partitioned into two subsets of white and black points. A pair of points with equal horizontal coordinates is said to be *vertical*, similarly, a pair of points with equal vertical coordinates is *horizontal*. If a configuration of points includes a pair of horizontal points p_1 and p_2 of different colors, then any consistent set of lines will include a vertical line L separating p_1 and p_2, which will be in the vertical strip with p_1 and p_2 on its boundaries. Such a strip is referred to as a *forcing strip*, and the line L as *forced* by points p_1 and p_2. Horizontal forcing strips and forced lines are defined similarly. The instance I_Φ has an underlying grid-like structure consisting of vertical and horizontal forcing strips. The rectangular regions inside the structure and consisting of points outside the strips are referred to as *f-rectangles* of the grid. The f-rectangles are arranged into rows and columns.

For each propositional variable p occurring in C use one special row and one special column of rectangles. In the f-rectangle that is at the intersection of the row and column place configuration R_p as depicted on Fig. 20.

Notice that R_p requires at least one horizontal and one vertical line to separate the white from the black points. If only one such vertical line occurs in a consistent set of lines, then it separates either the left or the right white point from the central black one, which we interpret as an assignment of the value *true* or *false* to p, accordingly.

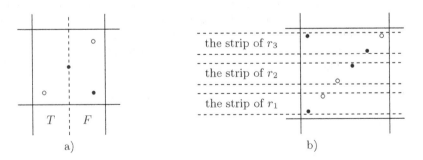

Fig. 20. a) Configuration R_p b) Configuration R_C

For each clause C in Φ use one special row and one special column of f-rectangles. Let C be of the form $C = r_1 \vee r_2 \vee r_3$, where the variables in the literals r_i are all different. Subdivide the row into three strips corresponding to the literals. For each such r_i place one black and one white points, of distinct vertical and horizontal coordinates, inside its strip in the column of the variable of r_i, in the 'true' vertical strip if $r_i = p$, and in the 'false' strip if $r_i = \neg p$. These two points are referred to as *configuration* $R_{C,i}$. In the region of the intersection of the row and column of C place configuration R_C as depicted in Fig. 20. Notice that R_C requires at least three lines to separate the white from the black points, and among them at least one vertical. The example of a fragment of this construction is depicted in Fig. 21. Column x_{p_i} and row x_{p_i} correspond to variable p_i, row y_C corresponds to clause C.

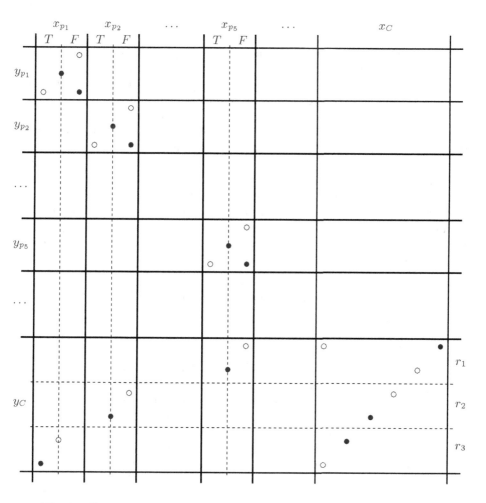

Fig. 21. Construction of configurations R_{p_i} and R_C for $C = p_1 \cdot \neg p_2 \cdot \neg p_5$

Let the underlying grid of f-rectangles be minimal to accommodate this construction. Add horizontal rows of f-rectangles, their number greater by 1 than the size of Φ. Suppose that a consistent set of lines W includes exactly one vertical and one horizontal line per each R_p, and exactly one vertical and two horizontal lines per each R_C, let L_1 be the set of all these lines. There is also the set L_2 of lines inside the forcing strips, precisely one line per each strip. We have $W = L_1 \cup L_2$. Let the number of horizontal lines in W be equal to l_h and vertical to l_v. These lines create $T = (l_h - 1) \cdot (l_v - 1)$ regions, and this number is the last component of I_Φ.

Next we show the correctness of the reduction. Suppose first that Φ is satisfiable, let us fix a satisfying assignment of logical values to the variables of Φ. The consistent set of lines is determined as follows. Place one line into each forcing strip. For each variable p place one vertical and one horizontal line to separate points in R_p, the vertical line determined by the logical value assigned to p. Each configuration R_C is handled as follows. Let C be of the form $C = r_1 \vee r_2 \vee r_3$. Since C is satisfied, at least one $R_{C,i}$, say $R_{C,1}$, is separated by the vertical line that separates also R_p, where p is the variable of r_1. Place two horizontal lines to separate the remaining $R_{C,2}$ nad $R_{C,3}$. They also separate two pairs of points in R_C. Add one vertical line to complete the separation of the points in R_C. All this means that there is a consistent set of lines which creates T regions.

On the other hand, suppose that there is a consistent set of lines for I_Φ, which determines at most T regions. The number T was defined in such a way that two lines must separate each R_p and three lines each R_C, in the latter case at least one of them is vertical. Notice that a horizontal line contributes fewer regions than a vertical one because the grid of splitting strips contains much more rows than columns. Hence, one vertical line and two horizontal lines separate each R_C, because changing horizontal to vertical would increase the number of regions beyond T. It follows that for each clause $C = r_1 \vee r_2 \vee r_3$ at least one $R_{C,i}$ is separated by a vertical line of R_p, where p is the variable of r_i, and this yields a satisfying truth assignment. ∎

6.4   Attribute Reduction vs. Discretization

We have presented two different concepts of data reduction, namely the attribute reduction and the discretization of real value attribute. Both concepts are useful as data preprocessing methods for knowledge discovery processes [32], particularly for rule-based classification algorithms. Attribute reduction eliminates redundant attributes and favours those attributes which are most relevant to the classification process. Discretization eliminates insignificant differences between real values by partition of real axis into intervals. Attribute reduction process can result in generation of short decision rules, while discretization process is helpful to obtain strong decision rules (supported by large number of objects).

There is a strong relationship between those concepts. If $\mathbf{C}$ is an optimal set of cuts for decision table $\mathbb{S}$ then the discretized decision table $\mathbb{S}|_{\mathbf{C}}$ is not reducible, i.e., it contains exactly one decision reduct. Every discretization process is associated with a loss of information, but for some rough set-based applications (e.g., dynamic reduct and dynamic rule methods [10]), where reducts are an important tool, the loss caused by optimal discretization is too large to obtain strong rules. In this situation we would like to search for a more excessive set of cuts to ensure bigger number of reducts of discretized decision table, and, at the same time, to keep some additional information.

In this section, we consider the problem of searching for the minimal set of cuts which preserves the discernibility between objects with respect to any subset of s attributes. One can show that this problem, called s-*optimal discretization problem* (s-OPTIDISC problem), is also NP-hard. Similarly to the case of OPTIDISC, we propose a solution based on approximate boolean reasoning approach for s-OPTIDISC problem.

Definition 19. *Let* $\mathbb{S} = (U, A \cup \{dec\})$ *be a given decision table and let* $1 \leq s \leq |A| = k$. *A set of cuts* $\mathbf{C}$ *is* s-*consistent with* $\mathbb{S}$ *(or* s-*consistent in short) iff for any subset of* s *attributes* B *(i.e.,* $|B| = s$*)* $\mathbf{C}$ *is consistent with subtable* $\mathbb{S}|_B = (U, B \cup \{dec\})$.

Any 1-consistent set of cuts is called *locally consistent* and any k-consistent set of cuts, where $k = |A|$, is called *globally consistent*.

Definition 20. *An* s-*consistent set of cuts is called* s-*irreducible if neither of its proper subsets is* s-*consistent.*

Definition 21. *An* s-*consistent set of cuts* $\mathbf{C}$ *is called* s-*optimal if* $|C| \leq |Q|$ *for any* s-*consistent set of cuts* $\mathbf{Q}$.

When $s = k = |A|$, the definition of k-irreducible and k-optimal set of cuts is exactly the same as the definition of irreducible and optimal set of cuts. Thus the concept of s-irreducibility and s-optimality is a generalization of irreducibility and optimality from Definition 18.

The following proposition presents important properties of s-consistent set of cuts:

Proposition 8. *Let a set of cuts* $\mathbf{C}$ *be* s-*consistent with a given decision table* $\mathbb{S} = (U, A \cup \{d\})$. *For any subset of attributes* $B \subset A$ *such that* $|B| \geq s$, *if* B *is relative super-reduct of* $\mathbb{S}$ *then the set of discretized attributes* $B|_{\mathbf{C}}$ *is also a relative super-reduct of discretized decision table* $\mathbb{S}|_{\mathbf{C}}$.

Example 17. Let us illustrate the concept of s-optimal set of cuts on the decision table from Table 14. One can see that the set of all relative reducts of Table 14 is equal to $R = \{\{a_1, a_2\}, \{a_2, a_3\}\}$.

The set of all generic cuts is equal to $\mathbf{GCuts}_A = \mathbf{C}_{a_1} \cup \mathbf{C}_{a_2} \cup \mathbf{C}_{a_3}$, where

$$\mathbf{C}_{a_1} = \{(a_1, 1.5), (a_1, 2.5), (a_1, 3.5), (a_1, 4.5), (a_1, 5.5), (a_1, 6.5), (a_1, 7.5)\}$$
$$\mathbf{C}_{a_2} = \{(a_2, 1.5), (a_2, 3.5), (a_2, 5.5), (a_2, 6.5), (a_2, 7.5)\}$$
$$\mathbf{C}_{a_3} = \{(a_3, 2.0), (a_3, 4.0), (a_3, 5.5), (a_3, 7.0)\}$$

is illustrated in Fig. 22.

Table 14. An example decision table with ten objects, three attributes and three decision classes

A	a_1	a_2	a_3	d
u_1	1.0	2.0	3.0	0
u_2	2.0	5.0	5.0	1
u_3	3.0	7.0	1.0	2
u_4	3.0	6.0	1.0	1
u_5	4.0	6.0	3.0	0
u_6	5.0	6.0	5.0	1
u_7	6.0	1.0	8.0	2
u_8	7.0	8.0	8.0	2
u_9	7.0	1.0	1.0	0
u_{10}	8.0	1.0	1.0	0

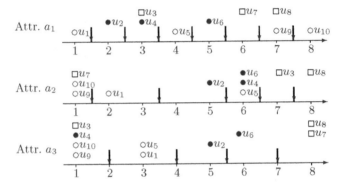

Fig. 22. Illustration of cuts on the table $\mathbb{S}$. Objects are marked by three labels with respect to their decision values.

Some examples of s-optimal sets of cuts for $s = 1, 2, 3$ are as follows:

$$\mathbf{C}_1 = \{(a_1, 1.5), (a_1, 2.5), (a_1, 3.5), (a_1, 4.5), (a_1, 5.5), (a_1, 6.5), (a_1, 7.5)\}$$
$$\cup \{(a_2, 1.5), (a_2, 3.5), (a_2, 5.5), (a_2, 6.5)\}$$
$$\cup \{(a_3, 2.0), (a_3, 4.0), (a_3, 7.0)\} \tag{42}$$
$$\mathbf{C}_2 = \{(a_1, 3.5), (a_1, 4.5), (a_1, 5.5), (a_1, 6.5)\} \cup \{(a_2, 3.5), (a_2, 6.5)\}$$
$$\cup \{(a_3, 2.0), (a_3, 4.0)\} \tag{43}$$
$$\mathbf{C}_3 = \{(a_1, 3.5)\} \cup \{(a_2, 3.5), (a_2, 6.5)\} \cup \{(a_3, 4.0)\} \tag{44}$$

Thus $\mathbf{C}_1$ is the smallest locally consistent set of cuts and $\mathbf{C}_3$ is the global optimal set of cuts. One can see that the table $\mathbb{S}|_{\mathbf{C}_3}$ has only one reduct: $\{a_1|_{\mathbf{C}_3}, a_2|_{\mathbf{C}_3}, a_3|_{\mathbf{C}_3}\}$, while both tables $\mathbb{S}|_{\mathbf{C}_1}$ and $\mathbb{S}|_{\mathbf{C}_2}$ still have two reducts.

The following theorem characterizes the complexity of s-OPTIDISC problem.

Table 15. The 2-optimal discretized table $\mathbb{S}|_{\mathbf{C_2}}$ still has two reducts: $\{a_1|_{\mathbf{C_2}}, a_2|_{\mathbf{C_2}}\}$ and $\{a_2|_{\mathbf{C_2}}, a_3|_{\mathbf{C_2}}\}$

$\mathbb{S}|_{\mathbf{C_2}}$	$a_1|_{\mathbf{C_2}}$	$a_2|_{\mathbf{C_2}}$	$a_3|_{\mathbf{C_2}}$	d
u_1	0	0	1	0
u_2	0	1	2	1
u_3	0	2	0	2
u_4	0	1	0	1
u_5	1	1	1	0
u_6	2	1	2	1
u_7	3	0	2	2
u_8	4	2	2	2
u_9	4	0	0	0
u_{10}	4	0	0	0

Theorem 15. *For a given decision table* $\mathbb{S} = (U, A \cup \{d\})$ *and an integer* s, *the problem of searching for* s-*optimal set of cuts is* $DTIME(kn \log n)$ *for* $s = 1$ *and is NP-hard for any* $s \geq 2$.

Proof: The first part is obvious. The proof of the second part follows from the NP-hardness of DISCSIZE2D (optimal discretization for two attributes), see Theorem 13 from Sect. 6.3. ∎

The following fact states that s-consistency is monotone with respect to s. In particular, it implies that one can reduce the s-optimal set of cuts to obtain $(s + 1)$-optimal set of cuts.

Theorem 16. *For any decision table* $\mathbb{S} = (U, A \cup \{d\})$, $card(A) = k$, *and for any integer* $s \in \{1, \ldots, k - 1\}$, *if the set of cuts* **P** *is* s-*consistent with* $\mathbb{S}$, *then* **P** *is also* $(s + 1)$-*consistent with* $\mathbb{S}$.

Proof: We assume that the set of cuts **C** is s-consistent and not $(s+1)$-consistent. Then there exists a set of $(s + 1)$ attributes $B = \{b_1, \ldots, b_s, b_{s+1}\}$, such that **C** is not consistent with subtable $\mathbf{B} = (U, B \cup \{d\})$. Hence, there are two objects u_i, u_j such that $d(u_i) \neq d(u_j)$ and $(u_i, u_j) \notin IND(B)$ (i.e., $\exists_{b \in B} [b(u_i) \neq b(u_j)]$) but there is no cut on **C** which discerns u_i and u_j. Since $(u_1, u_2) \notin IND(B)$, then one can choose the subset $B' \subset B$ with s attributes such that $(u_i, u_j) \notin IND(B')$. Therefore, **C** is not consistent with the subtable $\mathbf{B} = (U, B' \cup \{d\})$ and in the consequence **C** is not s-consistent which is a contradiction. ∎

6.4.1 Boolean Reasoning Approach to s-OPTIDISC

Consider a decision table $\mathbb{S} = (U, A \cup \{dec\})$ where $U = \{u_1, u_2, \ldots, u_n\}$ and $A = \{a_1, \ldots, a_k\}$. We encode the s-OPTIDISC problem by a Boolean function in a similar way as in Sect. 6.2:

boolean variables: Let $\mathbf{C}_{a_m}$ be the set of candidate cuts on the attribute a_m for $m = 1, \ldots, k$. We denote by $\mathbf{P}_{a_m} = \{p_1^{a_m}, \ldots, p_{n_m}^{a_m}\}$ the set of boolean variables corresponding to cuts from $\mathbf{C}_{a_m}$. Thus the set of all boolean variables is denoted by

$$\mathbf{P} = \bigcup_{m=1}^{k} \mathbf{P}_{a_m}.$$

Encoding function: In Sect. 6.2, for any objects $u_i, u_j \in U$, we denoted by $\mathbf{X}_{i,j}^a$ the set of cuts from $\mathbf{C}_a$ discerning u_i and u_j, i.e.

$$\mathbf{X}_{i,j}^a = \{(a; c_k^a) \in \mathbf{C}_a : (a(u_i) - c_k^a)(a(u_j) - c_k^a) < 0\}.$$

For any subset of attributes $B \subset A$, the B-discernibility function for u_i and u_j is defined as a disjunction of boolean variables corresponding to the cuts from B discerning u_i and u_j:

$$\psi_{i,j}^B = \sum_{a \in B} \Sigma \mathbf{X}_{i,j}^a = \Sigma \mathbf{X}_{i,j}^B$$

where $\mathbf{X}_{i,j}^B = \bigcup_{a \in B} \mathbf{X}_{i,j}^a$. The (Boolean) *discernibility function* for the set of attributes B is defined by:

$$\Phi_B = \prod_{d(u_i) \neq d(u_j)} \psi_{i,j}^B.$$

The encoding function for s-optimal discretization problem is defined as follows:

$$\Phi_s = \prod_{|B|=s} \Phi_B = \prod_{|B|=s} \prod_{d(u_i) \neq d(u_j)} \psi_{i,j}^B.$$

The construction of Φ_s enables us to prove of the following theorem.

Theorem 17. *A set of cuts $\mathbf{C}$ is s-optimal with a given decision table if and only if the corresponding boolean monomial $\Pi_\mathbf{C}$ is minimal prime implicant of Φ_s.*

One can see that the function Φ_s is a conjunction of N clauses of form $\psi_{i,j}^B$ where

$$N = \binom{k}{s} \cdot \underbrace{|\{(u_i, u_j) : d(u_i) \neq d(u_j)\}|}_{=conflict(\mathbb{S})} = O\left(\binom{k}{s} \cdot n^2\right)$$

Thus, any greedy algorithm of searching for minimal prime implicant of the function Φ_s needs at least $O\left(n^2 \cdot \binom{k}{s}\right)$ steps to compute the quality of a given cut (i.e., the number of clauses satisfied by this cut). Let us discuss some properties of the function Φ_s which are useful when solving this problem.

Recall that the discernibility function for the reduct problem was constructed by the discernibility matrix

$$\mathbf{M}(\mathbb{S}) = [M_{i,j}]_{i,j=1}^n$$

where $M_{i,j} = \{a \in A : a(u_i) \neq a(u_j)\}$ is the set of attributes discerning u_i and u_j. The relationship between reduct and discretization problems is expressed by the following technical lemma:

Lemma 1. *For any pair of objects $u_i, u_j \in U$ and for any subset of attributes B*

$$\prod_{|B|=s} \psi_{i,j}^B \geq \prod_{a \in M_{i,j}} \psi_{i,j}^a.$$

The equality holds if $|M_{i,j}| \leq k - s + 1$.

Proof: Firstly, from the definition of $M_{i,j}$, we have $\psi_{i,j}^B = \psi_{i,j}^{B \cap M_{i,j}}$. Thus

$$\psi_{i,j}^B = \sum_{a \in B \cap M_{i,j}} \psi_{i,j}^a \geq \prod_{a \in B \cap M_{i,j}} \psi_{i,j}^a \geq \prod_{a \in M_{i,j}} \psi_{i,j}^a.$$

Hence

$$\prod_{|B|=s} \psi_{i,j}^B \geq \prod_{a \in M_{i,j}} \psi_{i,j}^a.$$

On the other hand, if $|M_{i,j}| \leq k - s + 1$, then $|A - M_{i,j}| \geq s - 1$. Let $C \subset A - M_{i,j}$ be a subset of $s - 1$ attributes. For each attribute $a \in M_{i,j}$, we have $|\{a\} \cup C| = s$ and

$$\psi_{i,j}^{\{a\} \cup C} = \psi_{i,j}^a$$

Thus

$$\prod_{|B|=s} \psi_{i,j}^B = \left(\prod_{a \in M_{i,j}} \psi_{i,j}^a \right) \cdot \psi' \leq \prod_{a \in M_{i,j}} \psi_{i,j}^a.$$

∎

This lemma allows us to simplify many calculations over the function Φ_s. A pair of objects is called *conflicting* if $d(u_i) \neq d(u_j)$, i.e., u_i and u_j are from distinct decision classes. For any set of cuts $\mathbf{C}$, we denote by $A_{i,j}(\mathbf{C})$ the set of attributes for which there is at least one cut from $\mathbf{C}$ discerning objects u_i, u_j, thus

$$A_{i,j}(\mathbf{C}) = \{a \in A : \exists_{c \in \mathbb{R}}((a; c) \in \mathbf{C}) \wedge [(a(u_i) - c)(a(u_j) - c) < 0]\}.$$

It is obvious that

$$A_{i,j}(\mathbf{C}) \subseteq M_{i,j} \subseteq A.$$

A set of cuts $\mathbf{C}$ is consistent (or k-consistent) if and only if $|A_{i,j}(\mathbf{C})| \geq 1$ for any pair of conflicting objects u_i, u_j. We generalize this observation by showing that a set of cuts $\mathbf{C}$ is s-consistent if and only if the sets $A_{i,j}(\mathbf{C})$ are sufficiently large, or equivalently, the difference between A and $A_{i,j}(\mathbf{C})$ must be sufficiently small.

Theorem 18. *For any set of cuts $\mathbf{C}$, the following statements are equivalent:*

a) $\mathbf{C}$ *is s-consistent;*
b) *The following inequality*

$$|A_{i,j}(\mathbf{C})| \geq k_{i,j} = \min\{|M_{i,j}|, k - s + 1\} \tag{45}$$

holds for any pair of conflicting objects $u_i, u_j \in U$.

Proof: The function Φ_s can be rewritten as follows:

$$\Phi_s = \prod_{d(u_i) \neq d(u_j)} \prod_{|B|=s} \psi_{i,j}^B$$

Let $\mathcal{P}_s(A) = \{B \subset A : |B| = s\}$. Theorem 17 states that $\mathbf{C}$ is s-consistent if and only if $\mathbf{C} \cap \mathbf{X}_{i,j}^B \neq \varnothing$ for any pair of conflicting objects $u_i, u_j \in U$ and for any $B \in \mathcal{P}_s(A)$ such that $\mathbf{X}_{i,j}^B \neq \varnothing$.

Therefore, it is enough to prove that, for any pair of conflicting objects $u_i, u_j \in U$, the following statements are equivalent

    a) $|A_{i,j}(\mathbf{C})| \geq k_{i,j};$ (46)

    b) $\forall_{B \in \mathcal{P}_s(A)}(\mathbf{X}_{i,j}^B \neq \varnothing) \implies \mathbf{C} \cap \mathbf{X}_{i,j}^B \neq \varnothing$ (47)

To do so, let us consider two cases:

1. $|M_{i,j}| \leq k - s + 1$: in this case $k_{i,j} = |M_{i,j}|$. We have

$$|A_{i,j}(\mathbf{C})| \geq k_{i,j}; \iff \qquad\qquad A_{i,j}(\mathbf{C}) = M_{i,j}$$
$$\iff \qquad\qquad \forall_{a \in M_{i,j}} \mathbf{C} \cap \mathbf{X}_{i,j}^a \neq \varnothing$$

By previous lemma, we have

$$\prod_{|B|=s} \psi_{i,j}^B = \prod_{a \in M_{i,j}} \psi_{i,j}^a.$$

Thus a) $\iff$ b).

2. $|M_{i,j}| > k - s + 1$: in this case we have $k_{i,j} = k - s + 1$ and the condition $|A_{i,j}(\mathbf{C})| \geq k - s + 1$ is equivalent to

$$|A - A_{i,j}(\mathbf{C})| < s.$$

Consequently, any set $B \in \mathcal{P}_s(A)$ has non-empty intersection with $A_{i,j}(\mathbf{C})$, thus $\mathbf{C} \cap \mathbf{X}_{i,j}^B \neq \varnothing$.

We have shown in both cases that a) $\iff$ b).   ■

The MD-heuristic was presented in previous section as greedy algorithm for the Boolean function Φ_k. The idea was based on a construction and an the analysis of a new table $\mathbb{S}^* = (U^*, A^*)$, where

- $U^* = \{(u_i, u_j) \in U^2 : d(u_i) \neq d(u_j)\}$
- $A^* = \{c : c$ is a cut on $\mathbb{S}\}$, where $c((u_i, u_j)) = \begin{cases} 1 & \text{if } c \text{ discerns } u_i, u_j \\ 0 & \text{otherwise} \end{cases}$

This table consists of $O(nk)$ attributes (cuts) and $O(n^2)$ objects (see Table 16).

We denote by $Disc(a, c)$ the *discernibility degree* of the cut $(a; c)$ which is defined as the number of pairs of objects from different decision classes (or number of objects in table $\mathbb{S}^*$) discerned by c. The *MD-heuristic* is searching

Table 16. Temporary table $\mathbb{S}^*$ constructed for the decision table from Table 15

$\mathbb{S}^*$	a_1							a_2				a_3			$k_{i,j}$
	1.5	2.5	3.5	4.5	5.5	6.5	7.5	1.5	3.5	5.5	6.5	2.0	4.0	7.0	
(u_1,u_2)	1							1					1		2
(u_1,u_3)	1	1						1	1	1		1			2
(u_1,u_4)	1	1						1	1			1			2
(u_1,u_6)	1	1	1	1				1	1				1		2
(u_1,u_7)	1	1	1	1	1			1					1	1	2
(u_1,u_8)	1	1	1	1	1	1		1	1	1			1	1	2
(u_2,u_3)	1								1	1	1	1			2
(u_2,u_5)	1	1								1			1		2
(u_2,u_7)	1	1	1	1				1	1					1	2
(u_2,u_8)	1	1	1	1	1					1	1			1	2
(u_2,u_9)	1	1	1	1	1			1	1			1	1		2
(u_2,u_{10})	1	1	1	1	1	1		1	1			1	1		2
(u_3,u_4)											1				1
(u_3,u_5)			1								1	1			2
(u_3,u_6)			1	1							1	1	1		2
(u_3,u_9)			1	1	1	1		1	1	1	1				2
(u_3,u_{10})			1	1	1	1	1	1	1	1	1				2
(u_4,u_5)			1										1		2
(u_4,u_7)			1	1	1			1	1	1		1	1	1	2
(u_4,u_8)			1	1	1	1					1	1	1	1	2
(u_4,u_9)			1	1	1	1		1	1	1					2
(u_4,u_{10})			1	1	1	1	1	1	1	1					2
(u_5,u_6)			1										1		2
(u_5,u_7)			1	1									1	1	2
(u_5,u_8)			1	1	1						1		1	1	2
(u_6,u_7)				1				1	1	1				1	2
(u_6,u_8)				1	1									1	2
(u_6,u_9)				1	1			1	1	1		1	1		2
(u_6,u_{10})				1	1	1		1	1	1		1	1		2
(u_7,u_9)				1								1	1	1	2
(u_7,u_{10})				1	1							1	1	1	2
(u_8,u_9)								1	1	1	1	1	1	1	2
(u_8,u_{10})							1	1	1	1	1	1	1	1	2

for a cut $(a;c) \in A^*$ with the largest discernibility degree $Disc(a,c)$. Then we move the cut c from A^* to the result set of cuts $\mathbf{P}$ and remove from U^* all pairs of objects discerned by c. Our algorithm terminates if $U^* = \varnothing$. We have shown that MD-heuristic is quite efficient, since it determines the best cut in $O(kn)$ steps using $O(kn)$ space only.

One can modify this algorithm for the need of s-optimal discretization problem by applying Theorem 18. At the beginning, we confer *required cut number* $k_{i,j}$ and *set of discerning attributes* $A_{i,j} := \varnothing$ upon every pair of objects

$(u_i, u_j) \in U^*$ (see Theorem 18). Next we search for a cut $(a; c) \in A^*$ with the largest discernibility degree $Disc(a, c)$ and remove $(a; c)$ from A^* to the result set of cuts $\mathbf{P}$. Then we insert the attribute a into lists of attributes of all pairs of objects discerned by $(a; c)$. We also delete from U^* such pairs (u_i, u_j) that $|A_{i,j}| = k_{i,j}$. This algorithm is continued until $U^* = \varnothing$.

Algorithm 5. MD-heuristic for s-optimal discretization problem

---

Input: Decision table $\mathbb{S} = (U, A, dec)$
Output: The semi s-optimal set of cuts;
begin
> Construct the table $\mathbb{S}^* = (U^*, A^*)$ from $\mathbb{S}$;
> $\mathbf{C} := \varnothing$;
> **for** *each pair of conflicting objects* $(u_i, u_j) \in U^*$ **do**
> > Set $k[i, j] := \min\{|M_{i,j}|, k - s + 1\}$;
> > Set $A[i, j] := \varnothing$;
>
> **end**
> **while** $U^* \neq \varnothing$ **do**
> > Select the the cut (a, c) with the maximal number of occurrences of 1's in $\mathbb{S}^*$;
> > $\mathbf{C} := \mathbf{C} \cup \{(a, c)\}$;
> > Delete from $\mathbb{S}^*$ the column corresponding to (a, c);
> > **for** *each pair of conflicting objects* $(u_i, u_j) \in U^*$ **do**
> > > $A[i, j] := A[i, j] \cup \{a\}$;
> > > **if** $|A_{i,j}| \geq k_{i,j}$ **then**
> > > > Delete (u_i, u_j) from U^*;
> > >
> > > **end**
> >
> > **end**
>
> **end**
> Return $\mathbf{C}$;
end

In case of decision table from Example 17, the temporary table $\mathbb{S}^*$ consists of 33 pairs of objects from different decision classes (see Table 16). For $s = 2$, the required numbers of cuts $k_{i,j}$ for all $(u_i, u_j) \in U^*$ (see Theorem 18) are equal to 2 except $k_{3,4} = 1$. Our algorithm begins by choosing the best cut $(a_3, 4.0)$ discerning 20 pairs of objects from $\mathbb{S}$. In the next step the cut $(a_1, 3.5)$ will be chosen because of 17 pairs of objects discerned by this cut. After this step one can remove 9 pairs of objects from U^*, e.g., $(u_1, u_6), (u_1, u_7), (u_1, u_8), (u_2, u_5), \ldots$ because they are discerned by two cuts on two different attributes. If the algorithm stops, one can eliminate some superfluous cuts to obtain the set of cuts $\mathbf{P}_2$ as it was presented in Eqn. (43).

6.5   Bibliography Notes

The classification of discretization methods into three dimensions, i.e., local vs. global, dynamic vs. static and supervised vs. unsupervised has been introduced in [28]. Liu et al. [62] summarize the existing discretization methods and identify some issues yet to solve. They point out also future research for discretization. Below we describe some well-known discretization techniques with respect to this classification schema.

6.5.1   Equal Width and Equal Frequency Interval Binning

These are probably the simplest discretization methods. The first method called *equal width interval discretization* involves determining the domain of observed values of an attribute $a \in A$ (i.e., $[v^a_{min}, v^a_{max}]$) and dividing this interval into k_a equally sized intervals where $k_a \in \mathbf{N}$ is a parameter supplied by the user. One can compute the interval width: $\delta = \frac{v^a_{max} - v^a_{min}}{k_a}$ and construct interval boundaries (cut points): $c^a_i = v^a_{min} + i \cdot \delta$, where $i = 1, \ldots, k_a - 1$.

The second method called *equal frequency interval discretization* sorts the observed values of an attribute a (i.e., $v^a_1 < v^a_2 < \cdots < v^a_{n_a}$) and divides them into k_a intervals (k_a is also a parameter supplied by the user) where each interval contains $\lambda = \left\lceil \frac{n_a}{k_a} \right\rceil$ sequential values. The cut points are computed by $c_i = \frac{v_{i \cdot \lambda} + v_{i \cdot \lambda + 1}}{2}$ for $i = 1, \ldots, k_a - 1$.

These methods are global and applied to each continuous attribute independently, so they are static. They are also unsupervised discretization methods because they make no use of decision class information.

The described methods are efficient from the point of view of time and space complexity. However, because of the discretization of each attribute independently, decision rules generated over discretized data will not give us satisfiable quality of classification for unseen (so far) objects.

6.5.2   OneR Discretizer

Holte (1993)[48] proposed an error-based approach to discretization which is a global, static and supervised method and is known as **OneR (*One Rule*) Discretizer**. Each attribute is sorted into ascending order and a greedy algorithm that divides the feature into intervals where each of them contains a strong majority of objects from one decision class is used. There is an additional constraint that each interval must include at least some prespecified number of values (the user has to fix some constant M, which is a minimal number of observed values in intervals).

Given a minimal size M, each discretization interval is initiated to contain M consequent values and it is made as pure as possible by moving a partition boundary (cut) to add an observed value until the count of the dominant decision class in that interval will increase. Empirical analysis [48] suggests a minimal bin size of $M = 6$ performs the best.

6.5.3   Statistical Test Methods

Any cut $c \in \mathbf{C}_a$ splits the set of values (l_a, r_a) of the attribute a into two intervals: $L_c = (l_a, c)$ and $R_c = (c, r_a)$. Statistical tests allow to check the probabilistic independence between the object partition defined by decision attribute and by the cut c. The independence degree is estimated by χ^2 test described by

$$\chi^2 = \sum_{i=1}^{2} \sum_{j=1}^{r} \frac{(n_{ij} - E_{ij})^2}{E_{ij}}$$

where: r – number of decision classes,

n_{ij} – number of objects from j^{th} class in i^{th} interval,

R_i – number of objects in i^{th} interval $\left(= \sum_{j=1}^{r} n_{ij}\right)$,

C_j – number of objects in the jth class $\left(= \sum_{i=1}^{2} n_{ij}\right)$,

n – total number of objects $\left(\sum_{i=1}^{2} R_i\right)$,

E_{ij} – expected frequency of A_{ij} $\left(= \frac{R_i \times C_j}{n}\right)$. If either R_i or C_j is 0, E_{ij} is set to 0.1.

Intuitively, if the partition defined by c does not depend on the partition defined by the decision d then:

$$P(C_j) = P(C_j|L_c) = P(C_j|R_c) \tag{48}$$

for any $i \in \{1, \ldots, r\}$. The condition (48) is equivalent to $n_{ij} = E_{ij}$ for any $i \in \{1, 2\}$ and $j \in \{1, \ldots, r\}$, hence we have $\chi^2 = 0$. In the opposite case, if there exists a cut c which properly separates objects from different decision classes the value of χ^2 test for c is very high.

Discretization methods based on χ^2 test are choosing only cuts with large value of this test (and delete the cuts with small value of χ^2 test).

There are different versions of this method (see, e.g., ChiMerge system for discretization – Kerber (1992) [56], StatDisc- Richeldi & Rossotto (1995), [125], Chi2 (1995) [64]).

6.5.4   Entropy Methods

A number of methods based on entropy measure established the strong group of the research in the discretization domain. This concept uses class-entropy as a criterion to evaluate a list of best cuts which together with the attribute domain induce the desired intervals.

The class information entropy of the partition induced by a cut point c on attribute a is defined by

$$E(a; c; U) = \frac{|U_1|}{n} Ent(U_1) + \frac{|U_2|}{n} Ent(U_2),$$

where n is a number of objects in U and U_1, U_2 are the sets of objects on the left side (right side) of the cut c.

For a given feature A, the cut $c_{\min}$ which minimizes the entropy function over all possible cuts is selected. This method can be applied recursively to both object sets U_1, U_2 induced by $c_{\min}$ until some stopping condition is achieved.

There is a number of methods based on information entropy [113], [17], [31], [21], [114].

Different methods use different stopping criteria. As an example, we mention one of them which has been proposed by Fayyad and Irani [31].

Fayyad and Irani used the *Minimal Description Length Principle* [126] [127] to determine a stopping criteria for their recursive discretization strategy. First they defined the *Gain* of the cut $(a; c)$ over the set of objects U by:

$$Gain\,(a; c; U) = Ent\,(U) - E\,(a; c; U)$$

and the recursive partitioning within a set of objects U stops iff

$$Gain\,(a; c; U) < \frac{\log_2\,(n-1)}{n} + \frac{\Delta\,(a; c; U)}{n}$$

where

$$\Delta\,(a; c; U) = \log_2\,(3^r - 2) - [r \cdot Ent\,(U) - r_1 \cdot Ent\,(U_1) - r_2 \cdot Ent\,(U_2)],$$

and r, r_1, r_2 are the numbers of decision class labels represented in the sets U, U_1, U_2, respectively.

6.5.5   Decision Tree Based Methods

Decision tree is not only a useful tool for classification task but it can be treated as feature selection as well as discretization method. Information gain measure can be used to determine the threshold value the gain ratio is greatest in order to partition the data [124]. A divide and conquer algorithm is then successively applied to determine whether to split each partition into smaller subsets at each iteration.

6.5.6   Boolean Reasoning Based Methods

The general idea of applying Boolean reasoning methodology to the discretization problem was proposed in [95]. The global discretization method based on MD-heuristics and improvements were presented lately in [102], [79], [93]. The local discretization method based on MD-heuristics was proposed in [8].

The NP-hardness of general discretization problem was shown in [95] and [79]. The stronger results related to NP-hardness of discretization in two-dimensional space were presented in [98]. The discretization method that preserves some reducts of a given decision table was presented in [104].

7   Approximate Boolean Reasoning Approach to Decision Tree Induction

In this section, we consider another classification method called *decision tree*. The name of this method derives from the fact that it can be represented by an oriented tree structure, where each *internal node* is labeled by a *test* on an information vector, each *branch* represents an outcome of the test, and *leaf nodes* represent decision classes or class distributions.

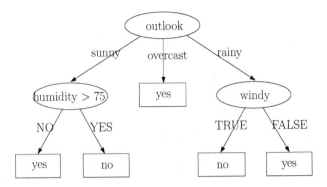

Fig. 23. An exemplary decision tree for the weather data from Table 13

Fig. 23 represents an example of "weather" decision table and corresponding decision tree for this decision table [123].

Usually, tests in decision trees are required to have small number of possible outcomes. In Fig. 23 two types of tests are presented. The first type is simply based on taking one of existing attributes (in Fig. 23 this type of tests occurs in nodes labeled by *outlook* and *windy*). The second type is defined by cuts on real value attributes (humidity > 75). In general, the following types of tests are considered in the literature:

1. **Attribute-based tests:** This type consists of tests defined by symbolic attributes, i.e., for each attribute $a \in A$ we define a test t_a such that for any object u from a universe, $t_a(u) = a(u)$;
2. **Value-based tests:** This type consists of binary tests defined by a pair of an attribute and its value, i.e., for each attribute $a \in A$ and for each value $v \in V_a$ we define a test $t_{a=v}$ such that for any object u from a universe,

$$t_{a=v}(u) = \begin{cases} 1 & \text{if } a(u) = v \\ 0 & \text{otherwise;} \end{cases}$$

3. **Cut-based tests:** Tests of this type are defined by cuts on real value attributes. For each attribute $a \in A$ and for each value $c \in \mathbb{R}$ we define a test $t_{a>c}$ such that for any object u from a universe,

$$t_{a>c}(u) = \begin{cases} 1 & \text{if } a(u) > c \\ 0 & \text{otherwise;} \end{cases}$$

4. **Value set-based tests:** For each attribute $a \in A$ and for each set of values $S \subset V_a$ we define a test $t_{a \in S}$ such that for any object u from a universe,

$$t_{a \in S}(u) = \begin{cases} 1 & \text{if } a(u) \in S \\ 0 & \text{otherwise;} \end{cases}$$

This is a generalization of previous types.

5. **Hyperplane-based tests:** Tests of this type are defined by linear combinations of continuous attributes. A test $t_{w_1 a_1 + \cdots + w_k a_k > w_0}$, where $a_1, \ldots, a_k$ are continuous attributes and $w_0, w_1, \ldots, w_k$ are real numbers, is defined as follows:

$$t_{w_1 a_1 + \cdots + w_k a_k > w_0}(u) = \begin{cases} 1 & \text{if } w_1 a_1(u) + \cdots + w_k a_k(u) > w_0 \\ 0 & \text{otherwise;} \end{cases}$$

A decision tree is called *binary* if it is labeled by binary tests only. In fact, binary decision tree is a classification algorithm defined by a nested "IF – THEN – ELSE –" instruction. More precisely, let decision table $\mathbb{S} = (U, A \cup \{dec\})$ be given, where $V_{dec} = \{1, \ldots, d\}$, each decision tree for $\mathbb{S}$ is a production of the following grammar system:

```
decision_tree := dec_class|
     <IF> test <THEN> decision_tree <ELSE> decision_tree;
dec_inst := <dec=1>|<dec=2>|...|<dec=d>;
test := t_1|t_2|\dots |t_m
```

where $\{t\_1, \ldots, t\_m\}$ is a given set of m binary tests. Similarly, non-binary decision trees can be treated as nested CASE instructions.

Decision tree is one of the most favorite types of templates in data mining, because of its simple representation and easy readability. Analogously to other classification algorithms, there are two issues related to the decision tree approach, i.e., how to classify new unseen objects using decision tree and how to construct an optimal decision tree for a given decision table.

In order to classify an unknown example, the information vector of this example is tested against the decision tree. The path is traced from the root to a leaf node that holds the class prediction for this example.

A decision tree is called *consistent* with a given decision table $\mathbb{S}$ if it properly classifies all objects from $\mathbb{S}$. A given decision table may have many consistent decision trees. The main objective is to build a decision tree of high prediction accuracy for a given decision table. This requirement is realized by a philosophical principle called Occam's Razor.

This principle thought up a long time ago by William Occam while shaving states that the shortest hypothesis, or solution to a problem, should be the one we should prefer (over longer, more complicated ones). This is one of the fundamental tenets of the way western science works and has received much debate and controversy. The specialized version of this principle applied to the decision trees can be formulated as follows:

"The world is inherently simple. Therefore, the smallest decision tree that is consistent with the samples is the one that is most likely to identify unknown objects correctly."

Unfortunately, the problem of searching for shortest tree for a decision table has shown to be NP-hard. It means that the no computer algorithm can solve this in a feasible amount of time in the general case. Therefore, only heuristic algorithms have been developed to find a good tree, usually very close to the best.

In the next section, we summarize the most popular decision tree induction methods.

7.1   Decision Tree Induction Methods

The basic heuristic for construction of decision tree (for example ID3 or later C4.5 – see [124], [123], CART [15]) is based on the top-down recursive strategy described as follows:

1. It starts with a tree with one node representing the whole training set of objects.
2. If all objects have the same decision class, the node becomes leaf and is labeled with this class.
3. Otherwise, the algorithm selects the best test t_{Best} from the set of all possible tests.
4. The current node is labeled by the selected test t_{Best} and it is branched accordingly to values of t_{Best}. Also, the set of objects is partitioned and assigned to new created nodes.
5. The algorithm uses the same processes (steps 2, 3, 4) recursively for each new nodes to form the whole decision tree.
6. The partitioning process stops when either all examples in a current node belong to the same class, or no test function has been selected in Step 3.

Developing decision tree induction methods (see [31], [123]) we should define some *heuristic measures (or heuristic quality functions)* to estimate the quality of tests. In tree induction process, the optimal test t_{Best} with respect to the function $\mathcal{F}$ is selected as the result of Step 3.

More precisely, let $\mathcal{T} = \{t_1, t_2, \ldots, t_m\}$ be a given set of all possible tests, heuristic measure is a function

$$\mathcal{F} : \mathcal{T} \times \mathcal{P}(U) \to \mathbb{R}$$

where $\mathcal{P}(U)$ is the family of all subsets of U. The value $\mathcal{F}(t, X)$, where $t \in \mathcal{T}$ and $X \subset U$, should estimate the chance that t_i labels the root of the optimal decision tree for X. Usually, the value $\mathcal{F}(t, X)$ depends on how the test t splits the set of objects X.

Definition 22. *A counting table w.r.t. the decision attribute dec for the set of objects $X \subset U$ – denoted by* **Count**$(X; dec)$ *– is the array of integers $(x_1, \ldots, x_d)$, where $x_k = |X \cap DEC_k|$ for $k \in \{1, \ldots, d\}$.*

We will drop the decision attribute *dec* from the notation of counting table, just for simplicity. Moreover, if the set of objects X is defined by a logical formula ϕ, i.e.,

$$X = \{u \in U : \phi(u) = \textbf{true}\}$$

then the counting table for X can be denoted by **Count**(ϕ). For example, by **Count**$(age \in (25, 40))$ we denote the counting table for the set of objects $X = \{x \in U : age(x) \in (25, 40)\}$.

Any test $t \in \mathcal{T}$ defines a partition of X into disjoint subsets of objects $X_{v_1}, \ldots, X_{v_{n_t}}$, where $V_t = \{v_1, \ldots, v_{n_t}\}$ is the domain of test t and $X_{v_i} = \{u \in X : t(u) = v_i\}$. The value $\mathcal{F}(t, X)$ of an arbitrary heuristic measure $\mathcal{F}$ is defined by counting tables **Count**$(X; dec)$, **Count**$(X_{v_1}; dec)$, $\ldots$, **Count**$(X_{v_{n_t}}; dec)$.

7.1.1   Entropy Measure

This is one of the most well-known heuristic measures, and it has been used in the famous C4.5 system for decision tree induction system [123].

This concept uses class-entropy as a criterion to evaluate the partition induced by a test. Precisely, the class information entropy of a set X with counting table $(x_1, \ldots, x_d)$, where $x_1 + \cdots + x_d = N$, is defined by

$$Ent(X) = -\sum_{j=1}^{d} \frac{x_j}{N} \log \frac{x_j}{N}.$$

The class information entropy of the partition induced by a test t is defined by

$$E(t, X) = \sum_{i=1}^{n_t} \frac{|X_{v_i}|}{|X|} Ent(X_{v_i}),$$

where $\{X_{v_1}, \ldots, X_{v_{n_t}}\}$ is the partition of X defined by t.

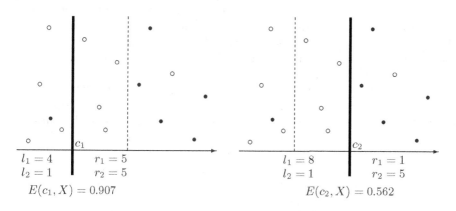

$$l_1 = 4 \qquad r_1 = 5$$
$$l_2 = 1 \qquad r_2 = 5$$
$$E(c_1, X) = 0.907$$

$$l_1 = 8 \qquad r_1 = 1$$
$$l_2 = 1 \qquad r_2 = 5$$
$$E(c_2, X) = 0.562$$

Fig. 24. Geometrical interpretation of the entropy measure. The set X consists of 15 objects, where $Ent(X) = 0.971$. The cut c_2 is preferred by entropy measure.

In the decision tree induction process, the test $t_{\min}$ that maximizes the information gain defined by

$$Gain(t, X) = Ent(X) - E(t, X)$$

or minimizes the entropy function $E(., X)$ is selected. There is a number of methods based on information entropy theory reported in [17], [31], [21], [123].

Information gain measure tends to favor those tests with larger numbers of outcomes. An obvious way to negate the bias or "greediness" of information gain is to take into account the number of values of an attribute. A new, improved calculation for test t over set of objects X is:

$$Gain\_Ratio(t, X) = \frac{Gain(t, X)}{IV(t)},$$

where

$$IV(t) = \sum (-\log_2 |X_i|/|X|)$$

Fig. 24 illustrates the entropy method on the set X containing 15 objects, where $\mathbf{Count}(X) = (9, 6)$. Comparing two cuts c_1 and c_2, one can see that c_2 is intuitively better because each set of the induced partition has an almost homogenous distribution. This observation is confirmed by the entropy measure.

7.1.2   Pruning Techniques

The overfitting is the phenomenon that a learning algorithm adapts so well to a training set, that the random disturbances in the training set are included in the model as meaningful. Consequently (as these disturbances do not reflect the underlying distribution), the performance on the test set (with its own, but definitively other, disturbances) will suffer from techniques that learn too well [130].

This is also the case in decision tree approach. We want our decision tree to generalise well, but unfortunately if we build a decision tree until all the training data has been classified perfectly and all leaf nodes are reached, there is a chance that we'll have a lot of misclassifications when we try using it. In response to the problem of overfitting nearly all modern decision tree algorithms adopt a pruning strategy of some sort.

Many algorithms use a technique known as postpruning or backward pruning. This essentially involves growing the tree from a dataset until all possible leaf nodes have been reached (i.e., purity) and then removing particular substrees (e.g., see "Reduced Error Pruning" method by Quinlan [124]). Studies have shown that post-pruning will result in smaller and more accurate trees by up to 25%. Among many pruning techniques developed and compared in several papers, it has been found that there is not much variation in terms of their performance (e.g., see Mingers [76] Esposito et al. [30]).

Various other pruning methods exist, including strategies that convert the tree to rules before pruning. Recent work is involved to incorporate some overfitting-prevention bias into the splitting part of the algorithm. One example of this is based on the minimum-description length principle [127] that states that the best hypothesis is the one that minimises length of encoding of the hypothesis and data. This has been shown to produce accurate trees with small size (e.g., see Mehta et al. [71]).

7.2   MD Algorithm

In Boolean reasoning approach to discretization, qualities of cuts were evaluated by their *discernibility properties*. In this section, we present an application of discernibility measure in induction of decision tree. This method of decision tree induction is called *the Maximal-Discernibility Algorithm*, or shortly *MD algorithm*.

MD algorithm uses discernibility measure to evaluate the quality of tests. Intuitively, a pair of objects is said to be *conflict* if they belong to different decision classes. An *internal conflict* of a set of objects $X \subset U$ is defined by the number of conflict pairs of objects from X. Let $(n_1, \ldots, n_d)$ be a counting table of X, then $conflict(X)$ can be computed by

$$conflict(X) = \sum_{i<j} n_i n_j.$$

If a test t determines a partition of a set of objects X into $X_1, X_2, \ldots, X_{n_t}$, then discernibility measure for t is defined by

$$Disc(t, X) = conflict(X) - \sum_{i=1}^{n_t} conflict(X_i). \tag{49}$$

Thus the more pairs of objects are separated by the test t the larger is the chance that t labels the root of the optimal decision tree for X.

MD algorithm is using two kinds of tests depending on attribute types. In case of symbolic attributes $a_j \in A$, *test functions defined by sets of values*, i.e.,

$$t_{a_j \in V}(u) = 1 \iff [a_j(u) \in V]$$

where $V \subset V_{a_j}$, are considered. For numeric attributes $a_i \in A$, only *test functions defined by cuts*:

$$t_{a_i > c}(u) = True \iff [a_i(u) \le c] \iff [a_i(u) \in (-\infty; c))]$$

where c is a *cut* in V_{a_i}, are considered.

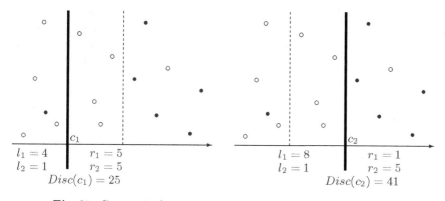

$l_1 = 4$	$r_1 = 5$	$l_1 = 8$	$r_1 = 1$
$l_2 = 1$	$r_2 = 5$	$l_2 = 1$	$r_2 = 5$
$Disc(c_1) = 25$		$Disc(c_2) = 41$	

Fig. 25. Geometrical interpretation of the discernibility measure

Usually, decision tree induction algorithms are described as recursive functions. Below we present the non-recursive version of MD algorithm (Algorithm 6), which is longer but more convenient for further consideration. During the construction, we additionally use some object sets to label nodes of the decision tree. This third kind of labels will be removed at the end of construction process.

Algorithm 6. MD algorithm for decision tree construction

---

1 **begin**
2   Initialize a decision tree **T** with one node labeled by the set of all objects U;
3   **Q** := [**T**]; // *Initialize a FIFO queue* **Q** *containing* **T**
4   **while Q** *is not empty* **do**
5          $N :=$ **Q**.$head()$; // *Get the first element of the queue*
6          $X := N.Label$;
7   **if** *the major class of X is pure enough* **then**
8              $N.Label := major\_class(X)$;
9   **else**
10             $t := ChooseBestTest(X);$ $N.Label := t$;
               // *Search for the best test of form $t_{a\in V}$ for $V \subset V_a$ with respect to $Disc(.,X)$*
11             $X_L = \{u \in X : t(u) = 0\};$ and $X_R = \{u \in X : t(u) = 1\}$;
12   **Create** two successors of the current node N_L and N_R and label them by X_L and X_R;
13   **Q**.$insert(N_L)$; **Q**.$insert(N_R)$;
               // *Insert N_L and N_R into* **Q**
14   **end**
15   **end**
16 **end**

In general, MD algorithm does not differ very much from other existing tree induction methods. However, there are some specific details, e.g., avoiding of overfitting (line 7), efficient searching for best tests (line 10) or creating soft decision tree (line 12), that distinguish this method. We will discuss those issues lately in the next sections.

7.3   Properties of the Discernibility Measure

In this section, we study the most important properties of discernibility measure that result in efficiency of the process of searching for best tests as well as in accuracy of the constructed tree.

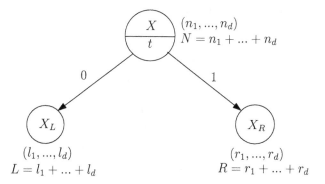

Fig. 26. The partition of the set of objects U defined by a binary test

To simplify the notation, we will use following notations for binary tests:

- d – the number of decision classes;
- $X_L = \{x \in X : t(x) = 0\}$ and $X_R = \{x \in U : t(x) = 1\}$;
- **Count**$(X) = (n_1, \ldots, n_d)$ – the counting table for X;
- **Count**$(X_L) = (l_1, \ldots, l_d)$ and **Count**$(X_R) = (r_1, \ldots, r_d)$ – the counting tables for U_L and U_R (obviously $n_j = l_j + r_j$ for $j \in \{1, \ldots, d\}$);
- $L = \sum_{j=1}^{d} l_j$, $R = \sum_{j=1}^{d} r_j$, $N = \sum_{i=1}^{d} n_j = L + R$ – total numbers of objects of X_L, X_R, X;

Fig. 26 illustrates the binary partition made by a cut on an attribute.

With those notations the discernibility measure for binary tests can be also computed as follows:

$$Disc(t, X) = conflict(X) - conflict(X_1) - conflict(X_2)$$

$$= \frac{1}{2} \sum_{i \neq j} n_i n_j - \frac{1}{2} \sum_{i \neq j} l_i l_j - \frac{1}{2} \sum_{i \neq j} r_i r_j$$

$$= \frac{1}{2} \left(N^2 - \sum_{i=1}^{d} n_i^2 \right) - \frac{1}{2} \left(L^2 - \sum_{i=1}^{d} l_i^2 \right) - \frac{1}{2} \left(R^2 - \sum_{i=1}^{d} r_i^2 \right)$$

$$= \frac{1}{2} \left(N^2 - L^2 - R^2 \right) - \frac{1}{2} \sum_{i=1}^{d} (n_i^2 - l_i^2 - r_i^2)$$

$$= \frac{1}{2} \left[(L + R)^2 - L^2 - R^2 \right] - \frac{1}{2} \sum_{i=1}^{d} [(l_i + r_i)^2 - l_i^2 - r_i^2]$$

$$= LR - \sum_{i=1}^{d} l_i r_i$$

One can show that in case of binary tests, the discernibility measure can be also computed by

$$Disc(t, X) = LR - \sum_{i=1}^{d} l_i r_i \qquad (50)$$

$$= \sum_{i=1}^{d} l_i \sum_{i=1}^{d} r_i - \sum_{i=1}^{d} l_i r_i$$

$$= \sum_{i \neq j} l_i r_j \qquad (51)$$

Thus the discernibility measure can be calculated using either Equation (50) or Equation (51). In the next sections, depending on the situation, we will use one of those forms to calculate the discernibility measure.

7.3.1   Searching for Binary Partition of Symbolic Values

Let us consider a nonnumeric (symbolic) attribute a of a given decision table $\mathbb{S}$. Let $P = (V_1, V_2)$ be a binary disjoint partition of V_a. A pair of objects (x, y) is said to be *discerned* by P if $d(x) \neq d(y)$ and either $(a(x) \in V_1)$ and $(a(y) \in V_2)$ or $(a(y) \in V_1)$ and $(a(x) \in V_2)$.

For a fixed attribute a and an object set $X \subset U$, we define the *discernibility degree* of a partition $P = (V_1, V_2)$ as follows

$$Disc_a(P|X) = Disc(t_{a \in V_1}, X)$$
$$= \left| \{(x, y) \in X^2 : x, y \text{ are discerned by } P\} \right|$$

In the MD algorithm as described above, we have considered the problem of searching for optimal binary partition with respect to discernibility. This problem, called *MD partition*, can be described as follows:

MD-PARTITION:

input: A set of objects X and an symbolic attribute a.

output: A binary partition P of V_a such that $Disc_a(P|X)$ is maximal.

We will show that the MD-PARTITION problem is NP-hard with respect to the size of V_a. The proof will suggest some natural searching heuristics for the optimal partition. We have applied those heuristics to search for best tests on symbolic attributes in the MD algorithm.

To prove the NP-hardness of the MD-PARTITION problem we consider the corresponding decision problem called the *binary partition problem* described as follows:

BINPART:
  input: A value set $V = \{v_1, \ldots, v_n\}$, two functions: $s_1, s_2 : V \to \mathbf{N}$
    and a positive integer K.
  question: Is there a binary partition of V into two disjoint subsets
    $P(V) = \{V_1, V_2\}$ such that the discernibility degree of P defined
    by
$$Disc\,(P) = \sum_{i \in V_1, j \in V_2} [s_1\,(i) \cdot s_2\,(j) + s_2\,(i) \cdot s_1\,(j)]$$

One can see that each instance of BINPART is a special case of MD-PARTITION. Indeed, let us consider a decision table with two decision classes, i.e., $V_{dec} = \{1, 2\}$. Assume that $V_a = \{v_1, \ldots, v_n\}$, we denote by $\mathbf{s}(v_i) = (s_1(v_i), s_2(v_i))$ the counting table of the set $X_{v_i} = \{x \in X : a(x) = v_i\}$. In this case, according to Equation (51), the discernibility degree of a partition P is expressed by

$$Disc\,(P|Z) = l_1 r_2 + l_2 r_1$$
$$= \sum_{v \in V_1} s_1(v) \sum_{w \in V_2} s_2(w) + \sum_{v \in V_1} s_2(v) \sum_{w \in V_2} s_1(w)$$
$$= \sum_{v \in V_1; w \in V_2} [s_1\,(v) \cdot s_2\,(w) + s_2\,(v) \cdot s_1\,(w)]$$

Thus, if BINPART problem is NP-complete, then MD-PARTITION problem is NP hard.

Theorem 19. *The binary partition problem is NP-complete.*

Proof: It is easy to see that the BINPART problem is in NP. The NP-completeness of the BINPART problem can be shown by polynomial transformation from *Set Partition Problem (SPP)*, which is defined as the problem of checking whether there is a partition of a given finite set of positive integers $S = \{n_1, n_2, \ldots, n_k\}$ into two disjoint subsets S_1 and S_2 such that $\sum_{i \in S_1} i = \sum_{j \in S_2} j$.

It is known that the SPP is NP-complete [35]. We will show that SPP is polynomially transformable to BINPART. Let $S = \{n_1, n_2, \ldots, n_k\}$ be an instance of SPP. The corresponding instance of the BINPART problem is as follows:

  − $V = \{1, 2, \ldots, k\}$;
  − $s_0\,(i) = s_1\,(i) = n_i$ for $i = 1, \ldots, k$;
  − $K = \frac{1}{2} \left(\sum_{i=1}^{k} n_i \right)^2$.

One can see that for any partition P of the set V_a into two disjoint subsets V_1 and V_2 the discernibility degree of P can be expressed by:

$$Disc\,(P) = \sum_{i \in V_1; j \in V_2} [s_0(i) \cdot s_1(j) + s_1(i) \cdot s_0(j)] =$$

$$= \sum_{i \in V_1; j \in V_2} 2n_i n_j = 2 \cdot \sum_{i \in V_1} n_i \cdot \sum_{j \in V_2} n_j =$$

$$\leq \frac{1}{2} \left(\sum_{i \in V_1} n_i + \sum_{j \in V_2} n_j \right)^2 = \frac{1}{2} \left(\sum_{i \in V} n_i \right)^2 = K,$$

i.e., for any partition P we have the inequality $Disc\,(P) \leq K$ and the equality holds if, and only if $\sum_{i \in V_1} n_i = \sum_{j \in V_2} n_j$. Hence, P is a good partition of V (into V_1 and V_2) for the BINPART problem iff it defines a good partition of S (into $S_1 = \{n_i\}_{i \in V_1}$ and $S_2 = \{n_j\}_{j \in V_2}$) for the SPP problem. Therefore, the BINPART problem is NP-complete and the MD-PARTITION problem is NP hard. ∎

Now we are going to describe some approximate solutions for MD-PARTITION problem, which can be treated as a 2-mean clustering problem over the set $V_a = \{v_1, \ldots, v_m\}$ of symbolic values, where distance between those values is defined by discernibility measure. Let $\mathbf{s}(v_i) = (n_1(v_i), n_2(v_i), \ldots, n_d(v_i))$ denote the counting table of the set $X_{v_i} = \{x \in X : a(x) = v_i\}$. The distance between two symbolic values $v, w \in V_a$ is determined as follows:

$$\delta_{disc}(v, w) = Disc(v, w) = \sum_{i \neq j} n_i(v) \cdot n_j(w)$$

One can generalize the definition of distance function by

$$\delta_{disc}(V_1, V_2) = \sum_{v \in V_1, w \in V_2} \delta_{disc}(v, w).$$

It is easy to observe that the distance function δ_{disc} is additive and symmetric, i.e.:

$$\delta_{disc}\,(V_1 \cup V_2, V_3) = \delta_{disc}\,(V_1, V_3) + \delta_{disc}\,(V_2, V_3) \tag{52}$$
$$\delta_{disc}\,(V_1, V_2) = \delta_{disc}\,(V_2, V_1) \tag{53}$$

for arbitrary sets of values V_1, V_2, V_3.

Example 18. Consider a decision table with two symbolic attributes in Fig. 27 (left). The counting tables and distance graphs between values of those attributes are presented in Fig. 27 (right).

We have proposed the following heuristics for MD-PARTITION problem:

1. The *grouping by minimizing conflict* algorithm starts with the most detailed partition $\mathbf{P}_a = \{\{v_1\}, \ldots, \{v_m\}\}$. Similarly to agglomerative hierarchical clustering algorithm, in every step the two nearest sets V_1, V_2 of $\mathbf{P}_a$ with

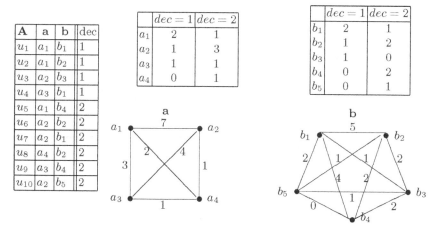

A	a	b	dec
u_1	a_1	b_1	1
u_2	a_1	b_2	1
u_3	a_2	b_3	1
u_4	a_3	b_1	1
u_5	a_1	b_4	2
u_6	a_2	b_2	2
u_7	a_2	b_1	2
u_8	a_4	b_2	2
u_9	a_3	b_4	2
u_{10}	a_2	b_5	2

	$dec = 1$	$dec = 2$
a_1	2	1
a_2	1	3
a_3	1	1
a_4	0	1

	$dec = 1$	$dec = 2$
b_1	2	1
b_2	1	2
b_3	1	0
b_4	0	2
b_5	0	1

Fig. 27. An exemplary decision table with two symbolic attributes

respect to the function $\delta_{disc}(V_1, V_2)$ is selected and replaced by their union set $V = V_1 \cup V_2$. Distances between sets in the partition $\mathbf{P}_a$ are also updated according to Eqn. (52). The algorithm repeats this step until $\mathbf{P}_a$ contains two sets only. An illustration of this algorithm is presented in Fig. 28.

2. The second technique is called *grouping by maximizing discernibility*. The algorithm also starts with a family of singletons $\mathbf{P}_a = \{\{v_1\}, \ldots, \{v_m\}\}$, but first we look for two singletons with the largest discernibility degree to create kernels of two groups; let us denote them by $V_1 = \{v_1\}$ and $V_2 = \{v_2\}$. For any symbolic value $v_i \notin V_1 \cup V_2$ we compare the distances $Disc(\{v_i\}, V_1)$ and $Disc(\{v_i\}, V_2)$ and attach it to the group with a smaller discernibility

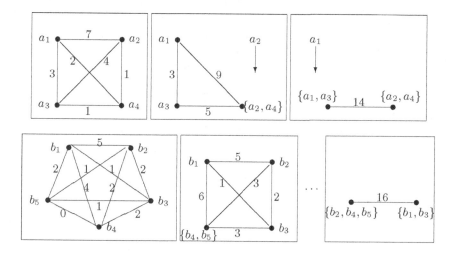

Fig. 28. Illustration of grouping by minimizing conflict algorithm

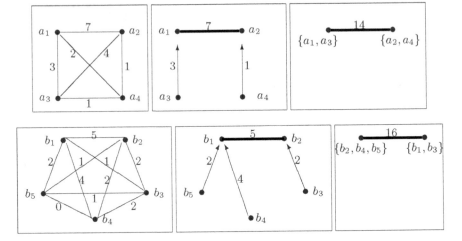

Fig. 29. Illustration of grouping by maximizing discernibility algorithm

degree for v_i. This process ends when all the values in V_a are drawn out. Fig. 29 presents an illustration of this method. For the considered example, both grouping methods give the same results on each attribute, but it is not true in general.

7.3.2   Incomplete Data

Now we consider a data table with incomplete value attributes. The problem is how to guess unknown values in a data table to guarantee maximal discernibility of objects in different decision classes.

The idea of grouping values proposed in the previous sections can be used to solve this problem. We have shown how to extract patterns from data by using discernibility of objects in different decision classes. Simultaneously, the information about values in one group can be used to guess the unknown values. Below we define the searching problem for unknown values in an incomplete decision table.

The decision table $\mathbb{S} = (U, A \cup \{d\})$ is called *"incomplete"* if attributes in A are defined as functions $a : U \rightarrow V_a \cup \{*\}$ where for any $u \in U$ by $a(u) = *$ we mean an *unknown value* of the attribute a. All values different from $*$ are called *fixed values*.

We say that a pair of objects $x, y \in U$ is *inconsistent* if $d(x) \neq d(y) \wedge \forall_{a \in A}[a(x) = * \vee a(y) = * \vee a(x) = a(y)]$. We denote by $Conflict(\mathbb{S})$ the number of inconsistent pairs of objects in the decision table $\mathbb{S}$.

The problem is to search for possible fixed values which can be substituted for the $*$ value in the table $\mathbb{S}$ in such a way that the number of conflicts $Conflict\left(\mathbb{S}'\right)$ in the new table $\mathbb{S}'$ (obtained by changing entries $*$ in table $\mathbb{S}$ into fixed values) is minimal.

The main idea is to group values in the table $\mathbb{S}$ so that discernibility of objects in different decision classes is maximized. Then we replace $*$ by a value depending on fixed values belonging to the same group.

To group attribute values we can use heuristics proposed in the previous sections. We assume that all the unknown values of attributes in A are pairwise different and different from the fixed values. Hence, we can label the unknown values by distinct indices before applying algorithms proposed in the previous sections. This assumption allows to create the discernibility matrix for an incomplete table as in the case of complete tables and we can then use the *Global Partition* method presented in Sect. 7.3 for grouping unknown values.

The function $Disc(V_1, V_2)$ can also be computed for all pairs of subsets which may contain unknown values. Hence, we can apply both heuristics of Dividing and Conquer methods for grouping unknown values.

After the grouping step, we assign to the unknown value one (or all) of the fixed values in the same group which contains the unknown one. If there is no fixed value in the group we choose an arbitrary value (or all possible values) from an attribute domain, that does not belong to other groups. If such values do not exist either, we can say that these unknown values have no influence on discernibility in a decision table and we can assign to them an arbitrary value from the domain.

7.3.3 Searching for Cuts on Numeric Attributes

In this section, we discuss some properties of the best cuts with respect to discernibility measure. Let us fix a continuous attribute a and, for simplification, we will denote the discernibility measure of a cut c on the attribute a by $Disc(c)$ instead of $Disc(a, c)$.

Let us consider two cuts $c_L < c_R$ on attribute a. The following formula shows how to compute the difference between the discernibility measures of c_L and c_R using information about class distribution in intervals defined by these cuts.

Lemma 2. *The following equation holds:*

$$Disc(c_R) - Disc(c_L) = \sum_{i=1}^{d} \left[(R_i - L_i) \sum_{j \neq i} M_j \right] \tag{54}$$

where $(L_1, \ldots, L_d)$, $(M_1, \ldots, M_d)$ *and* $(R_1, \ldots, R_d)$ *are the counting tables of intervals* $(-\infty; c_L)$, $[c_L; c_R)$ *and* $[c_R; \infty)$, *respectively (see Fig. 30).*

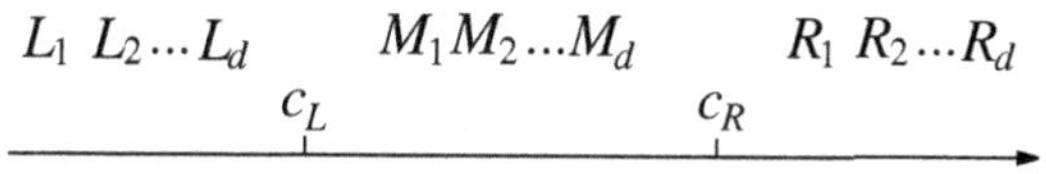

$$L_1\ L_2 \ldots L_d \qquad M_1 M_2 \ldots M_d \qquad R_1\ R_2 \ldots R_d$$

$$\underset{c_L}{\underline{\qquad\qquad\qquad\qquad\qquad\qquad\qquad}}\underset{c_R}{\qquad\qquad}\longrightarrow$$

Fig. 30. The counting tables defined by cuts c_L, c_R

Proof: According to Eqn. (50) we have

$$Disc(c_L) = \sum_{i=1}^{d} L_i \sum_{i=1}^{d}(M_i + R_i) - \sum_{i=1}^{d} L_i(M_i + R_i)$$

$$= \sum_{i=1}^{d} L_i \sum_{i=1}^{d} M_i + \sum_{i=1}^{d} L_i \sum_{i=1}^{d} R_i - \sum_{i=1}^{d} L_i(M_i + R_i)$$

Analogously

$$Disc(c_R) = \sum_{i=1}^{d}(L_i + M_i) \sum_{i=1}^{d} R_i - \sum_{i=1}^{d}(L_i + M_i)R_i$$

$$= \sum_{i=1}^{d} L_i \sum_{i=1}^{d} R_i + \sum_{i=1}^{d} M_i \sum_{i=1}^{d} R_i - \sum_{i=1}^{d} R_i(M_i + L_i)$$

Hence,

$$Disc(c_R) - Disc(c_L) = \sum_{i=1}^{d} M_i \sum_{i=1}^{d}(R_i - L_i) - \sum_{i=1}^{d} M_i(R_i - L_i)$$

$$= \sum_{i,j=1}^{d} M_i(R_j - L_j) - \sum_{i=1}^{d} M_i(R_i - L_i)$$

Simplifying the last formula we obtain (54). ∎

7.3.4 Boundary Cuts

Let $C_a = \{c_1, \ldots, c_N\}$ be a set of consecutive candidate cuts on attribute a such that $c_1 < c_2 < \cdots < c_N$.

Definition 23. *The cut $c_i \in C_a$, where $1 < i < N$, is called the* **boundary cut** *if there exist at least two such objects $u_1, u_2 \in U$ that $a(u_1) \in [c_{i-1}, c_i)$, $a(u_2) \in [c_i, c_{i+1})$ and $dec(u_1) \neq dec(u_2)$.*

The notion of *boundary cut* has been introduced by Fayyad et al. [31], who showed that best cuts with respect to entropy measure can be found among boundary cuts. We will show the similar result for discernibility measure, i.e., it is enough to restrict the search to the set of boundary cuts.

Theorem 20. *The cut c_{Best} maximizing the function $Disc(a, c)$ can be found among boundary cuts.*

Proof: Assume that c_a and c_b are consecutive boundary cuts. Then the interval $[c_a, c_b)$ consists of objects from one decision class, say $CLASS_i$. Let $(L_1, \ldots, L_d)$ and $(R_1, \ldots, R_d)$ are the counting tables of intervals $(-\infty; c_a)$ and $[c_a; \infty)$, respectively

For arbitrary cuts c_L and c_R such that $c_a \leq c_L < c_R \leq c_b$. According to the notation in Fig. 30, we have $M_i \neq 0$ and $\forall_{j \neq i} M_j = 0$. Then (54) has a form

$$Disc(c_R) - Disc(c_L) = M_i \sum_{j \neq i}(R_j - L_j).$$

Thus, function $Disc(c)$ is monotone within the interval $[c_a, c_b)$ because $\sum_{j \neq i}(R_j - L_j)$ is constant for all sub intervals of $[c_a, c_b)$. More precisely, for any cut $c \in [c_a, c_b)$

$$Disc(c) = Disc(c_a) + A \cdot x$$

where $A = \sum_{j \neq i}(R_j - L_j)$ and $x > 0$ is the number of objects lying between c_a and c. ∎

Theorem 20 makes it possible to look for optimal cuts among boundary cuts only. *This fact allows us to save time and space in the MD-heuristic because one can remove all non-boundary points from the set of candidate cuts.*

7.4   The Properties of MD Decision Trees

The decision trees, which are built by MD-heuristics (using discernibility measures), are called *MD decision trees*. In this section, we study some properties of MD decision tress.

A real number $v_i \in a(U)$ is called *single value* of an attribute a if there is exactly one object $u \in U$ such that $a(u) = v_i$. The cut $(a; c) \in \mathbf{Cut}_{\mathbb{S}}(a)$ is called *the single cut* if c is lying between two single values v_i and v_{i+1}. We have the following theorem related to single cuts:

Theorem 21. *In case of decision tables with two decision classes, any single cut c_i, which is a local maximum of the function $Disc$, resolves more than a half of conflicts in the decision table, i.e.*

$$Disc(c_i) \geq \frac{1}{2} \cdot conflict(\mathbb{S}).$$

Proof: Let c_{i-1} and c_{i+1} be the nearest neighboring cuts to c_i (from the left and the right hand sides). The cut c_i is a local maximum of the function W if and only if

$$Disc(c_i) > \max\{Disc(c_{i-1}), Disc(c_{i+1})\}.$$

Because c_i is a single cut, one can assume that there are only two objects u and v such that

$$a(u) \in (c_{i-1}; c_i) \quad \text{and} \quad a(v) \in (c_i; c_{i+1}).$$

Theorem 20 allows us to conclude that c_i is a boundary cut. One can assume, without loss of generality, $u \in CLASS_1$ and $v \in CLASS_2$.

Let (L_1, L_2) and (R_1, R_2) be counting tables of intervals $(-\infty; c_i)$ and $(c_i; \infty)$. We have

$$Disc(c_i) = L_1 R_2 + L_2 R_1.$$

From definition of *conflict* measure, we have

$$con flict(U) = (L_1 + R_1)(L_2 + R_2).$$

Because $a(u) \in (c_{i-1}; c_i)$ and $u \in CLASS_1$, hence after applying Lemma 2 we have

$$Disc(c_i) - Disc(c_{i-1}) = R_2 - L_2.$$

Similarly, we have

$$Disc(c_{i+1}) - Disc(c_i) = R_1 - L_1;$$

Then we have the following inequality:

$$(R_1 - L_1)(R_2 - L_2) = [Disc(c_i) - Disc(c_{i-1})][Disc(c_{i+1}) - Disc(c_i)] \leq 0.$$

Thus

$$(R_1 - L_1)(R_2 - L_2) = R_1 R_2 + L_1 L_2 - L_1 R_2 - R_1 L_2 \leq 0$$
$$\Leftrightarrow R_1 R_2 + L_1 L_2 + L_1 R_2 + R_1 L_2 \leq 2(L_1 R_2 + R_1 L_2)$$
$$\Leftrightarrow (L_1 + R_1)(L_2 + R_2) \leq 2(L_1 R_2 + R_1 L_2)$$
$$\Leftrightarrow conflict(\mathbb{S}) \leq 2W(c_i)$$

Therefore, $Disc(c_i) \geq \frac{1}{2} \cdot conflict(\mathbb{S})$, what completes the proof. ∎

The single cuts defining local maxima w.r.t. the discernibility measure can be found when (for example) the feature $a : U \rightarrow \mathbb{R}$ is a "$1 - 1$" mapping. If the original attributes of a given decision table are not "$1 - 1$" mapping, we can try to create new features by taking linear combination of the existing attributes. The cuts on new attributes being linear combinations of the existing ones are called hyperplanes.

This fact will be useful in the proof of an upper-bound on the height of decision tree in the following part.

Our heuristic method aims to minimize the conflict function using possibly small number of cuts. The main subject of decision tree algorithms is to minimize the number of leaves (or rules) in decision tree. In this section we will show that the height of the decision tree generated by MD algorithm is quite small.

Theorem 22. *Let a cut c_i satisfy conditions of Theorem 21 and let c_i divide $\mathbb{S}$ into two decision tables $\mathbb{S}_1 = (U_1, A \cup \{d\})$ and $\mathbb{S}_2 = (U_2, A \cup \{d\})$ such that*

$$U_1 = \{u \in U : a(u) < c_i\} \ and \ U_2 = \{u \in U : a(u) > c_i\},$$

then $con flict(\mathbb{S}_1) + con flict(\mathbb{S}_2) \leq \frac{1}{2}conflict(\mathbb{S})$.

Proof: This fact is obtained directly from Theorem 21 and the observation that

$$con flict(\mathbb{S}_1) + con flict(\mathbb{S}_2) + Disc(c_i) = con flict(\mathbb{S})$$

for any cut c_i. ∎

Theorem 23. *In case of decision table with two decision classes and n objects, the height of the MD decision tree using hyperplanes is not larger than $2\log n - 1$.*

Proof: Let $conflict(h)$ be a sum of $conflict(N_h)$ for all nodes N_h on the level h. From Theorem 22 we have

$$conflict(h) \geq 2conflict(h+1).$$

Let n be the number of objects in given decision table and n_1, n_2 the numbers of decision classes (we assumed that there are only two decision classes). From Proposition 2 we can evaluate the $conflict$ of the root of generated decision tree by:

$$conflict(0) = conflict(\mathbb{A}) = n_1 n_2 \leq \left(\frac{n_1 + n_2}{2}\right)^2 = \frac{n^2}{4}.$$

Let $h(\mathbf{T})$ be the height of the decision tree $\mathbf{T}$, we have $conflict(h(\mathbf{T})) = 0$. Therefore:

$$conflict(0) \geq 2^{h(\mathbf{T})-1} \Rightarrow$$

$$h(\mathbf{T}) \leq \log_2(conflict(0)) + 1 \leq \log_2\left(\frac{n^2}{4}\right) + 1 = 2\log n - 1$$

∎

Let us assume that any internal node N of the constructed decision tree satisfies the condition:

$$conflict(N_L) = conflict(N_R)$$

or more generally:

$$\max(conflict(N_L), conflict(N_R)) \leq \frac{1}{4}conflict(N),$$

where N_L, N_R are left and right sons of N. In a similar way we can prove that

$$h(\mathbf{T}) \leq \log_4(conflict(\mathbb{A})) + 1 \leq \log_4\left(\frac{n^2}{4}\right) + 1 = \log n.$$

7.5   The Accuracy of MD-Algorithm

Experiments for classification methods have been carried over decision tables using two techniques called "train-and-test" and "n-fold-cross-validation". In Table 17 we present some experimental results obtained by testing the proposed methods for classification quality on well-known data tables from the "UC Irvine Machine Learning repository"[6] and execution times. Similar results obtained by alternative methods are reported in [33]. It is interesting to compare those results with regard to both classification quality and execution time.

---

[6] http://www1.ics.uci.edu/~mlearn/MLRepository.html

Table 17. The quality comparison between decision tree methods. MD: MD-heuristics; MD-G: MD-heuristics with symbolic value partition.

Names of	Classification accuracies			
Tables	S-ID3	C4.5	MD	MD-G
Australian	78.26	85.36	83.69	84.49
Breast (L)	62.07	71.00	69.95	69.95
Diabetes	66.23	70.84	71.09	76.17
Glass	62.79	65.89	66.41	69.79
Heart	77.78	77.04	77.04	81.11
Iris	96.67	94.67	95.33	96.67
Lympho	73.33	77.01	71.93	82.02
Monk-1	81.25	75.70	100	93.05
Monk-2	69.91	65.00	99.07	99.07
Monk-3	90.28	97.20	93.51	94.00
Soybean	100	95.56	100	100
TicTacToe	84.38	84.02	97.7	97.70
Average	78.58	79.94	85.48	87.00

7.6   Bibliographical Notes

The MD-algorithm for decision tree (i.e., using discernibility measure to construct decision tree from decision table) and properties of such decision trees was described in [92]. The idea of symbolic value grouping was presented in [103].

7.6.1   Other Heuristic Measures

In next sections we recall some other well-known measures for decision tree induction. To simplify the notation, we will consider only binary tests with values from $\{0, 1\}$. In this case, we will use the same notations as in Sect. 7.3 (page 434).

1. **Statistical test:** Statistical tests are applied to check the probabilistic independence between the object partition defined by a test t. The independence degree is estimated by the χ^2 test given by

$$\chi^2(t, X) = \sum_{j=1}^{d} \frac{(l_j - E(X_{L,j}))^2}{E(X_{L,j})} + \sum_{j=1}^{d} \frac{(r_j - E(X_{R,j}))^2}{E(X_{R,j})},$$

where $E(X_{L,j}) = L \cdot \frac{n_j}{N}$ and $E(X_{R,j}) = R \cdot \frac{n_j}{N}$ are the expected numbers of objects from j^{th} class which belong to X_L and and X_R, respectively.

   Intuitively, if the partition defined by t does not depend on the partition defined by the decision attribute dec then one can expect that counting tables for X_L and X_R are proportional to the counting table of X, that is

$$(l_1, \ldots, l_d) \simeq \left(L \frac{n_1}{N}, \ldots, L \frac{n_d}{N} \right) \text{ and } (r_1, \ldots, r_d) \simeq \left(R \frac{n_1}{N}, \ldots, R \frac{n_d}{N} \right),$$

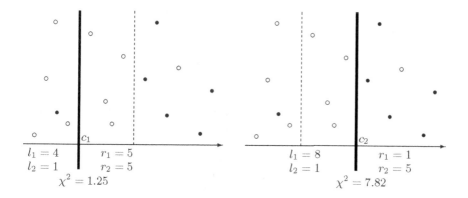

Fig. 31. Geometrical interpretation of χ^2 method

thus we have $\chi^2(c) = 0$. In the opposite case if the test t properly separates objects from different decision classes, the value of χ^2 test for t is maximal. Fig. 31 illustrates the χ^2 method on the set X containing 15 objects, where **Count**$(X) = (9, 6)$. Comparing two cuts c_1 and c_2, one can see that the more counting tables of X_L and X_R from the counting table of X differ, the larger is the value of χ^2 test.

2. **Gini's index:**

$$Gini(t, X) = \frac{L}{N}\left(1 - \sum \frac{l_i^2}{L^2}\right) + \frac{R}{N}\left(1 - \sum \frac{r_i^2}{R^2}\right)$$

3. **Sum-Minority:**

$$Sum\_Minority\,(t, X) = \min_{i=1,\ldots,d}\{l_i\} + \min_{i=1,\ldots,d}\{r_i\}$$

4. **Max-Minority:**

$$Max\_Minority\,(t, X) = \max\left\{\min_{i=1,\ldots,d}\{l_i\},\ \min_{i=1,\ldots,d}\{r_i\}\right\}$$

5. **Sum-Impurity:**

$$Sum\_impurity\,(t, X) = \sum_{i=1}^{d} l_i \cdot (i - avg_L)^2 + \sum_{i=1}^{d} r_i \cdot (i - avg_R)^2$$

where

$$avg_L = \frac{\sum_{i=1}^{d} i \cdot l_i}{L} \text{ and } avg_R = \frac{\sum_{i=1}^{d} i \cdot r_i}{R}$$

are averages of decision values of objects of the left set and the right set of the partition defined by the test t (respectively). Usually, this measure is applied for decision tables with two decision classes. In fact, Sum-Impurity is a sum of variations of both sides of t and it is minimal if t separates the set of objects correctly.

8   Approximate Boolean Reasoning Approach to Feature Extraction Problem

We have presented so far some applications of rough sets and ABR in many issues of data mining like feature selection, rule generation, discretization, and decision tree generation. Discretization of numeric attributes can be treated not only as the data reduction process (in which some of original attributes are replaced by the discretized ones), but also as the feature extraction method since it defines a new set of attributes. In this section, we consider some extensions of discretization methods in the mean of feature extraction problem. Particularly, we consider the problem of searching for new features defined either by linear combinations of attributes (hyperplanes) or by sets of symbolic values.

8.1   Grouping of Symbolic Values

We have considered the real value attribute discretization problem as a problem of searching for a partition of real values into *intervals*. The efficiency of discretization algorithms is based on the existence of the natural linear order "$<$" in the real axis $\mathbb{R}$. In case of symbolic value attributes (i.e., without any pre-assumed order in the value sets of attributes) the problem of searching for partitions of value sets into a "small" number of subsets is more complicated than for continuous attributes.

Once again, we will apply the Boolean reasoning approach to construct a partition of symbolic value sets into small number of subsets.

Let us consider a decision table $\mathbb{S} = (U, A \cup \{d\})$. By *grouping of symbolic values* from the domain V_{a_i} of an attribute $a_i \in A$ we mean an arbitrary mapping $P : V_{a_i} \to \{1, \ldots, m_i\}$. Two values $x, y \in V_{a_i}$ are in the same group if $P(x) = P(y)$. One can see that the notion of partition of attribute domain is a generalized concept of discretization and it can be used for both continuous and symbolic attributes. Intuitively, the mapping $P : V_{a_i} \to \{1, \ldots, m_i\}$ defines a partition of V_{a_i} into disjoint subsets of values as follows:

$$V_{a_i} = V_1(P) \cup \cdots \cup V_{m_i}(P),$$

where $V_j(P) = \{v \in V_{a_i} : P(v) = j\}$.

Thus any grouping of symbolic values $P : V_{a_i} \to \{1, \ldots, m_i\}$ defines a new attribute $a_i|_P = P \circ a_i : U \to \{1, \ldots, m_i\}$ where

$$a_i|_P (u) = P (a_i (u))$$

for any object $u \in U$. By *rank of a partition* P on a_i we denote the number of non-empty subsets occurring in its partition, i.e.,

$$rank (P) = |P (V_{a_i})|$$

Similarly to the discretization problem, grouping of symbolic values can reduce some superfluous data but it is also associated with a loss of some significant information. We are interested in those groupings which guarantee the high quality of classification.

Let $B \subset A$ be an arbitrary subset of attributes. A family of partitions $\{P_a\}_{a \in B}$ on B is called $B-consistent$ if and only if it maintains the discernibility relation $DISC(B, d)$ between objects, i.e., for any $u, v \in U$,

$$[d(u) \neq d(v) \wedge inf_B(u) \neq inf_B(v)] \implies \exists_{a \in B} [P_a(a(u)) \neq P_a(a(v))] \quad (55)$$

We consider the following optimization problem called *the symbolic value partition problem:*

SYMBOLIC VALUE PARTITION PROBLEM:
  input: a given decision table $\mathbb{S} = (U, A \cup \{d\})$, and a set $B \subseteq A$ of nominal attributes in $\mathbb{S}$.
  output: minimal $B-consistent$ family of partitions (i.e., B-consistent family $\{P_a\}_{a \in B}$ with the minimal value of $\sum_{a \in B} rank(P_a)$).

This concept is useful when we want to reduce attribute domains with large cardinalities. The discretization problem can be derived from the partition problem by adding the monotonicity condition for the family $\{P_a\}_{a \in A}$ such that

$$\forall_{v_1, v_2 \in V_a} (v_1 \leq v_2) \Rightarrow (P_a(v_1) \leq P_a(v_2)).$$

In the next sections, we present three solutions for this problem, namely the *local partition method,* the *global partition method* and the *"divide and conquer"* method. The first approach is based on grouping the values of each attribute independently whereas the second approach is based on grouping of attribute values simultaneously for all attributes. The third method is similar to the decision tree techniques: the original data table is divided into two subtables by selecting the *"best binary partition of some attribute domain"* and this process is continued for all subtables until some stop criterion is satisfied.

8.1.1   Local Partition

The local partition strategy is very simple. For any fixed attribute $a \in A$, we search for a partition P_a that preserves the consistency condition (55) for the attribute a (i.e., $B = \{a\}$).

For any partition P_a the equivalence relation $\approx_{P_a}$ is defined by:

$$v_1 \approx_{P_a} v_2 \Leftrightarrow P_a(v_1) = P_a(v_2)$$

for all $v_1, v_2 \in V_a$. We consider the relation $\mathbf{UNI}_a$ defined on V_a as follows:

$$v_1 \mathbf{UNI}_a v_2 \Leftrightarrow \forall_{u, u' \in U} (a(u) = v_1 \wedge a(u') = v_2) \Rightarrow d(u) = d(u'). \quad (56)$$

It is obvious that the relation $\mathbf{UNI}_a$ defined by Eqn. (56) is an equivalence relation. One can show [103] that the equivalence relation $\mathbf{UNI}_a$ defines a minimal $a-consistent$ partition on a, i.e., if P_a is a-consistent then $\approx_{P_a} \subseteq \mathbf{UNI}_a$.

8.1.2 Divide and Conquer Approach to Partition

A partition of symbolic values can be also obtained from MD-decision tree algorithm (see Sect. 7). Assume that $\mathbf{T}$ is the decision tree constructed by MD-decision tree method for decision table $\mathbb{S} = (U, A \cup \{d\})$.

For any symbolic attribute $a \in A$, let $P_1, P_2, \ldots, P_k$ be the binary partitions on V_a which are presented in $\mathbf{T}$. The partition P_a of symbolic values on V_a can be defined as follows:

$$P_a(v) = P_a(v') \Leftrightarrow \forall_i P_i(v) = P_i(v')$$

This method has been implemented in RSES system[7].

8.1.3 Global Partition Method Based on ABR

In this section, we present the ABR approach to the symbolic value partition problem. Let us describe the basic steps of this solution:

Problem modeling: We can encode the problem as follows:

Let us consider the discernibility matrix $\mathbf{M}(\mathbb{S}) = [m_{i,j}]_{i,j=1}^n$ (see [143]) of the decision table $\mathbb{S}$, where $m_{i,j} = \{a \in A : a(u_i) \neq a(u_j)\}$ is the set of attributes discerning two objects u_i, u_j. Observe that if we want to discern an object u_i from another object u_j we need to preserve one of the attributes in $m_{i,j}$. To put it more precisely: *for any two objects u_i, u_j there exists an attribute $a \in m_{i,j}$ such that the values $a(u_i), a(u_j)$ are discerned by P_a.*

Hence, instead of cuts as in the case of continuous values (defined by pairs (a_i, c_j)), we consider boolean variables corresponding to triples (a_i, v, v') called *constraints*, where $a_i \in A$ for $i = 1, \ldots, k$ and $v, v' \in V_{a_i}$. Obviously two tripes (a_i, v, v') and (a_i, v', v) represents the same constraint and are treated as identical.

The Boolean function that encodes this problem is constructed as follows:

$$f_{\mathbb{S}} = \prod_{\substack{u_i, u_j \in U : \\ dec(u_i) \neq dec(u_j)}} \psi_{i,j}, \tag{57}$$

where

$$\psi_{i,j} = \sum_{a \in A} (a, a(u_i), a(u_j)).$$

Development: Searching for prime implicants;

We can build a new decision table $\mathbb{S}^+ = (U^+, A^+ \cup \{d^+\})$ assuming $U^+ = U^*; d^+ = d^*$ and $A^+ = \{(a, v_1, v_2) : (a \in A) \wedge (v_1, v_2 \in V_a)\}$. Once again, the greedy heuristic can be applied to A^+ to search for a minimal set of constraints discerning all pairs of objects in different decision classes.

---

[7] Rough Set Exploration System: http://logic.mimuw.edu.pl/~rses/

Reasoning: Unlike previous applications of Boolean reasoning approach, it is not trivial to decode the result of the previous step to obtain a direct solution for the symbolic partition problem. The minimal (or semi-minimal) prime implicant of a Boolean function f_S (Eqn. (57)) describes the minimal set of constraints for the target partition. Thus the problem is how to convert the minimal set of constraints into a low rank partition.

Let us notice that our problem can be solved by efficient heuristics of "*graph $k-$colorability*" problem which is formulated as a problem of checking whether, for a given graph $G = (V, E)$ and an integer k, there exists a function $f : V \rightarrow \{1, \ldots, k\}$ such that $f(v) \neq f(v')$ whenever $(v, v') \in E$.

This graph $k-$colorability problem is solvable in polynomial time for $k = 2$, but is NP-complete for any $k \geq 3$. However, similarly to the discretization problem, some efficient heuristic searching for the optimal graph coloring determining optimal partitions of attribute value sets can be applied.

For any attribute a_i in a semi-minimal set X of constraints returned from the above heuristic we construct a graph $\Gamma_{a_i} = \langle V_{a_i}, E_{a_i} \rangle$, where E_{a_i} is the set of all constraints of the attribute a_i in X. Any coloring of all the graphs Γ_{a_i} defines an A-consistent partition of value sets. Hence, heuristic searching for minimal graph coloring returns also sub-optimal partitions of attribute value sets.

The corresponding boolean formula has $O(knl^2)$ variables and $O(n^2)$ clauses, where l is the maximal value of $card(V_a)$ for $a \in A$. When prime implicants of boolean formula have been constructed, a heuristic for graph coloring should be applied to generate new features.

Example 19. Let us consider the decision table presented in Fig. 32 and the reduced form of its discernibility matrix.

Firstly, we have to find a shortest prime implicant of the Boolean function f_S with boolean variables of the form $a_{v_1}^{v_2}$ (corresponding to the constraints (a, v_1, v_2)). For the considered example, the minimal prime implicant encodes the following set of constraints:

$$\{\mathbf{a}_{a_2}^{a_1}, \mathbf{a}_{a_3}^{a_2}, \mathbf{a}_{a_4}^{a_1}, \mathbf{a}_{a_4}^{a_3}, \mathbf{b}_{a_4}^{a_1}, \mathbf{b}_{a_4}^{a_2}, \mathbf{b}_{a_3}^{a_2}, \mathbf{b}_{a_3}^{a_1}, \mathbf{b}_{a_5}^{a_3}\}$$

and it is represented by graphs of constraints for each attribute (Fig. 32).

Next we apply a heuristic to color vertices of those graphs as it is shown in Fig. 32. The colors are corresponding to the partitions:

$$P_{\mathbf{a}}(a_1) = P_{\mathbf{a}}(a_3) = 1; \quad P_{\mathbf{a}}(a_2) = P_{\mathbf{a}}(a_4) = 2;$$
$$P_{\mathbf{b}}(b_1) = P_{\mathbf{b}}(b_2) = P_{\mathbf{b}}(b_5) = 1; P_{\mathbf{b}}(b_3) = P_{\mathbf{b}}(b_4) = 2$$

and at the same time one can construct the new decision table (see Fig. 32). The following set of decision rules can be derived from the table $\mathbb{S}^P$:

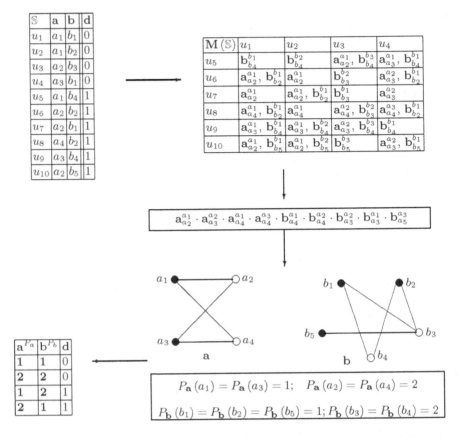

Fig. 32. The decision table and the corresponding discernibility matrix. Coloring of attribute value graphs and the reduced table.

if $a(u) \in \{a_1, a_3\}$ and $b(u) \in \{b_1, b_2, b_5\}$ then $d = 0$

                                                              (supported by u_1, u_2, u_4)

if $a(u) \in \{a_2, a_4\}$ and $b(u) \in \{b_3, b_4\}$ then $d = 0$

                                                             (supported by u_3)

if $a(u) \in \{a_1, a_3\}$ and $b(u) \in \{b_3, b_4\}$ then $d = 1$

                                                             (supported by u_5, u_9)

if $a(u) \in \{a_2, a_4\}$ and $b(u) \in \{b_1, b_2, b_5\}$ then $d = 1$

                                                           (supported by u_6, u_7, u_8, u_{10})

8.2 Searching for New Features Defined by Oblique Hyperplanes

In Sect. 6, we have introduced the optimal discretization problem as the problem of searching for the minimal set of cuts. Every cut (a, c) on an attribute a can be interpreted as a linear $(k-1)$-dimensional surface that divides the affine space $\mathbb{R}^k$ into two half-spaces.

In this section, we consider the problem of searching for the optimal set of oblique hyperplanes which is a generalization of the problem of searching for the minimal set of cuts.

Let $\mathbb{S} = (U, A \cup \{dec\})$ be a decision table where $U = \{u_1, \ldots, u_n\}$, $A = \{a_1, \ldots, a_k\}$, $a_i : U \to \mathbb{R}$ is a real function from universe U for any $i \in \{1, \ldots, k\}$ and $d : U \to \{1, \ldots, r\}$ is a *decision*.

Any set of objects described by real value attributes $a_1, \ldots, a_k \in A$ can be treated as a set of points in k-dimensional real *affine space* $\mathbb{R}^k$. In fact, the object $u_i \in U$ is represented by the point

$$P_i = (a_1(u_i), a_2(u_i), \ldots, a_k(u_i)) \text{ for } i \in \{1, 2, \ldots, n\}.$$

Any hyperplane can be defined as a set of points by the linear equation

$$H = \left\{ \mathbf{x} \in \mathbb{R}^k : \mathcal{L}(\mathbf{x}) = 0 \right\},$$

where $\mathcal{L} : \mathbb{R}^k \to \mathbb{R}$ is a given linear function defined by

$$\mathcal{L}(x_1, \ldots, x_k) = \sum_{i=1}^{k} \alpha_i \cdot x_i + \alpha_0.$$

Any hyperplane $\mathbf{H}$ defined by a linear function $\mathcal{L}$ divides the space $\mathbb{R}^k$ into *the left half-space* $\mathbf{H}^L$ and *the right half-space* $\mathbf{H}^R$ of $\mathbf{H}$ by

$$\mathbf{H}^L = \left\{ \mathbf{x} \in \mathbb{R}^k : \mathcal{L}(\mathbf{x}) < 0 \right\} \quad \text{and}$$
$$\mathbf{H}^R = \left\{ \mathbf{x} \in \mathbb{R}^k : \mathcal{L}(\mathbf{x}) > 0 \right\}.$$

We say that the hyperplane $\mathbf{H}$ discerns a pair of objects u_i, u_j if and only if the points P_i, P_j corresponding to u_i, u_j, respectively, are in different half-spaces of the hyperplane $\mathbf{H}$. This condition is expressed by:

$$\mathcal{L}(u_i) \cdot \mathcal{L}(u_j) < 0.$$

Any hyperplane $\mathbf{H}$ defines a new feature (attribute) $a_{\mathbf{H}} : U \to \{0, 1\}$ by

$$a_{\mathbf{H}}(u) = \begin{cases} 0 \text{ if } L(u) < 0 \\ 1 \text{ if } L(u) \geq 0 \end{cases}$$

($a_{\mathbf{H}}$ is the characteristic function of the right half-space $\mathbf{H}^R$).

The discretization concept defined by cuts (attribute-value pairs) can be generalized by using oblique hyperplanes. In fact, normal cuts are the special hyperplanes which are orthogonal (parallel) to axes. A set of hyperplanes $\mathcal{H} = \{\mathbf{H}_1, \ldots, \mathbf{H}_m\}$ is said to be compatible with a given decision table $\mathbb{S} = (U, A \cup \{dec\})$ if and only if for any pair of objects $u_i, u_j \in U$ such that $dec(u_i) \neq$

$dec(u_j)$ there exists a hyperplane $\mathbf{H} \in \mathcal{H}$ discerning u_i and u_j whenever $inf_A(u_i) \neq inf_A(u_j)$

In this section, we consider the problem of searching for minimal compatible set of hyperplanes.

Similarly to the problems of optimal discretization and optimal symbolic value partition, this problem can be solved by the Boolean reasoning approach. The idea is as following:

- **boolean variables:** each candidate hyperplane $\mathbf{H}$ is associated with a boolean variable $v_{\mathbf{H}}$.
- **Encoding Boolean function:**

$$f = \prod_{\substack{u_i, u_j \in U : \\ dec(u_i) \neq dec(u_j)}} \sum_{\mathbf{H} \text{ discerns } u_i, u_j} v_{\mathbf{H}}$$

All searching strategies as well as heuristic measures for the optimal discretization problem can be applied to the corresponding problem for hyperplane. Unfortunately, the problem of searching for the best hyperplane with respect to a given heuristic measure usually shows to be very hard. The main difficulty is caused by the large number, i.e., $O\left(n^k\right)$, of possible candidate hyperlanes. For example, Heath [47] has shown that the problem of searching for the hyperplane with minimal energy with respect to Sum-Minority measure is NP-hard.

8.2.1 Hyperplane Searching Methods

Usually, because of the high complexity, the local search strategy – using decision tree as a data structure – is employed to extract a set of hyperplanes from data. In this section, we put a special emphasis on the problem of searching for the best single hyperplane.

Let us mention three approximate solutions of this problem: the *simulated annealing based method* [47], *OC1 method* [78] and *genetic algorithm based method* [82].

Simulated annealing based method: Heath et al. [47] have presented an interesting technique by applying the notion of *annealing process*[8] in material sciences. In previous section, we presented some heuristic measures for partitions of object set defined by cuts or symbolic values. We can use one of those measures to define the energy of hyperplanes in such a way that the energy of the optimal hyperpane is minimal.

The simulated annealing algorithm starts with randomly initial hyperplane, because the choice of the first hyperplane is not important for this method. In particular, one can choose the hyperplane passing through the points where

---

[8] **anneal:** *to make (as glass or steel) less brittle by subjecting to heat and then cooling.* (according to Webster Dictionary).

$x_i = 1$ and all other $x_j = 0$, for each dimension i. This hyperplane is defined by the linear equation:

$$x_1 + x_2 + \cdots + x_k - 1 = 0,$$

i.e., $\alpha_i = 1$ for $i = 1, \ldots, k$ and $\alpha_0 = -1$.

Next, the perturbation process of the hyperplane H is repeated until some stop criteria hold. The perturbation algorithm is based on random picking of one coefficient α_i and adding to it a uniformly chosen random variable in the range $[-0.5, 0.5)$. The energy of the new hyperplane and the change of energy ΔE should be computed.

If $\Delta E < 0$, then the energy has decreased and the new hyperplane becomes the current hyperplane. In general, the probability of replacing the current hyperplane by the new hyperplane is defined by

$$P = \begin{cases} 1 & \text{if } \Delta E < 0 \\ e^{\frac{-\Delta E}{T}} & \text{otherwise,} \end{cases}$$

where T is a *temperature of the system* (in practice the temperature of the system can be given by any decreasing function with respect to the number of iterations). Once the probability described above is larger than some threshold we replace the current hyperplane by the new hyperplane.

The process will be continued until keeping the hyperplane with the lowest energy seen so far at the current state (i.e., if the energy of the system does not change for a number of iterations).

OC1 method: Murthy et al. [78] proposed another method called OC1 to search for hyperplanes. This method combines the Heath's randomize strategy with the decision tree method proposed by Breiman et al. [15]. This method also starts with an arbitrary hyperplane H defined by linear function

$$\mathcal{L}(x_1, \ldots, x_k) = \sum_{i=1}^{k} \alpha_i \cdot x_i + \alpha_0$$

and next it perturbs the coefficients of H one at a time.

If we consider the coefficient α_m as a variable, and all other coefficients as constants then we can define a linear projection p_m of any object u_j onto the the real axis as follows:

$$p_m(u_j) = \frac{\alpha_m a_m(u_j) - \mathcal{L}(u_j)}{a_m(u_j)}$$

(the function p_m does not depend on coefficient α_m). One can note that the object u_j is above H if $\alpha_m > p_m(u_i)$, and below otherwise. By fixing the values of all other coefficients we can obtain n constraints on the value of α_m defined by $p_m(u_1), p_m(u_2), \ldots, p_m(u_n)$ (assuming no degeneracies). Let α_m^* be the best univariate split point (with respect to the impurity measure) of those constraints. One can obtain a new hyperplane by changing α_m to α_m^*.

Murthy at al [78] proposed different strategies of deciding the order of coefficient perturbation, but he has observed that the perturbation algorithm stops when the hyperplane reaches a local minimum. In such situation OC1 tries to jump out of local minima by using some randomization strategies.

8.2.2   Genetic Algorithm Based Method for Hyperplanes

A general method of searching for the *optimal set of hyperplanes* with respect to an arbitrary measure was proposed in [93], [82]. This method was based on evolution strategy and the main problem was related to chromosome representation of hyperplanes. The representation scheme should be efficient, i.e., it should represent different hyperplanes using as small number of bits as possible. Moreover, the complexity of the fitness function should be taken into account.

Algorithm 7. Searching for hyperplanes

---

1   **begin**
2 | Initialize a new table $\mathbf{B} = (U, B \cup \{d\})$ such that $B = \varnothing$;
3 | **while** $\partial_\mathbf{B} \neq \partial_\mathbb{s}$ **do**
    | | // //Search for the best hyperplane
4 | | **for** $i := 1$ *to* k **do**
5 | | | Search for the best hyperplane H_i attached to the axis x_i using genetic algorithm;
6 | | **end**
7 | | $H :=$ Best hyperplane from the set $\{H_1, H_2, \ldots, H_k\}$;
8 | | $B := B \cup \{Test_H\}$;
9 | **end**
10 **end**

In the presented above algorithm, the main effort should concentrate on searching for the best hyperplane attached to each of axes. Let us describe the GA-based method for this problem, assuming that we are searching for hyperplanes attached to x_1.

Chromosomes : Let us fixe an integer b. In each two-dimensional plane $\mathbf{L}(x_1, x_i)$ we select 2^b vectors $v_1^i, v_2^i, \ldots, v_{2^b}^i$ of the form:

$$v_j^i = \left[\alpha_j^i, 0, \ldots, 0, \overset{\substack{i\text{-th position}}}{1}, 0, \ldots, 0 \right] \text{ for } i = 2, \ldots, k \text{ and } j = 1, \ldots, 2^b$$

These vectors, which are not parallel to x_1, can be selected by one of the following methods:

1. Random choice of 2^b values: $\alpha_1^i, \alpha_2^i, \ldots, \alpha_{2^b}^i$.
2. The values $\alpha_1^i, \alpha_2^i, \ldots, \alpha_{2^b}^i$ are chosen in such a way that all angles between successive vectors are equal, i.e., $\alpha_j^i = \cot\left(j\frac{\pi}{1+2^i}\right)$

3. The sequence $\alpha_1^i, \alpha_2^i, \ldots, \alpha_{2^b}^i$ is an arithmetical progression (e.g., $\alpha_j^i = j - 2^{l-1}$).

Any chromosome is a bit vector of the length $b(k-1)$ containing the $(k-1)$ blocks of length b. The ith block (for $i = 1, 2, \ldots, k-1$) encodes an integer $j_{i+1} \in \{1, \ldots, 2^b\}$ corresponding to one of the vectors of the form $v_{j_{i+1}}^{i+1}$. Thus any chromosome represents an array of $(k-1)$ integers $[j_2, j_3, \ldots, j_k]$ and can be interpreted as a linear subspace $L = Lin(v_{j_2}^2, v_{j_3}^3, \ldots, v_{j_k}^k)$. Let f_L be the projection parallel to L onto the axis x_1. The function f_L can be treated as a new attribute as follows:

$$f_L(u) := a_1(u) - \alpha_{j_2}^2 a_2(u) - \alpha_{j_3}^3 a_3(u) - \cdots - \alpha_{j_k}^k a_k(u)$$

for each object $u \in U$.

Operators: Let us consider two examples of chromosomes (assuming $b = 4$):

$$chr_1 = \underset{1}{0010}\ \underset{2}{1110} \ldots \underset{i}{0100} \ldots \underset{k-1}{1010}$$

$$chr_2 = \underset{1}{0000}\ \underset{2}{1110} \ldots \underset{i}{1000} \ldots \underset{k-1}{0101}$$

The genetic operators are defined as follows:

1. Mutation and selection are defined in standard way [72]. Mutation of chr_1 is realized in two steps; first one block, say i-th, is randomly chosen and next its contents (in our example "0100") are randomly changed into a new block, e.g., "1001". The described example of mutation is changing the chromosome chr_1 into chr_1', where:

$$chr_1' = \underset{1}{0010}\ \underset{2}{1110} \ldots \underset{i}{\underline{1001}} \ldots \underset{k-1}{1010} .$$

2. Crossover is done by the exchange of the whole fragments of chromosome corresponding to one vector. The result of crossover of two chromosomes is realized in two steps as well; first the block position i is randomly chosen and next the contents of ith blocks of two chromosomes are exchanged. For example, if crossover is performed on chr_1, chr_2 and i^{th} block position is randomly chosen then we obtain their offspring:

$$chr_1' = \underset{1}{0010}\ \underset{2}{1110} \ldots \underset{i}{\underline{1000}} \ldots \underset{k-1}{1010}$$

$$chr_2' = \underset{1}{0000}\ \underset{2}{1110} \ldots \underset{i}{\underline{0100}} \ldots \underset{k-1}{0101}$$

Fitness function: The fitness of any chromosome χ representing a linear subspace $L = Lin(v_{j_2}^2, v_{j_3}^3, \ldots, v_{j_k}^k)$ is calculated by the quality of the best cut on the attribute f_L. Moreover, together with the best cut on f_L, the chromosome determines the best hyperplane parallel to L. In fact, any cut $p \in \mathbb{R}$ on f_L defines the hyperplane $\mathbf{H} = H\left(p, v_{j_2}^2, v_{j_3}^3, \ldots, v_{j_k}^k\right)$ as follows:

$$\mathbf{H} = (p, 0, \ldots, 0) \oplus \mathbf{L} = \left\{ \mathbf{P} \in \mathbb{R}^k : \overrightarrow{\mathbf{P_0 P}} \in \mathbf{L} \right\}$$

$$= \left\{ (x_1, x_2, \ldots, x_k) \in \mathbb{R}^k : [x_1 - p, x_2, \ldots, x_k] = b_2 v_{j_2}^2 + b_3 v_{j_3}^3 + \cdots + b_k v_{j_k}^k \right\}$$

$$\text{for some } b_2, \ldots, b_k \in \mathbb{R}$$

$$= \left\{ (x_1, x_2, \ldots, x_k) \in \mathbb{R}^k : x_1 - p = \alpha_{j_2}^2 x_2 + \alpha_{j_3}^3 x_3 + \cdots + \alpha_{j_k}^k x_k \right\}$$

$$= \left\{ (x_1, x_2, \ldots, x_k) \in \mathbb{R}^k : x_1 - \alpha_{j_2}^2 x_2 - \alpha_{j_3}^3 x_3 - \cdots - \alpha_{j_k}^k x_k - p = 0 \right\}.$$

The hyperplane quality and, in consequence, the fitness of the chromosome can be calculated using different measures introduced in previous section. In [93] we have proposed to evaluate the quality of chromosome using two factors, i.e., discernibility function as an award factor and indiscernibility function as a penalty. Thus the fitness of chromosome $chi = [j_2, \ldots, j_k]$ is defined by

$$fitness(\chi) = power(\mathbf{H})$$
$$= F(award(\mathbf{H}), penalty(\mathbf{H})).$$

where $\mathbf{H}$ is the best hyperplane parallel to the linear subspace spanning on base vectors $(v_{j_2}^2, v_{j_3}^3, \ldots, v_{j_k}^k)$ and $F(.,.)$ is a two-argument function which is increasing w.r.t. the first argument and decreasing w.r.t. the second argument.

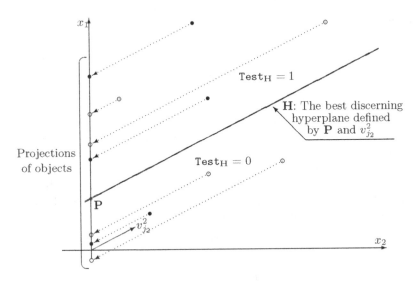

Fig. 33. Interpretation of the projection function in two-dimensional space

8.2.3   Searching for Optimal Surfaces

In the previous section we considered a method of searching for semi-optimal hyperplanes. Below, we present a natural way to generate a semi-optimal set of *high degree surfaces (curves)* applying the existing methods for hyperplanes.

Let us note that any i-th *degree surface in* $\mathbb{R}^k$ can be defined as follows:

$$S = \left\{ (x_1, \ldots, x_k) \in \mathbb{R}^k : P(x_1, \ldots, x_k) = 0 \right\},$$

where $P(x_1, \ldots, x_k)$ is an arbitrary i^{th} degree polynomial over k variables.

Any i^{th} degree polynomial is a linear combination of monomials, each of degree not greater than i. By $\eta(i, k)$ we denote the number of k-variable monomials of degrees $\leq i$. Then, instead of searching for i^{th} degree surfaces in k-dimensional affine real space $\mathbb{R}^k$, one can search for hyperplanes in space $\mathbb{R}^{\eta(i,k)}$.

It is easy to see that the number of jth degree monomials built from k variables is equal to $\binom{j+k-1}{k-1}$. Then we have

$$\eta(i, k) = \sum_{j=1}^{i} \binom{j+k-1}{k-1} = O(k^i). \tag{58}$$

As we can see, applying the above surfaces we have better chance to discern objects from different decision classes with smaller number of "cuts". This is because higher degree surfaces are more flexible than normal cuts. This fact can be shown by applying the VC (Vapnik-Chervonenkis) dimension for corresponding set of functions [154].

To search for an optimal set of ith degree surfaces discerning objects from different decision classes of a given decision table $\mathbb{S} = (U, A \cup \{d\})$ one can construct a new decision table $\mathbb{S}^i = (U, A^i \cup \{d\})$ where A^i is a set of all monomials of degree $\leq i$ built on attributes from A. Any hyperplane found for the decision table $\mathbb{S}^i$ is a surface in the original decision table $\mathbb{S}$. The cardinality of A^i is estimated by the formula (58).

Hence, for the better solution, we must pay with the increase of space and time complexity.

9   Rough Sets and Association Analysis

In this section, we consider a well-known and famous nowadays data mining technique, called association rules [3], to discover useful patterns in transactional databases. The problem is to extract all associations and correlations among data items where the presence of one set of items in a transaction implies (with a certain degree of confidence) the presence of other items. Besides market basket data, association analysis is also applicable to other application domains such as customer relationship management (CRM), bioinformatics, medical diagnosis, Web mining, and scientific data analysis.

We will point out also the contribution of rough sets and approximate Boolean reasoning approach in association analysis, as well as the correspondence between the problem of searching for approximate reduct and the problem of generating association rules from frequent item sets.

9.1 Approximate Reducts

Let $\mathbb{S} = (U, A \cup \{dec\})$ be a given decision table, where $U = \{u_1, u_2, \ldots, u_n\}$ and $A = \{a_1, \ldots, a_k\}$. Discernibility matrix of $\mathbb{S}$ was defined as the $(n \times n)$ matrix $\mathbf{M}(\mathbb{S}) = [M_{i,j}]_{i,j=1}^n$ where

$$M_{i,j} = \begin{cases} \{a_m \in A : a_m(x_i) \neq a_m(x_j)\} & \text{if } dec(x_i) \neq dec(x_j) \\ \varnothing & \text{otherwise.} \end{cases} \tag{59}$$

Let us recall that a set $B \subset A$ of attributes is "consistent with dec" (or dec-consistent) if B has non-empty intersection with each non-empty set $M_{i,j}$, i.e.,

$$B \text{ is consistent with } dec \quad \text{iff} \quad \forall_{i,j}(C_{i,j} = \varnothing) \vee (B \cap C_{i,j} \neq \varnothing).$$

Minimal (with respect to inclusion) dec-consistent sets of attributes are called decision reducts.

In some applications (see [138], [120]), instead of reducts we prefer to use their approximations called α-reducts, where $\alpha \in [0, 1]$ is a real parameter. A set of attributes is called α-reduct if it is minimal (with respect to inclusion) among the sets of attributes B such that

$$\frac{disc(B)}{conflict(\mathbb{S})} = \frac{|\{M_{i,j} : B \cap M_{i,j} \neq \varnothing\}|}{|\{C_{i,j} : C_{i,j} \neq \varnothing\}|} \geq \alpha.$$

If $\alpha = 1$, the notions of an α-reduct and a (normal) reduct coincide. One can show that for a given α, problems of searching for shortest α-reducts and for all α-reducts are also NP-hard [96].

9.2 From Templates to Optimal Association Rules

Let $\mathbb{S} = (U, A)$ be an information table. By *descriptors* (or simple descriptors) we mean the terms of the form $(a = v)$, where $a \in A$ is an attribute and $v \in V_a$ is a value in the domain of a (see [98]). By *template* we mean the conjunction of descriptors:

$$\mathbf{T} = D_1 \wedge D_2 \wedge \ldots \wedge D_m,$$

where $D_1, \ldots D_m$ are either simple or generalized descriptors. We denote by $length(\mathbf{T})$ the number of descriptors being in $\mathbf{T}$.

For the given template with length m:

$$\mathbf{T} = (a_{i_1} = v_1) \wedge \ldots \wedge (a_{i_m} = v_m)$$

the object $u \in U$ is said to satisfy the template $\mathbf{T}$ if and only if $\forall_j a_{i_j}(u) = v_j$. In this way the template $\mathbf{T}$ describes the set of objects having the common property: "*values of attributes* $a_{i_1}, \ldots, a_{i_m}$ are equal to $v_1, \ldots, v_m$, respectively". In this sense one can use templates to describe the regularity in data, i.e., *patterns* - in data mining or *granules* - in soft computing.

Templates, except for length, are also characterized by their support. The *support* of a template **T** is defined by

$$support(\mathbf{T}) = |\{u \in U : u \text{ satisfies } \mathbf{T}\}|.$$

From descriptive point of view, we prefer long templates with large support.

The templates that are supported by a predefined number (say *min_support*) of objects are called *the frequent templates*. This notion corresponds exactly to the notion of *frequent itemsets* for transaction databases [1]. Many efficient algorithms for frequent itemset generation has been proposed in [1], [3], [2], [161] [44]. The problem of frequent template generation using rough set method has been also investigated in [98], [105]. In Sect. 5.4 we considered a special kind of templates called *decision templates* or *decision rules*. Almost all objects satisfying a decision template should belong to one decision class.

Let us assume that the template **T**, which is supported by at least s objects, has been found (using one of existing algorithms for frequent templates). We assume that **T** consists of m descriptors i.e.

$$\mathbf{T} = D_1 \wedge D_2 \wedge \cdots \wedge D_m$$

where D_i (for $i = 1, \ldots, m$) is a descriptor of the form $(a_i = v_i)$ for some $a_i \in A$ and $v_i \in V_{a_i}$. We denote the set of all descriptors occurring in the template **T** by $DESC(\mathbf{T})$, i.e.,

$$DESC(\mathbf{T}) = \{D_1, D_2, \ldots, D_m\}.$$

Any set of descriptors $\mathbf{P} \subseteq DESC(\mathbf{T})$ defines an association rule

$$\mathcal{R}_\mathbf{P} =_{def} \left(\bigwedge_{D_i \in \mathbf{P}} D_i \implies \bigwedge_{D_j \notin \mathbf{P}} D_j \right).$$

The *confidence* factor of the association rule $\mathcal{R}_\mathbf{P}$ can be redefined as

$$confidence\,(\mathcal{R}_\mathbf{P}) =_{def} \frac{support(\mathbf{T})}{support(\bigwedge_{D_i \in \mathbf{P}} D_i)},$$

i.e., the ratio of the number of objects satisfying **T** to the number of objects satisfying all descriptors from **P**. The *length* of the association rule $\mathcal{R}_\mathbf{P}$ is the number of descriptors from **P**.

In practice, we would like to find as many association rules with satisfactory confidence as possible (i.e., $confidence\,(\mathcal{R}_\mathbf{P}) \geq c$ for a given $c \in (0; 1)$). The following property holds for the confidence of association rules:

$$\mathbf{P}_1 \subseteq \mathbf{P}_2 \quad \implies \quad confidence\,(\mathcal{R}_{\mathbf{P}_1}) \leq confidence\,(\mathcal{R}_{\mathbf{P}_2}). \qquad (60)$$

This property says that if the association rule $\mathcal{R}_\mathbf{P}$ generated from the descriptor set **P** has satisfactory confidence then the association rule generated from any superset of **P** also has satisfactory confidence.

For a given confidence threshold $c \in (0; 1]$ and a given set of descriptors $\mathbf{P} \subseteq DESC(\mathbf{T})$, the association rule $\mathcal{R}_\mathbf{P}$ is called *c-representative* if

1. $confidence\,(\mathcal{R}_{\mathbf{P}}) \geq c$;
2. for any proper subset $\mathbf{P'} \subset \mathbf{P}$ we have $confidence\,(\mathcal{R}_{\mathbf{P'}}) < c$.

From Eqn. (60) one can see that instead of searching for all association rules, it is enough to find all c-representative rules. Moreover, every c-representative association rule covers a family of association rules. The shorter the association rule $\mathcal{R}$ is, the bigger is the set of association rules covered by $\mathcal{R}$. First of all, we show the following theorem:

Theorem 24. *For a fixed real number $c \in (0; 1]$ and a template $\mathbf{T}$, the optimal c-association rules problem – i.e., searching for the shortest c-representative association rule from $\mathbf{T}$ in a given table $\mathbb{A}$ – is NP-hard.*

Proof: Obviously, the Optimal c–Association Rules Problem belongs to NP. We show that the Minimal Vertex Covering Problem (which is NP-hard, see e.g. [35]) can be transformed to the Optimal c-Association Rules Problem.

Let the graph $\mathbf{G} = (V, E)$ be an instance of the Minimal Vertex Cover Problem, where $V = \{v_1, v_2, \ldots v_n\}$ and $E = \{e_1, e_2, \ldots e_m\}$. We assume that every edge e_i is represented by two-element set of vertices, i.e., $e_i = \{v_{i_1}, v_{i_2}\}$. We construct the corresponding information table (or transaction table) $\mathbb{A}(\mathbf{G}) = (U, A)$ for the Optimal c-Association Rules Problem as follows:

1. The set U consists of m objects corresponding to m edges of the graph G and $k + 1$ objects added for some technical purpose, i.e.,

$$U = \{x_1, x_2, \ldots, x_k\} \cup \{x^*\} \cup \{u_{e_1}, u_{e_2}, \ldots, u_{e_m}\},$$

where $k = \left\lfloor \frac{c}{1-c} \right\rfloor$ is a constant derived from c.

2. The set A consists of n attributes corresponding to n vertices of the graph G and an attribute a^* added for some technical purpose, i.e.,

$$A = \{a_{v_1}, a_{v_2}, \ldots, a_{v_n}\} \cup \{a^*\}.$$

The value of attribute $a \in A$ over the object $u \in U$ is defined as follows:
(a) if $u \in \{x_1, x_2, \ldots, x_k\}$ then

$$a(x_i) = 1 \text{ for any } a \in A.$$

(b) if $u = x^*$ then for any $j \in \{1, \ldots, n\}$:

$$a_{v_j}(x^*) = 1 \qquad \text{and} \qquad a^*(x^*) = 0.$$

(c) if $u \in \{u_{e_1}, u_{e_2}, \ldots, u_{e_m}\}$ then for any $j \in \{1, \ldots, n\}$:

$$a_{v_j}(u_{e_i}) = \begin{cases} 0 & \text{if } v_j \in e_i \\ 1 & \text{otherwise} \end{cases} \qquad \text{and} \qquad a^*(u_{e_i}) = 1.$$

Example

Let us consider the Optimal c-Association Rules Problem for $c = 0.8$. We illustrate the proof of Theorem 24 by the graph $\mathbf{G} = (V, E)$ with five vertices $V = \{v_1, v_2, v_3, v_4, v_5\}$ and six edges $E = \{e_1, e_2, e_3, e_4, e_5, e_6\}$. First we compute $k = \left\lfloor \frac{c}{1-c} \right\rfloor = 4$. Hence, the information table $\mathbb{A}(\mathbf{G})$ consists of six attributes $\{a_{v_1}, a_{v_2}, a_{v_3}, a_{v_4}, a_{v_5}, a^*\}$ and $(4 + 1) + 6 = 11$ objects $\{x_1, x_2, x_3, x_4, x^*, u_{e_1}, u_{e_2}, u_{e_3}, u_{e_4}, u_{e_5}, u_{e_6}\}$. The information table $\mathbb{A}(\mathbf{G})$ constructed from the graph $\mathbf{G}$ is presented in the figure below.

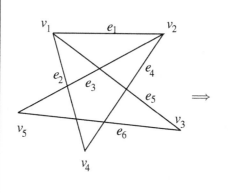

$\mathbb{A}(\mathbf{G})$	a_{v_1}	a_{v_2}	a_{v_3}	a_{v_4}	a_{v_5}	a^*
x_1	1	1	1	1	1	1
x_2	1	1	1	1	1	1
x_3	1	1	1	1	1	1
x_4	1	1	1	1	1	1
x^*	1	1	1	1	1	0
u_{e_1}	0	0	1	1	1	1
u_{e_2}	0	1	1	0	1	1
u_{e_3}	1	0	1	1	0	1
u_{e_4}	1	0	1	0	1	1
u_{e_5}	0	1	0	1	1	1
u_{e_6}	1	1	0	1	0	1

Fig. 34. The construction of the information table $\mathbb{A}(\mathbf{G})$ from the graph $\mathbf{G} = (V, E)$ with five vertices and six edges for $c = 0.8$

The illustration of our construction is presented in Fig. 34.

We will show that any set of vertices $W \subseteq V$ is a minimal covering set for the graph $\mathbf{G}$ if and only if the set of descriptors

$$\mathbf{P}_W = \{(a_{v_j} = 1) : \text{ for } v_j \in W\}$$

defined by W encodes the shortest c-representative association rule for $\mathbb{A}(\mathbf{G})$ from the template

$$\mathbf{T} = (a_{v_1} = 1) \wedge \cdots \wedge (a_{v_n} = 1) \wedge (a^* = 1).$$

The first implication ($\Rightarrow$) is obvious. We show that implication ($\Leftarrow$) also holds.

The only objects satisfying $\mathbf{T}$ are $x_1, \ldots, x_k$ hence we have $support(\mathbf{T}) = k$. Let $\mathbf{P} \Rightarrow \mathbf{Q}$ be an optimal c-confidence association rule derived from $\mathbf{T}$. Then we have $\frac{support(\mathbf{T})}{support(\mathbf{P})} \geq c$, hence

$$support(\mathbf{P}) \leq \frac{1}{c} \cdot support(\mathbf{T}) = \frac{1}{c} \cdot k = \frac{1}{c} \cdot \left\lfloor \frac{c}{1-c} \right\rfloor \leq \frac{1}{1-c} = \frac{c}{1-c} + 1.$$

Because $support(\mathbf{P})$ is an integer number, we have

$$support(\mathbf{P}) \leq \left\lfloor \frac{c}{1-c} + 1 \right\rfloor = \left\lfloor \frac{c}{1-c} \right\rfloor + 1 = k + 1.$$

Thus, there is at most one object from the set $\{x^*\}\cup\{u_{e_1}, u_{e_2}, \dots, u_{e_m}\}$ satisfying the template $\mathbf{P}$. We consider two cases:

1. *The object x^* satisfies $\mathbf{P}$*: then the template $\mathbf{P}$ cannot contain the descriptor $(a^* = 1)$, i.e.,

$$\mathbf{P} = (a_{v_{i_1}} = 1) \cdots\cdots (a_{v_{i_t}} = 1)$$

   and there is no object from $\{u_{e_1}, u_{e_2}, \dots, u_{e_m}\}$ which satisfies $\mathbf{P}$, i.e., for any edge $e_j \in E$ there exists a vertex $v_i \in \{v_{i_1}, \dots, v_{i_t}\}$ such that $a_{v_i}(u_{e_j}) = 0$ (which means that $v_i \in e_j$). Hence, the set of vertices $W = \{v_{i_1}, \dots, v_{i_t}\} \subseteq V$ is a solution of the Minimal Vertex Cover Problem.

2. *An object u_{e_j} satisfies $\mathbf{P}$*: then $\mathbf{P}$ consists of the descriptor $(a^* = 1)$; thus

$$\mathbf{P} = (a_{v_{i_1}} = 1) \cdots\cdots (a_{v_{i_t}} = 1) \cdot (a^* = 1).$$

   Let us assume that $e_j = \{v_{j_1}, v_{j_2}\}$. We consider two templates $\mathbf{P_1}, \mathbf{P_2}$ obtained from $\mathbf{P}$ by replacing the last descriptor by $(a_{v_{j_1}} = 1)$ and $(a_{v_{j_2}} = 1)$, respectively, i.e.

$$\mathbf{P_1} = (a_{v_{i_1}} = 1) \cdots\cdots (a_{v_{i_t}} = 1) \cdot (a_{v_{j_1}} = 1)$$

$$\mathbf{P_2} = (a_{v_{i_1}} = 1) \cdots\cdots (a_{v_{i_t}} = 1) \cdot (a_{v_{j_2}} = 1).$$

   One can prove that both templates are supported by exactly k objects: $x_1, x_2, \dots, x_t$ and x^*. Hence, similarly to the previous case, the two sets of vertices $W_1 = \{v_{i_1}, \dots, v_{i_t}, v_{j_1}\}$ and $W_2 = \{v_{i_1}, \dots, v_{i_t}, v_{j_2}\}$ establish the solutions of the Minimal Vertex Cover Problem.

We showed that any instance I of the Minimal Vertex Cover Problem can be transformed to the corresponding instance I' of the Optimal c–Association Rule Problem in polynomial time and any solution of I can be obtained from solutions of I'. Our reasoning shows that the Optimal c–Association Rules Problem is NP-hard. ∎

Since the problem of searching for the shortest representative association rules is NP-hard, the problem of searching for all association rules must be also as least NP-hard because this is a more complex problem. Having all association rules one can easily find the shortest representative association rule. Hence, we have the following:

Theorem 25. *The problem of searching for all (representative) association rules from a given template is at least NP-hard unless $P = NP$.*

The NP-hardness of presented problems forces us to develop efficient approximate algorithms solving them. In the next section we show that they can be developed using rough set methods.

9.3   Searching for Optimal Association Rules by Rough Set Methods

To solve the presented problem, we show that the problem of searching for optimal association rules from a given template is equivalent to the problem of searching for local α-reducts for a decision table, which is a well-known problem in rough set theory. We propose the Boolean reasoning approach for association rule generation.

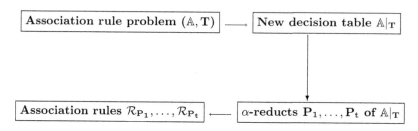

Fig. 35. The Boolean reasoning scheme for association rule generation

We construct a new decision table $\mathbb{A}|_\mathbf{T} = (U, A|_\mathbf{T} \cup d)$ from the original information table $\mathbb{A}$ and the template $\mathbf{T}$ as follows:

– $A|_\mathbf{T} = \{a_{D_1}, a_{D_2}, \ldots, a_{D_m}\}$ is a set of attributes corresponding to the descriptors of the template $\mathbf{T}$

$$a_{D_i}(u) = \begin{cases} 1 \text{ if the object } u \text{ satisfies } D_i, \\ 0 \text{ otherwise}; \end{cases} \quad (61)$$

– the decision attribute d determines whether a given object satisfies the template $\mathbf{T}$, i.e.,

$$d(u) = \begin{cases} 1 \text{ if the object } u \text{ satisfies } \mathbf{T}, \\ 0 \text{ otherwise}. \end{cases} \quad (62)$$

The following theorems describe the relationship between association rules problem and reduct searching problem.

Theorem 26. *For a given information table $\mathbb{A} = (U, A)$ and a template $\mathbf{T}$, the set of descriptors $\mathbf{P}$ is a reduct in $\mathbb{A}|_\mathbf{T}$ if and only if the rule*

$$\bigwedge_{D_i \in \mathbf{P}} D_i \Rightarrow \bigwedge_{D_j \notin \mathbf{P}} D_j$$

is 100%-representative association rule from $\mathbf{T}$.

Proof: Any set of descriptors $\mathbf{P}$ is a reduct in the decision table $\mathbb{A}|_\mathbf{T}$ if and only if every object u with decision 0 is discerned from objects with decision 1 by one of the descriptors from $\mathbf{P}$ (i.e., there is at least one 0 in the information vector $inf_{\mathbf{P}}(u)$). Thus u does not satisfy the template $\bigwedge_{D_i \in \mathbf{P}} D_i$. Hence

$$support\left(\bigwedge_{D_i \in \mathbf{P}} D_i\right) = support(\mathbf{T}).$$

The last equality means that

$$\bigwedge_{D_i \in \mathbf{P}} D_i \Rightarrow \bigwedge_{D_j \notin \mathbf{P}} D_j$$

is 100%-confidence association rule for table $\mathbb{A}$.   ■

Analogously, one can show the following fact:

Theorem 27. *For a given information table* $\mathbb{A} = (U, A)$, *a template* $\mathbf{T}$, *a set of descriptors* $\mathbf{P} \subseteq DESC(\mathbf{T})$, *the rule*

$$\bigwedge_{D_i \in \mathbf{P}} D_i \Rightarrow \bigwedge_{D_j \notin \mathbf{P}} D_j$$

is a c-representative association rule obtained from $\mathbf{T}$ *if and only if* $\mathbf{P}$ *is a α-reduct of* $\mathbb{A}|_{\mathbf{T}}$, *where* $\alpha = 1 - \frac{\frac{1}{c} - 1}{\frac{n}{s} - 1}$, *n is the total number of objects from U and* $s = support(\mathbf{T})$. *In particular, the problem of searching for optimal association rules can be solved using methods for α-reduct finding.*

Proof: Assume that $support(\bigwedge_{D_i \in \mathbf{P}} D_i) = s + e$, where $s = support(\mathbf{T})$. Then we have

$$confidence\left(\bigwedge_{D_i \in \mathbf{P}} D_i \Rightarrow \bigwedge_{D_j \notin \mathbf{P}} D_j\right) = \frac{s}{s+e} \geq c.$$

This condition is equivalent to

$$e \leq \left(\frac{1}{c} - 1\right) s.$$

Hence, one can evaluate the discernibility degree of $\mathbf{P}$ by

$$disc\_degree(\mathbf{P}) = \frac{e}{n-s} \leq \frac{\left(\frac{1}{c} - 1\right) s}{n-s} = \frac{\frac{1}{c} - 1}{\frac{n}{s} - 1} = 1 - \alpha.$$

Thus

$$\alpha = 1 - \frac{\frac{1}{c} - 1}{\frac{n}{s} - 1}.$$
          ■

Searching for minimal α-reducts is a well-known problem in the rough set theory. One can show that the problem of searching for shortest α-reducts is NP-hard [96] and the problem of searching for the all α-reducts is at least NP-hard. However, there exist many approximate algorithms solving the following problems:

1. Searching for shortest reduct (see [143]);
2. Searching for a number of short reducts (see, e.g., [158]);
3. Searching for all reducts (see, e.g., [7]).

The algorithms for the first two problems are quite efficient from computational complexity point of view. Moreover, in practical applications, the reducts generated by them are quite closed to the optimal one.

In Sect. 9.3.1, we present some heuristics for these problems in terms of association rule generation.

9.3.1 Example

The following example illustrates the main idea of our method. Let us consider the information table $\mathbb{A}$ (Table 18) with 18 objects and 9 attributes.

Assume that the template

$$\mathbf{T} = (a_1 = 0) \wedge (a_3 = 2) \wedge (a_4 = 1) \wedge (a_6 = 0) \wedge (a_8 = 1)$$

has been extracted from the information table $\mathbb{A}$. One can see that $support(\mathbf{T}) = 10$ and $length(\mathbf{T}) = 5$. The new decision table $\mathcal{A}|_{\mathbf{T}}$ is presented in Table 19.

The discernibility function for decision table $\mathbb{A}|_{\mathbf{T}}$ is as follows

$$\begin{aligned}
f(D_1, D_2, D_3, D_4, D_5) = {} & (D_2 \vee D_4 \vee D_5) \wedge (D_1 \vee D_3 \vee D_4) \wedge (D_2 \vee D_3 \vee D_4) \\
& \wedge (D_1 \vee D_2 \vee D_3 \vee D_4) \wedge (D_1 \vee D_3 \vee D_5) \\
& \wedge (D_2 \vee D_3 \vee D_5) \wedge (D_3 \vee D_4 \vee D_5) \wedge (D_1 \vee D_5)
\end{aligned}$$

Table 18. The example of information table $\mathbb{A}$ and template $\mathbf{T}$ with support 10

$\mathbb{A}$	a_1	a_2	a_3	a_4	a_5	a_6	a_7	a_8	a_9
u_1	0	*	1	1	*	2	*	2	*
u_2	0	*	2	1	*	0	*	1	*
u_3	0	*	2	1	*	0	*	1	*
u_4	0	*	2	1	*	0	*	1	*
u_5	1	*	2	2	*	1	*	1	*
u_6	0	*	1	2	*	1	*	1	*
u_7	1	*	1	2	*	1	*	1	*
u_8	0	*	2	1	*	0	*	1	*
u_9	0	*	2	1	*	0	*	1	*
u_{10}	0	*	2	1	*	0	*	1	*
u_{11}	1	*	2	2	*	0	*	2	*
u_{12}	0	*	3	2	*	0	*	2	*
u_{13}	0	*	2	1	*	0	*	1	*
u_{14}	0	*	2	2	*	2	*	2	*
u_{15}	0	*	2	1	*	0	*	1	*
u_{16}	0	*	2	1	*	0	*	1	*
u_{17}	0	*	2	1	*	0	*	1	*
u_{18}	1	*	2	1	*	0	*	2	*
$\mathbf{T}$	0	*	2	1	*	0	*	1	*

Table 19. The new decision table $\mathbb{A}|_{\mathbf{T}}$ constructed from $\mathbb{A}$ and template $\mathbf{T}$

$\mathbb{A}|_{\mathbf{T}}$	D_1 $a_1 = 0$	D_2 $a_3 = 2$	D_3 $a_4 = 1$	D_4 $a_6 = 0$	D_5 $a_8 = 1$	d
u_1	1	0	1	0	0	0
u_2	1	1	1	1	1	1
u_3	1	1	1	1	1	1
u_4	1	1	1	1	1	1
u_5	0	1	0	0	1	0
u_6	1	0	0	0	1	0
u_7	0	0	0	0	1	0
u_8	1	1	1	1	1	1
u_9	1	1	1	1	1	1
u_{10}	1	1	1	1	1	1
u_{11}	0	1	0	1	0	0
u_{12}	1	0	0	1	0	0
u_{13}	1	1	1	1	1	1
u_{14}	1	1	0	0	0	0
u_{15}	1	1	1	1	1	1
u_{16}	1	1	1	1	1	1
u_{17}	1	1	1	1	1	1
u_{18}	0	1	1	1	0	0

After the condition presented in Table 20 is simplified, we obtain six reducts for the decision table $\mathbb{A}|_{\mathbf{T}}$.

$$f(D_1, D_2, D_3, D_4, D_5) = (D_3 \wedge D_5) \vee (D_4 \wedge D_5) \vee (D_1 \wedge D_2 \wedge D_3) \vee$$
$$(D_1 \wedge D_2 \wedge D_4) \vee (D_1 \wedge D_2 \wedge D_5) \vee (D_1 \wedge D_3 \wedge D_4)$$

Thus, we have found from the template $\mathbf{T}$ six association rules with (100%)-confidence (see Table 20).

For $c = 90\%$, we would like to find α-reducts for the decision table $\mathbb{A}|_{\mathbf{T}}$, where

$$\alpha = 1 - \frac{\frac{1}{c} - 1}{\frac{n}{s} - 1} = 0.86.$$

Hence, we would like to search for a set of descriptors that covers at least

$$\lceil (n - s)(\alpha) \rceil = \lceil 8 \wedge 0.86 \rceil = 7$$

elements of discernibility matrix $\mathbb{M}(\mathbb{A}|_{\mathbf{T}})$. One can see that the following sets of descriptors:

$$\{D_1, D_2\}, \{D_1, D_3\}, \{D_1, D_4\}, \{D_1, D_5\}, \{D_2, D_3\}, \{D_2, D_5\}, \{D_3, D_4\}$$

have non-empty intersection with exactly 7 members of the discernibility matrix $\mathbb{M}(\mathbb{A}|_{\mathbf{T}})$. Table 20 presents all association rules achieved from those sets.

Table 20. The simplified version of the discernibility matrix $M(A|_\mathbf{T})$; representative association rules with (100%)-confidence and representative association rules with at least (90%)-confidence

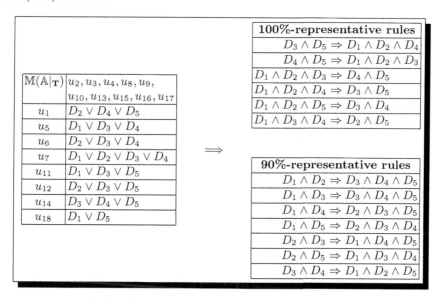

In Fig. 36, we present the set of all 100%–association rules (light gray region) and 90%–association rules (dark gray region). The corresponding representative association rules are represented in bold frames.

9.3.2   The Approximate Algorithms

From the previous example it follows that the searching problem for the representative association rules can be considered as a searching problem in the lattice of attribute subsets (see Fig. 36). In general, there are two searching strategies: bottom–up and top–down. The *top–down strategy* starts with the whole descriptor set and tries to go down through the lattice. In every step, we reduce the most superfluous subsets keeping the subsets which most probably can be reduced in the next step. Almost all existing methods realize this strategy (e.g., Apriori algorithm [2]). The advantage of these methods is as follows:

1. They generate all association rules during searching process.
2. It is easy to implement them for either parallel or concurrent computer.

But this process can take very long computation time because of NP-hardness of the problem (see Theorem 25).

The rough set based method realizes the *bottom–up strategy*. We start with the empty set of descriptors. Here we describe the modified version of greedy heuristics for the decision table $A|_\mathbf{T}$. In practice, we do not construct this additional decision table. The main problem is to compute the occurrence number of descriptors in the discernibility matrix $M(A|_\mathbf{T})$. For any descriptor D, this

Algorithm 8. Searching for shortest representative association rule

Input: Information table $\mathbb{A}$, template $\mathbf{T}$, minimal confidence c.
Output: Short c-representative association rule

```
1  begin
2  │  Set P := ∅; U_P := U ;
3  │  min_support := |U| − 1/c · support(T);
4  │  Select the descriptor D from DESC(T) \ P which is satisfied by the smallest
   │  number of objects from U_P;
5  │  Set P := P ∪ {D};
6  │  U_P := satisfy(P);
   │  // i.e., set of objects satisfying all descriptors from P
7  │  if |U_P| > min_support then
8  │  │  GOTO Step 4;
9  │  else
10 │  │  STOP;
11 │  end
12 end
```

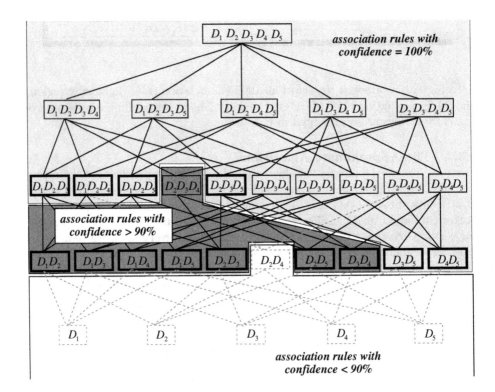

Fig. 36. The illustration of 100% and 90% representative association rules

number is equal to the number of "0" occurring in the column a_D represented by this descriptor and it can be computed using simple SQL queries of the form

SELECT COUNT ...   WHERE ...

We present two algorithms: the first (Algorithm 8) finds *almost the shortest c-representative association rule*. The presented algorithm does not guarantee that the descriptor set **P** is c-representative. But one can achieve it by removing from **P** (which is in general small) all unnecessary descriptors.

The second algorithm (Algorithm 9) finds k *short c-representative association rules* where k and c are parameters given by the user. This algorithm makes use of the beam search strategy which evolves k most promising nodes at each depth of the searching tree.

Algorithm 9. Searching for k short representative association rules

---

Input: Information table $\mathbb{A}$, template **T**, minimal confidence c, number of
representative rules $k \in \mathbb{N}$.
Output: k short c-representative association rules $\mathcal{R}_{\mathbf{P}_1}, \ldots, \mathcal{R}_{\mathbf{P}_k}$.

```
 1  begin
 2      for i := 1 to k do
 3          Set P_i := ∅;
 4          U_{P_i} := U;
 5      end
 6      Set min_support := |U| − 1/c · support(T);
 7      Result_set := ∅;
 8      Working_set := {P_1,..., P_k};
 9      Candidate_set := ∅;
10      for (each P_i ∈ Working_set) do
11          Select k descriptors D^i_1,..., D^i_k from DESC(T) \ P_i which is satisfied by
            the smallest number of objects from U_{P_i};
12          Insert P_i ∪ {D^i_1},..., P_i ∪ {D^i_k} to the Candidate_set;
13      end
14      Select k descriptor sets P'_1,..., P'_k from the Candidate_set (if exist) which
        are satisfied by smallest number of objects from U;
15      Set Working_set := {P'_1,..., P'_k};
16      for (P_i ∈ Working_set) do
17          Set U_{P_i} := satisfy(P_i);
18          if |U_{P_i}| < min_support then
19              Move P_i from Working_set to the Result_set;
20          end
21          if (|Result_set| > k or Working_set is empty) then
22              STOP;
23          else
24              GOTO Step 9;
25          end
26      end
27  end
```

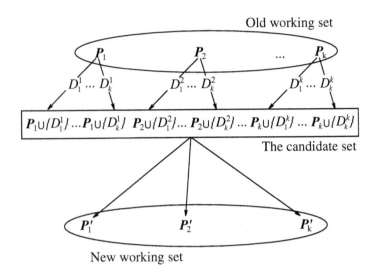

Fig. 37. The illustration of the "k short representative association rules" algorithm

10   Rough Set and Boolean Reasoning Approach to Mining Large Data Sets

Mining large data sets is one of the biggest challenges in KDD. In many practical applications, there is a need of data mining algorithms running on terminals of a client–server database system where the only access to database (located in the server) is enabled by SQL queries.

Unfortunately, the proposed so far data mining methods based on rough sets and Boolean reasoning approach are characterized by high computational complexity and their straightforward implementations are not applicable for large data sets. The critical factor for time complexity of algorithms solving the discussed problem is the number of simple SQL queries like

```
SELECT COUNT FROM aTable WHERE aCondition
```

In this section, we present some efficient modifications of these methods to solve out this problem. We consider the following issues:

- Searching for short reducts from large data sets;
- Induction of rule based rough classifier from large data sets;
- Searching for best partitions defined by cuts on continuous attributes;
- Soft cuts: a new paradigm for discretization problem.

10.1   Searching for Reducts

The application of ABR approach to reduct problem was described in Sect. 5. We have shown (see Algorithm 2 on page 389) that the greedy heuristic for minimal reduct problem uses only two functions:

− $disc(B)$ = number of pairs of objects discerned by attributes from B;
− $isCore(a)$ = check whether a is a core attribute.

In this section, we will show that this algorithm can be efficiently implemented in DBMS using only simple SQL queries.

Let $\mathbb{S} = (U, A \cup \{dec\})$ be a decision table. Recall that by "*counting table*" of a set of objects $X \subset U$ we denoted the vector:

$$CountTable(X) = \langle n_1, \ldots, n_d \rangle,$$

where $n_k = card(X \cap CLASS_k)$ is the number of objects from X belonging to the kth decision class. We define a conflict measure of X by

$$conflict(X) = \sum_{i<j} n_i n_j = \frac{1}{2} \left[\left(\sum_{k=1}^{d} n_k \right)^2 - \sum_{k=1}^{d} n_k^2 \right].$$

In other words, $conflict(X)$ is the number of pairs of different class objects.

By *counting table* of a set of attributes B we mean the two-dimensional array $Count(B) = [n_{v,k}]_{v \in INF(B), k \in V_{dec}}$, where

$$n_{v,k} = card(\{x \in U : inf_B(x) = v \text{ and } dec(x) = k\}).$$

Thus $Count(B)$ is a collection of counting tables of equivalence classes of the indiscernibility relation $IND(B)$. It is clear that the complexity time for the construction of counting table is $O(nd \log n)$, where n is the number of objects and d is the number of decision classes. It is clear that counting table can be easily constructed in data base management systems using simple SQL queries.

The discernibility measure of a set of attributes B can be easily calculated from the counting table as follows:

$$disc_{dec}(B) = \frac{1}{2} \sum_{v \neq v', k \neq k'} n_{v,k} \cdot n_{v',k'}.$$

The disadvantage of this equation relates to the fact that it requires $O(S^2)$ operations, where S is the size of the counting table $Count(B)$.

The discernibility measure can be understood as a number of unresolved (by the set of attributes B) conflicts. One can show that:

$$disc_{dec}(B) = conflict(U) - \sum_{[x] \in U/IND(B)} conflict([x]_{IND(B)}). \tag{63}$$

Thus, the discernibility measure can be determined in $O(S)$ time:

$$disc_{dec}(B) = \frac{1}{2} \left(n^2 - \sum_{k=1}^{d} n_k^2 \right) - \frac{1}{2} \sum_{v \in INF(B)} \left[\left(\sum_{k=1}^{d} n_{v,k} \right)^2 - \sum_{k=1}^{d} n_{v,k}^2 \right], \tag{64}$$

where $n_k = |CLASS_k| = \sum_v n_{v,k}$ is the size of kth decision class.

Moreover, one can show that attribute a is a core attribute of decision table $\mathbb{S} = (U, A \cup \{dec\})$ if and only if

$$disc_{dec}(A - \{a\}) < disc_{dec}(A).$$

Thus both operations $disc_{dec}(B)$ and $isCore(a)$ can be performed in linear time with respect to the counting table.

Example 20. The counting table for a_1 is as follows:

$Count(a_1)$	$dec = no$	$dec = yes$
$a_1 = sunny$	3	2
$a_1 = overcast$	0	3
$a_1 = rainy$	1	3

We illustrate Eqn. (64) by inserting some additional columns to the counting table:

$Count(a_1)$	$dec = no$	$dec = yes$	$\sum$	$conflict(.)$
$a_1 = sunny$	3	2	5	$\frac{1}{2}(5^2 - 2^2 - 3^2) = 6$
$a_1 = overcast$	0	3	3	$\frac{1}{2}(3^2 - 0^2 - 3^2) = 0$
$a_1 = rainy$	1	3	4	$\frac{1}{2}(4^2 - 1^2 - 3^2) = 3$
U	4	8	12	$\frac{1}{2}(12^2 - 8^2 - 4^2) = 32$

Thus $disc_{dec}(a_1) = 32 - 6 - 0 - 3 = 23$.

10.2   Induction of Rough Classifiers

Decision rules play an important role in KDD and data mining. Rule-based classifiers establish an accurate and interpretable model for data.

As it has been mentioned before (Sect. 5), any rule-based classification method consists of three steps: (1) rule generation, (2) rule selection and (3) decision making (e.g., by voting). The general framework for rule based classification methods was presented in Fig. 11. In machine learning, this approach is called *eager (or laborious) learning* methodology.

In *lazy learning* methods new objects are classified without the generalization step. For example, in kNN (k Nearest Neighbors) classifiers, the decision of new object x can be made by taking a vote between k nearest neighbors of x. In lazy decision tree method, we try to reconstruct the path $p(x)$ of the "imaginable decision tree" that can be applied to x.

In this section, we present a lazy learning approach to rule-based classification methods. The proposed method can be applied to solve the classification problem on large data sets.

10.2.1   Induction of Decision Rules by Lazy Learning
Lazy learning methods need more time complexity for the classification step, i.e., the answer time for the question about decision of a new object is longer

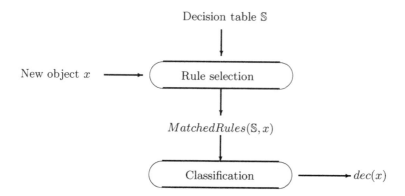

Fig. 38. The lazy rule-based classification system

than in eager classification methods. But lazy classification methods are *well scalable*, i.e., they can be realized for larger decision table using distributed computer system [151]. The scalability property is also very advisable in data mining. Unfortunately, the eager classification methods are weakly scalable. As we recalled before, the time and memory complexity of existing algorithms does not make possible to apply rule base classification methods for very large decision table.

The most often approach, which is placed for rough set-based methods, relates to the lack of scalability. We will show that some classification methods based on rough set theory can be modified using lazy learning algorithms to make them more scalable. The lazy rule-based classification diagram is presented in Fig. 38.

In other words, we will extract the set of decision rules covering the object x directly from data without learning process. The large decision table must be stroed in a data base system and the main problem is to minimize the number SQL queries used in the algorithm. We show that this diagram can work for the classification method described in Sect. 5 using the set of decision rules from $MinRules(\mathbb{S}, \lambda_{\max}, \sigma_{\min}, \alpha_{\min})$. Formally, the problem is formulated as follows: *given a decision table $\mathbb{S} = (U, A \cup \{dec\})$ and a new object x, find all (or almost all) decision rules from the set*

$$MatchRules(\mathbb{S}, x) = \{\mathbf{r} \in MinRules(\mathbb{S}, \lambda_{\max}, \sigma_{\min}, \alpha_{\min}) : x \text{ satisfies } \mathbf{r}\}.$$

In the case of too large number of such rules, one can find as many rules from $MatchRules(\mathbb{S}, x)$ as required.

Let $Desc(x) = \{d_1, d_2, \ldots, d_k\}$, where $d_i := (a_i = a_i(x))$, be the set of all descriptors derived from x. We denote by $\mathbf{P}_i = \{S \subset Desc(x) : |S| = i\}$ the family of sets consisting of exactly i descriptors, and let $\mathbf{P} = \bigcup_{i=1}^{k} \mathbf{P}_i$. One can see that every decision rule $\mathbf{r} \in MatchRules(\mathbb{S}, x)$ should have the form

$$d_{i_1} \wedge \cdots \wedge d_{i_m} \Rightarrow (dec = k)$$

for some $d_{i_1}, \ldots, d_{i_m} \in Desc(x)$. The problem of searching for $MatchRules(\mathbb{S}, x)$ is equivalent to the problem of searching for corresponding families of subsets from **P** using minimal number of basic SQL queries to the database. We will show that the set $MatchRules(\mathbb{S}, x)$ can be found by modified Apriori algorithm (see [2]).

Let $S \in \mathbf{P}$ be an arbitrary set of descriptors from $Desc(x)$. The support of S can be defined by

$$support(S) = |\{u \in U : u \text{ satisfies all descriptors from } S\}|.$$

Let $(s_1, \ldots, s_d)$ be the counting table of the set of objects satisfying all descriptors from S (the *class distribution of S*), i.e.,

$$s_i = |\{u \in U : (u \in DEC_i) \text{ and } (u \text{ satisfies } \bigwedge S)\}|.$$

It is obvious that $support(S) = s_1 + \cdots + s_d$.

We assume that the function $GetClassDistribution(S)$ returns the class distribution of S. One can see that this function can be computed using simple SQL query of the form SELECT COUNT FROM ... WHERE ... GROUP BY ...

Algorithm 10. Rule selection method based on Apriori algorithm

---

Input: The object x, the maximal length $\lambda_{\max}$, the minimal support $\sigma_{\min}$, and the minimal confidence $\alpha_{\min}$.

Output: $MatchRules(\mathbb{S}, x)$: decision rules from $MinRules(\mathbb{S}, \lambda_{\max}, \sigma_{\min}, \alpha_{\min})$ covering x.

1 **begin**
2     $\mathbf{C}_1 := \mathbf{P}_1$; $i := 1$;
3   **while** $((i \leq \lambda_{\max})$ *AND* $(\mathbf{C}_i$ is not empty$))$ **do**
4        $\mathbf{F}_i := \varnothing$;
5        $\mathbf{R}_i := \varnothing$;
6   **for** $C \in \mathbf{C}_i$ **do**
7           $(s_1, \ldots, s_d) := GetClassDistribution(C)$;
8           $support = s_1 + \cdots + s_d$;
9   **if** $support \geq \sigma_{\min}$ **then**
10   **if** $(\max\{s_1, \ldots, s_d\} \geq \alpha_{\min} * support)$ **then**
11                 $\mathbf{R}_i := \mathbf{R}_i \cup \{C\}$;
12   **else**
13                 $\mathbf{F}_i := \mathbf{F}_i \cup \{C\}$;
14   **end**
15   **end**
16   **end**
17        $\mathbf{C}_{i+1} := AprGen(\mathbf{F}_i)$; $i := i + 1$;
18   **end**
19   Return $\bigcup_i \mathbf{R}_i$
20 **end**

Table 21. The "weather" decision table $\mathbb{S}$ and the object x to be classified

$\mathbb{S}$	a_1	a_2	a_3	a_4	dec
ID	outlook	temperature	humidity	windy	play
1	sunny	hot	high	FALSE	no
2	sunny	hot	high	TRUE	no
3	overcast	hot	high	FALSE	yes
4	rainy	mild	high	FALSE	yes
5	rainy	cool	normal	FALSE	yes
6	rainy	cool	normal	TRUE	no
7	overcast	cool	normal	TRUE	yes
8	sunny	mild	high	FALSE	no
9	sunny	cool	normal	FALSE	yes
10	rainy	mild	normal	FALSE	yes
11	sunny	mild	normal	TRUE	yes
12	overcast	mild	high	TRUE	yes
13	overcast	hot	normal	FALSE	yes
14	rainy	mild	high	TRUE	no
x	sunny	mild	high	TRUE	?

The algorithm consists of k iterations where k is the number of attributes. In the ith iteration all decision rules containing i descriptors (length $= i$) are extracted. For this purpose, we compute three families $\mathbf{C}_i$, $\mathbf{R}_i$ and $\mathbf{F}_i$ of subsets of descriptors in the ith iteration:

- The family $\mathbf{C}_i \subset \mathbf{P}_i$ consists of "candidate sets" of descriptors and it can be generated without any database operation.
- The family $\mathbf{R}_i \subset \mathbf{C}_i$ consists of such candidates which contains descriptors (from the left hand side) of some decision rules from $MatchRules(\mathbb{S}, x)$.
- The family $\mathbf{F}_i \subset \mathbf{C}_i$ consists of such candidates which are supported by more than $\sigma_{\min}$ (frequent subsets).

In the algorithm, we apply the function $AprGen(\mathbf{F}_i)$ to generate the family $\mathbf{C}_{i+1}$ of candidate sets from $\mathbf{F}_i$ (see [2]) using the following observations:

1. Let $S \in \mathbf{P}_{i+1}$ and let $S_1, S_2, \ldots, S_{i+1}$ be subsets formed by removing from S one descriptor, we have $support(S) \leq \min\{support(S_j)$, for any $j = 1, \ldots, j+1$. This means that if $S \in \mathbf{R}_{i+1}$ then $S_j \in \mathbf{F}_i$ for $j = 1, \ldots, i+1$. Hence, if $S_j \in \mathbf{F}_i$ for $j = 1, \ldots, i+1$, then S can be inserted to $\mathbf{C}_{i+1}$;
2. Let $s_1^{(j)}, \ldots, s_d^{(j)}$ be the class distribution of S_j and let $s_1, \ldots, s_d$ be the class distribution of S, we have $s_k \leq \min\{s_k^{(1)}, \ldots, s_k^{(i+1)}\}$, for $k = 1, \ldots, d$. This means that if $\max_k\{\min\{s(1)_k, \ldots, s(i+1)_k\}\} \leq \alpha_{\min} \cdot \sigma_{\min}$, then we can remove S from $\mathbf{C}_{i+1}$;

Example 21. Let us illustrate the idea by the well-known for us example of *weather* decision table. Assume that we have to classify a new unseen object (Table 21):

$$x = [sunny, mild, high, TRUE].$$

Table 22. The set of all minimal decision rules

No	$MinConsRules(\mathbb{S})$	supp.
1	outlook(overcast)$\Rightarrow$play(yes)	4
2	humidity(normal) AND windy(FALSE)$\Rightarrow$play(yes)	4
3	outlook(sunny) AND humidity(high)$\Rightarrow$play(no)	3
4	outlook(rainy) AND windy(FALSE)$\Rightarrow$play(yes)	3
5	outlook(sunny) AND temperature(hot)$\Rightarrow$play(no)	2
6	outlook(rainy) AND windy(TRUE)$\Rightarrow$play(no)	2
7	outlook(sunny) AND humidity(normal)$\Rightarrow$play(yes)	2
8	temperature(cool) AND windy(FALSE)$\Rightarrow$play(yes)	2
9	temperature(mild) AND humidity(normal)$\Rightarrow$play(yes)	2
10	temperature(hot) AND windy(TRUE)$\Rightarrow$play(no)	1
11	outlook(sunny) AND temperature(mild) AND windy(FALSE)$\Rightarrow$play(no)	1
12	outlook(sunny) AND temperature(cool)$\Rightarrow$play(yes)	1
13	outlook(sunny) AND temperature(mild) AND windy(TRUE)$\Rightarrow$play(yes)	1
14	temperature(hot) AND humidity(normal)$\Rightarrow$play(yes)	1

$i=1$				$i=2$				$i=3$			
$\mathbf{C_1}$	check	$\mathbf{R_1}$	$\mathbf{F_1}$	$\mathbf{C_2}$	check	$\mathbf{R_2}$	$\mathbf{F_2}$	$\mathbf{C_3}$	check	$\mathbf{R_3}$	$\mathbf{F_3}$
$\{d_1\}$	(3,2)		$\{d_1\}$	$\{d_1,d_2\}$	(1,1)		$\{d_1,d_2\}$	$\{d_1,d_3,d_4\}$	(0,1)	$\{d_1,d_3,d_4\}$	
$\{d_2\}$	(4,2)		$\{d_2\}$	$\{d_1,d_3\}$	(3,0)	$\{d_1,d_3\}$					
$\{d_3\}$	(4,3)		$\{d_3\}$	$\{d_1,d_4\}$	(1,1)		$\{d_1,d_4\}$	$\{d_2,d_3,d_4\}$	(1,1)		$\{d_2,d_3,d_4\}$
$\{d_4\}$	(3,3)		$\{d_4\}$	$\{d_2,d_3\}$	(2,2)		$\{d_2,d_3\}$				
				$\{d_2,d_4\}$	(1,1)		$\{d_2,d_4\}$				
				$\{d_3,d_4\}$	(2,1)		$\{d_3,d_4\}$				

$MatchRules(\mathbb{S}, x) = \mathbf{R_2} \cup \mathbf{R_3}$:

(outlook = sunny) AND (humidity = high) $\Rightarrow$ *play = no*

(outlook = sunny) AND (temperature = mild) AND (windy = TRUE) $\Rightarrow$ *play = yes*

Fig. 39. The illustration of algorithm for $\lambda_{max}=3; \sigma_{\min}=1; \alpha_{\min}=1$

We will compare the standard approach with the proposed method based on lazy learning approach.

The set $MinConsRules(\mathbb{S})$ for this decision table consists of 14 of all minimal consistent rules and is presented in Table 22.

One can see that the set $MatchRules(\mathbb{S}, x)$ consists of two rules:

(outlook = sunny) AND (humidity = high) $\Rightarrow$ *play = no* (rule 3)

(outlook = sunny) AND (temperature = mild) AND (windy = TRUE) $\Rightarrow$ *play = yes* (rule 13)

Fig. 39 illustrates the main steps of Algorithm 10. One can see that both decision rules from $MatchRules(\mathbb{S}, x)$ are discovered by the proposed algorithm.

S	a_1	a_2	a_3	a_4	dec
ID	outlook	temperature	humidity	windy	play
1	sunny	hot	high	FALSE	no
2	sunny	hot	high	TRUE	no
3	overcast	hot	high	FALSE	yes
4	rainy	mild	high	FALSE	yes
5	rainy	cool	normal	FALSE	yes
6	rainy	cool	normal	TRUE	no
7	overcast	cool	normal	TRUE	yes
8	sunny	mild	high	FALSE	no
9	sunny	cool	normal	FALSE	yes
10	rainy	mild	normal	FALSE	yes
11	sunny	mild	normal	TRUE	yes
12	overcast	mild	high	TRUE	yes
13	overcast	hot	normal	FALSE	yes
14	rainy	mild	high	TRUE	no
x	sunny	mild	high	TRUE	?

$\Longrightarrow$

$\mathbb{S}\|_x$	d_1	d_2	d_3	d_4	dec
ID	$a_1\|_x$	$a_2\|_x$	$a_3\|_x$	$a_4\|_x$	dec
1	1	0	1	0	no
2	1	0	1	1	no
3	0	0	1	0	yes
4	0	1	1	0	yes
5	0	0	0	0	yes
6	0	0	0	1	no
7	0	0	0	1	yes
8	1	1	1	0	no
9	1	0	0	0	yes
10	0	1	0	0	yes
11	1	1	0	1	yes
12	0	1	1	1	yes
13	0	0	0	0	yes
14	0	1	1	1	no

Fig. 40. A decision table $\mathbb{S}$, object x, new decision table $\mathbb{S}\|_x$

To explain the essence of Algorithm 10, let us define a new decision table $\mathbb{S}\|_x = (U, A\|_x \cup \{dec\})$, where $A\|_x = \{a_1\|_x, \ldots, a_k\|_x\}$ is a new set of binary attributes defined as follows

$$a_i\|_x(u) = \begin{cases} 1 \text{ if } a_i(u) = a_i(x) \\ 0 \text{ otherwise.} \end{cases}$$

In fact, the decision table $\mathbb{S}\|_x$ is a tabular form of the Boolean function f_x encoding the problem of searching for minimal consistent decision rules covering the object x (see Eqn. (35), page 396).

The table $\mathbb{S}\|_x$ can be treated as a special type of transaction data set because it consists of a decision attribute. In Fig. 40, we present the decision table $\mathbb{S}\|_x$ for the "weather" decision table from the previous example. Table $\mathbb{S}\|_x$ is a useful construction for proving the correctness of the proposed algorithm. One can show that each decision rule from $MatchRules(\mathbb{S}, x)$ can be derived from $\mathbb{S}\|_x$.

One can see that if $x \in U$, then the presented algorithm can generate the object oriented reducts for x. Hence, the proposed method can be applied also for eager learning. This method can be used for adaptive rule generation system where data is growing up in time.

10.3 Searching for Best Cuts

Searching for the best partitions is a common problem for discretization and decision tree methods. We consider the problem of searching for optimal cuts of real value attributes assuming that the decision table is large and is stored in a relational database management system (DBMS).

Usually, when developing of decision tree induction methods [31], [123] and some supervised discretization methods [17], [28], [95], [93] it is necessary to use a *measure* (or *quality function*) to estimate the quality of candidate cuts. Definitions of basic measures were discussed in Sect. 5 and Sect. 6.

Let us assume that a set of candidate cuts $\mathbf{C}_a = \{c_1, \ldots, c_N\}$ on an attribute a and the quality measure

$$\mathcal{F} : \mathbf{C}_a \to \mathbb{R}^+$$

are given. The *straightforward algorithm* should compute the values of F for all cuts: $\mathcal{F}(c_1), \ldots, \mathcal{F}(c_N)$, and returns the cut c_{Best} which maximizes or minimizes the value of function $\mathcal{F}$ as a result of the searching process. Thus the algorithm of searching for best cuts from $\mathbf{C}_a$ with aspect to measure $\mathcal{F}$ requires at least $O(N + n)$ steps, where n is the number of objects in the decision table. In the case of large data tables which are stored in relational databases, for every cut $c_i \in \mathbf{C}_a$ the algorithm should draw out the counting tables $(L_1, \ldots, L_d)$ and $(R_1, \ldots, R_d)$ (for intervals $(-\infty, c_i)$ and $[c_i, \infty)$, respectively) to compute the value of $\mathcal{F}(c_i)$. Hence, the straightforward algorithm requires at least $O(Nd)$ simple queries to search for the best cut on each attribute.

Thus, in million object data bases, the number of simple queries amounts to millions and the time complexity of algorithm becomes unacceptable. Of course, some simple queries can be wrapped in packages or replaced by complex queries, but the DBMS still has to transfer millions class distributions from the server to the client.

The most popular strategy for mining large data tables is the sampling technique [4], i.e., building a model (decision tree or discretization) for small, randomly selected subset of data, and then evaluate the quality of the constructed model on the whole data. If the quality of generated model is not sufficient, it is necessary to refine the existing model or to construct a new model using new random sample (see [52]). Another strategy for mining large data tables is the parallelization technique [5], [134] using computer network architecture.

In this section, we would like to present an alternative solution to the sampling technique.

10.3.1 Algorithm Acceleration Methods for Discernibility Measure
In this section, we present some properties for Boolean reasoning approach. They make it possible to induce decision trees and perform discretization of real value attributes directly from large data bases.

Tail cuts: The following property is interesting in application of MD heuristics for large data tables as it allows for the quick elimination of a large number of cuts.

Firstly, let us recall the definition of *median* which is a well-known in statistics.

Definition 24. *Median of kth decision class is the middle point of its distribution. In other words, if we denote by $L_k(c)$ and $R_k(c)$ the number of objects from k^{th} decision class that are on the left side of c and the right side of c, respectively, then the median of the kth decision class is defined by*

$$Median(k) = \arg \max_{c \in \mathbf{C}_a} \left(\min\{L_k(c) - R_k(c), 0\} \right).$$

Intuitively, for any cut c, if it is on the left hand side of the median of k-decision class, i.e., $c \leq Median(k)$, we have $L_k(c) \leq R_k(c)$, otherwise, i.e., $c > Median(k)$, we have $L_k(c) > R_k(c)$.

Let $c_1 < c_2 \cdots < c_N$ be the set of consecutive candidate cuts, and let

$$c_{min} = \min_i \{Median(i)\} \text{ and } c_{max} = \max_i \{Median(i)\}.$$

Then we have the following theorem.

Theorem 28. *The quality function*

$$Disc : \{c_1, \ldots, c_N\} \to \mathbb{N}$$

defined over the set of cuts is increasing in $\{c_1, \ldots, c_{min}\}$ and decreasing in $\{c_{max}, \ldots, c_N\}$. Hence

$$c_{Best} \in \{c_{min}, \ldots, c_{max}\}.$$

Proof: Let us consider two cuts $c_L < c_R < c_{min}$. Using Eqn. (54) we have

$$Disc(c_R) - Disc(c_L) = \sum_{i=1}^{d} \left[(R_i - L_i) \sum_{j \neq i} M_j \right].$$

Because $c_L < c_R < c_{min}$, hence $R_i - L_i \geq 0$ for any $i = 1, \ldots d$. Thus $Disc(c_R) \geq Disc(c_L)$.

Analogously, one can show that for $c_{max} < c_L < c_R$ we have $Disc(c_R) \leq Disc(c_L)$ ∎

The theorem states that one can reduce the searching space using $O(d \log N)$ SQL queries to determine the medians of decision classes (by applying the Binary Search Algorithm). Let us also observe that if all decision classes have similar medians then almost all cuts can be eliminated.

Efficient localization of the best cut: The idea is to apply the *"divide and conquer"* strategy to determine the best cut $c_{Best} \in \{c_1, \ldots, c_N\}$ with respect to a given quality function.

First we divide the interval containing all possible cuts into k intervals (e.g., $k = 2, 3, \ldots$). Then we choose the interval that most probably contains the best cut. We will use some *approximate discernible measures* to predict the interval which most probably contains the best cut with respect to discernibility measure. This process is repeated until the considered interval consists of one cut. Then the best cut can be chosen between all visited cuts.

The problem arises how to define the measure evaluating the quality of the interval $[c_L; c_R]$ having class distributions: $(L_1, \ldots, L_d)$ in $(-\infty, c_L)$; $(M_1, \ldots, M_d)$ in $[c_L, c_R)$; and $(R_1, \ldots, R_d)$ in $[c_R, \infty)$ (see Fig. 30). This measure should estimate the quality of the best cut among those belonging to the interval $[c_L, c_R]$.

We consider two specific probabilistic models for distribution of objects in the interval $[c_L, c_R]$.

Let us consider a random cut c lying between c_L and c_R and let $(x_1, x_2, \ldots, x_d)$ be the counting table of the set of objects belonging to the interval $[c_L, c]$. Let us assume that $x_1, x_2, \ldots, x_d$ are independent random variables with uniform distribution over sets $\{0, \ldots, M_1\}, \ldots, \{0, \ldots, M_d\}$, respectively. This assumption is called "*fully independent assumption*". Under this assumption we have

$$E(x_i) = \frac{M_i}{2} \text{ and } D^2(x_i) = \frac{M_i(M_i + 2)}{12}$$

for all $i \in \{1, \ldots, d\}$. The following theorem [88] [83] characterizes the quality of the random cut $c \in [c_L, c]$.

Theorem 29. *Let $X = disc(c)$ be a random variable defined by the discernibility measure for the random cut $c \in [c_L, c_R]$. The mean and the standard deviation of X can be calculated as follows:*

$$E(X) = E(disc(c)) = \frac{disc(c_L) + disc(c_R) + conflict([c_L, c_R])}{2}, \qquad (65)$$

where $conflict([c_L, c_R]) = \sum_{i \neq j} M_i M_j$, and

$$D^2(X) = \sum_{i=1}^{n} \left[\frac{M_i(M_i + 2)}{12} \left(\sum_{j \neq i} (R_j - L_j) \right)^2 \right]. \qquad (66)$$

Proof: Let us consider any random cut c lying between c_L and c_R. The situation is shown in Fig. 41.

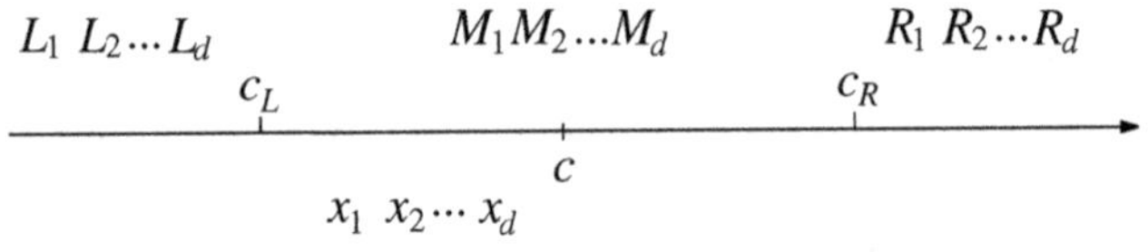

Fig. 41. Random cut c and random class distribution $x_1, \ldots, x_d$ induced by c

$$X - disc(c_L) = \sum_{i=1}^{d} \left[(R_i + M_i - x_i - L_i) \sum_{j \neq i} x_j \right]$$

$$= \sum_{i=1}^{d} \left[(R_i - L_i) \sum_{j \neq i} x_j + (M_i - x_i) \sum_{j \neq i} x_j \right]$$

$$X - disc(c_R) = \sum_{i=1}^{d} \left[(L_i + x_i - R_i) \sum_{j \neq i} (M_j - x_j) \right]$$

$$= \sum_{i=1}^{d} \left[(R_i - L_i) \sum_{j \neq i} (x_j - M_j) + x_i \sum_{j \neq i} (M_j - x_j) \right].$$

Hence

$$X = \frac{disc(c_L) + disc(c_R)}{2} + \sum_{i \neq j} x_i(M_j - x_j) + \sum_{i=1}^{d}\left[(R_i - L_i)\sum_{j \neq i}\left(x_j - \frac{M_j}{2}\right)\right].$$

Then we have

$$E(X) = \frac{disc(c_L) + disc(c_R)}{2} + \sum_{i \neq j} E(x_i)(M_j - E(x_j))$$

$$+ \sum_{i=1}^{d}\left[(R_i - L_i)\sum_{j \neq i}\left(E(x_j) - \frac{M_j}{2}\right)\right]$$

$$= \frac{disc(c_L) + disc(c_R)}{2} + \frac{1}{4}\sum_{i \neq j} M_i M_j.$$

In the consequence we have

$$E(X) = \frac{disc(c_L) + disc(c_R) + conflict(c_L; c_R)}{2};$$

$$X - E(X) = \sum_{i \neq j}\left(x_i - \frac{M_i}{2}\right)\left[(R_j - L_j) - \left(x_j - \frac{M_j}{2}\right)\right].$$

Thus

$$D^2(X) = E\left([X - E(X)]^2\right)$$

$$= \sum_{i=1}^{n}\left[\frac{M_i(M_i + 2)}{12}\left(\sum_{j \neq i}(R_j - L_j)\right)^2\right]$$

which complete the proof.   ∎

One can use formulas (65) and (66) to construct a measure estimating quality of the best cut in $[c_L, c_R]$

$$Eval\left([c_L, c_R], \alpha\right) = E(W(c)) + \alpha\sqrt{D^2(W(c))}, \tag{67}$$

where the real parameter α from $[0, 1]$ can be tuned in the learning process. The details of this method are presented in Algorithm 11.

One can see that to determine the value $Eval\left([c_L, c_R], \alpha\right)$ we need to have the class distributions $(L_1, \ldots, L_d)$, $(M_1, \ldots, M_d)$ and $(R_1, \ldots, R_d)$ of the attribute a in $(-\infty, c_L)$, $[c_L, c_R)$ and $[c_R, \infty)$. This requires only $O(d)$ simple SQL queries of the form:

```
SELECT COUNT
FROM DecTable
WHERE (attribute_a BETWEEN value_1 AND value_2) AND (dec = i)
```

Algorithm 11. Localization of optimal cuts

Input: Attribute a and a set of candidate cuts $\mathbf{C}_a = \{c_1, \ldots, c_N\}$ on a;
Two parameters: $k \in \mathbb{N}$ and $\alpha \in [0, 1]$.
Output: Semi-optimal cut $c \in \mathbf{C}_a$.

1 **begin**
2 $\quad\quad c_{i_{\min}} := \min_k\{Median(k)\};$
3 $\quad\quad c_{i_{\max}} := \max_k\{Median(k)\};$
4 $\quad\quad Left := i_{\min}; \ Right := i_{\max};$
5 $\quad\quad$ **while** *(Left < Right)* **do**
6 $\quad\quad\quad\quad$ Divide $[Left; Right]$ into k intervals with equal length by $(k+1)$
$\quad\quad\quad\quad$ boundary points, i.e.,

$$p_i = Left + i * \frac{Right - Left}{k};$$

$\quad\quad\quad\quad$ for $i = 0, \ldots, k$.
7 $\quad\quad\quad\quad$ **for** $i = 1, \ldots, k$ **do**
8 $\quad\quad\quad\quad\quad\quad$ compute $Eval([c_{p_{i-1}}, c_{p_i}], \alpha)$ using Formula (67);
9 $\quad\quad\quad\quad$ **end**
10 $\quad\quad\quad\quad [p_{j-1}; p_j] :=$ the interval with maximal value of $Eval(.);$
11 $\quad\quad\quad\quad Left := p_{j-1}; \ Right := p_j;$
12 $\quad\quad$ **end**
13 $\quad\quad$ **Return** the cut $c_{Left};$
14 **end**

Hence, the number of queries required for running our algorithm is of order O $(dk \log_k N)$. In practice we set $k = 3$ because the function $f(k) = dk \log_k N$ over positive integers is taking minimum for $k = 3$. For $k > 2$, instead of choosing the best interval $[p_{i-1}, p_i]$, the algorithm can select the best union $[p_{i-m}, p_i]$ of m consecutive intervals in every step for a predefined parameter $m < k$. The modified algorithm needs more – but still of order $O(\log N)$ – simple questions only.

10.3.2   Examples

We consider a data table consisting of 12000 records. Objects are classified into 3 decision classes with the distribution $(5000, 5600, 1400)$, respectively. One real value attribute has been selected and $N = 500$ cuts on its domain has generated class distributions (histograms) as shown in Fig. 42.

The medians of classes are c_{166}, c_{414} and c_{189}, respectively. The median of every decision class has been determined by the *binary search algorithm* using $\log N = 9$ simple queries. Applying Theorem 28 we conclude that it is enough to consider only cuts from $\{c_{166}, \ldots, c_{414}\}$. In this way 251 cuts have been eliminated by using 27 simple queries only.

In Fig. 43 we show the graph of $W(c_i)$ for $i \in \{166, \ldots, 414\}$ and we illustrated the outcome of application of our algorithm to reduce the set of cuts for $k = 2$ and $\alpha = 0$.

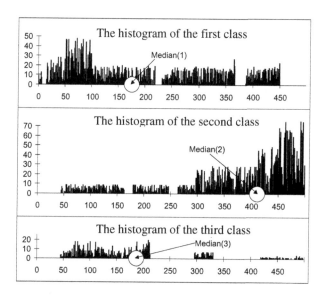

Fig. 42. Histograms of the first, the second and the third decision classes, respectively

First, the cut c_{290} is chosen and it is necessary to determine to which of the intervals $[c_{166}, c_{290}]$ and $[c_{290}, c_{414}]$ the best cut belongs. The values of function *Eval* on these intervals are computed:

$$Eval([c_{166}, c_{290}], 0) = 23927102, \quad Eval([c_{290}, c_{414}], 0) = 24374685.$$

Hence, the best cut is predicted to belong to $[c_{290}, c_{414}]$ and the search process is reduced to the interval $[c_{290}, c_{414}]$. The above procedure is repeated recursively until the selected interval consists of single cut only. For our example, the best cut c_{296} has been successfully selected by our algorithm. In general, the cut selected by the algorithm is not necessarily the best one. However, numerous experiments on different large data sets showed that the cut c^* returned by the algorithm is close to the best cut c_{Best} (i.e., $\frac{W(c^*)}{W(c_{Best})} \cdot 100\%$ is about 99.9%).

10.3.3 Local and Global Search

The algorithm presented above is called also *"local search strategy"*. In local search algorithm, we are looking for the best cuts for each attribute separately. Next, we compare all obtained best cuts to find out the global best one. This is a typical search strategy for decision tree construction (see [124]).

The approximate measure makes possible to construct *"global search strategy"* for best cuts. This strategy becomes helpful if we want to control the computation time, because it performs both attribute selection and cut selection processes at the same time.

The global strategy is searching for the best cut over all attributes. At the beginning, the best cut can be relative to every attribute, hence for each attribute

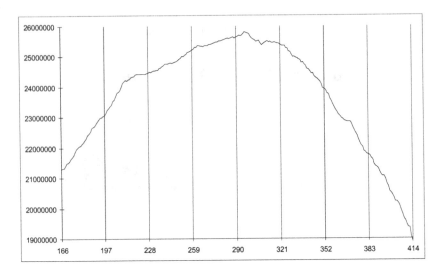

Fig. 43. Graph of $W(c_i)$ for $i \in \{166, \ldots, 414\}$

we keep the interval in which the best cut can be found (see Theorem 28), i.e., we have a collection of all potential intervals

$$\mathbf{Interval\_Lists} = \{(a_1, l_1, r_1), (a_2, l_2, r_2), \ldots, (a_k, l_k, r_k)\}.$$

Next we iteratively run the following procedure:

- *remove the interval $I = (a, c_L, c_R)$ having highest probability of containing the best cut (using Formula 67);*
- *divide interval I into smaller ones $I = I_1 \cup I_2 \cdots \cup I_k$;*
- *insert $I_1, I_2, \ldots, I_k$ to* **Interval_Lists**.

These iterative steps can be continued until we have one–element interval or the time limit of searching algorithm is exhausted. This strategy can be simply implemented using priority queue to store the set of all intervals, where the priority of intervals is defined by Formula (67).

10.3.4   Approximate Measures

We presented the approximate discernibility measure with respect to the fully independent assumption, i.e., distribution of objects from each decision class in $[c_L, c_R]$ is independent of the others.

In this section, we consider the problem of searching for approximation of discernibility measure under *"fully dependent assumption"* as well as approximate entropy measure under both independent and dependent assumptions.

The full dependency is based on the assumption that the values $x_1, \ldots, x_d$ are proportional to $M_1, \ldots, M_d$, i.e.,

$$\frac{x_1}{M_1} \simeq \frac{x_2}{M_2} \simeq \cdots \simeq \frac{x_d}{M_d}.$$

Let $x = x_1 + \cdots + x_d$ and let $t = \frac{x}{M}$, we have

$$x_1 \simeq M_1 \cdot t; \qquad x_2 \simeq M_2 \cdot t; \qquad \cdots \qquad x_d \simeq M_d \cdot t, \qquad (68)$$

where t is a real number from $[0, 1]$.

Approximate discernibility measure. The following theorem has been proved in [84]:

Theorem 30. *Under fully independent assumption, the quality of the interval $[c_R, c_L]$ can be evaluated by*

$$Eval([c_L, c_R]) = \frac{W(c_L) + W(c_R) + conflict([c_L, c_R])}{2} + \frac{[W(c_R) - W(c_L)]^2}{8 \cdot conflict([c_L, c_R])}$$

if $|W(c_R) - W(c_L)| < 2 \cdot conflict([c_L, c_R])$.
Otherwise, it is evaluated by $\max\{W(c_L), W(c_R)\}$.

One can see that under both dependent and independent assumptions, the discernibility measure of best cut in the interval $[c_R, c_L]$ can be evaluated by the same component

$$\frac{W(c_L) + W(c_R) + conflict([c_L, c_R])}{2}$$

and it is extended by the second component Δ, where

$$\Delta = \frac{[W(c_R) - W(c_L)]^2}{8 \cdot conflict([c_L; c_R])} \qquad \text{(under fully dependent assumption)}$$

$$\Delta = \alpha \cdot \sqrt{D^2(W(c))} \qquad \text{(under fully independent assumption)}$$

for some $\alpha \in [0, 1]$.

Moreover, under fully dependent assumption, one can predict the placement of the best cut. This observation is very useful in the construction of efficient algorithms.

Approximate entropy measures: Recall that in the standard entropy-based methods for decision tree induction (see [123]) we need the following notions:

1. *Information measure* of the set of objects U

$$Ent(U) = -\sum_{j=1}^{d} \frac{N_j}{N} \log \frac{N_j}{N} = -\sum_{j=1}^{d} \frac{N_j}{N} (\log N_j - \log N)$$

$$= \log N - \frac{1}{N} \sum_{j=1}^{d} N_j \log N_j = \frac{1}{N} \left(N \log N - \sum_{j=1}^{d} N_j \log N_j \right)$$

$$= \frac{1}{N} \left(h(N) - \sum_{j=1}^{d} h(N_j) \right),$$

where $h(x) = x \log x$.

2. *Information Gain* over the set of objects U received by the cut (a, c) is defined by

$$Gain(a, c; U) = Ent(U) - \left(\frac{|U_L|}{|U|} Ent(U_L) + \frac{|U_R|}{|U|} Ent(U_R)\right),$$

where $\{U_L, U_R\}$ is a partition of U defined by c. We have to choose such a cut (a, c) that maximizes the *information gain* $Gain(a, c; U)$ or minimizes the *entropy induced by this cut*.

$$Ent(a, c; U) = \frac{|U_L|}{|U|} Ent(U_L) + \frac{|U_R|}{|U|} Ent(U_R)$$

$$= \frac{L}{N}\left[\frac{1}{L}\left(h(L) - \sum_{j=1}^{d} h(L_j)\right)\right] + \frac{R}{N}\left[\frac{1}{R}\left(h(R) - \sum_{j=1}^{d} h(R_j)\right)\right]$$

$$= \frac{1}{N}\left[h(L) - \sum_{j=1}^{d} h(L_j) + h(R) - \sum_{j=1}^{d} h(R_j)\right],$$

where $(L_1, \ldots, L_d), (R_1, \ldots, R_d)$ are class distribution of U_L and U_R, respectively.

Analogously to the discernibility measure case, the main goal is to predict the quality of the best cut (in sense of the entropy measure) among those from the interval $[c_L, c_R]$, i.e., $Ent(a, c; U) = \frac{1}{N} f(x_1, \ldots, x_d)$, where

$$f(x_1, \ldots, x_d) = h(L + x) - \sum_{j=1}^{d} h(L_j + x_j) + h(R + M - x) - \sum_{j=1}^{d} h(R_j + M_j - x_j).$$

We have presented in [84] the approximate entropy measure under both independent and dependent assumptions.

- **Approximate entropy measure under fully independent assumption:** is defined by the average value of entropy of cuts $c \in (c_L, c_R)$. This value can be evaluated by

$$Ent(c_L, c_R) = H(L, M) - \sum_{j=1}^{d} H(L_j, M_j) + H(R, M) - \sum_{j=1}^{d} H(R_j, M_j),$$

where

$$H(a, b) = \frac{(a + b)h(a + b) - ah(a)}{2b} - \frac{2a + b}{4 \ln 2}$$

- **Approximate entropy measure under fully dependent assumption:** can be defined by the minimum of the following function

$$f(t) = h(L + M \cdot t) - \sum_{j=1}^{d} h(L_j + M_j \cdot t) + h(R + M - M \cdot t) - \sum_{j=1}^{d} h(R_j + M_j - M_j \cdot t)$$

for $t \in [0, 1]$. It has been shown in [84] that $f'(t)$ is increasing in $[0, 1]$. This fact can be used to find the value t_0 for which $f'(t_0) = 0$. If such t_0 exists, the function f achieves maximum at t_0. Hence, one can predict the the entropy measure of the best cut in the interval $[c_L, c_R]$ (under assumption about strong dependencies between classes) as follows:

- If $f'(1) \geq 0$ then $f'(t) > 0$ for any $t \in (0; 1)$, i.e., $f(t)$ is increasing function. Hence, c_R is the best cut.
- If $f'(0) \leq 0$ then $f'(t) \leq 0$ for any $t \in (0; 1)$, i.e., $f(t)$ is decreasing function. Hence, c_L is the best cut.
- If $f'(0) < 0 < f'(1)$ then locate the root t_0 of $f'(t)$ using "Binary Search Strategy". Then the best cut in $[c_L, c_R]$ can be estimated by $\frac{1}{N} f(t_0)$.

10.4   Soft Cuts and Soft Decision Trees

The standard discretization methods and decision tree methods are working with crisp partitions defined by cuts, which are partitioning the real axis into disjoint intervals.

Except the computational problem that occurs in large data tables, there is another ideological problem related to the usage of crisp cuts to object discerning. This can lead to misclassification of new objects which are very close to the cut points, and this fact can result in low quality of new object classification.

In this paper, we propose a novel approach based on *soft cuts* which makes it possible to overcome the second problem. Decision trees using soft cuts as test functions are called *soft decision trees*. The new approach leads to new efficient strategies in searching for the best cuts (both soft and crisp cuts) using the whole data. We show some properties of considered optimization measures that allows us to reduce the size of the searching space. Moreover, we prove that using only $O(\log N)$ simple queries, one can construct soft partitions that are very close to the optimal one.

In this section, we introduce *soft cuts* discerning two given values if those values are far enough from the cut. The formal definition of a soft cut is the following:

A soft cut is any triple $p = \langle a, l, r \rangle$, where $a \in A$ is an attribute, $l, r \in \Re$ are called the left and right bounds of p ($l \leq r$); the value $\varepsilon = \frac{r-l}{2}$ is called the uncertainty radius of p. We say that a soft cut p discerns pair of objects x_1, x_2 that $a(x_1) < a(x_2)$ if $a(x_1) < l$ and $a(x_2) > r$.

The intuitive meaning of $p = \langle a, l, r \rangle$ is that there is a real cut somewhere between l and r. So we are not sure where one can place the real cut in the interval $[l, r]$. Hence, for any value $v \in [l, r]$ we are not able to check if v is either on the left side or on the right side of the real cut. Then, we say that the interval $[l, r]$ is an uncertain interval of the soft cut p. Any normal cut can be treated as soft cut of radius equal to 0.

Any set of soft cuts splits the real axis into intervals of two categories: the intervals corresponding to new nominal values and the intervals of uncertain

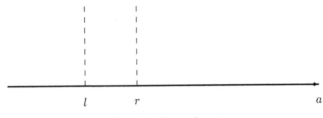

Fig. 44. The soft cut

values called boundary regions. The problem of searching for the minimal set of soft cuts with a given uncertainty radius can be solved in a similar way to the case of sharp cuts. Some heuristics for this problem are described in the next section. The problem becomes more complicated if we want to obtain as small as possible set of soft cuts with the largest radius. We will discuss this problem in the future. Now we recall some existing rule induction methods for real value attribute data and their modifications using soft cuts.

Instead of sharp cuts (see previous sections), the soft cuts determine additionally some uncertainty regions. Assume that $\mathbf{P} = \{p_1, p_2, \ldots, p_k\}$ is a set of soft cuts on attribute $a \in A$, where $p_i = (a, l_i, r_i); l_i \leq r_i$ and $r_i < l_{i+1}$ for $i = 1, \ldots, k-1$. The set of soft cuts $\mathbf{P}$ defines on $\Re$ a partition

$$\Re = (-\infty, l_1) \cup [l_1, r_1] \cup (r_1, l_2) \cup \cdots \cup [l_k, r_k] \cup (r_k, +\infty)$$

and at the same time defines a new nominal attribute $a^{\mathbf{P}} : U \to \{0, 1, \ldots, k\}$, such that $a^{\mathbf{P}}(x) = i$ if and only if $a(x) \in (r_i, l_{i+1}); i = 1, \ldots, k$. In the following section we are proposing some possible classification methods using soft discretization. These methods are based on fuzzy set approach, rough set approach, clustering approach and decision tree approach.

10.4.1 Soft Decision Tree

The test functions defined by traditional cuts can be replaced by soft cuts. We have proposed two strategies being modifications of the standard classification method for decision tree with soft cuts [94]. They are called *fuzzy decision tree* and *rough decision tree*.

In fuzzy decision tree method instead of checking the condition $a(u) > c$ we have to check how strong is the hypothesis that u is on the left or the right side of the cut (a, c). This condition can be expressed by $\mu_L(u)$ and $\mu_R(u)$, where μ_L and μ_R are the membership function of the left and the right interval (respectively). The values of these membership functions can be treated as a probability distribution of u in the node labeled by soft cut $(a, c - \varepsilon, c + \varepsilon)$. Then one can compute the probability of the event that object u is reaching a leaf. The decision for u is equal to decision labeling the leaf with the largest probability.

In the case of rough decision tree, when we are not able to decide if to turn left or right (the value $a(u)$ is too close to c) we do not distribute the probability to the children of considered node. We have to compare their answers taking into account the numbers of objects supported by them. The answer with most number of supported objects is the decision of test object.

Overfitting is the situation when the model fits exactly the data but it has a poor performance on unseen instances. Usually, overfitting is caused by the presence of noise in data. Most decision tree algorithms perform a MDL[9] based pruning phase after the building phase in which nodes are iteratively pruned to prevent overfitting.

Soft decision tree is another method to prevent the overfitting problem. This concept allows efficient abandoning small noise in data.

10.4.2   Searching for Soft Cuts

Recall that we have presented an efficient algorithm for searching for best cuts using divide and conquer strategy (Algorithm 11). One can modify this algorithm to determine "soft cuts" in large data bases. The modification is based on changing the stop condition. In every iteration of Algorithm 11, the current interval $[Left; Right]$ is divided equally into k smaller intervals and the best smaller interval will be chosen as the current interval. In the modified algorithm, one can either select one of smaller intervals as the current interval or stop the algorithm and return the current interval as a result.

Intuitively, the divide and conquer algorithm is stopped and returns the interval $[c_L; c_R]$ as a result if the following conditions hold:

- The class distribution in $[c_L; c_R]$ is stable, i.e., there is no sub-interval of $[c_L; c_R]$ which is considerably better than $[c_L; c_R]$ itself;
- The interval $[c_L; c_R]$ is sufficiently small, i.e., it contains a small number of cuts;
- The interval $[c_L; c_R]$ does not contain too much objects (because the large number of uncertain objects cans result in larger decision tree and then the time of decision tree construction prolongs).

10.4.3   Accuracy of Soft Decision Tree

In this section, we present the accuracy evaluation of soft decision tree approach. The main goal is to compare the classification accuracy of soft decision tree built from semi-optimal cuts with other decision tree techniques.

We have implemented three decision tree algorithms called "ENT" (based on entropy measure, similar to C4.5 [123]), "MD" (based on discernibility measure [82]) and "MD*"(the soft tree constructed by approximate discernibility measure). All experiments are done on "small" data set (from STATLOG project [75]) only, since the first two algorithms handle the data sets that fit in memory. We also recall the experiment results achieved by SLIQ algorithm (see [70]).

In our experiments, the standard algorithms based on entropy and discernibility measures are implemented without pruning step. The MD* algorithm is based on approximate discernibility measure (see Formula ((67))), where Δ was set to 0, i.e.

$$Eval\left([c_L; c_R]\right) = \frac{W(c_L) + W(c_R) + conflict([c_L; c_R])}{2}.$$

---

[9] MDL: "Minimal Description Length" principle.

Table 23. The experimental results of different decision tree algorithms on benchmark data

Data sets	#objects× #attr.		SLIQ	ENT	MD	MD*
Australian	690	× 14	84.9	85.2	86.2	86.2
German	1000	× 24	-	70	69.5	70.5
Heart	270	× 13	-	77.8	79.6	79.6
Letter	20000	× 16	84.6	86.1	85.4	83.4
SatImage	6435	× 36	86.3	84.6	82.6	83.9
Shuttle	57000	× 9	99.9	99.9	99.9	98.7

We used fuzzy decision tree classification method (described in Sect. 10.4) to classify new objects. From experimental results we can see that, even MD* algorithm constructs decision tree from approximate measure, its accuracy is still comparable with other exact measures.

10.4.4   Other Applications of Soft Cuts

Fuzzy Set Approach: In the fuzzy set approach, one can treat the interval $[l_i, r_i]$ for any $i \in \{1, \ldots, k\}$ as a kernel of some fuzzy set Δ_i. The membership function $f_{\Delta_i} : \Re \to [0, 1]$ is defined as follows:

1. $f_{\Delta_i}(x) = 0$ for $x < l_i$ or $x > r_{i+1}$.
2. $f_{\Delta_i}(x)$ increases from 0 to 1 for $x \in [l_i, r_i]$.
3. $f_{\Delta_i}(x)$ decreases from 1 to 0 for $x \in [l_{i+1}, r_{i+1}]$.
4. $f_{\Delta_i}(x) = 1$ for $x \in (r_i, l_{i+1})$.

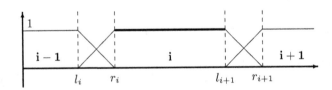

Fig. 45. Membership functions of intervals

Having defined membership function, one can use the idea of *fuzzy graph* [160] to represent the discovered knowledge.

Rough Set Approach: The boundary interval $[l_i, r_i]$ can be treated as uncertainty region for a real sharp cut. Hence, using rough set approach the intervals (r_i, l_{i+1}) and $[l_i, r_{i+1}]$ are treated as the lower and the upper approximations of any set X. Hence, we use the following notation $\mathbf{L}_a(X_i) = (r_i, l_{i+1})$ and $\mathbf{U}_a(X_i) = [l_i, r_{i+1}]$ such that $(r_i, l_{i+1}) \subseteq X \subseteq [l_i, r_{i+1}]$.

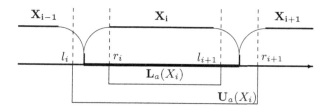

Fig. 46. Illustration of soft cuts

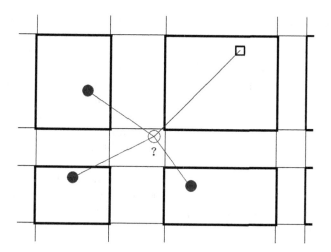

Fig. 47. Clustering approach

Having approximations of nominal values of all attributes, we can generate an upper and lower approximation of decision classes by taking a Cartesian product of rough sets. For instance, for a set X given by its rough representation $[\mathbf{L}_B(X), \mathbf{U}_B(X)]$ and for a set Y given by the representation $[\mathbf{L}_C(Y), \mathbf{U}_C(Y)]$, and let $B \cap C = \varnothing$. One can define a rough representation of $X \times Y$ by $[\mathbf{L}_{B \cup C}(X \times Y), \mathbf{U}_{B \cup C}(X \times Y)]$ where

$$\mathbf{L}_{B \cup C}(X \times Y) = \mathbf{L}_B(X) \times \mathbf{L}_C(Y)$$

and

$$\mathbf{U}_{B \cup C}(X \times Y) = \mathbf{U}_B(X) \times \mathbf{U}_C(Y).$$

Clustering Approach: Any set of soft cuts $\mathbf{P}$ defines a partition of real values of attributes into disjoint intervals, which determines a natural equivalence relation $IND(\mathbf{P})$ over the set of objects. New objects belonging to the boundary regions can be classified by applying the rough set membership function to test the hypothesis that the new object belongs to certain decision class.

One can also apply the idea of clustering. Any set of soft cuts defines a partition of $\Re^k$ into k-dimensional cubes. Using rough set approach one can classify some of these cubes to be in the lower approximation of a certain set, and they can be treated as clusters. To classify a new object belonging to any boundary cube one can compare distances from this object to the centers of adjacent clusters (see Fig. 47).

11   Conclusions

We have presented the approximate Boolean reasoning methodology as an extension of the original Boolean reasoning scheme. Theoretical foundations of the proposed method as well as many applications of rough set theory including attribute selection (calculation of reducts), decision rule induction, discretization and feature extraction have been developed. We also presented some efficient data mining algorithms based on the approximate Boolean reasoning approach. Thus, the results of this paper may be considered as belonging to the intersection of three research domains: the Boolean reasoning methods, rough sets and data mining.

We would like to emphasize the fact that approximate Boolean reasoning approach is not only a concrete method for problem solving, but it is a general methodology for development of concept approximation heuristics and data mining solutions. The secret is embedded in the first step where the investigated problem is encoded by a Boolean function. The encoding function creates the basis for designing different approximate solutions for the same problem. In many applications there is no need to construct the encoding function, but the satisfactory knowledge about it facilitates to develop appropriate approximate algorithms that fulfill some predefined requirements about the quality and complexity.

The general evaluation of data mining solutions depends on their accuracy, complexity, description length, interpretability and some other factors. Usually, the function expressing the dependency is not exactly specified by the expert and it can be approximated through an interaction. The presented data mining methods (based on the approximate Boolean reasoning methodology) make the interaction with experts easier by tuning the quality through other features. This possibility is essential for KDD which is an iterative and interactive process of identifying valid, novel, potentially useful, and ultimately understandable patterns in data [32].

This property has also a close connection with the minimal description length (MDL) principle, which was introduced by Jorma Rissanen [126], [127] as a method for inductive reasoning where the success in finding such regularities can be measured by the model length with which the data can be described. The methods proposed in this paper suggest a possibility to generalize the MDL idea by evaluating data mining methods with respect to a combined optimization criterion over such factors as the accuracy, understandability, description length, complexity, etc.

Moreover, the flexibility in modeling and designing approximate solutions for complex concepts is also a valuable property that can be applied to challenging problems like "reasoning from sensor measures to perception", "granular computing" or "computing with words".

The investigations on approximate Boolean reasoning refer to a new direction in computational complexity theory. We have noticed a regular dependency between hardness in developing accurate approximate algorithms and complexity of the encoding function. We have presented different heuristics for the same problem for which the better solution requires the heuristic with larger time and space complexity. The proposed methods can be treated as a step forward in developing of methods for checking which heuristics can be applied in a given situation, e.g., using the current possibilities of computer systems, limitations of the computing time and other available resources.

Acknowledgement

Firstly, I would like to express my deepest gratitude to my mentor and supervisor, Professor Dr hab. Andrzej Skowron. I can find no words to thank him for his excellent guidance, patience and advice through almost 20 years of our cooperation. Without his motivation and moral support, comments and criticisms, this work would not have been possible.

I am grateful to my dear daughter, Mai Lan, and my wife, Sinh Hoa, for their sincere love, unconditional and unreserving support. My greatest thanks go to my wife as a best colleague for the numerous scientific discussions, for her professional and energetic support in experiment preparation and execution, for many our common papers. Thanks.

Many of my colleagues have assisted with and offered valuable comments on the draft of my research. However I would like to thank especially Adam Grabowski and James F. Peters for proofreading this paper and suggesting improvements. All mistakes that remain are, of course, my errors.

Many of the results presented here were joint research between myself and coauthors including Jan Bazan, Chi Lang Ngo, Thanh Trung Nguyen, Tuan Trung Nguyen, Sinh Hoa Nguyen, James Peters, Andrzej Skowron, Dominik Ślęzak, Jarosław Stepaniuk, Marcin Szczuka, Hui Wang, Jan Komorowski, Marta and Karina Łuksza, Ewa Mąkosa. Here I wish to thank each of these colleagues for pleasant and rewarding collaboration.

I would like to thank colleagues from the School of Computing and Mathematics at University of Ulster, the Linnaeus Centre for Bioinformatics at Uppsala University and the College of Technology at the Vietnam National University for their support and cooperation during my visits to these places.

Financial support for the study was provided by the grant 3T11C00226 from Ministry of Scientific Research and Information Technology of the Republic of Poland.

My real gratitude is greatly indebted to my parents and my parents-in-law for their invaluable love, inestimable patience and innumerable sacrifices.

Bibliography

[1] R. Agrawal, T. Imielinski, and A. N. Swami. Mining Association Rules between Sets of Items in Large Databases. In P. Buneman and S. Jajodia, editors, *ACM SIGMOD International Conference on Management of Data*, pages 207–216, Washington, DC, 26-28 1993.

[2] R. Agrawal, H. Mannila, R. Srikant, H. Toivonen, and A. I. Verkamo. Fast Discovery of Association Rules. In *Advances in Knowledge Discovery and Data Mining*, pages 307–328, Menlo Park, CA, 1996. AAAI Press/The MIT Press.

[3] R. Agrawal and R. Srikant. Fast Algorithms for Mining Association Rules. In J. B. Bocca, M. Jarke, and C. Zaniolo, editors, *Twentieth International Conference on Very Large Data Bases VLDB*, pages 487–499. Morgan Kaufmann, May 12-15 1994.

[4] K. Alsabti, S. Ranka, and V. Singh. CLOUDS: A Decision Tree Classifier for Large Datasets. In *Knowledge Discovery and Data Mining*, pages 2–8, 1998.

[5] H. Andrade, T. Kurc, A. Sussman, and J. Saltz. Decision tree construction for data mining on clusters of shared-memory multiprocessors. Technical Report CS-TR-4203 and UMIACS-TR-2000-78. University of Maryland, Department of Computer Science and UMIACS, Dec. 2000.

[6] M. Anthony and N. Biggs. *Computational learning theory: an introduction*, volume 30 of *Cambridge Tracts in Theoretical Computer Science*. Cambridge University Press, 1992.

[7] J. Bazan. A Comparison of Dynamic and non-Dynamic Rough Set Methods for Extracting Laws from Decision Tables. In L. Polkowski and A. Skowron, editors, *Rough Sets in Knowledge Discovery 1: Methodology and Applications*, volume 18 of *Studies in Fuzziness and Soft Computing*, chapter 17, pages 321–365. Springer, Heidelberg, Germany, 1998.

[8] J. Bazan, H. S. Nguyen, S. H. Nguyen, P. Synak, and J. Wróblewski. Rough set algorithms in classification problems. In Polkowski et al. [116], pages 49–88.

[9] J. Bazan, H. S. Nguyen, A. Skowron, and M. Szczuka. A View on Rough Set Concept Approximations. In G. Wang, Q. Liu, Y. Yao, and A. Skowron, editors, *Rough Sets, Fuzzy Sets, Data Mining, and Granular Computing. Proceedings of RSFDGrC 2003*, volume 2639 of *Lecture Notes in Artificial Intelligence*, pages 181–188, Chongqing, China, 2003. Springer.

[10] J. Bazan, A. Skowron, and P. Synak. Dynamic reducts as a tool for extracting laws from decision tables. In *International Symposium on Methodologies for Intelligent Systems ISMIS*, volume 869 of *Lecture Notes in Artificial Intelligence*, pages 346–355, Charlotte, NC, October 16-19 1994. Springer.

[11] J. Bazan and M. Szczuka. RSES and RSESlib – A collection of tools for rough set computations. In W. Ziarko and Y. Yao, editors, *Proceedings of the 2nd International Conference on Rough Sets and Current Trends in Computing (RSCTC'2000), Banff, Canada, October 16-19, 2000*, volume 2005 of *Lecture Notes in Artificial Intelligence*, pages 106–113, Heidelberg, Germany, 2001. Springer.

[12] A. Blake. *Canonical Expressions in Boolean Algebra*. PhD thesis, University of Chicago, 1937.

[13] G. Boole. *Mathematical Analysis of Logic*. London, 1847.

[14] G. Boole. *The Law of Thought*. MacMillan (also Dover Publications, New-York), 1854.

[15] L. Breiman, J. Friedman, R. Olshen, and C. Stone. *Classification and Regression Trees*. Wadsworth and Brooks, Monterey, CA, 1984.

[16] F. Brown. *Boolean Reasoning*. Kluwer Academic Publishers, Dordrecht, Germany, 1990.

[17] J. Catlett. On Changing Continuous Attributes into Ordered Discrete Attributes. In Y. Kodratoff, editor, *European Working Session on Learning, Machine Learning - EWSL-91*, volume 482 of *Lecture Notes in Computer Science*, pages 164–178. Springer, 1991.

[18] A. K. Chandra and G. Markowsky. On the number of prime implicants. *Discrete Mathematics*, 24(1):7–11, 1978.

[19] C.-L. Chang and R. C.-T. Lee. *Symbolic Logic and Mechanical Theorem Proving*. Academic Press, London, 1973.

[20] B. S. Chlebus and S. H. Nguyen. On Finding Optimal Discretizations for Two Attributes. In *First International Conference on Rough Sets and Soft Computing RSCTC'1998*, pages 537–544.

[21] M. R. Chmielewski and J. W. Grzymała-Busse. Global discretization of continuous attributes as preprocessing for machine learning. *Int. J. Approx. Reasoning*, 15(4):319–331, 1996.

[22] P. Clark and R. Boswell. Rule induction with CN2: Some recent improvements. In *Proc. Fifth European Working Session on Learning*, pages 151–163, Berlin, 1991. Springer.

[23] P. Clark and T. Niblett. The CN2 induction algorithm. *Machine Learning*, 3(4):261–283, 1989.

[24] W. W. Cohen. Fast Effective Rule Induction. In *Proceedings of the Twelfth International Conference on Machine Learning (ICML-95)*, pages 115–123, San Francisco, CA, 1995.

[25] W. W. Cohen. Learning Trees and Rules with Set-Valued Features. In *Proceedings of the Thirteenth National Conference on Artificial Intelligence (AAAI-96)*, pages 709–716, Portland, OR, Aug. 1996.

[26] M. Davis, G. Logemann, and D. Loveland. A machine program for theorem proving. *Communications of the ACM*, 5(7):394–397, July 1962.

[27] M. Davis and H. Putnam. A computing procedure for quantification theory. *Journal of the ACM*, 7(3):201–215, July 1960.

[28] J. Dougherty, R. Kohavi, and M. Sahami. Supervised and Unsupervised Discretization of Continuous Features. In *ICML*, pages 194–202, 1995.

[29] I. Duentsch, G. Gediga, and H. S. Nguyen. Rough Set Data Analysis in the KDD Process. In *Proc. of 8th International Conference IPMU'2000*, pages 220–226, Madrid, Spain, July 2000. Universidad Politechnica de Madrid.

[30] F. Esposito, D. Malerba, and G. Semeraro. A Comparative Analysis of Methods for Pruning Decision Trees. *IEEE Transactions on Pattern Analysis and Machine Intelligence*, 19(5):476–491, 1997.

[31] U. M. Fayyad and K. B. Irani. Multi-Interval Discretization of Continuous-Valued Attributes for Classification Learning. In *IJCAI*, pages 1022–1029, 1993.

[32] U. M. Fayyad, G. Piatetsky-Shapiro, P. Smyth, and R. Uthurusamy, editors. *Advances in Knowledge Discovery and Data Mining*. The AAAI Press/The MIT Press, Cambridge, MA, 1996.

[33] J. H. Friedman, R. Kohavi, and Y. Yun. Lazy Decision Trees. In *Thirteenth National Conference on Artificial Intelligence and Eighth Innovative Applications of Artificial Intelligence Conference, AAAI/IAAI 96, Vol. 1*, pages 717–724, 1996.

[34] H. Gallaire and J. Minker, editors. *Logic and Databases*, New York, 1978. Plenum Press.

[35] M. R. Garey and D. S. Johnson. *Computers and Intractability: A Guide to the Theory of NP-Completeness*. W.H. Freeman & Co., New York, NY, 1979.

[36] E. Goldberg and Y. Novikov. BerkMin: A fast and robust SAT-solver. In *Proceedings of DATE-2002*, pages 142–149, 2002.

[37] J. Goldsmith, M. Hagen, and M. Mundhenk. Complexity of DNF and Isomorphism of Monotone Formulas. In J. Jedrzejowicz and A. Szepietowski, editors, *Mathematical Foundations of Computer Science 2005, Proceedings of 30th International Symposium, MFCS 2005, Gdansk, Poland*, volume 3618 of *Lecture Notes in Computer Science*, pages 410–421. Springer, 2005.

[38] S. Greco, B. Matarazzo, and R. Słowiński. Data mining tasks and methods: Classification: multicriteria classification. In W. Kloesgen and J. Żytkow, editors, *Handbook of KDD*, pages 318–328. Oxford University Press, Oxford, 2002.

[39] J. W. Grzymała-Busse. LERS – A System for Learning from Examples based on Rough Sets. In Słowiński [147], pages 3–18.

[40] J. W. Grzymała-Busse. LERS – A Data Mining System. In Maimon and Rokach [67], pages 1347–1351.

[41] J. W. Grzymała-Busse. Rule Induction. In Maimon and Rokach [67], pages 277–294.

[42] J. W. Grzymała-Busse and W. Ziarko. Data Mining and Rough Set Theory. *Communications of the ACM*, 43:108–109, 2000.

[43] J. Han and M. Kamber. *Data Mining: Concepts and Techniques*. Morgan Kaufmann Publishers Inc., San Francisco, CA, USA, 2000.

[44] J. Han, J. Pei, and Y. Yin. Mining frequent patterns without candidate generation. In W. Chen, J. Naughton, and P. A. Bernstein, editors, *2000 ACM SIGMOD Intl. Conference on Management of Data*, pages 1–12. ACM Press, 05 2000.

[45] D. Hand, H. Mannila, and P. Smyth. *Principles of Data Mining*. MIT Press, 2001.

[46] F. Hayes-Roth, D. A. Waterman, and D. B. Lenat. An Overview of Expert Systems. In F. Hayes-Roth, D. A. Waterman, and D. B. Lenat, editors, *Building Expert Systems*, pages 3–29. Addison-Wesley, London, 1983.

[47] D. G. Heath, S. Kasif, and S. Salzberg. Induction of Oblique Decision Trees. In *IJCAI*, pages 1002–1007, 1993.

[48] R. C. Holte. Very simple classification rules perform well on most commonly used datasets. *Machine Learning*, 11:63–91, 1993.

[49] E. V. Huntington. Boolean algebra. A correction. *Transactions of AMS*, 35:557–558, 1933.

[50] R. Jensen, Q. Shen, and A. Tuso. Finding rough set reducts with SAT. In D. Ślęzak, G. Wang, M. Szczuka, I. Düntsch, and Y. Yao, editors, *Proceedings of the 10th International Conference on Rough Sets, Fuzzy Sets, Data Mining, and Granular Computing (RSFDGrC'2005), Regina, Canada, August 31-September 3, 2005, Part I*, volume 3641 of *Lecture Notes in Artificial Intelligence*, pages 194–203. Springer, Heidelberg, Germany, 2005.

[51] R. G. Jeroslow. *Logic-Based Decision Support. Mixed Integer Model Formulation.* Elsevier, Amsterdam, 1988.

[52] G. H. John and P. Langley. Static Versus Dynamic Sampling for Data Mining. In E. Simoudis, J. Han, and U. M. Fayyad, editors, *Proc. 2nd Int. Conf. Knowledge Discovery and Data Mining, KDD*, pages 367–370. AAAI Press, 2-4 1996.

[53] H. A. Kautz and B. Selman. Planning as Satisfiability. In *Proceedings of the Tenth European Conference on Artificial Intelligence (ECAI'92)*, pages 359–363, 1992.

[54] H. A. Kautz and B. Selman. Pushing the Envelope : Planning, Propositional Logic, and Stochastic Search. In *Proceedings of the Twelfth National Conference on Artificial Intelligence (AAAI'96)*, pages 1194–1201, 1996.

[55] R. Keefe. *Theories of Vagueness.* Cambridge Studies in Philosophy, Cambridge, UK, 2000.

[56] R. Kerber. ChiMerge: Discretization of Numeric Attributes. In *Proceedings of the Tenth National Conference on Artificial Intelligence*, pages 123–128, San Jose, CA, 1992. AAAI Press.

[57] W. Kloesgen and J. Żytkow, editors. *Handbook of Knowledge Discovery and Data Mining.* Oxford University Press, Oxford, 2002.

[58] J. Komorowski, Z. Pawlak, L. Polkowski, and A. Skowron. Rough sets: a tutorial. In S. K. Pal and A. Skowron, editors, *Rough Fuzzy Hybridization: A New Trend in Decision-Making*, pages 3–98. Springer, Singapore, 1999.

[59] R. A. Kowalski. *Logic for problem solving.* North Holland, New York, 1980.

[60] M. Kryszkiewicz. Maintenance of reducts in the variable precision rough set model. In T. Y. Lin and N. Cercone, editors, *Rough Sets and Data Mining – Analysis of Imperfect Data*, pages 355–372. Kluwer Academic Publishers, Boston, USA, 1997.

[61] M. Kryszkiewicz and K. Cichoń. Towards Scalable Algorithms for Discovering Rough Set Reducts. In J. F. Peters, A. Skowron, D. Dubois, J. W. Grzymała-Busse, M. Inuiguchi, and L. Polkowski, editors, *Transactions on Rough Sets II. Rough sets and fuzzy sets: Journal Subline*, volume 3135 of *Lecture Notes in Computer Science*, pages 120–143, Heidelberg, Germany, 2004. Springer.

[62] H. Liu, F. Hussain, C. L. Tan, and M. Dash. Discretization: An Enabling Technique. *Data Mining Knowledge Discovery*, 6(4):393–423, 2002.

[63] H. Liu and H. Motoda, editors. *Feature Selection for Knowledge Discovery and Data Mining*. Kluwer Academic Publishers, 1999.

[64] H. Liu and R. Setiono. Chi2: Feature Selection and Discretization of Numeric Attributes. In *TAI '95: Proceedings of the Seventh International Conference on Tools with Artificial Intelligence*, page 88, Washington, DC, USA, 1995. IEEE Computer Society.

[65] D. W. Loveland. *Automated Theorem Proving. A Logical Basis*, volume 6 of *Fundamental Studies in Computer Science*. North-Holland, 1978.

[66] Y. Z. M. Moskewicz, C. Madigan and L. Zhang. Chaff: Engineering and efficient SAT solver. In *Proceedings of 38th Design Automation Conference (DAC2001)*, June 2001.

[67] O. Maimon and L. Rokach, editors. *The Data Mining and Knowledge Discovery Handbook*. Springer, Heidelberg, Germany, 2005.

[68] V. M. Manquinho, P. F. Flores, J. P. M. Silva, and A. L. Oliveira. Prime Implicant Computation Using Satisfiability Algorithms. In *9th International Conference on Tools with Artificial Intelligence (ICTAI '97)*, pages 232–239, 1997.

[69] J. P. Marques-Silva and K. A. Sakallah. GRASP – A New Search Algorithm for Satisfiability. In *Proceedings of IEEE/ACM International Conference on Computer-Aided Design*, pages 220–227, November 1996.

[70] M. Mehta, R. Agrawal, and J. Rissanen. SLIQ: A Fast Scalable Classifier for Data Mining. In *Extending Database Technology*, pages 18–32, 1996.

[71] M. Mehta, J. Rissanen, and R. Agrawal. MDL-Based Decision Tree Pruning. In *Proceedings of the First International Conference on Knowledge Discovery and Data Mining (KDD'95)*, pages 216–221, 1995.

[72] Z. Michalewicz. *Genetic Algorithms + Data Structures = Evolution Programs*. Springer Verlag, New York, NY, 1994.

[73] R. Michalski. Discovery Classification Rules Using Variable-valued Logic System VL1. In *Proceedings of the Third International Conference on Artificial Intelligence*, pages 162–172. Stanford University, 1973.

[74] R. Michalski, I. Mozetič, J. Hong, and N. Lavrač. The multi-purpose incremental learning system AQ15 and its testing application on three medical domains. In *Proc. Fifth National Conference on Artificial Intelligence*, pages 1041–1045, San Mateo, CA, 1986. Morgan Kaufmann.

[75] D. Michie, D. J. Spiegelhalter, and C. Taylor. *Machine learning, Neural and Statistical Classification*. Ellis Horwood, New York, NY, 1994.

[76] J. Mingers. An empirical comparison of pruning methods for decision tree induction. *Machine Learning*, 4(2):227–243, 1989.

[77] T. Mitchell. *Machine Learning*. Mc Graw Hill, 1998.

[78] S. K. Murthy, S. Kasif, and S. Salzberg. A System for Induction of Oblique Decision Trees. *Journal of Artificial Intelligence Research*, 2:1–32, 1994.

[79] H. S. Nguyen. *Discretization of Real Value Attributes, Boolean Reasoning Approach*. PhD thesis, Warsaw University, Warsaw, Poland, 1997.

[80] H. S. Nguyen. Rule induction from Continuous Data: New Discretization Concepts. In *Proc. of the III Joint Conference on Information Sciences*, volume 3, pages 81–84, NC, USA, March 1997. Duke University.

[81] H. S. Nguyen. Discretization Problems for Rough Set Methods. In L. Polkowski and A. Skowron, editors, *New Directions in Rough Sets, Data Mining and Granular-Soft Computing (Proc. of RSCTC'98, Warsaw, Poland)*, LNAI 1424, pages 545–552. Springer, Berlin Heidelberg, 1998.

[82] H. S. Nguyen. From Optimal Hyperplanes to Optimal Decision Trees. *Fundamenta Informaticae*, 34(1–2):145–174, 1998.

[83] H. S. Nguyen. Efficient SQL-Querying Method for Data Mining in Large Data Bases. In *Proc. of Sixteenth International Joint Conference on Artificial Intelligence, IJCAI-99*, pages 806–811, Stockholm, Sweden, 1999. Morgan Kaufmann.

[84] H. S. Nguyen. On Efficient Handling of Continuous Attributes in Large Data Bases. *Fundamenta Informaticae*, 48(1):61–81, 2001.

[85] H. S. Nguyen. On Exploring Soft Discretization of Continuous Attributes. In S. K. Pal, L. Polkowski, and A. Skowron, editors, *Rough-Neuro-Computing Techniques for Computing with Words*, pages 333–350. Springer-Verlag New York, Inc., 2002.

[86] H. S. Nguyen. Scalable Classification Method Based on Rough Sets. In J. J. Alpigini, J. F. Peters, A. Skowron, and N. Zhong, editors, *Rough Sets and Current Trends in Computing (Proc. RSCTC'2002)*, LNAI 2475, pages 433–440. Springer, Berlin Heidelberg, 2002.

[87] H. S. Nguyen. A Soft Decision Tree. In M. A. Kłopotek, S. Wierzchoń, and M. Michalewicz, editors, *Intelligent Information Systems 2002 (Proc. IIS'2002)*, Advanced in Soft Computing, pages 57–66. Springer, Berlin Heidelberg, 2002.

[88] H. S. Nguyen. On Exploring Soft Discretization of Continuous Attributes. In S. K. Pal, L. Polkowski, and A. Skowron, editors, *Rough-Neural Computing: Techniques for Computing with Words*, Cognitive Technologies, pages 333–350. Springer Verlag, 2003.

[89] H. S. Nguyen. On the Decision Table with Maximal Number of Reducts. In A. Skowron and M. Szczuka, editors, *Electronic Notes in Theoretical Computer Science*, volume 82. Elsevier, 2003.

[90] H. S. Nguyen. Approximate Boolean Reasoning Approach to Rough Sets and Data Mining. In D. Ślęzak, J. Yao, J. F. Peters, W. Ziarko, and X. Hu, editors, *Rough Sets, Fuzzy Sets, Data Mining, and Granular Computing, 10th International Conference, RSFDGrC 2005, Regina, Canada, August 31 – September 3, 2005, Proceedings, Part II*, LNCS 3642, pages 12–22. Springer-Verlag, 2005.

[91] H. S. Nguyen. *Rough Sets Approach to Learning in MAS – A tutorial*. Compiegne University, Compiegne, France, tutorial notes of the 2005 IEEE/WIC/ACM international conference on web intelligence and intelligent agent technology edition, 2005.

[92] H. S. Nguyen and S. H. Nguyen. From Optimal Hyperplanes to Optimal Decision Trees. In S. Tsumoto, S. Kobayashi, T. Yokomori, H. Tanaka, and A. Nakamura, editors, *Proceedings of the Fourth International Workshop on Rough Sets, Fuzzy Sets, and Machine Discovery (RSFD'96)*, pages 82–88, Tokyo, Japan, November 6-8 1996. The University of Tokyo.

[93] H. S. Nguyen and S. H. Nguyen. Discretization Methods for Data Mining. In L. Polkowski and A. Skowron, editors, *Rough Sets in Knowledge Discovery*, pages 451–482. Springer, Heidelberg New York, 1998.

[94] H. S. Nguyen and S. H. Nguyen. Fast split selection method and its application in decision tree construction from large databases. *International Journal of Hybrid Intelligent Systems.*, 2(2):149–160, 2005.

[95] H. S. Nguyen and A. Skowron. Quantization of Real Values Attributes, Rough set and Boolean Reasoning Approaches. In *Proc. of the Second Joint Conference on Information Sciences*, pages 34–37, Wrightsville Beach, NC, USA, October 1995.

[96] H. S. Nguyen and D. Ślęzak. Approximate Reducts and Association Rules – Correspondence and Complexity Results. In A. Skowron, S. Ohsuga, and N. Zhong, editors, *New Directions in Rough Sets, Data Mining and Granular-Soft Computing (Proc. of RSFDGrC'99, Yamaguchi, Japan)*, volume 1711 of *LNAI 1711*, pages 137–145. Springer, Heidelberg, Germany, 1999.

[97] H. S. Nguyen and M. Szczuka. *Rough Sets In Knowledge Discovery and Data Mining – A tutorial.* VAST-Vietnam academy of Science and Technology, Tutorial Notes of the 9th Pacific-Asia Conference on Knowledge Discovery and Data Mining, 18 may 2005, Hanoi, Wietnam, Takashi Washio (ed.) edition, 2005.

[98] S. H. Nguyen. Regularity analysis and its applications in Data Mining. In L. Polkowski, T. Y. Lin, and S. Tsumoto, editors, *Rough Set Methods and Applications: New Developments in Knowledge Discovery in Information Systems*, volume 56 of *Studies in Fuzziness and Soft Computing*, pages 289–378. Springer, Heidelberg, Germany, 2000.

[99] S. H. Nguyen. *Regularity Analysis and Its Applications in Data Mining.* PhD thesis, Warsaw University, Warsaw, Poland, 2000.

[100] S. H. Nguyen, J. Bazan, A. Skowron, and H. S. Nguyen. Layered Learning for Concept synthesis. In J. F. Peters, A. Skowron, J. W. Grzymala-Busse, B. Kostek, R. W. Swiniarski, and M. S. Szczuka, editors, *Transactions on Rough Sets I*, volume LNCS 3100 of *Lecture Notes on Computer Science*, pages 187–208. Springer, 2004.

[101] S. H. Nguyen and H. S. Nguyen. Some Efficient Algorithms for Rough Set Methods. In *Sixth International Conference on Information Processing and Management of Uncertainty on Knowledge Based Systems IPMU'1996*, volume III, pages 1451–1456, Granada, Spain, July 1-5 1996.

[102] S. H. Nguyen and H. S. Nguyen. Some Efficient Algorithms for Rough Set Methods. In *Proceedings of the Conference of Information Processing and Management of Uncertainty in Knowledge-Based Systems IPMU'96*, pages 1451–1456, Granada, Spain, July 1996.

[103] S. H. Nguyen and H. S. Nguyen. Pattern Extraction from Data. *Fundamenta Informaticae*, 34(1–2):129–144, 1998.

[104] S. H. Nguyen and H. S. Nguyen. Pattern Extraction from Data. In *Proceedings of the Conference of Information Processing and Management of Uncertainty in Knowledge-Based Systems IPMU'98*, pages 1346–1353, Paris, France, July 1998.

[105] S. H. Nguyen, A. Skowron, and P. Synak. Discovery of data patterns with applications to decomposition and classification problems. In L. Polkowski and A. Skowron, editors, *Rough Sets in Knowledge Discovery 2: Applications, Case Studies and Software Systems*, volume 19 of *Studies in Fuzziness and Soft Computing*, chapter 4, pages 55–97. Springer, Heidelberg, Germany, 1998.

[106] A. Øhrn, J. Komorowski, A. Skowron, and P. Synak. The ROSETTA software system. In L. Polkowski and A. Skowron, editors, *Rough Sets in Knowledge Discovery 2. Applications, Case Studies and Software Systems*, number 19 in Studies in Fuzziness and Soft Computing, pages 572–576. Springer, Heidelberg, Germany, 1998.

[107] Z. Pawlak. Classification of objects by means of attributes. *Research Report PAS 429, Institute of Computer Science, Polish Academy of Sciences, ISSN 138-0648*, 1981.

[108] Z. Pawlak. Information systems – theoretical foundations. *Information Systems*, 6:205–218, 1981.

[109] Z. Pawlak. Rough sets. *International Journal of Computer and Information Sciences*, 11:341–356, 1982.

[110] Z. Pawlak. Rough Classification. *International Journal of Man-Machine Studies*, 20(5):469–483, 1984.

[111] Z. Pawlak. *Rough Sets: Theoretical Aspects of Reasoning about Data*, volume 9 of *System Theory, Knowledge Engineering and Problem Solving*. Kluwer Academic Publishers, Dordrecht, The Netherlands, 1991.

[112] Z. Pawlak. Some Issues on Rough Sets. *Transactions on Rough Sets*, 1:1–58, 2004.

[113] Z. Pawlak, S. K. M. Wong, and W. Ziarko. Rough Sets: Probabilistic Versus Deterministic Approach. In B. Gaines and J. Boose, editors, *Machine Learning and Uncertain Reasoning Vol. 3*, pages 227–242. Academic Press, London, 1990.

[114] B. Pfahringer. Compression-Based Discretization of Continuous Attributes. In *Proceedings of the 12th International Conference on Machine Learning*, pages 456–463, 1995.

[115] C. Pizzuti. Computing Prime Implicants by Integer Programming. In *Eighth International Conference on Tools with Artificial Intelligence (ICTAI '96)*, pages 332–336, 1996.

[116] L. Polkowski, T. Y. Lin, and S. Tsumoto, editors. *Rough Set Methods and Applications: New Developments in Knowledge Discovery in Information Systems*, volume 56 of *Studies in Fuzziness and Soft Computing*. Springer, Heidelberg, Germany, 2000.

[117] L. Polkowski and A. Skowron, editors. *Rough Sets in Knowledge Discovery 2: Applications, Case Studies and Software Systems*, volume 19 of *Studies in Fuzziness and Soft Computing*. Springer, Heidelberg, Germany, 1998.

[118] P. Prosser. Hybrid algorithms for the constraint satisfaction problem. *Computational Intelligence*, 9(3):268–299, August 1993.

[119] F. J. Provost and T. Fawcett. Analysis and Visualization of Classifier Performance: Comparison under Imprecise Class and Cost Distributions. In *Knowledge Discovery and Data Mining*, pages 43–48, 1997.

[120] M. Quafafou. α-RST: A generalization of rough set theory. *Information Sciences*, 124(1-4):301–316, 2000.

[121] W. V. O. Quine. On cores and prime implicants of truth functions. *American Mathematical – Monthly*, 66:755–760, 1959.

[122] W. V. O. Quine. *Mathematical Logic*. Harward University Press, Camb-Mass, 1961.

[123] J. Quinlan. *C4.5 – Programs for Machine Learning*. Morgan Kaufmann, 1993.

[124] R. Quinlan. Induction of Decision Trees. *Machine Learning*, 1:81–106, 1986.

[125] M. Richeldi and M. Rossotto. Class-driven statistical discretization of continuous attributes. In *Proceedings of the 8th European Conference on Machine Learning (ECML-95), Heraclion, Crete, Greece, April 25-27, 1995*, volume 912 of *Lecture Notes in Computer Science*, pages 335–338. Springer, 1995.

[126] J. Rissanen. Modeling by shortes data description. *Automatica*, 14:465–471, 1978.

[127] J. Rissanen. *Minimum-description-length principle*, pages 523–527. John Wiley & Sons, New York, NY, 1985.

[128] S. Rudeanu. *Boolean Functions and Equations*. North-Holland/American Elsevier, Amsterdam, 1974.

[129] L. O. Ryan. Efficient algorithms for clause-learning SAT solvers. Master's thesis, Simon Fraser University, Burnaby, Canada, 2004.

[130] W. Sarle. Stopped training and other remedies for overfitting. In *In Proceedings of the 27th Symposium on Interface*, 1995.

[131] B. Selman, H. A. Kautz, and D. A. McAllester. Ten Challenges in Propositional Reasoning and Search. In *Proceedings of Fifteenth International Joint Conference on Artificial Intelligence*, pages 50–54, 1997.

[132] B. Selman, H. Levesque, and D. Mitchell. A new method for solving hard satisfiability problems. In *Proceedings of the Tenth National Conference on Artificial Intelligence (AAAI'92)*, pages 459–465, 1992.

[133] S. Sen. Minimal cost set covering using probabilistic methods. In *SAC '93: Proceedings of the 1993 ACM/SIGAPP symposium on Applied computing*, pages 157–164, New York, NY, USA, 1993. ACM Press.

[134] J. C. Shafer, R. Agrawal, and M. Mehta. SPRINT: A Scalable Parallel Classifier for Data Mining. In T. M. Vijayaraman, A. P. Buchmann, C. Mohan, and N. L. Sarda, editors, *Proc. 22nd Int. Conf. Very Large Databases, VLDB*, pages 544–555. Morgan Kaufmann, 3–6 1996.

[135] C. E. Shannon. A symbolic analysis of relay and switching circuits. *Transactions of AIEE*, (57):713–723, 1938.

[136] C. E. Shannon. *A symbolic analysis of relay and switching circuits*. MIT, Dept. of Electrical Engineering, 1940.

[137] A. Skowron. Boolean Reasoning for Decision Rules generation. In J. Komorowski and Z. W. Raś, editors, *Seventh International Symposium for Methodologies for Intelligent Systems ISMIS*, volume 689 of *Lecture Notes in Artificial Intelligence*, pages 295–305, Trondheim, Norway, June 15-18 1993. Springer.

[138] A. Skowron. Synthesis of Adaptive Decision Systems from Experimental Data. In A. Aamodt and J. Komorowski, editors, *Fifth Scandinavian Conference on Artificial Intelligence SCAI'1995*, volume 28 of *Frontiers in Artificial Intelligence and Applications*, pages 220–238, Trondheim, Norway, May 29-31 1995. IOS Press.

[139] A. Skowron. Rough sets in KDD – plenary talk. In Z. Shi, B. Faltings, and M. Musen, editors, *16-th World Computer Congress (IFIP'2000): Proceedings of Conference on Intelligent Information Processing (IIP'2000)*, pages 1–14. Publishing House of Electronic Industry, Beijing, 2000.

[140] A. Skowron and H. S. Nguyen. Boolean Reasoning Schema with some Applications in Data Mining. In J. Żytkow and J. Rauch, editors, *Principple of Data Mining and Knowledge Discovery (Proc. of PKDD'1999, Praga, Czech Republic)*, LNAI 1704, pages 107–115. Springer, Berlin Heidelberg, 1999.

[141] A. Skowron and S. K. Pal, editors. *Special volume: Rough sets, pattern recognition and data mining*, volume 24(6) of *Pattern Recognition Letters*. 2003.

[142] A. Skowron, Z. Pawlak, J. Komorowski, and L. Polkowski. A rough set perspective on data and knowledge. In W. Kloesgen and J. Żytkow, editors, *Handbook of KDD*, pages 134–149. Oxford University Press, Oxford, 2002.

[143] A. Skowron and C. Rauszer. The Discernibility Matrices and Functions in Information Systems. In Słowiński [147], chapter 3, pages 331–362.

[144] A. Skowron and J. Stepaniuk. Tolerance Approximation Spaces. *Fundamenta Informaticae*, 27(2-3):245–253, 1996.

[145] D. Ślęzak. Various approaches to reasoning with frequency-based decision reducts: a survey. In Polkowski et al. [116], pages 235–285.

[146] D. Ślęzak. Approximate Entropy Reducts. *Fundamenta Informaticae*, 53:365–387, 2002.

[147] R. Słowiński, editor. *Intelligent Decision Support – Handbook of Applications and Advances of the Rough Sets Theory*. Kluwer Academic Publishers, Dordrecht, Netherlands, 1992.

[148] R. Słowiński and D. Vanderpooten. Similarity Relation as a Basis for Rough Approximations. In P. Wang, editor, *Advances in Machine Intelligence and Soft Computing Vol. 4*, pages 17–33. Duke University Press, Duke, NC, 1997.

[149] J. Stefanowski and A. Tsoukiàs. Incomplete Information Tables and Rough Classification. *International Journal of Computational Intelligence*, 17(3):545–566, 2001.

[150] J. Stepaniuk. Approximation spaces, reducts and representatives. In Polkowski and Skowron [117], chapter 6, pages 109–126.

[151] P. Stolorz and R. Musick, editors. *Scalable High Performance Computing for Knowledge Discovery and Data Mining.* Kluwer Academic Publishers, Norwell, MA, USA, 1998.

[152] M. H. Stone. The theory of representations for Boolean algebras. *Transactions of AMS*, 40:37–111, 1936.

[153] C. Umans. The minimum equivalent dnf problem and shortest implicants. *Journal of Computer and System Sciences*, 63(4):597–611, 2001.

[154] V. Vapnik. *Statistical Learning Theory.* John Wiley & Sons, New York, NY, 1998.

[155] I. H. Witten and E. Frank. *Data Mining: Practical Machine Learning Tools and Techniques with Java Implementations.* Morgan Kaufmann, 2005.

[156] J. Wnek and R. S. Michalski. Hypothesis-Driven Constructive Induction in AQ17-HCI: A Method and Experiments. *Machine Learning*, 14:139–168, 1994.

[157] J. Wróblewski. Theoretical Foundations of Order-Based Genetic Algorithms. *Fundamenta Informaticae*, 28(3-4):423–430, 1996.

[158] J. Wróblewski. Genetic Algorithms in Decomposition and Classification Problem. In Polkowski and Skowron [117], pages 471–487.

[159] J. Wróblewski. *Adaptive Methods of Object Classification, Ph. D. Thesis.* Warsaw University, Warsaw, 2002.

[160] L. Zadeh. Fuzzy logic and the calculi of fuzzy rules, fuzzy graphs, and fuzzy probabilities. *Comput. Math. Appl.*, 37(11-12), 1999.

[161] M. Zaki. Efficient Enumeration of Frequent Sequences. In *Seventh International Conference on Information and Knowledge Management*, pages 68–75, Washington DC, 1998.

[162] W. Ziarko. The Discovery, Analysis and Representation of Data Dependencies in Databases. In G. Piatetsky-Shapiro and W. J. Frawley, editors, *Knowledge Discovery in Databases*, pages 195–212. AAAI/MIT Press, Palo Alto, CA, 1991.

[163] W. Ziarko. Variable Precision Rough Set Model. *Journal of Computer and System Sciences*, 46:39–59, 1993.

[164] W. Ziarko, editor. *Rough Sets, Fuzzy Sets and Knowledge Discovery: Proceedings of the Second International Workshop on Rough Sets and Knowledge Discovery (RSKD'93), Banff, Alberta, Canada, October 12–15 (1993).* Workshops in Computing. Springer–Verlag & British Computer Society, London, Berlin, 1994.

Author Index

Lecture Notes in Computer Science

For information about Vols. 1–4143

please contact your bookseller or Springer

Vol. 4190: R. Larsen, M. Nielsen, J. Sporring (Eds.), Medical Image Computing and Computer-Assisted Intervention – MICCAI 2006, Part I. XXXVVVIII, 949 pages. 2006.

Vol. 4189: D. Gollmann, J. Meier, A. Sabelfeld (Eds.), Computer Security – ESORICS 2006. XI, 548 pages. 2006.

Vol. 4188: P. Sojka, I. Kopeček, K. Pala (Eds.), Text, Speech and Dialogue. XIV, 721 pages. 2006. (Sublibrary LNAI).

Vol. 4187: J.J. Alferes, J. Bailey, W. May, U. Schwertel (Eds.), Principles and Practice of Semantic Web Reasoning. XI, 277 pages. 2006.

Vol. 4186: C. Jesshope, C. Egan (Eds.), Advances in Computer Systems Architecture. XIV, 605 pages. 2006.

Vol. 4185: R. Mizoguchi, Z. Shi, F. Giunchiglia (Eds.), The Semantic Web – ASWC 2006. XX, 778 pages. 2006.

Vol. 4184: M. Bravetti, M. Núñez, G. Zavattaro (Eds.), Web Services and Formal Methods. X, 289 pages. 2006.

Vol. 4183: J. Euzenat, J. Domingue (Eds.), Artificial Intelligence: Methodology, Systems, and Applications. XIII, 291 pages. 2006. (Sublibrary LNAI).

Vol. 4182: H.T. Ng, M.-K. Leong, M.-Y. Kan, D. Ji (Eds.), Information Retrieval Technology. XVI, 684 pages. 2006.

Vol. 4180: M. Kohlhase, OMDoc – An Open Markup Format for Mathematical Documents [version 1.2]. XIX, 428 pages. 2006. (Sublibrary LNAI).

Vol. 4179: J. Blanc-Talon, W. Philips, D. Popescu, P. Scheunders (Eds.), Advanced Concepts for Intelligent Vision Systems. XXIV, 1224 pages. 2006.

Vol. 4178: A. Corradini, H. Ehrig, U. Montanari, L. Ribeiro, G. Rozenberg (Eds.), Graph Transformations. XII, 473 pages. 2006.

Vol. 4177: R. Marín, E. Onaindía, A. Bugarín, J. Santos (Eds.), Current Topics in Artificial Intelligence. XIII, 621 pages. 2006. (Sublibrary LNAI).

Vol. 4176: S.K. Katsikas, J. Lopez, M. Backes, S. Gritzalis, B. Preneel (Eds.), Information Security. XIV, 548 pages. 2006.

Vol. 4175: P. Bücher, B.M.E. Moret (Eds.), Algorithms in Bioinformatics. XII, 402 pages. 2006. (Sublibrary LNBI).

Vol. 4174: K. Franke, K.-R. Müller, B. Nickolay, R. Schäfer (Eds.), Pattern Recognition. XX, 773 pages. 2006.

Vol. 4173: S. El Yacoubi, B. Chopard, S. Bandini (Eds.), Cellular Automata. XV, 734 pages. 2006.

Vol. 4172: J. Gonzalo, C. Thanos, M. F. Verdejo, R.C. Carrasco (Eds.), Research and Advanced Technology for Digital Libraries. XVII, 569 pages. 2006.

Vol. 4169: H.L. Bodlaender, M.A. Langston (Eds.), Parameterized and Exact Computation. XI, 279 pages. 2006.

Vol. 4168: Y. Azar, T. Erlebach (Eds.), Algorithms – ESA 2006. XVIII, 843 pages. 2006.

Vol. 4167: S. Dolev (Ed.), Distributed Computing. XV, 576 pages. 2006.

Vol. 4166: J. Górski (Ed.), Computer Safety, Reliability, and Security. XIV, 440 pages. 2006.

Vol. 4165: W. Jonker, M. Petković (Eds.), Secure, Data Management. X, 185 pages. 2006.

Vol. 4163: H. Bersini, J. Carneiro (Eds.), Artificial Immune Systems. XII, 460 pages. 2006.

Vol. 4162: R. Královič, P. Urzyczyn (Eds.), Mathematical Foundations of Computer Science 2006. XV, 814 pages. 2006.

Vol. 4161: R. Harper, M. Rauterberg, M. Combetto (Eds.), Entertainment Computing - ICEC 2006. XXVII, 417 pages. 2006.

Vol. 4160: M. Fisher, W.v.d. Hoek, B. Konev, A. Lisitsa (Eds.), Logics in Artificial Intelligence. XII, 516 pages. 2006. (Sublibrary LNAI).

Vol. 4159: J. Ma, H. Jin, L.T. Yang, J.J.-P. Tsai (Eds.), Ubiquitous Intelligence and Computing. XXII, 1190 pages. 2006.

Vol. 4158: L.T. Yang, H. Jin, J. Ma, T. Ungerer (Eds.), Autonomic and Trusted Computing. XIV, 613 pages. 2006.

Vol. 4156: S. Amer-Yahia, Z. Bellahsène, E. Hunt, R. Unland, J.X. Yu (Eds.), Database and XML Technologies. IX, 123 pages. 2006.

Vol. 4155: O. Stock, M. Schaerf (Eds.), Reasoning, Action and Interaction in AI Theories and Systems. XVIII, 343 pages. 2006. (Sublibrary LNAI).

Vol. 4154: Y.A. Dimitriadis, I. Zigurs, E. Gómez-Sánchez (Eds.), Groupware: Design, Implementation, and Use. XIV, 438 pages. 2006.

Vol. 4153: N. Zheng, X. Jiang, X. Lan (Eds.), Advances in Machine Vision, Image Processing, and Pattern Analysis. XIII, 506 pages. 2006.

Vol. 4152: Y. Manolopoulos, J. Pokorný, T. Sellis (Eds.), Advances in Databases and Information Systems. XV, 448 pages. 2006.

Vol. 4151: A. Iglesias, N. Takayama (Eds.), Mathematical Software - ICMS 2006. XVII, 452 pages. 2006.

Vol. 4150: M. Dorigo, L.M. Gambardella, M. Birattari, A. Martinoli, R. Poli, T. Stützle (Eds.), Ant Colony Optimization and Swarm Intelligence. XVI, 526 pages. 2006.

Vol. 4149: M. Klusch, M. Rovatsos, T.R. Payne (Eds.), Cooperative Information Agents X. XII, 477 pages. 2006. (Sublibrary LNAI).

Vol. 4148: J. Vounckx, N. Azemard, P. Maurine (Eds.), Integrated Circuit and System Design. XVI, 677 pages. 2006.

Vol. 4147: M. Broy, I.H. Krüger, M. Meisinger (Eds.), Automotive Software – Connected Services in Mobile Networks. XIV, 155 pages. 2006.

Vol. 4146: J.C. Rajapakse, L. Wong, R. Acharya (Eds.), Pattern Recognition in Bioinformatics. XIV, 186 pages. 2006. (Sublibrary LNBI).

Vol. 4145: L. Moreau, I. Foster (Eds.), Provenance and Annotation of Data and Processes. XI, 288 pages. 2006.

Vol. 4144: T. Ball, R.B. Jones (Eds.), Computer Aided Verification. XV, 564 pages. 2006.